Pharmaceutical Water

Pharmaceutical Water

System Design, Operation, and Validation

Second Edition

William V. Collentro

New York   London

First published in 1998 by Interpharm/CRC.
This edition published in 2011 by Informa Healthcare, Telephone House, 69-77 Paul Street, London EC2A 4LQ, UK.

Simultaneously published in the USA by Informa Healthcare, 52 Vanderbilt Avenue, 7th Floor, New York, NY 10017, USA.

Informa Healthcare is a trading division of Informa UK Ltd. Registered Office: 37–41 Mortimer Street, London W1T 3JH, UK. Registered in England and Wales number 1072954.

©2011 Informa Healthcare, except as otherwise indicated

No claim to original U.S. Government works

Reprinted material is quoted with permission. Although every effort has been made to ensure that all owners of copyright material have been acknowledged in this publication, we would be glad to acknowledge in subsequent reprints or editions any omissions brought to our attention.

All rights reserved. No part of this publication may be reproduced, stored in a retrieval system, or transmitted, in any form or by any means, electronic, mechanical, photocopying, recording, or otherwise, unless with the prior written permission of the publisher or in accordance with the provisions of the Copyright, Designs and Patents Act 1988 or under the terms of any licence permitting limited copying issued by the Copyright Licensing Agency, 90 Tottenham Court Road, London W1P 0LP, UK, or the Copyright Clearance Center, Inc., 222 Rosewood Drive, Danvers, MA 01923, USA (http://www.copyright.com/ or telephone 978-750-8400).

Product or corporate names may be trademarks or registered trademarks, and are used only for identification and explanation without intent to infringe.

This book contains information from reputable sources and although reasonable efforts have been made to publish accurate information, the publisher makes no warranties (either express or implied) as to the accuracy or fitness for a particular purpose of the information or advice contained herein. The publisher wishes to make it clear that any views or opinions expressed in this book by individual authors or contributors are their personal views and opinions and do not necessarily reflect the views/opinions of the publisher. Any information or guidance contained in this book is intended for use solely by medical professionals strictly as a supplement to the medical professional's own judgement, knowledge of the patient's medical history, relevant manufacturer's instructions and the appropriate best practice guidelines. Because of the rapid advances in medical science, any information or advice on dosages, procedures, or diagnoses should be independently verified. This book does not indicate whether a particular treatment is appropriate or suitable for a particular individual. Ultimately it is the sole responsibility of the medical professional to make his or her own professional judgements, so as appropriately to advise and treat patients. Save for death or personal injury caused by the publisher's negligence and to the fullest extent otherwise permitted by law, neither the publisher nor any person engaged or employed by the publisher shall be responsible or liable for any loss, injury or damage caused to any person or property arising in any way from the use of this book.

A CIP record for this book is available from the British Library.

Library of Congress Cataloging-in-Publication Data available on application

ISBN-13: 9781420077827

Orders may be sent to: Informa Healthcare, Sheepen Place, Colchester, Essex CO3 3LP, UK
Telephone: +44 (0)20 7017 5540
Email: CSDhealthcarebooks@informa.com
Website: http://informahealthcarebooks.com/

For corporate sales please contact: CorporateBooksIHC@informa.com
For foreign rights please contact: RightsIHC@informa.com
For reprint permissions please contact: PermissionsIHC@informa.com

Typeset by MPS Limited, a Macmillan Company

*This book is dedicated to my family—my wife Caroline;
our children Andrew, Christopher, Michael, and Emily;
and our grandchildren Isabella, Julia, Joseph, Drew, Nicholas, Chase,
Anna, Jackson, Adelyn, and Samantha*

Preface to the second edition

The preface of the first edition of *Pharmaceutical Water* discusses the numerous seminars, conferences, and training sessions conducted each year that specifically discuss water purification and compendial water systems. It is interesting to note how much the world has changed in the last 12 years. Perhaps, the Internet explosion with virtual instant access to material coupled with budget restriction explains the limited number of events with fewer attendees. Despite this, the relative number of citations associated with compendial water systems continues to grow. Perhaps, it is time to go "back to basics," acknowledging that compendial water systems consist of multiple unit operations with demanding design, installation, operation, validation, and preventative maintenance requirements. Further, unique feed water properties for different systems significantly influence these parameters. In summary, while some may imply, there is no standard system or "magic bullet."

During preparation of this second edition, several enhancements were performed. The original text has been completely reviewed and edited to incorporate changes in technology, feed water quality, and regulatory requirements. Significant expansion and entirely new sections have been added discussing the following:

- Expanded use of chloramines in raw water supplies
- Use of ozone for microbial control in Purified Water systems
- Discussion of biofilm theory and operating experience
- Chemical sanitization material selection and execution
- The effects of disinfection byproducts in raw water supplies
- Reverse osmosis membrane performance and system design
- Continuous electrodeionization enhancement
- System documentation
- System commissioning

Twelve more years of experience has provided a significant number of observations. Many of these observations have been included as "case histories," with data inserted into the text at several locations. While theory, design, operation, validation, and maintenance considerations are discussed, the second edition reflects actual observations and expands on the "hands-on" presentation philosophy of the original text.

References have been added or updated throughout the text. These should be used by the reader as an expansion of the material presented in the text. On many occasions, "older" references are retained. These are like a good wine, improving with age, often proving information of significant importance.

Knowledge is good, but sharing that knowledge and experience is excellent.

Preface to the first edition

Each year, I am fortunate enough to present several workshops, seminars, and courses associated with pharmaceutical water systems. At the beginning of my presentation, mostly for sessions lasting longer than a few hours, I generally discuss some ground rules. I explain how I will do everything possible to avoid speaking from the podium. Often, I remove the microphone and simply walk in front of or around the attendees. It is important to emphasize that the material being presented is an attempt to share my personal experience associated with pharmaceutical water systems on a daily basis. I encourage questions since they demonstrate that the attendees have been properly stimulated by the subject matter. I have approached the preparation of this text in a manner similar to my approach to workshops, with the thought that it provided me with a forum to discuss topics without a time limit.

I am constantly amazed by the number of presentations offered each year regarding pharmaceutical water systems. I rarely review a brochure for a major conference without noting a session on water systems. Certainly, the number of seminars presented for pharmaceutical water systems over the past several years is disproportional to the balance of technical emphasis for all other systems, components, and functions at a facility. It is my belief that this situation has evolved due to improper "horizontal-vertical" integration of disciplines and management with regard to information associated with pharmaceutical water systems. Horizontal integration is extremely important since it relates to interdisciplinary exchange of information associated not only with pharmaceutical water systems but also the regulatory, operating, maintenance, and similar issues associated with these systems. From an organizational viewpoint, it is critical that all appropriate disciplines be involved in aspects of the pharmaceutical water system. This role should not be limited to an engineering department or, more specifically, a facilities engineering department. Regulatory, quality control, analytical, manufacturing, and other input is vital to the success of design, operation, validation, and maintenance of any pharmaceutical water system. Of equal importance is a vertical integration of knowledge. This entails a transfer of items such as the details associated with routine operation of a system from operating personnel, through supervisory personnel, all the way to senior management personnel. It is impossible for senior managers to determine proper system design and related budgets or to address regulatory concerns without "hands-on" knowledge of detailed system operation. When proper horizontal and vertical integration are performed, all the necessary tools required to weave interdisciplinary input through a project is achieved.

In preparing this text, I elected to personally write all of the material myself. This appears to contradict the weaving concept indicated above. However, I believe that the contrary is true. When multiple individuals attempt to prepare a text addressing all aspects of pharmaceutical water systems, the lack of proper horizontal and vertical integration becomes very obvious. Continuity of the presentation is affected. This text deals with pharmaceutical water systems by addressing the compendial requirements, the nature of raw water supplies, specification preparation, and validation documentation. Occasionally, it may appear that certain items are repeated. I am extremely sensitive to the fact that an individual focusing on a particular topic may review a certain section of this text without reviewing another section, which may discuss related items in greater detail. To avoid this situation, some repetition, coupled with a table of contents and a complete index, ensures that all aspects associated with a particular topic have been reviewed.

Finally, I have attempted to simplify the somewhat overpowering nature of engineering aspects associated with specific water purification unit operations by presenting a brief overview of the theory and application of the technology, a discussion of design considerations, and a discussion of operating and maintenance considerations that incorporates extensive field experience and "case histories." Throughout the entire text, regulatory and related issues are woven into the presentation. The resulting tapestry may occasionally present opinions. These are clearly designated by indicating that I am suggesting, based on experience associated with over 400 pharmaceutical and related water purification systems, that my opinion should be considered. If this generates a difference of opinion, my objective has still been met since it has stimulated the reader to consider a different view of a particular topic.

Acknowledgements

I would like to thank the numerous individuals who provided assistance during the preparation of this text. Special thanks to Tahar El-Korchi and Worcester Polytechnic Institute for encouragement and understanding, and Amanda Bowden, Christian Peterson, and Kostas Avgiris for assistance with preparation of figures. Finally, I would like to thank the staff at Informa Healthcare for their professional support, particularly Aimee Laussen.

Contents

Preface to the second edition vi
Preface to the first edition vii
Acknowledgements ix

1. Introduction 1
2. Impurities in raw water 13
3. Pretreatment techniques 34
4A. Ion removal techniques—reverse osmosis 93
4B. Ion removal techniques—ion exchange 133
4C. Additional ion removal techniques 159
5. Distillation and Pure Steam generation 181
6. Storage systems and accessories 221
7. Ozone systems and accessories 256
8. Polishing components 280
9. Distribution systems—design, installation, and material selection 317
10. Controls and instrumentation 367
11. System design and specification guidelines 387
12. System installation, start-up, and commissioning 411
13. System validation 419

Index 463

1 | Introduction

DEFINITION OF PHARMACEUTICAL WATERS

- Water and steam used in the pharmaceutical industry and related disciplines are classified by various pharmacopeias. The U.S. Pharmacopeia (USP) classifies compendial waters as follows:
  - Water for Injection
  - Bacteriostatic Water for Injection
  - Sterile Water for Inhalation
  - Sterile Water for Injection
  - Sterile Water for Irrigation
  - Purified Water
  - Sterile Purified Water
  - Water for Hemodialysis
  - Pure Steam
  - Drinking Water (indirectly)
- With the exception of Drinking Water, USP Purified Water, USP Water for Injection, and USP Pure Steam, the classifications listed above refer to "packaged water" (USP, 2010(a)). Drinking Water, USP Purified Water, and USP Water for Injection are the primary waters used for most pharmaceutical applications, and are the primary topic of this book. Validation is required for all compendial water systems producing USP Purified Water or USP Water for Injection, with the exception of Drinking Water. Drinking Water used in a specific application generally requires "commissioning/qualification" to an "internal" specification, verifying that the quality of the product water, from both a chemical and microbiological standpoint, does not vary from established internal specifications with time. This qualification process is often used not only to maintain control of product water but also to document the nature of the system by preparing and executing documents similar to those used for compendial water systems. Obviously, the internal specifications established for a qualified system may parallel a particular USP official monograph specification, such as that for Purified Water. Finally, certain applications may expand the USP requirements for a particular grade of water. As an example, many biotechnology water specifications require "low bacterial endotoxin" Purified Water. For such application, the biotechnology company would validate the system as a USP Purified Water system and incorporate an internal bacterial endotoxin specification. Chemical, bacteria, bacterial endotoxin, and other parameters associated with each of the pharmaceutical grades of water identified above are addressed individually in this chapter.
- USP is prepared and published by The United States Pharmacopeial Convention, a private organization. The material within USP is established by "Expert Committees," circulated to the general public for comment and review, and revised after acceptance. The Expert Committees as well as the review processes include U.S. Food and Drug Administration comment, review, and approval. Since new volumes of USP are published periodically, it is suggested that reference to USP states the number of the most recent addition and/or most recent edition including all "Supplements."

DEFINITION OF PHARMACEUTICAL WATERS—EP, JP, BP, etc.

- As indicated, water and steam used in the pharmaceutical industry and related disciplines are also classified by other pharmacopeias, including the European

Pharmacopeia (EP), Japanese Pharmacopeia (JP), and the British Pharmacopeia (BP). Over the past several years, there have been many attempts to "harmonization" descriptions, specifications, and method of production for compendial waters. While significant progress has been achieved, specific differences of importance will be addressed within this chapter.

CHEMICAL SPECIFICATIONS
Drinking Water
- From a chemical standpoint, water classified as Drinking Water, for applications such as some initial rinsing operations and active pharmaceutical ingredient manufacturing operations, must meet the U.S. Environmental Protection Agency's (EPA) National Primary Drinking Water Regulations (NPDWR), or comparable regulations of the European Union, Japan, and/or World Health Organization, as applicable, for "Drinking Water." This would include but not be limited to the parameters presented in Table 1.1 for U.S. EPA Drinking Water (EPA, 2010). It is important to note that the NPDWR *will* change with time, incorporating additional parameters or changing regulated item concentrations. It should be emphasized that all validated USP systems, as well as systems using Drinking Water, should have access to correspondence identifying changes to these regulations.
- As discussed further in subsequent chapters of this book, it is highly recommended that supplemental analysis for Drinking Water, including feedwater to a USP Purified Water or USP Water for Injection system, be considered. The nature and type of analyses are dictated by the intended use of the Drinking Water. For example, if groundwater is used for an initial rinsing step during applications such as "clean-in-place" (CIP) or the production of an active pharmaceutical ingredient, it may be appropriate to treat the water through a particulate removal filter and/or water softening system. If water softening is used, the presence of high molecular weight multivalent cations, such as barium, strontium, and aluminum, in the feedwater should be identified. As discussed in chapter 3, these compounds will affect the Standard Operating Procedures (SOPs), specifically the regeneration salt dosing and concentration, during regeneration of the water softening system. Multivalent cations, such as calcium and magnesium, are *not* included in the NPDWR, but affect the performance of the system.
- Other specific components are critical to different water purification unit operations. Another example is the level of naturally occurring organic material (NOM) in a surface water supply to a USP Purified Water system. Both anion resin and reverse osmosis (RO) membranes will foul with organic material. The level of the NOM in feedwater will not only dictate the nature of pretreatment equipment but also establish an analytical monitoring program clearly demonstrating that the selected pretreatment operations "protect" the anion resin within the ion exchange system or RO membranes from fouling.

Purified Water
- Chemical specifications for USP Purified Water are outlined in the *Official Monograph* by referencing *Physical Tests* chapters for conductivity and total organic carbon (TOC). *Physical Tests* Section <643> provides the TOC specification, capability of the TOC analyzer, "system suitability" requirements, and calibration requirements. The section does *not* set forth requirements for online measurement versus "grab" sampling and laboratory analysis. Further, the section does not state the frequency of analysis. The TOC limit for USP Purified Water is 0.50 mg/L. The specification agrees with the current EP specification.
- USP *Physical Tests* Section <645> outlines the specification for conductivity, method of determination, instrument (meter and probe), calibration requirements, etc. This section outlines a three-stage test method that compensates for the presence of carbon dioxide and pH. The most restrictive specification, "Stage 1," is 1.3 µS/cm at 25°C or

Table 1.1 U.S. EPA Drinking Water Regulated Contaminants

Contaminant	Potential long-term health effect—concentration above NPDWR	Source of contaminant in Drinking Water
Category: Microorganisms		
Cryptosporidium	Gastrointestinal illness	Human and animal fecal waste
Giardia lamblia	Gastrointestinal illness	Human and animal fecal waste
Legionella	Legionnaire's disease	Naturally occurring—heating/cooling systems
Total coliform	Numerous, if confirmed as *Escherichia coli* and/or fecal coliform	Coliform confirmed as *Escherichia coli* or fecal coliform from human or animal fecal waste
Turbidity	(Indicator of presence of waterborne disease)	Soil runoff
Viruses (enteric)	Gastrointestinal illness	Human and animal fecal waste
Category: Disinfection by-products		
Bromate	Increased cancer risk	Ozone disinfection of Drinking Water
Chlorite	Anemia; nervous system effects	Chlorine dioxide disinfection of Drinking Water
Haloacetic acids (HAA5)	Increased cancer risk	Chlorine disinfection of Drinking Water
Total trihalomethanes	Liver, kidney, or central nervous system problems; increased cancer risk	Chlorine disinfection of Drinking Water
Category: Disinfecting agents		
Chloramines	Eye/nose irritation; stomach discomfort; anemia	Water disinfecting agent
Chlorine	Eye/nose irritation; stomach discomfort	Water disinfecting agent
Chlorine dioxide	Anemia; nervous system effects	Water disinfecting agent
Category: Inorganic chemicals		
Antimony	Increase in blood cholesterol; decrease in blood sugar	Waste from refineries, fire retardants, ceramics, and solder
Arsenic	Skin damage, circulatory system problems, possible increased cancer risk	Erosion of natural deposits, runoff from orchards, runoff from glass and electronic production waste
Asbestos (fibers > 10 μm)	Increase risk of developing benign intestinal polyps	Decay of asbestos cement in water mains; erosion of natural deposits
Barium	Increase in blood pressure	Discharge of drilling waste, discharge from metal refineries; erosion of natural deposits
Beryllium	Intestinal lesions	Discharge from metal refineries and coal burning factories; discharge from electrical, aerospace, and defense industries
Cadmium	Kidney damage	Corrosion of galvanized pipes; erosion of natural deposits; discharge from metal refineries; runoff from waste batteries and paints
Total chromium	Allergic dermatitis	Discharge from steel and pulp mills; erosion of natural deposits
Copper	Gastrointestinal distress; liver or kidney damage	Corrosion of household plumbing; erosion of natural deposits
"Free" cyanide	Nerve damage or thyroid problems	Discharge from steel/metal factories; discharge from plastic and fertilizer factories
Fluoride	Bone diseases; mottled teeth	Water additive to promote strong teeth; discharge from fertilizer and aluminum factories
Lead	Children: delay in physical or mental development Adults: kidney problems and high blood pressure	Corrosion of household plumbing systems; erosion of natural deposits

(Continued)

Table 1.1 U.S. EPA Drinking Water Regulated Contaminants (*Continued*)

Contaminant	Potential long-term health effect—concentration above NPDWR	Source of contaminant in Drinking Water
Mercury	Kidney damage	Erosion of natural deposits; discharge from refineries and factories; runoff from landfills and croplands
Nitrate (as nitrogen)	Infants: serious illness, shortness of breath, blue baby syndrome	Runoff from fertilizer use. Leaching from septic tanks; sewage; erosion of natural deposits
Nitrite (as nitrogen)	Infants: serious illness, shortness of breath, blue baby syndrome	Runoff from fertilizer use; leaching from septic tanks; sewage; erosion of natural deposits
Selenium	Hair or fingernail loss; numbness in fingers or toes; circulatory problems	Discharge from petroleum refineries; erosion of natural deposits; discharge from mines
Thallium	Hair loss; changes in blood; kidney, intestine, or liver problems	Leaching from ore-processing sites; discharge from electronics, glass, and drug factories
Category: Organic chemicals		
Acrylamide	Nervous system or blood problems; increased risk of cancer	Added to water during sewage/wastewater treatment
Alachlor	Eye, liver, kidney, or spleen problems; anemia; increased risk of cancer	Runoff from herbicide used on row crops
Atrazine	Cardiovascular system or reproductive system problems	Runoff from herbicide used on row crops
Benzene	Anemia; decrease in blood platelets; increased risk of cancer	Discharge from factories; leaching from gas storage tanks and landfills
Benzo(*a*)pyrene (PAHs [polycyclic aromatic hydrocarbons])	Reproductive difficulties; increased risk of cancer	Leaching from linings of water storage tanks and distribution lines
Carbofuran	Problems with blood, nervous system, or reproductive system	Leaching of soil fumigant used on rice and alfalfa
Carbon tetrachloride	Liver problems; increased risk of cancer	Discharge from chemical plants and other industrial factories
Chlordane	Liver or nervous system problems; increased risk of cancer	Residue of banned termiticide
Chlorobenzene	Liver or kidney problems	Discharge from chemical and agricultural chemical factories
2,4-D (2,4-dichlorophenoxyacetic acid)	Kidney, liver, or adrenal gland problems	Runoff from herbicide used on row crops
Dalapon	Minor kidney changes	Runoff from herbicide used on rights of way
1,2-Dibromo-3-chloropropane (DBCP)	Reproductive difficulties; increased risk of cancer	Runoff/leaching from soil fumigant used on soybeans, cotton, pineapples, and orchards
o-Dichlorobenzene	Liver, kidney, or circulatory system problems	Discharge from industrial chemical factories
p-Dichlorobenzene	Anemia; liver, kidney, or spleen damage; changes in blood	Discharge from industrial chemical factories
1,2-Dichloroethane	Increased risk of cancer	Discharge from industrial chemical factories
1,1-Dichloroethylene	Liver problems	Discharge from industrial chemical factories
cis-1,2-Dichloroethylene	Liver problems	Discharge from industrial chemical factories
trans-1,2-Dichloroethylene	Liver problems	Discharge from industrial chemical factories

Table 1.1 (*Continued*)

Contaminant	Potential long-term health effect—concentration above NPDWR	Source of contaminant in Drinking Water
Dichloromethane	Liver problems; increased risk of cancer	Discharge from industrial chemical factories
1,2-Dichloropropane	Increased risk of cancer	Discharge from industrial chemical factories
Di(2-ethylhexyl) adipate	Weight loss, liver problems, or possible reproductive difficulties	Discharge from chemical factories
Di(2-ethylhexyl)phthalate	Reproductive difficulties; liver problems; increased risk of cancer	Discharge from rubber and chemical factories
Dinoseb	Reproductive difficulties	Runoff from herbicide used on soybeans and vegetables
Dioxin (2,3,7,8-TCDD [2,3,7,8-tetrachloridibenzo-p-dioxin])	Reproductive difficulties; increased risk of cancer	Emissions from waste incineration and other combustion; discharge from chemical factories
Diquat	Cataracts	Runoff from herbicide use
Endothall	Stomach and intestinal problems	Runoff from herbicide use
Endrin	Liver problems	Residue of banned insecticide
Epichlorohydrin	Increased cancer risk and stomach problems	Discharge from industrial chemical factories; an impurity of some water treatment chemicals
Ethylbenzene	Liver or kidney problems	Discharge from petroleum refineries
Ethylene dibromide	Problems with liver, stomach, reproductive system, or kidneys; increased risk of cancer	Discharge from petroleum refineries
Glyphosate	Kidney problems; reproductive difficulties	Runoff from herbicide use
Heptachlor	Liver damage; increase risk of cancer	Residue of banned termiticide
Heptachlor epoxide	Liver damage; increase risk of cancer	Breakdown of heptachlor
Hexachlorobenzene	Liver or kidney problems; reproductive difficulties; increased risk of cancer	Discharge from metal refineries and agricultural chemical factories
Hexachlorocyclopentadiene	Kidney or stomach problems	Discharge from chemical factories
Lindane	Liver or kidney problems	Runoff/leaching from insecticide used on cattle, lumber, gardens
Methoxychlor	Reproductive difficulties	Runoff/leaching used on fruits, vegetables, alfalfa, livestock
Oxamyl (Vydate)	Slight nervous system effects	Runoff/leaching from insecticides used on apples, potatoes, and tomatoes
Polychlorinated biphenyls (PCBs)	Skin changes; thymus gland problems; immune deficiencies; reproductive or nervous system difficulties; increased risk of cancer	Runoff from landfills; discharge of waste chemicals
Pentachlorophenol	Liver or kidney problems; increased cancer risk	Discharge from wood preserving factories
Picloram	Liver problems	Herbicide runoff
Simazine	Problems with blood	Herbicide runoff
Styrene	Liver, kidney, or circulatory system problems	Discharge from rubber and plastic factories; leaching from landfills
Tetrachloroethylene	Liver problems; increased cancer risk	Discharge from factories and dry cleaners
Toluene	Nervous system, kidney, or liver problems	Discharge from petroleum factories

(*Continued*)

Table 1.1 U.S. EPA Drinking Water Regulated Contaminants (*Continued*)

Contaminant	Potential long-term health effect—concentration above NPDWR	Source of contaminant in Drinking Water
Toxaphene	Kidney, liver, or thyroid problems; increased risk of cancer	Runoff/leaching from insecticide used on cotton and cattle
2,4,5-TP ([(2,4,5-trichlorophenoxy) propionic acid] Silvex)	Liver problems	Residual of banned herbicide
1,2,4-Trichlorobenzene	Changes in adrenal glands	Discharge from textile finishing factories
1,1,1-Trichloroethane	Liver, nervous system, or circulatory problems	Discharge from metal degreasing sites and other factories
1,1,2-Trichloroethane	Liver, kidney, or immune system problems	Discharge from industrial chemical factories
Trichloroethylene	Liver problems; increased risk of cancer	Discharge from metal decreasing sites and other factories
Vinyl chloride	Increased risk of cancer	Leaching from PVC pipes; discharge from plastic factories
Xylenes (total)	Nervous system damage	Discharge from petroleum factories; discharge from chemical factories
Category: Radionuclides		
Alpha particles	Increased risk of cancer	Erosion of natural deposits of certain minerals
Beta particles and photon emitters	Increased risk of cancer	Decay of natural and man-made deposits of certain materials
Radium 226 and Radium 228	Increased risk of cancer	Erosion of natural material
Uranium	Increased risk of cancer; kidney toxicity	Erosion of natural deposits

Abbreviation: NPDWR, National Primary Drinking Water Regulations.

- 1.1 μS/cm at 20°C, in agreement with the EP specification. The section does *not* set forth requirements for online measurement versus grab sampling and laboratory analysis. Further, the section does not state the frequency of analysis.
- Other pharmacopeias *may* have additional testing requirements. As an example, the EP contains a nitrate specification of 0.2 mg/L, maximum. It is important to review the various pharmacopeial requirements for countries where products will be sold.
- As indicated previously, certain systems may require supplemental sampling and monitoring for important contaminants not addressed specifically within the pharmacopeia. As an example, for a Purified Water system using ozone for microbial control (storage and distribution system), control of residual disinfecting by-products such as trihalomethanes (chap. 2) may be critical. Ozone will oxidize trihalomethane compounds to carbon dioxide, which will react with water yielding the hydronium and bicarbonate ion, increasing the conductivity of Purified Water.

USP Sterile Purified Water
- USP Sterile Purified Water is USP Purified Water that is sterilized and suitably packaged. It contains *no* antimicrobial agents. The chemical specifications for USP Sterile Purified Water are defined in the *Official Monograph* and currently include the following:
  - Oxidizable substances
  - Conductivity
- Unlike Drinking Water, USP Purified Water, or USP Water for Injection, packaged waters must meet laboratory-type chemical tests.

USP Water for Injection
- The chemical specifications for USP Water for Injection are identical to the chemical specifications for USP Purified Water.

- The EP chemical specification for Water for Injection includes a nitrate specification (0.2 mg/L, maximum) (EP, 2010(a)).

USP Sterile Water for Injection
- USP Sterile Water for Injection is USP Water for Injection that has been sterilized and suitably packaged. It contains *no* antimicrobial agents or added substances. It must pass the physical and chemical tests set forth in the current edition of USP with all supplements.

USP Bacteriostatic Water for Injection
- Bacteriostatic Water for Injection is USP Water for Injection sterilized and suitably packaged. It contains one or more suitable antimicrobial agents. It must pass the physical and chemical tests set forth in the current edition of USP with all supplements.

USP Sterile Water for Irrigation
- USP Sterile Water for Irrigation is USP Water for Injection sterilized and suitably packaged. It contains *no* antimicrobial agents or other added substances. It must pass the physical and chemical tests set forth in the current edition of USP with all supplements.

USP Sterile Water for Inhalation
- USP Sterile Water for Inhalation is USP Water for Injection sterilized and suitably packaged. It contains *no* antimicrobial agents, except when used in humidifiers or other similar devices that may be liable to contamination over a period of time, or other added substances. It must pass the physical and chemical tests set forth in the current edition of USP with all supplements.

BACTERIA ACTION AND ALERT LIMITS
- It is important to note that, with the exception of the reference to sterility for packaged waters, there are no bacteria specifications contained in the USP *Official Monographs* for pharmaceutical waters. Drinking Water has been defined. The EPA's NPDWR references specific guidelines related to the absence of coliform bacteria. However, the USP Purified Water and Water for Injection *Official Monographs* merely state that the feedwater to the water purification system producing Purified Water or Water for Injection must meet the NPDWR (or other appropriate European, Japanese, or World Health Organization standards). Subsequently, it is extremely important to point out that information associated with bacterial levels is outlined in USP *General Information* Section <1231> "Water for Pharmaceutical Purposes." While this section provides valuable information to support the *Official Monographs*, material contained within this section, as implied, is for "general information."
- Suggested maximum total viable bacteria levels as well as the enumeration method are included directly in the EP Purified Water and Water for Injection *Monographs* (EP, 2010(b)).
- In reviewing the USP *General Information* section with regard to "Alert" and "Action" levels for bacteria (USP, 2010(b)), the following definitions are provided:

  > **Alert Levels** are events or levels that, when they occur or are exceeded, indicate that a process may have drifted from its normal operating condition. Alert Level excursions constitute a warning and do not necessarily require a corrective action. However, alert level excursions usually lead to the alerting of personnel involved in water

system operation as well as QA. Alert level excursions may also lead to additional monitoring with more intense scrutiny of resulting and neighboring data as well as other process indicators.

Action Levels are events or higher levels that, when they occur or are exceeded, indicate that a process is probably drifting from its normal operating range. Examples of kinds of Action Level "events" include exceeding the Alert Levels repeatedly; or in multiple simultaneous locations, a single occurrence of exceeding a higher microbial level; or the individual or repeated recovery of specific objectionable microorganisms. Exceeding an Action Level should lead to immediate notification of both QA and personnel involved in water system operations so that corrective actions can immediately be taken to bring the process back into its normal operating range.

While it should be indicated that the definition stated above are directly from the USP *General Information* section, a "guideline" to support the balance of the material in USP, they provide an excellent description of Alert and Action Level definition and corrective action.

The Action and Alert Levels are further defined with regard to product purity as follows:

A specification excursion may trigger an extensive finished product impact investigation, substantial remedial actions within the water system that may include a complete shutdown, and possible product rejection.

- The selection of Alert and Action Levels must consider the product being manufactured as well as the method of manufacturing the product. While the levels should be extremely conservative, ensuring the safety of the public, it should be pointed out that regulatory authorities generally consider exceeding an Action Level as an "out-of-specification" (OOS) condition, requiring an evaluation of the potential effects on the quality of the product (from a microbiological standpoint) and an "Incident Evaluation Report."
- The USP *General Information* section indicates that the maximum suggested Action Levels are as follows:
  - Drinking Water: 500 cfu/mL
  - Purified Water: 100 cfu/mL
  - Water for Injection: 10 cfu/100 mL

  The EP suggested maximum limits for Purified Water and Water for Injection are similar but the indicated microbial enumeration methodologies are different. As discussed within this text, actual bacteria Alert and Action Limits for Purified Water and Water for Injection systems are generally lower than the indicated values.
- The recommended USP enumeration methodologies in the *General Information* section are as follows (USP, 2010(c)):
  - *Drinking Water*:
    - Method: pour plate or membrane filtration
    - Minimum sample volume: 1 mL
    - Growth medium: plate count agar
    - Incubation time: 42–72 hours minimum
    - Incubation temperature: 30–35°C
  - *Purified Water*:
    - Method: pour plate or membrane filtration
    - Minimum sample volume: 1 mL
    - Growth medium: plate count agar
    - Incubation time: 48–72 hours minimum
    - Incubation temperature: 30–35°C
  - *Water for Injection*:
    - Method: membrane filtration
    - Minimum sample volume: 100 mL

- Growth medium: plate count agar
 - Incubation time: 48–72 hours minimum
 - Incubation temperature: 30–35°C
- Other pharmacopeias, such as EP, reference different bacteria enumeration methods than USP. The EP *Monograph* requirements are indicated as follows:
  - *Purified Water*:
    - Method: membrane filtration
    - Minimum sample volume: chosen in relation to the expected results
    - Growth medium: R2A agar
    - Incubation time: 5 days
    - Incubation temperature: 30–35°C
  - *Water for Injection*:
    - Method: membrane filtration
    - Minimum sample volume: 200 mL, minimum
    - Growth medium: R2A agar
    - Incubation time: 5 days
    - Incubation temperature: 30–35°C
- While the USP *General Information* section and the EP *Monograph* contain specific methods for enumeration of bacteria, it is suggested that the selected method, including culture media, incubation time period, and incubation temperature be established for the specific system, process, product, and operating conditions. Ideally, selection should provide a result as quickly as possible. However, bacteria, particularly in a low-nutrient environment, may require a number of days to culture. Chapter 13 of this text addresses performance qualification testing to evaluate options for bacteria determination. As indicated, rapid microbial techniques may be coupled with conventional techniques to meet the indicated requirements.
- SOPs associated with the methodology used for bacterial monitoring should clearly state that the analyst inspect "plates" every 24 hours (as a minimum) until the recommended incubation time period is reached. This would allow the analyst to report excursions in bacterial levels as quickly as possible, minimizing the amount of product manufactured with bacterial levels exceeding the Action Limit. If performed in an adequately "air-filtered" environment, the effects of atmospheric bacteria, during observation of colonies prior to the specified incubation time, should be minimal.

BACTERIAL ENDOTOXIN SPECIFICATION

- Bacterial endotoxin specifications for USP Waters (bulk and packaged) are presented as follows:
  - Drinking Water: none
  - USP Purified Water: none
  - USP Sterile Purified Water: none
  - USP Water for Injection: <0.25 EU/mL
  - USP Sterile Water for Injection: <0.25 EU/mL
  - USP Bacteriostatic Water for Injection: <0.5 EU/mL
  - USP Sterile Water for Irrigation: <0.25 EU/mL
  - USP Sterile Water for Inhalation: <0.5 EU/mL
  - Water for Hemodialysis: <2 EU/mL

FEEDWATER PARAMETERS

- For systems producing Drinking Water, there is no defined feedwater requirement (chemical or microbial) for the system. However, the product water must meet the chemical and microbial attributes for EPA's NPDWR (or comparable regulations of the European Union, Japan, or World Health Organization), be free of coliform bacteria, and have a total viable bacteria value of <500 cfu/mL. These product water parameters apply to feedwater for a USP Purified Water or USP Water for Injection system. Obviously, since packaged waters refer back to the USP Purified Water or Water for Injection specifications, they are also produced from a system with feedwater meeting the NPDWR.

ADDED SUBSTANCES

- The *General Notices* section of USP defines an "added substance." An added substance is defined as:

    > Official substances may contain only the specific Added Substances that are permitted by the Official Monograph.

    Both the USP Purified Water and Water for Injection *Official Monographs* indicated that:

    > It contains no added substance.

- For USP Purified Water systems, there is an *exception* to the added substance *rule* that should be discussed in detail. Many USP Purified Water systems use ozone for microbial control (chap. 7). Briefly, these systems consist of a storage system, distribution pump, dissolved ozone destruct inline ultraviolet units, and a distribution loop. Ozone rapidly decomposes to oxygen. When ozone is considered as an oxidizing agent for the destruction of bacteria, or during its decomposition to produce oxygen, it is not considered an added substance. It is important, however, to consider the reaction between ozone and organic material, principally NOM (chap. 2), which may be present in USP Purified Water treated by ozone. The literature contains numerous references of undesirable, partially oxidized, NOM by-products of the ozonation process, as well as "stable" compounds, such as formaldehyde and chloroform, associated with ozonation of NOM (Faroog et al., 1977; Frank, 1987; Gurol, 1984; Legube et al., 1989; Westerhoff et al., 1998; Siddigui and Amy, 1993). Obviously, the dissolved ozone concentration, contact time, and TOC concentration of makeup water to the Purified Water storage tank are important parameters that should be addressed as discussed in chapter 7. If the Purified Water generating system does not contain a process for removing NOM (or provides only the partial removal of NOM), it is quite possible that the undesirable fragmented compounds of NOM may be produced. With the possible exception of a noted decrease in pH and an increase in carbon dioxide associated with the reaction of NOM and ozone, it is impossible to determine the rate and extent of the reaction of the NOM with ozone. Subsequently, for systems using ozone, particularly when RO is not included as a unit operation within the Purified Water generating system, periodic analysis for fragmented organic material, a potential added substance, must be performed.
- USP Purified Water system employing ozone, supplemental analysis *may* be required to identify added substances associated with the situation described above. These analyses may use gas chromatography (GC)/mass spectrometry (MS). Sample collection, transport, handling, and analysis are difficult. Quite often, the compounds produced by the reaction are not in the "library" of compounds maintained by the analytical laboratory used. These compounds are often identified as a "peak" with an estimated molecular weight and/or chemical structure. The concentration and chemical profile of fragmented organic material associated with this phenomenon will vary with seasonal and climatic changes in the water supply to the USP Purified Water generating system. Unlike the *General Information* section of USP, the *General Notices* section is an "enforceable" extension of the *Official Monographs* for USP Purified Water and USP Water for Injection.

FOREIGN SUBSTANCES AND IMPURITIES

- References and a definition of "foreign substances and impurities" are contained in the *General Notices* section of the USP. This is, perhaps, one of the most important sections of the official monographs, particularly as it relates to an "extension" of the monograph. Within this definition, the following is stated:

   > Tests for the presence of impurities and foreign substances are provided to limit such substances to amounts that are unobjectionable under conditions in which the article is customarily used.

- Any material introduced during the Purified Water generating or Water for Injection process must be considered as a foreign substance and impurity unless it is clearly demonstrated that it complies with the above quoted requirement or is fully removed subsequent to injection. For example, polymers and antiscalants introduced in some RO-based USP Purified Water generating systems, and even regenerant chemicals used for deionization units within a Purified Water generating system, must be analyzed to verify that impurities are not present that could affect final product water quality. Certificates of Analysis for material introduced, "prerelease" analytical programs conducted internally for materials introduced into a USP Purified Water generating system, and a responsive analytical monitoring program are required to ensure that foreign substances and impurities are not present in final product water.

METHOD OF PRODUCTION—WATER FOR INJECTION

- The "method of production" for Water for Injection may vary with pharmacopeias. An example is EP and USP Water for Injection production methods.
- The USP Water for Injection *Official Monograph* states that:

   > Water for Injection is water produced by distillation or a purification process that is equivalent to or superior to distillation in the removal of chemicals and microorganisms.

- The EP Water for Injection *Monograph* states that:

   > Water for Injection in bulk is obtained from water that complies with the regulations on water intended for human consumption laid down by the competent authority or from purified water by distillation in an apparatus of which the parts in contact with the water are of neutral glass, quartz, or suitable metal and which is fitted with an effective device to prevent entrainment of droplets. The correct maintenance of the apparatus is essential. The first portion of the distillate obtained when the apparatus begins to function is discarded and the distillate is collected.
   >
   > Formulation of product for consumption in countries or regions complying with EP must employ distillation for production of Water for Injection.

In all other chapters, there will be frequent references to this chapter because it has presented the defining regulations for pharmaceutical waters.

REFERENCES

United States EPA. United States Environmental Protection Agency, "Drinking Water Contaminants—National Primary Drinking Water Regulations," 2010.

EP 2010(a). European Pharmacopeia. Vol. 2, 7th ed. 2010a:3221.

EP 2010(b). European Pharmacopeia. Vol. 2, 7th ed. 2010b:3219–3220, 3224–3225.

Faroog S, Chian ESK, Engelbrecht RS. Basic concepts in disinfection with ozone. J Water Pollut Control Fed 1977; 50:1818–1831.

Frank CA. Destruction of volatile organic contaminants in drinking water by ozone treatment. Ozone Sci Eng 1987; 9(3):265–288.

Gurol MC. Factors Controlling the Removal of Organic Pollutants in an Ozone Reactor. Proceedings of the American Water Works Associates Annual Conference, Dallas, Texas, June 10–14, 1984.

Legube B, Crové JP, DeLaat J, et al. Ozonation of an extracted aquatic fulvic acid: theoretical and practical aspects. Ozone Sci Eng 1989; 11(1):69–92.
Siddigui MS, Amy GL. Factor affecting DBP formation during ozone-bromide reactions. J Am Water Works Assoc 1993; 85(1):63–72.
USP(a). United States Pharmacopeia 33—National Formulary 28, Reissue, *Official Monographs*, The United States Pharmacopeial Convention, Rockville, MD, 2010a:3870–3873.
USP(b). United States Pharmacopeia 33—National Formulary 28, Reissue, *General Information* Chapter <1231>, "Water for Pharmaceutical Purposes," The United States Pharmacopeial Convention, Rockville, MD, 2010b:740.
USP(c). United States Pharmacopeia 33—National Formulary 28, Reissue, *Official Monographs*, The United States Pharmacopeial Convention, Rockville, MD, 2010c:741.
Westerhoff P, Song R, Amy G, et al. NOM's role in bromine and bromate formation during ozonation. J Am Water Works Assoc 1998; 90(2):82–94.

2 | Impurities in raw water

INTRODUCTION

It would be inappropriate to discuss the various unit operations configured in a water purification system without discussing the impurities present in raw water. The nature, type, and concentration of various impurities should define the water purification system. This is particularly true for USP Purified Water systems and systems used for the production of active pharmaceutical ingredients, where there is limited, if any, definition of water purification, techniques required to produce the desired water quality. Raw water supplies to pharmaceutical water purification systems must, as discussed in chapter 1, meet the National Primary Drinking Water Regulation (NPDWR) as defined by the U.S. EPA or an appropriate similar agency. The requirements for feedwater to a compendial water system are set forth in the current applicable USP *Official Monograph*. It is important to recognize the myriad of different potential sources of feedwater to facilities. Raw feedwater may be from a municipal (private or public) supply or a dedicated private supply for the facility. It should be obvious that the nature and concentration of various impurities in raw water supplies will be a function of the ultimate "source" of the water. Sources may include reservoirs, lakes, streams, rivers, and groundwater. Groundwater supplies may be influenced by surface water supplies depending on the topography and depth of the aquifer. In the Appendix A of this chapter, general characteristics are presented for various impurities present in surface waters, groundwaters influenced by a surface supply, and groundwaters not influenced by a surface supply.

The U.S. EPA National Primary and Secondary Drinking Water Regulations are defined (40CFR141 and 40CFR142). The regulations are periodically updated to reflect available treatment technology, analytical detection capability, and ongoing research associated with the health effects of impurities. The Surface Water Treatment Rule (U.S. EPA, 2006b) defines specific requirements for municipal facility treatment of both surface water supplies and groundwater influenced by surface water supplies. The Disinfection Byproducts Rule (U.S. EPA, 2006c) outlines specific limits for undesirable by-products produced by the reaction of disinfecting agents with naturally occurring organic material (NOM). While impurities in raw water are discussed in this chapter, specific limits for impurities are defined by the U.S. EPA and may change periodically.

PARTICULATE MATTER

Particulate matter refers to nonsoluble inorganic or organic impurities that may exist in a water supply. Particulate matter can be considered as the residue present on a filter disk after water is passed through the filter. Particulate matter can be related to total suspended solids determined by a method similar to that outlined by Eaton et al. (2005)—Standard Methods, Procedure No. 2540D. It is important to emphasize the fact that total suspended solids should not be confused with the definition of "Total Solids" prior to the fifth supplement of USP 23. The outdated USP Total Solids measurement, performed with an evaporative technique, measured both dissolved and nondissolved impurities, with the potential exception of low molecular weight volatile impurities that are primarily organic in nature.

Any definition of particulate matter in water must reference a size. Can a nondissolved inorganic or organic material be classified as particulate in nature if it is not visible in a water solution to the human eye? In general, particles become nonvisible when the size is less than 40 µm. For water purification applications, it appears appropriate to define particulate matter as the no dissolved inorganic and/or organic material that will be removed by a particulate removal filter. While chapter 3 discusses pretreatment techniques, including particulate removal filters, a properly operating, backflushable particulate removal filter will remove particles greater than or equal to 10 µm in size (Collentro, 1994; Coulter, 1996). Subsequently, it is appropriate to establish a definition for particulate matter as nondissolved inorganic or

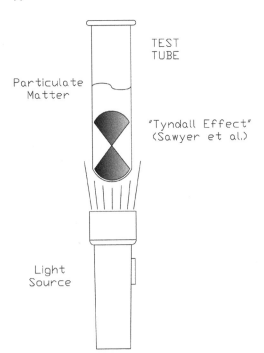

Figure 2.1 Tyndall effect. *Source*: From Sawyer et al. (1994).

organic material with a size greater than 10 μm. This material could include dirt; iron oxides or hydroxides; other metallic oxides or hydroxides; tiny organisms such as slime, algae, or iron-rich species of bacteria such as *Leptothrix* and *Clonothrix* (AWWA, 2004); and dissolved inorganic ions that react with precipitate generation cations and anions (e.g., calcium, magnesium, barium, sulfate, and carbonate).

In certain cases, particulate matter may exist as a very fine material. This is generally true for nondissolved inorganic compounds. If a sample of water-containing fine particulate matter is dispersed into a test tube with a round bottom, a darkened room and flashlight can be used to visualize this particulate matter. This technique is demonstrated in Figure 2.1. While reference to the total suspended solids measurement is related to the concentration of particulate matter, a review of a raw water analysis, particularly from a municipality, may provide an important additional analytical method as the turbidity of particulate matter. Higher total suspended solid concentrations and/or turbidity levels are an indication of high levels of particulate matter in water.

IONIC MATERIAL

Ionic material is classified as any material that both dissolves in water and produces positive ions (cations) and negative ions (anions). There are two important processes to consider: solubility and ionization. Before a substance can produce ions, it must dissolve in water. All compounds have a characteristic "solubility product," which is associated with the extent of saturation of the material in a water solution. A highly soluble material, for example, is sodium chloride (NaCl), common table salt. It readily dissolves in water, producing a sodium ion with a +1 charge and a chloride ion with a −1 charge:

$$NaCl + \xrightarrow{H_2O} Na^+ + Cl^-$$

It is important to consider that even very soluble sodium chloride can reach a saturation limit; it becomes particulate matter when its saturation value is exceeded. To verify, one can slowly pour salt into a glass of water. During this procedure, it will become obvious that the saturation point is reached when the sodium chloride no longer dissolves in the water, but settles to the bottom of the glass. Compounds such as calcium carbonate ($CaCO_3$) and magnesium carbonate ($MgCO_3$) are much less soluble in water than sodium chloride. In general, salts of monovalent, low molecular weight cations (e.g., Na) and monovalent, low molecular weight anions (e.g., Cl) are relatively soluble.

It would be inappropriate to conclude a discussion of solubility and subsequent ionization in water without addressing the relationship of ionic removal to the "dissolving" process. This concept can be explained by relating it to a water purification unit operation-water softening with a regenerant salt storage tank. When "wet" salt storage is employed within the brine tank, salt pellets remain below the level of water. When the water softener is regenerated, saturated salt solution is withdrawn from the salt storage tank. The saturated brine solution is diluted with water and regenerates the cation resin in the water softener (see chap. 3). Prior to the next regeneration cycle, "makeup water" is introduced into the salt storage tank. Previously nondissolved salt dissolves in the makeup water. The salt level in the salt storage tank decreases each time the unit is regenerated. Thus, increased solubility occurs by "dilution" of the water solution in the salt tank.

Ions are produced when certain compounds are added to water and dissolved to produce an electrically neutral solution of positive and negative ions. Ionization examples for various compounds are as follows:

$$Na_2SO_4 \xrightarrow{H_2O} 2Na^+ + SO_4^{2-}$$

$$KCl \xrightarrow{H_2O} K^+ + Cl^-$$

$$K_2SO_4 \xrightarrow{H_2O} 2K^+ + SO_4^{2-}$$

The ionic material produced can be related to a single atom (e.g., Na^+) or a radical (e.g., SO_4^-). For purposes of defining ions for water purification, it is relatively unimportant to refer to molecular weights and the associated charge (monovalent, divalent, etc.). Obviously, the ability of a compound to produce ions is extremely important to certain unit operations (e.g., ion exchange). As ions enter a water solution, the ability to conduct an electrical current increases. An indication of the concentration of ions in a water solution is determined by measuring the conductance or resistance to an applied electric current associated with a probe immersed in a water solution containing the ions. Prior to discussing the measurement of the ionic strength of a water solution, it is important to indicate that the "equivalent conductance" of all positive and negative ions is, in general, approximately equal, with the exception of the hydronium (H_3O^+) and hydroxyl (OH^-) ions (Table 2.1). This important ionic property, for water solutions with relatively neutral pH values (e.g., 5.0–9.0), can be used to approximate the total dissolved solid (TDS) level of a solution.

While most conductivity or resistivity cells used with water purification systems have a complex configuration, the basic principles associated with the measurement can be explained by referring to a classical "dip-type" cell, as opposed to a flow-through cell (Fig. 2.2). Within the cell, two parallel plates are positioned. The effective area of each plate, A, is expressed in cm^2. The length (l) between the plates is expressed in cm. A cell constant is generated by

Table 2.1 Equivalent Conductivity for Select Ions at Infinite Dilution

Ion	Molar conductivity
½Ca^{2+}	59.47
H_3O^+	349.65
½ Mg^{2+}	53.0
Na^+	50.08
½ Mn^{2+}	53.5
HCO_3^-	44.5
NO_3^-	48.3
OH^-	198
½SO_4^{-2}	80.0

Source: Handbook of Chemistry and Physics.

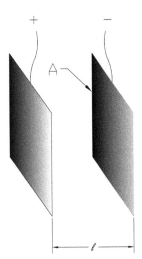

Figure 2.2 Classical conductivity "Dip Cell."

Table 2.2 Resistivity (MΩ-cm) Vs. Conductivity (μS/cm)

Resistivity	Conductivity[a]
0.01	100
0.1	10
0.2	5
0.5	2
1.0	1.0
2.0	0.5
4.0	0.25
5.0	0.20
8.0	0.125
10.0	0.100
15.0	0.067
18.2	0.055

[a]Conductivity = 1/Resistivity

dividing the length between the plates by the effective area of each plate (l/A). The conductance or resistance between the two plates is measured. The specific conductance, or conductivity, is determined by multiplying the measured conductance by the cell constant. Conversely, the specific resistance, or resistivity, is determined by dividing the resistance by the cell constant. As a result, conductivity or resistivity cannot be expressed without a dimensional term (generally cm). For the vast majority of water purification applications, conductivity is expressed a microsiemens (μS)/cm. Resistivity is expressed as Ω-cm, kΩ-cm, or, for high purity water, MΩ-cm (megohms-cm). Since water temperature increases the mobility of ions, conductivity and resistivity should reference a measurement temperature such as 25°C or be corrected to 25°C by using appropriate correction curves. The units for conductivity and resistivity are direct reciprocals of each other. Table 2.2 provides a comparison of resistivity and conductivity values that demonstrates the indicated reciprocity. The maximum obtainable resistivity for theoretically "ion-free" water is 18.2 MΩ-cm (0.055 μS/cm conductivity) at 25°C. From a theoretical standpoint, the conductivity of a solution can be calculated by using the equivalent conductance of individual ions and multiplying by the concentration of the specific ions. An approximation of the TDS level of ions in solution can be obtained by multiplying the conductivity (at 25°C) by 0.5. Again, this calculation *only* provides an estimate of the TDS value.

The level of ionic material in water determines the purity of the water. Chapter 1 discussed the chemical specifications for pharmaceutical grades of water. It would be inappropriate, and misleading, to conclude a discussion of ionic material without discussing weakly ionized substances. For sodium chloride in water, the solubility and direct ionization characteristics were presented. For pharmaceutical water applications, one can assume that the sodium chloride that dissolves in solution is ionized as sodium and chloride ions. Some salts, acids, and bases do not completely ionize. These materials will dissolve (based on their solubility product), producing, in a simplified nomenclature, an unionized portion of the original material and an ionized portion of the material. Examples of weakly ionized salts, existing in equilibrium with a dissolved "unionized" portion, are shown as follows:

$$CaCO_3 \xrightarrow{H_2O} Ca^{2+} + CO_3^{2-}$$

$$MgCO_3 \xrightarrow{H_2O} Mg^{2+} + CO_3^{2-}$$

$$BaCO_3 \xrightarrow{H_2O} Ba^{2+} + CO_3^{2-}$$

The degree of ionization for each equilibrium equation is determined by an ionization constant. The larger the ionization constant, the greater the percentage of ions present when compared with the unionized portion. The lower the ionization constant, the greater the percentage of unionized material compared to ionized material. The kinetics of the equilibrium reaction, particularly as it relates to gaseous components, is a very critical factor in determining the product water purity from a single- or double-pass reverse osmosis (RO) unit. Subsequently, it is very important to remember that certain materials are weakly ionized when evaluating the use of RO for particular applications (e.g., the nature of the feedwater supply).

Finally, certain salts, which are weakly ionized, are referred to as "acid salts" or "basic salts." Adding sodium chloride to water has no effect on the pH of the solution. However, if magnesium chloride ($MgCl_2$) is added to water, a noticeable drop in pH will occur, consistent with the following reactions:

$$MgCl_2 + 4H_2O \leftrightarrow Mg(OH)_2 \downarrow + 2H_3O^+ + 2Cl^-$$

where $\downarrow$ indicates that $Mg(OH)_2$ is not dissolved (i.e., it is a precipitate). The pH drop is associated with the product H_3O^+ hydronium ion.

Basic salts consist of a "strongly ionized anion" and "weakly ionized cation." An acid salt consists of a "strongly ionized cation" and "weakly ionized anion." In general, strongly and weekly ionized ions relate to their associated acid or base. As an example, the chloride ion is associated with hydrochloric acid, a strong acid, and is subsequently a strongly ionized anion. Magnesium is associated with magnesium hydroxide, a weak base often used as a liquid dose for temporary relief of stomach or intestinal discomfort. For water purification purposes, a general understanding of these phenomena is important. Since raw water supplies consist of both weakly and strongly ionized cations and anions, the resulting pH is affected by numerous equilibriums. It is *unimportant* to know the nature of the equilibriums involved or the equilibrium (ionization) constants. However, it *is important* to recognize that the effect of this situation is exhibited by a pH value that can be either slightly acidic or slightly basic.

The equilibrium process and nature of ions must be considered when evaluating water exhibiting a pH value that may only slightly vary from a neutral value of 7.0. A particular example relates to pH adjustment prior to RO systems. For RO systems with infrequently used (pharmaceutical applications) cellulostic membranes, it is desired to provide feedwater with a pH in the range of 5.0 to 6.5 to minimize membrane-degrading hydrolysis. This situation will be discussed in more detail in chapter 4. For most pharmaceutical RO applications, thin-film composite polyamide membranes are employed. Depending on the nature of the downstream RO polishing components, RO pH feedwater adjustment to 8.0–9.0 may be performed to minimize the presence of carbon dioxide (CO_2) (in the gaseous state). Consistent with the information associated with weakly ionized substances and acid and basic salts, it is *inappropriate* to attempt to "calculate" the amount of acid or base required to obtain desired RO feedwater pH value. The "reservoir" of unionized material will affect, in some cases to a significant extent, the actual amount of acid or base required versus the calculated amount. The only method of effectively determining the required acid and basic volume to achieve the desired pH is to "titrate" a measured volume of RO feedwater with a "standard solution" of strong acid (e.g., sulfuric acid, H_2SO_4) or strong base (e.g., sodium hydroxide, NaOH), as appropriate (cellulostic or polyamide membrane material).

DISSOLVED GASES: NONREACTIVE

Another impurity present in raw water is dissolved gases. Nonreactive dissolved gases are gases, such as oxygen and nitrogen, that do not chemically react with water to produce compounds that will introduce ionic contaminants to the solution and affect the pH of the water. The solubility of both oxygen and nitrogen in water at a constant pressure is a function of temperature, with solubility decreasing with increasing temperature. Table 2.3 presents data associated with the dissolved oxygen concentration in water as a function of temperature. Table 2.4 presents the dissolved nitrogen concentration in water as a function of temperature. The presence of dissolved nonreactive gases in water does not have any significant effect on pharmaceutical water purification system performance. The presence of oxygen in product water for systems using stainless steel storage and distribution systems can oxidize stainless steel surfaces (see subsequent chapters of this book). However, with very minor exceptions, the removal of dissolved nonreactive gases in pharmaceutical water purification systems is seldom employed.

Table 2.3 Dissolved Oxygen Concentration in Water Vs. Temperature

Temperature (°C)	Dissolved oxygen concentration (mg/L)
10	12.84
20	10.47
30	8.84
40	7.85
50	7.14
60	6.68
70	6.33
80	6.11
90	6.01
100	5.99

Source: From Perry et al. (2007).

Table 2.4 Dissolved Nitrogen Concentration in Water Vs. Temperature

Temperature (°C)	Dissolved nitrogen concentration (mg/L)
10	18.40
20	15.28
30	13.30
40	11.82
50	10.88
60	10.24
70	9.83
80	9.75
90	9.75
100	9.75

Source: From Perry et al. (2007).

DISSOLVED GASES: REACTIVE

Since the mid-1980s, single-pass and double-pass RO systems have gained popularity for the production of USP Purified Water, certain active pharmaceutical ingredient manufacturing operations, and for pretreatment of feedwater to distillation units in Water for Injection systems. The presence of dissolved reactive gases can have a dramatic impact on the quality of product water. The primary dissolved reactive gases of concern are carbon dioxide and ammonia (NH_3). On the basis of their significant impact on state-of-the-art technology, it is appropriate to provide a comprehensive explanation of the source, anticipated concentration, and effect on pH of both gases.

Carbon dioxide is present in air and readily dissolves in water. Its solubility as a function of temperature is presented in Table 2.5. As a reactive gas, carbon dioxide undergoes equilibrium chemical reactions with water as follows:

$$(1) CO_2 + 2H_2O \leftrightarrow H_3O^+ + HCO_3^-$$

$$(2) HCO_3^- + H_2O \leftrightarrow H_3O^+ + CO_3^{2-}$$

The primary reaction encountered at anticipated pH values of raw water is the first equation. Obviously, equilibrium is affected by pH, the lower the pH value, the greater the hydronium ion concentration. As the hydronium ion concentration increases (pH decreases), the equilibrium reaction is forced to the "left," producing more carbon dioxide and less bicarbonate ion (HCO_3^-). Conversely, as the pH increases, the hydronium ion concentration decreases. Excess hydroxide ions, associated with the higher pH, will react with hydronium ions, producing water by the following reaction:

$$H_3O^+ + OH^- \leftrightarrow 2H_2O$$

As hydronium ions are removed, the equilibrium reaction is forced to the "right," resulting in a lower concentration of carbon dioxide and higher concentration of bicarbonate ions. The relationship between the hydronium ion concentration, pH, the carbon dioxide concentration, and bicarbonate ion concentration is presented in Table 2.6. When ion exchange is used as the

Table 2.5 Solubility of Carbon Dioxide in Water Vs. Temperature

Temperature (°C)	Total carbon dioxide solubility in water (mg/L)
10	0.73
20	0.53
30	0.40
40	0.31
50	0.24
60	0.18

Note: Does *not* consider the reaction of carbon dioxide with water.
Source: From Perry et al. (2007).

Table 2.6 Carbon Dioxide/Bicarbonate Equilibrium Concentration as a Function of pH[a]

pH	Free CO_2 in water (mg/L)	Bicarbonate ion (mg/L)
5.5	78.3	14.7
6.0	62.2	37.1
6.5	37.7	71.1
7.0	16.8	100
7.5	7.0	114
8.0	2.1	120

[a]Calculation assume a bicarbonate concentration of 100 mg/L (as bicarbonate ion) at pH 7.0. Values are calculated using an equilibrium constant of 4.3×10^{-7} at 25°C.

Table 2.7 Ammonia/Ammonium Equilibrium Concentrations as a Function of pH[a]

pH	Free NH$_3$ in water (mg/L)	Ammonium ion (mg/L)
6.0	0.0005	~1.0
6.5	0.002	~1.0
7.0	0.005	1.0
7.5	0.02	0.99
8.0	0.06	0.95
8.5	0.21	0.79
9.0	0.34	0.64
9.5	0.61	0.36
10.0	0.81	0.15

[a]Calculation assume an ammonium concentration of 1.0 mg/L (as ammonium ion) at pH 7.0. Values are calculated using an equilibrium constant of 1.77×10^{-5} at 25°C.

ion removal technique within a pharmaceutical water purification system, the equilibrium associated with carbon dioxide has minimal effect on the quality of the product water, since bicarbonate ions are rapidly and continuously removed during passage through anion resin, forcing the equilibrium reaction to the "right" and eliminating carbon dioxide by removing the bicarbonate ions. On the other hand, single- or double-pass RO systems will allow carbon dioxide to pass *directly* through the RO membrane as a gas. The product water carbon dioxide, which will reestablish equilibrium with bicarbonate and hydronium ions, affects the purity of the product water by increasing its conductivity.

Ammonia gas is generally present at much lower concentrations than carbon dioxide in raw water. Ammonia can be associated with chloramines (an alternate municipal distribution system disinfecting agent), which are produced by the reaction of chlorine and ammonia. Relatively high concentrations of residual chloramines, which are required to obtain the same or similar disinfection properties as chlorine, can be associated with the production of ammonia by the following equation:

$$2NH_2Cl \leftrightarrow NHCl_2 + NH_3$$

Ammonia is extremely soluble in water and will react with water to produce the ammonium (NH_4^+) and hydroxide ions as follows:

$$NH_3 + H_2O \leftrightarrow NH_4^+ + OH^-$$

Equilibrium is affected by pH. As the pH decreases, hydroxide ions react with hydronium ions, forcing the equilibrium reaction to the right, thus decreasing the concentration of ammonia gas and increasing the concentration of ammonium ion. As the pH increases, the concentration of hydroxide ions increases, forcing the equilibrium reaction to the left, thus increasing the ammonia gas concentration and decreasing the ammonium ion concentration. Table 2.7 presents the relative concentrations of ammonia gas, the ammonium ion, and the hydroxide ion as a function of pH. Ammonia gas passes directly through an RO membrane, decreasing product water conductivity. In ion exchange systems, ammonia gas is entirely eliminated as the ammonium ion passes through cation resin and the equilibrium equation is forced to the right.

While there are other "reactive" gases, carbon dioxide and ammonia are the two primary reactive gases of concern for pharmaceutical water purification systems. Considering the increased use of single- and double-pass RO units and the USP conductivity specification for Purified Water and Water for Injection, the concentration of carbon dioxide and ammonia is an important parameter.

MICROORGANISMS

Numerous microorganisms may exist in a raw water supply. Microorganisms can include viable bacteria, nonviable bacteria, slime, algae, viruses, and cysts. Chapter 9 discusses microbial monitoring techniques for pharmaceutical water systems.

Table 2.8 Raw Water Total Viable Bacteria Levels Determined by PCA and DBA

Sample no.	Culture media	Total viable bacteria (cfu/mL)
1	PCA	40
1	DBA	~4300
2	PCA	30
2	DBA	~13,000
3	PCA	4
3	DBA	470
4	PCA	190
4	DBA	~13,200
5	PCA	60
5	DBA	~5800

Notes: PCA results by plate count agar (tryptone, yeast extract, glucose, agar, and reagent-grade water), heterotrophic plate count of 1.0 mL (pour-plate) 22°C, and 48 hours incubation time period. DBA by diluted broth agar (peptone, agar, and reagent grade water), heterotrophic plate count (spread-plate), 25°C, and 10 to 14 days incubation time period.
Source: From van der Kooij (1992).

Table 2.9 Raw Water Total Viable Bacteria Levels Determined by PCA and R2A

Identification	Culture media: incubation time period (hr)	Total viable bacteria (cfu/mL)
EPA study: Buchanan, West Virginia (Cross, 1993)	PCA: 48	103
EPA study: Buchanan, West Virginia (Cross, 1993)	R2A: 48	~360,000
EPA study: Lake Havasu, CA (Cross, 1993)	PCA: 48	170
EPA study: Lake Havasu, CA (Cross, 1993)	R2A: 48	~36,000
Reasoner and Geldreich (1985)	PCA: 48	390
Reasoner and Geldreich (1985)	R2A: 48	~1500
Reasoner and Geldreich (1985)	PCA: 72	400
Reasoner and Geldreich (1985)	R2A: 72	~3400
Reasoner and Geldreich (1985)	PCA: 168	430
Reasoner and Geldreich (1985)	R2A: 168	~3800

Note: Cross data: Heterotrophic Plate Count per "Standard Methods" Section 9215, 35°C, 48-hour incubation time period.
Reasoner and Geldreich Data: Heterotrophic Plate Count per "Standard Methods" Section 9215, (spread-plate), at 35°C.

Measured bacterial levels in raw water supplies are significantly affected by the enumeration culture media. Historically, a heterotrophic plate count, as outlined in *Standard Methods for the Examination of Water and Wastewater* (Eaton et al., 2005, Section 9215), is used with standard plate count agar (PCA) to determine the total viable bacteria levels in raw water. Tables 2.8 and 2.9 indicate how highly sensitive media, such as R2A, are not generally appropriate for determining total viable bacteria levels in raw water. Observed total viable bacteria levels with R2A as compared to PCA are at least one order of magnitude greater (Bartdoni, 2006; Massa, 1998). The literature indicates that only a small fraction of bacteria in Drinking Water are detected by even the most selective culture media (Rosenfeldt, 2009). Further, while R2A culture media may indicate higher bacteria values than other media, it requires a much longer incubation time period that classical PCA (Reasoner and Geldreich, 1985). This inhibits the ability to detect a total viable bacteria problem in time to avoid public health issues. While total viable bacteria levels in drinking water are a concern, Total Coliform measurement provides a rapid method of excursions when augmented with tests for "other" regulated microorganisms. A maximum total viable bacteria Action Limit historically stated in the nonbinding *General Information* Section of the USP is 500 cfu/mL.

There are critical issues that should be addressed regarding the presence of bacteria and other microorganisms in raw water supplies. If the municipality employs chloramines as the residual disinfecting agent in the distribution system, it should be recognized that three chloramine compounds will exist. The concentration of each compound is a function of pH and affects the disinfecting properties. Large microorganisms, such as cysts, require time for total destruction (Table 2.10). Table 2.10 also verifies the fact that it is possible for viable bacteria and residual disinfecting agent to exist simultaneously. Municipalities will generally perform multiple chlorination steps through the treatment process. After final chlorination, municipalities generally prefer to establish minimum contact time intervals to the "first user" from the distribution system, as further demonstrated by the data for *Escherichia coli* in Table 2.11. Residual disinfectant should not be considered as a source operation for removing either *Giardia* or *Cryptosporidium* (EPA, 2009; Korich et al., 1990; Wickramanayake et al., 1984). The U. S. EPA Surface Water Treatment Rule provides specific guidelines for municipal plant treatment to insure complete removal of cysts. *Giardia* alone is attribute at least 100,000 to 2,500,000 million infections a year in the United States and is the most prevalent protozoan parasite in the world (Furness et al., 2000; Kucik, 2004).

Table 2.10 Contact Time Required for Chlorine Inactivation of Cysts and Viruses

Cyst: *Giardia lamblia*	
Residual chlorine concentration (mg/L)	Contact time (min)
0.1	400
0.2	200
1.0	90

Notes: Temperature = 5°C.
pH 6.0.
Inactivation ≥99%.
Source: From Hoff and Akin (1986).
U.S. EPA SWTR requires ≥99.9% removal. Filtration is generally coupled with filtration. Data do not imply that chlorination should be solely employed for removal of *Giardia lamblia* but demonstrates the significant contact time required for ≥99% inactivation at the stated conditions.

Poliovirus	
Residual chlorine concentration (mg/L)	Contact time (min)
0.1	2.5
0.2	1.5
1.0	0.6

Notes: Temperature = 2°C.
pH 6.0.
Inactivation ≥99%.
Source: From Haas and Karra (1984).

Table 2.11 Time Required for 99% Inactivation of *Escherichia coli*

Temperature (°C)	Residual chlorine concentration (mg/L)	Time (min)
5	0.1	7.0
5	0.2	4.0
5	1.0	0.6
20–25	0.1	4.5
20–25	0.2	1.5
20–25	1.0	0.6

Notes: pH 8.5.
Source: From Haas and Karra (1984).

Total Coliform bacteria should be absent in Drinking Water. For municipalities collecting fewer than 40 routine samples per month, no more than one sample can indicate "coliform-positive" per month (U.S. EPA, 2006b). Every sample that indicates positive for Total Coliform must be analyzed for either *Escherichia* or fecal coliform. If two consecutive Total Coliform-positive samples indicate the presence of *E. coli* or fecal coliform, the municipal system in violation of the U.S. EPA NPDWR. The presence of Fecal Coliform and *E. coli* indicate that the water is contaminated with human or animal waste containing pathogens. Procedures for determining both total coliforms, fecal coliform and *E. coli*, are presented by Eaton et al. (2005, Section 9221).

A myriad of types and species of bacteria can be present in raw water. The nature of bacteria present is related to available nutrients. Bacteria are extremely flexible in adapting to environmental conditions, principally the degree of organic and inorganic nutrients present. When residual disinfectant is added to a raw water supply, or attempts are made at removing inorganic and organic nutrients, the extremely low nutrient conditions produce a situation where the majority of the bacteria present are classified as "gram negative." These bacteria do not retain Gram's stain in a standard staining method. The presence of objectionable gram-negative species of bacteria is a concern for pharmaceutical and related water purification applications, from raw feedwater to product water. Gram-negative bacteria, adapting to low nutrient environments, can survive by maximizing the surface area to volume ratio. With the exception of specific microorganisms stated in the preceding text, the U.S. EPA NPDWR does not contain a definitive limit for total viable bacteria.

BACTERIAL ENDOTOXINS

Bacteria present in a raw water supply exists in a "dynamic" state. Depending on environmental conditions controlling the rate of growth, bacteria will continually proliferate, while other bacteria are "dying." Both gram-negative and gram-positive bacteria can be present in a water supply. Gram-positive bacteria have a single cell wall. Gram-negative bacteria, on the other hand, have an inner membrane and a second outer membrane. The presence of this outer membrane is extremely important. It maximizes the gram-negative bacteria's ability to receive nutrients from the environment into the cell. Further, the outer membrane is composed of phospholipid and, most importantly, lipopolysaccharide and protein (Gould, 1993). The lipopolysaccharide is extremely stable. It is considered an endotoxin because the toxin is synthesized as part of (or endogenous to) the bacteria cell structure. In other words, when gram-negative bacteria are destroyed, bacterial endotoxins are released. If bacterial endotoxins are introduced into the human bloodstream, a body's defense mechanism is activated. This defense mechanism may result in an elevated body temperature. The nature of the body's response to bacterial endotoxins (lipopolysaccharides) may result in death. On the basis of the body's response to bacterial endotoxins, they are often referred to as pyrogens, from the Greek word meaning "fire" due to the fever produced when injected into the human bloodstream.

Since it is fully anticipated that gram-negative bacteria will be found in raw water supplies, particularly supplies from a surface source or groundwater source influenced by a surface source, and since the raw water supplies will contain a mechanism for destroying bacteria, bacterial endotoxins may be present in raw water supplies. The removal of bacterial endotoxins from raw water supplies is critical to the production of any material that will be introduced or come in contact with the bloodstream. Further, the presence of bacterial endotoxins in non-USP Water for Injection systems (USP Purified Water) used for many research applications is also undesirable. Finally, bacterial endotoxins are associated with biofilm, as discussed in chapter 9.

ORGANIC MATERIAL

Organic material present in raw water supplies can be classified into two general categories: naturally occurring and pollutants. NOM will be present in raw water supplies from a surface water source or groundwater source influenced by surface water. In general, the concentration of NOM in a water supply from a groundwater source will be extremely low, which is dependent on the source of the groundwater, the nature of the aquifer, the depth of the aquifer yielding the raw water supply, and the physical characteristics and makeup of the soil between the ground surface and actual well water withdrawal point.

NOM present in raw water supplies can be classified into a light molecular weight fraction (fulvic acid) and a heavy molecular weight fraction (humic acid) (Aiken and Cotsaris, 1995; Owen et al., 1993; AWWRF, 1993). Theoretically, a series of chemical reactions is used to determine if a naturally occurring organic molecule is classified as fulvic acid or humic acid (Aiken, 1988; Owen et al., 1995). However, for purposes of defining these materials for water purification unit operation treatment techniques, molecular weight can be considered as an excellent indicator. The fulvic acid fraction of NOM contains an identifiable chemical structure and is, in general, removed and controlled within pharmaceutical water purification systems. On the other hand, humic acid exhibits "snowflake-like" properties with a large range of molecular weight molecules and varying chemical structure (Kunin, 1981; Amy et al., 1987; Logan and Jiang, 1990). The literature contains references to a proposed chemical structure for humic acid molecules (Kunin, 1981; Croue et al., 1993). While these chemical structures may represent a general description of humic acid molecules, it is impossible to define the nature of the molecule accurately, since humic acid has no single chemical structure but rather an infinite number of molecular structures. Humic acid is extremely important in pharmaceutical water purification systems. For example, its concentration and characteristics can determine the operating life of activated carbon media (for organic removal); fouling of anion exchange resins in deionization systems; fouling of membrane processes, such as ultrafiltration or, to a much greater extent, RO membranes. Thermal decomposition products of NOM include chloride ions that contribute to chloride stress corrosion and pitting of austenitic stainless steel surfaces in heated unit operations, such as distillation (Collentro, 1987; Dvorin and Zahn, 1987; D'Auria et al., 1987). In USP Purified Water systems with feedwater from a surface source (or groundwater source influenced by a surface source), water purification system without RO, and ozonation of storage and distribution system for microbial control, "Added Substances," as defined by the USP *Official Monograph* for Purified Water, may be produced. This can be attributed to incomplete oxidation of NOM when exposed to ozone (Glaze and Weinberg, 1993; Digiano et al., 2001). In fact, the reaction of NOM with residual disinfectant produces undesirable compounds (to be discussed later in this chapter).

The vast majority of pharmaceutical water purification systems employ RO as the primary ion removal technique. The presence and required removal of NOM is critical to the long-term successful operation of RO-based systems (see chap. 4). While organic fouling of membranes can occur, microbial fouling attributed to the uncontrolled proliferation of bacteria on membrane surfaces from nutrients provided by NOM is extremely important. Subsequently, while "delivered water" TOC measurement for both USP Purified Water and Water for Injection is specified within the individual *Official Monographs*, pharmaceutical water purification systems with feedwater containing NOM should include provisions for monitoring (and trending) TOC levels from the raw water inlet through each unit operation.

A second and increasingly important source of organic material present in a raw water supply is pollutants (e.g., industrial waste, fertilizers, and pesticides). While the number of potential raw water supplies containing pollutants would appear to be minimal, unlike NOM, pollutants may be noted in groundwater supplies. This situation is compounded by the fact that groundwater supplies feeding a pharmaceutical water purification system may be from a "private" source rather than a "municipal" source. The U.S. EPA has defined a significant number of organic compounds that are considered pollutants, with maximum concentrations for each (U.S. EPA, 2009). If the raw water supply to a facility is from a "private" groundwater source, appropriate analysis may not be performed with adequate frequency to identify the presence/absence of pollutants. Further, since groundwater generally exhibit minimal mixing in a vertical direction, seasonal and climatic conditions affecting the actual "depth" from the ground to the top of the aquifer (and, subsequently, to the draw-off point for wells) will have a pronounced effect on the representative nature of any measurement associated with potential pollutants.

COLLOIDS

Colloids are substances that are much larger than ionic material but much smaller than particulate matter. In general, colloids will exhibit a size of 0.01 to 0.1 µm or larger. On the basis of the ability of ultrafiltration membranes with stated "molecular weight cutoffs" to remove colloids, it would appear that colloids have a molecular weight in the range of 10,000 to 100,000 Da or higher. While colloids are commonly found in raw water supplies from a surface water source (or groundwater source influenced by a surface source), they can also be present in raw water supplies from a groundwater source. Further, an increasing number of municipal treatment facilities inject chemicals to convert both soluble and particulate forms of iron to colloidal iron to minimize staining of domestic facilities using water. For water purification systems, colloids of silica, iron, and aluminum are important. For raw water supplies from a surface source, or a groundwater source under the influence of a surface source, colloids are generally complexed with organic material (Fig. 2.3). Colloids exhibit a slight negative charge. In high-purity water systems, this negative charge may be balanced with a hydronium ion. The high equivalent conductance of hydronium ions in a high-purity water system, for even trace amounts of colloidal material, will inhibit the production of water with a conductivity equivalent to that of theoretically ion-free water. Experimental results from pilot studies demonstrating this situation are presented in Table 2.12. The presence of colloids in pharmaceutical water purification systems and support systems is important. For example, the presence of colloidal silica in the feedwater to a multiple-effect distillation unit may result

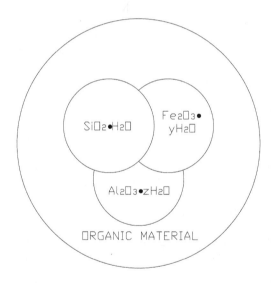

Figure 2.3 The universal colloid. *Source*: From Kunin (1980).

Table 2.12 Effect of Colloidal Silica Concentration on Water Purification System Product Water Quality

Sample no.	TOC concentration (µg/L)	Colloidal silica concentration (µg/L)	Conductivity (µS/cm at 25°C)
1	110	16	0.059
2	160	16	0.074
3	350	50	0.40
4	40	2	0.057

Notes: Water purification system includes pretreatment, two-bed deionization, and final mixed bed polisher. Sample Nos. 1, 2, and 3 are from the effluent of the mixed bed polisher. Sample 4 is from the effluent of an ultrafiltration unit positioned downstream of the mixed bed polisher. Raw feedwater is from a river source. Sample No. 3 was collected after a two-day period of heavy rain.
Source: From Collentro (1993).

in "volatilization" of the colloidal silica, which produces deposits on vapor-liquid disengaging sections positioned above the evaporator sections of the distillation units (see chap. 5). If the deposits are not removed during periodic maintenance, liquid "carryover" to the distilled product water may occur, potentially resulting in the presence of bacterial endotoxins in product water. The presence of colloids in feedwater to an RO unit can result in colloidal fouling of the membranes, requiring periodic chemical cleaning of the membranes (see chap. 4).

RESIDUAL DISINFECTANT

Residual disinfectant will be present in raw water municipal supplies from a surface source or groundwater source influenced by a surface source. The U.S. EPA Ground Water Rule (GWR) (U.S. EPA, 2006a) requires "community water systems" with potential introduction of viral and bacterial pathogens to conduct a "risk-targeting approach" that consists of four components, one of which is treatment to provide at least a 4-log inactivation or removal of viruses as well as compliance with the Total Coliform Rule (U.S. EPA 815-F-06-003). While raw water supplies from a municipal facility with a groundwater source may not exhibit bacteria at the treatment facility, it is highly likely that bacteria will be introduced during distribution. Subsequently, raw water supplies from a surface source, a groundwater source influenced by a surface source, and a groundwater source will be treated with residual disinfectant to destroy bacteria. In general, raw water supplies in large metropolitan areas in the United States are from a surface water source. Table 2.13 demonstrates that a high percentage of the U.S. population obtains raw water from a relatively small percentage of municipal water supplies. This table also demonstrates the high percentage of municipal water suppliers providing water to a relatively small population. These data are important since the selection of residual disinfectant is greatly enhanced by the source of feedwater to a treatment facility.

Chlorination of municipal water supplies in the United States began around 1908 (McGuire, 2006). Chlorine is an extremely effective disinfecting agent. When added to water, chlorine produces a mixture of hypochlorous acid and hypochloric acid, which then produces hypochlorite ion, as outlined by the following reactions:

$$Cl_2 + 2H_2O \leftrightarrow HOCl + H_3O^+ + Cl^-$$

$$HOCl + H_2O \leftrightarrow H_3O^+ + OCl^-$$

The ratio of hypochlorite ion to hypochlorous acid is a direct function of pH. For example, at a pH of 7.5, equal concentrations of hypochloric acid and hypochlorite ions will be present. However, at a pH of 9, approximately 95% of the chlorine "residual" occurs as hypochlorite ion, a very powerful oxidant. Unfortunately, research conducted in the late 1960s and 1970s indicated that chlorination of municipal water supplies containing NOM and certain inorganic matter resulted in the production of disinfection by-products—trihalomethanes (THMs) (Kruithof, 1986) and haloacetic acids (HAA5) (Arora et al., 1997). Research identified that the humic acid fraction of NOM, the trihalomethane "precursor," is the primary mechanism for production of the THMs (Reckhow and Singer, 1990; Edzwald et al., 1985; Young and Singer, 1979). In November 1979, the U.S. EPA established a maximum total THM level of 100 μg/L (U.S. EPA, 1979). The Stage 1 Disinfection Byproducts Rule reduced the maximum average total trihalomethane limit to 80μg/L. The Stage 2 Disinfection Byproduct

Table 2.13 Distribution of U.S. Community Water Systems by Size

System size (population served)	Percent of total number of systems	% of U.S. population served
<500	55.8	1.4
501–3300	26.9	5.3
3301–10,000	9.7	8.0
10,001–100,000	6.7	26.5
>100,000	1.0	58.8

Notes: Estimated total number of systems: 170,000.
Source: U.S. EPA (2002).

Table 2.14 Hypochlorous Acid-Ammonia Reactions Producing Chloramines; Chloramine Equilibrium Vs. pH

$$NH_3 + HOCl \rightarrow H_2O + NH_2Cl \text{ (monochloramine)}$$
$$NH_2Cl + HOCl \rightarrow H_2O + NHCl_2 \text{ (dichloramine)}$$
$$NHCl_2 + HOCl \rightarrow H_2O + NCl_3 \text{ (trichloramine)}$$

pH	% NH_2Cl	% $NHCl_2$	% NCl_3
4.0	5	25	70
5.0	65	35	0
6.0	80	20	0
7.0	95	5	0
8.0	100	0	0

Source: From Collentro (1985) and Glaze (1990).

Rule (U.S. EPA, 2006c) provides a phased distribution system monitoring program outlining enforcement dates based on the number of individual served by the municipal treatment facility.

In an attempt to provide excellent disinfecting properties while minimizing the production of THMs, many treatment facilities employ primary disinfection with chlorine and final disinfection, prior to distribution, with a mixture of chlorine and ammonia, producing an alternate disinfecting agent, chloramines. Hypochlorous acid produced by the reaction of chlorine with water will react with ammonia to form multiple species of chloramines (Table 2.14). This table also contains the percentage of monochloramine, dichloramine, or trichloramine present at various pH values. The disinfecting properties of trichloramines and dichloramines are poor when compared with monochloramines. As a result, the pH of systems using chloramines for disinfection is generally maintained at a value greater than 7–8. In certain larger and "older" municipalities, where many domestic "end users" may have lead piping, pH values are maintained at 9.0 or higher. While several articles have been published presenting arguments associated with the disinfecting properties of chloramines versus chlorine (Neden et al., 1992; Rosenfeldt et al., 2009), it appears, in general, that residual chloramines at a concentration of approximately 3.0 mg/L will produce the same disinfecting properties as residual chlorine at a concentration of 0.5 to 1.0 mg/L. Some researchers have questioned the ability of monochloramine to destroy certain viruses effectively, primarily viruses in the polio family (Esposito, 1974; Kelly and Sanderson, 1958, 1960).

The transition of residual disinfecting agents from chlorine to chloramines (monochloramine) significantly affects pharmaceutical water purification systems. This is particularly true for systems using thin-film composite polyamide RO membranes that are incapable of tolerating trace concentrations of residual disinfectant; the most popular membranes used in pharmaceutical water purification systems. For these systems, a conservative design of activated carbon units for residual disinfectant removal must be maintained. These design features are discussed in detail in Chapter 3. In addition, raw water supplies containing residual chloramines will also require more frequent replacement of activated carbon media, also discussed in chapter 3. Finally, for raw water supplies using chloramines as a residual disinfectant, the ability of *conventional* double-pass RO to produce USP Purified Water (conductivity specification) is extremely poor.

A limited number of municipal treatment facilities employ chlorine dioxide or ozone for primary disinfection. These agents produce limited disinfection by-products when compared to chlorine and are subsequently attractive primary disinfecting agents (Digiano et al., 2001).

It would be inappropriate to conclude any discussion of residual disinfectant without discussing general operating "trends" associated with seasonal and climatic changes. For systems using residual chlorine, THM production will continue to occur in distribution piping from the municipal treatment facility. Further, bacteria will also deplete the residual disinfectant at an increased rate as the temperature of the surface water increases with seasonal fluctuations, or if the water contains a higher degree of bacteria associated with severe climatic conditions. Some municipalities may elect to "shave" residual chlorine concentrations to ensure that the total THM level is not exceeded. Under these conditions, particularly during summer months, it is possible that extremely low residual disinfectant concentrations may be present in the feedwater to a facility. This situation may require initial injection of residual disinfectant to ensure that the feedwater supply to the pharmaceutical water purification system meets the U.S. EPA's NPDWR and the 500 cfu/mL total viable bacteria level historically indicated in the USP *General Information* section. Well water supplies influenced by a surface source treated with chloramines will also exhibit seasonal and climatic changes. In general, however, since the production of THMs is not a "controlling" factor, the increased demand for residual disinfectant is generally addressed by increasing the concentration of residual chloramines. Again, for systems using RO (or double-pass RO) as the ion removal technique, a decrease in product water quality (increase in RO unit product water conductivity) should be anticipated (for conventional RO units), since higher concentrations of ammonia associated with the increased chloramine concentration will affect the RO system product water quality. Alternative techniques to compensate for the presence of ammonia will be discussed in chapter 4.

DISINFECTION BY-PRODUCTS

As indicated earlier, organic and inorganic matter in source water will react with residual disinfecting agents to produce undesirable by-products. Table 2.15 presents a list of U.S. EPA NPDWR regulated disinfection by-products. Chlorine reacts with heavy molecular NOM and bromide producing THMS and HAA5. THMs consist of chloroform, bromodichloromethane, dibromomonochloromethane, and bromoform. These compounds are all carcinogens. The U.S. EPA annual average concentration limit for total trihalomethanes is 0.080 µg/L (U.S. EPA, 2006). The HAA5 compounds include monochloroacetic acid, dichloroacetic acid, trichloroacetic acid, monobromoacetic acid, and dibromoacetic acid. The U.S. EPA annual average concentration limit for HAA5 is 0.060 µg/L (US EPA, 2006).

Bromate is produced by the reaction of bromide and ozone when ozone is employed as a primary disinfecting agent. The U.S. EPA annual average concentration limit for Bromate is 0.010 µg/L (U.S. EPA, 2006).

Chlorite is produced by the reaction of chlorine dioxide with NOM and inorganic matter. The U.S. EPA annual average concentration limit for chlorite is 1.0 mg/L (U.S. EPA, 2006).

Toxicology studies with laboratory animals indicate that disinfection by-products such as chloroform, bromoform, brmodichloromethane, dichloroacetic acid, and bromate are

Table 2.15 Disinfection By-products and Source of By-products

Compound	Source
Bromate	Disinfection byproduct of ozone with naturally occurring bromide
Chlorite	Disinfection byproduct of chlorine dioxide
Chloroform (THM)	Disinfection byproduct of reaction of NOM, inorganic matter and disinfecting agents (chlorine and chloramines)
Bromodichloromethane (THM)	Disinfection byproduct of reaction of NOM, inorganic matter and disinfecting agents (chlorine and chloramines)
Dibromochloromethane (THM)	Disinfection byproduct of reaction of NOM, inorganic matter and disinfecting agents (chlorine and chloramines)
Bromoform (THM)	Disinfection byproduct of reaction of NOM, inorganic matter and disinfecting agents (chlorine and chloramines)
Monochloroacetic acid (HAA5)	Disinfection byproduct of reaction of NOM, inorganic matter and disinfecting agents (chlorine and chloramines)
Dichloroacetic acid (HAA5)	Disinfection byproduct of reaction of NOM, inorganic matter and disinfecting agents (chlorine and chloramines)
Trichloroacetic acid (HAA5)	Disinfection byproduct of reaction of NOM, inorganic matter and disinfecting agents (chlorine and chloramines)
Monobromoacetic acid (HAA5)	Disinfection byproduct of reaction of NOM, Inorganic matter and disinfecting agents (chlorine and chloramines)
Dibromoacetic acid (HAA5)	Disinfection byproduct of reaction of NOM, inorganic matter and disinfecting agents (chlorine and chloramines)

Source: U.S. EPA (2009).

carcinogens. Chlorite, bromodichloromethane, and certain haloacetic acids cause adverse reproductive and/or development effects in laboratory animals (U.S. EPA, 1998).

Chloroform is the predominant THM compound. Historically employed activated carbon has limited ability to remove chloroform. The presence of chloroform in final product water from a USP Purified Water system is common when THMs are present in feedwater. Further, for USP Water for Injection systems, chloroform, a volatile organic compound, may be present in distilled product water (Kroneld, 1991). Again, the concentration of chloroform in USP Water for Injection systems with raw feedwater from a surface source should also be determined periodically.

REFERENCES

Aiken GR. A critical evaluation of the use of macroporous resins for the isolation of aquatic humic substances. In: Frimmel FH, Christman RF, eds. Humic Substances and Their Role in the Environment. New York: John Wiley & Sons, 1988.

Aiken G, Cotsaris E. Soil and hydrology: their effect on NOM. J Am Water Works Assoc 1995; 87(1):36–45.

Amy GL, Collins MR, Kuo CJ, et al. Comparing gel permeation chromatography and ultrafiltration for the molecular weight characterization of aquatic organic matter. J Am Water Works Assoc 1987; 79(1): 43–49.

Arora H, LeChevallier MW, Dixon KL. DBP—occurrence survey. J Am Water Works Assoc 1997; 89(6):60–68.

AWWA. Problem Organisms in Water: Identification and Treatment. In: Manual of Water Supply Practices. 3rd ed. Denver: American Water Works Association, 2004:7–17.

AWWRF. Natural Organic Matter in Drinking Water—Origin, Characterization, and Removal (workshop proceedings). Denver: American Water Works Research Foundation/Chamonix, France: Lyonnaise des Eaux-Dumez, 19–22 September, 1993.

Bartolani JA, Porteous NB, Zarzabal LA. Measuring the validity of two in-office water test kits. J Am Dental Assoc 2006; 137:363–371.

Collentro WV. Pretreatment—activated carbon filtration, part I. Ultrapure Water 1985; 2(4):24–33.

Collentro WV. A new approach to the production of ultra high purity, low TOC water. Paper presented at the First Annual High Purity Water Conference and Exposition (12–15 April). Philadelphia: Tall Oaks Publishing, 1987: 20–22.

Collentro WV. An overview of present and future technologies for semiconductor, pharmaceutical and power applications. Ultrapure Water 1993; 10(6):20–31.

Collentro WV. U.S.P. purified water systems: discussion of pretreatment, part I. Pharm Technol 1994; 18 (4):38–46.

Coulter SL. Pretreatment—organics, TOC, color, turbidity, and SDI pretreatment for unit operations. Ultrapure Water 1996; 13(7):54–59.

Cross RA. Purification of drinking water with ultrafiltration. In: Proceedings of the Eleventh Annual Membrane Technology/Separations Planning Conference. Newton, Mass., USA: Business Communications Company, Inc./Filtrex, Inc., October 11–13, 1993.

Croue JP, Martin B, Dequin A, et al. Isolation and characterization of dissolved hydrophobic and hydrophilic organic substances of a reservoir water. Workshop Proceedings: Natural Organic Matter in Drinking Water—Origin, Characterization, and Removal. Denver: American Water Works Research Foundation/Chamonix, France: Lyonnaise Des Eaux-Dumez, 1993:73–81.

D'Auria G, Itteilag T, Pastrick R. The impact of reverse osmosis on make-up water chemistry at millstone unit two nuclear power station. Paper presented at the First Annual High Purity Water Conference and Exposition (12–15 April). Philadelphia: Tall Oaks Publishing, 1987:1–6.

Digiano FA, Singer PC, Parameswar C, et al. Biodegradation kinetics of ozonated NOM and aldehydes. J Am Water Works Assoc 2001; 93(8):92–104.

Dvorin R, Zahn J. Organic and inorganic removal by ultrafiltration. Paper presented at the First Annual High Purity Water Conference and Exposition (12–15 April). Philadelphia: Tall Oaks Publishing, 1987:7–9.

Eaton AD, Clesceri LS, Rice EW, et al. Standard Methods for the Examination of Water and Wastewater, 21st ed. American Public Health Association, American Water Works Association, and Water Environment Federation. Washington, DC: American Public Health Association, Section 9215, 2005:9-34–9-41.

Edzwald JK, Becker WC, Wattier KL. Surrogate parameters for monitoring organic matter and trihalomethane precursors in water treatment. J Am Water Works Assoc 1985; 77(4):122–132.

Esposito MP. The Inactivation of Viruses in Water by Dichloramine. Master's thesis. Cincinnati, Ohio, USA: University of Cincinnati, 1974.

Furness BW, Beach WC, Roberts JM. Giardiasis Surveillance, Morbidity and Mortality Weekly Report, 49(SS07). Surveillance Summary, Centers for Disease Control and Prevention, Atlanta, GA, USA, August 11, 2000, pp. 1–13.

Glaze WH. Chemical oxidation. In: Pontius F, ed. Water Quality and Treatment—A Handbook of Community Water Systems. 4th ed. Denver: American Water Works Association. New York: McGraw-Hill, Inc., 1990:747–779.

Glaze WH, Weinberg HS. Identification and Occurrence of Ozonation Byproducts in Drinking Water. Denver: American Water Works Association Research Foundation, 1993.

Gould MJ. Evaluation of microbial/endotoxin contamination using the LAL test. Ultrapure Water 1993; 10(6):43–47.

Haas CN, Karra SB. Kinetics of microbial inactivation by chlorine. Part II: kinetics in the process of chlorine demands. Water Res 1984; 18:1451.

Handbook of Chemistry and Physics. 87th ed. In: Lide DR, ed. Boca Raton, FL, USA: CRC Press, Inc., 2006–2007:5-90–5-91.

Hoff JC, Akin EW. Microbial resistance to disinfectants: mechanisms and significance. Environ Health Perspect 1986; 69:7–13.

Kelly SM, Sanderson WW. The effect of chlorine in water on enteric viruses. Am J Public Health 1958; 48 (10):1323–1334.

Kelly SM, Sanderson WW. The effect of chlorine in water on enteric viruses. Part 2: the effect of combined chlorine of poliomyelitis and coxsackie viruses. Am J Public Health 1960; 50(1):14–20.

Korich DG, Mead JR, Madore MS, et al. Effects of ozone, chlorine dioxide, chlorine, and monochloramine on *cryptosporidium parvum* oocyst viability. Appl Environ Microbiol 1990; 56(1):423–428.

Kroneld R. Volatile hydrocarbons in pharmaceutical solutions. J Parenter Sci Technol 1991; 45(4):200–203.

Kruithof JC. Chlorination By-products: Production and Control. Report from the Committee on the Side Effects of Chlorination. KIWA Communication No. 74. Denver: American Water Works Association Research Foundation, 1986.

Kucik C, Martin GL, Sortor BV. Common intestinal parasites. Am Fam Pract 2004; 69(5):1161–1168.

Kunin R (ed.). The role of silica in water treatment—part I. In Amber-Hi-Lites, No. 164. Philadelphia: Rohm and Haas Company, 1980.

Kunin R (ed.). The Role of Organic Matter in Water Treatment. In Amber-Hi-Lites, No. 167. Philadelphia: Rohm and Haas Company, 1981.

Logan BE, Jiang Q. Molecular size distribution of dissolved organic matter. J Environ Eng 1990; 116(6):1046.

Massa S, Caruso M, Trovatelli F, et al. Comparison of plate count agar and R2A medium for enumeration of heterotrophic bacteria in natural mineral water. World J Microbiol Biotechnol 1998; 14:727–730.

McGuire MJ. Eight revolutions in the history of U.S. Drinking water disinfection. J Am Water Works Assoc 2006; 98(3):129.

Neden DG, Jones RJ, Smith JR, et al. Comparing chlorination and chloramination for controlling bacterial growth. J Am Water Works Assoc 1992; 84(7):80–88.

Owen D, Amy G, Chowdhury Z. Characterization of Natural Organic Matter and Its Relationship to Treatability. Denver: American Water Works Association Research Foundation, 1993.

Owen DM, Amy GL, Chowdhury Z, et al. NOM characterization and treatability. J Am Water Works Assoc 1995; 87(1):46–63.

Perry RH, Green DW, Maloney JO. Perry's Chemical Engineer's Handbook. 8th ed. New York: McGraw-Hill, Inc., 2007:3-97–3-100; table 3-120.

Reasoner DJ, Geldreich EE. A new medium for the enumeration and subculture of bacteria from potable water. Appl Environ Microbiol January, 1985:1–7.

Reckhow DA, Singer PC. Chlorination by-products in drinking waters: from formation potentials to finished water concentrations. J Am Water Works Assoc 1990; 82(4):173–180.

Rosenfeldt EJ, Baeza C, Krappe DRU. Effect of free chlorine application on microbial quality of drinking water in chloraminated systems. J Am Water Works Assoc 2009; 101(10):60–70.

Sawyer CN, McCarty PL, Parkin GF. Chemistry for Environmental Engineering. 4th ed. New York: McGraw Hill, Inc., 1994:330.

U.S. EPA. National Interim Primary Drinking Water Regulations: Control of Trihalomethanes in Drinking Water; Final Rule. Federal Register 1979; 44:68624.

U.S. EPA. Stage 1 Disinfectants and Disinfection Byproducts Rule, Final Rule, Publication EPA 815-F-98-010, 12/98, 1998.

U.S. EPA. Community Water Systems Survey 2000, Volume 1, Overview. Office of Water (46OTM), Publication EPA 815:R-02-005A, 2002:7–9.

U.S. EPA. Final Ground Water Rule Fact Sheet, Publication: EPA 818-F-06-003, 2006.

U.S. EPA. National Primary Drinking Water Regulations: Ground Water Rule; Final Rule, 40CFR Parts 9, 141, and 142, Federal Register 71:216:65574, 2006a.

U.S. EPA. National Primary Drinking Water Regulations: Long Term 2 Enhanced Surface Water Treatment Rule; Final Rule, 40CFR Parts 9, 141, and 142, Federal Register 71:3:654, 2006b.

U.S. EPA. National Primary Drinking Water Regulations: Stage 2 Disinfectants and Disinfection Byproducts; Final Rule, 40CFR Parts 9, 141, and 142, Federal Register 71:2:388, 2006c.

U.S. EPA. Drinking Water Contaminants—List of Contaminants and Their MCLs. Publication: EPA 816-F-09-0004, 2009.

van der Kooij D. Assimilable organic carbon as an indicator of bacterial re-growth. J Am Water Works Assoc 1992; 84(2):57–65.

Wickramanayake GB, Rubin AJ, Sproul OJ. Inactivation of *Giardia lamblia* cysts with ozone. Appl Environ Microbiol 1984; 48(1):671–672.

Young JS, Singer PC. Chloroform formation in public water supplies: a case study. J Am Water Works Assoc 1979; 71(2):87–95.

APPENDIX A: CHARACTER OF WATER SUPPLIES
Groundwater
- Temperature generally constant (±5°C).
- Chemical changes subtle except during drought or flood conditions.
- Predominant cations are calcium and magnesium.
- Predominant anions are bicarbonate and sulfate.
- Silica concentration may be significant (>5 mg/L).
- Iron and manganese may be present.
- Iron bacteria may be present.
- NOM is generally not present.
- Trace organic pollutants may be present.
- Colloids generally not present.
- Bacteria generally low or not present.
- Bacterial endotoxin levels are generally low.

Groundwater Influenced by a Surface Water Supply
- Temperature may vary with seasonal and climatic changes.
- Chemical changes are highly affected by seasonal and climatic conditions.
- During drought conditions, primary cations will probably be calcium and magnesium. During heavy periods of rain, the primary cation may be sodium. Seasonal changes will produce ongoing fluctuations in cation profile.
- During drought conditions, primary anions will probably be bicarbonate and sulfate. During heavy periods of rain, the primary anion may be chloride. Seasonal changes will produce ongoing fluctuations in anion profile.
- Silica will be present at moderate to high levels, depending on seasonal and climatic changes.
- Iron and/or manganese will probably be present at moderate levels depending on seasonal and climatic conditions.
- Iron bacteria may be present.
- NOM will be present, with concentrations increasing with greater influences from the surface water source.
- Trace organic pollutants may be present. The type and nature of the pollutants will vary with the percentage of ground-to-surface water.
- Colloidal material will be present, with concentrations increasing with greater influences from the surface source. Concentration will generally "parallel" the increase in NOM.
- Bacteria will be present, with levels increasing with greater influences from the surface source.
- Bacterial endotoxins may be present

Surface Water
- Temperature will vary with both seasonal and climatic conditions.
- Significant changes in water quality attributes will occur with seasonal and climatic changes.
- Predominant cation is generally sodium.
- Predominant anion is generally chloride.
- Reactive silica, as well as colloidal silica, will be present.
- Iron and manganese concentrations are generally low, but iron concentration may be influenced by older iron pipe distribution systems and a low (negative, zero, or slightly positive) Langelier Saturation Index.
- The TDS level is generally lower than groundwater supplies.

- Numerous microorganisms may be present, including, but not limited to, bacteria, crustacea, zebra mussels, algae, and protozoa.
- The presence of protozoa, such as *Cryptosporidium parvum* and *Giardia lamblia* may require special treatment techniques, such as filtration, by the municipality.

Disinfection by-products (THMs) present a problem that is generally addressed by increasing the pH and adding chlorine and ammonia, thus producing residual chloramines (monochloramine) as a disinfecting agent.

- Significant amounts of NOM will be present, consisting of a light molecular weight fraction (fulvic acid) and a heavy molecular weight fraction (humic acid).
- Trace organic and inorganic pollutants may be present, particularly during periods of heavy runoff.
- Colloids of silica, aluminum, and iron will be present.
- Organic material may produce a light yellow/brown color.
- Cysts may be present in source water.
- Bacteria may be present in both source and facility feedwater.
- Bacterial endotoxins will be present.

3 | Pretreatment techniques

INTRODUCTION
Numerous water purification unit operations are employed as pretreatments to an ion removal technique. The selection of a pretreatment scheme is a function of several variables, including the nature of the feed water supply, the ion removal technique, the overall system design criteria, and other factors, including individual preferences of the water purification system designer or owner. These factors influence the design of pharmaceutical water purification feed water systems such that numerous unit operations may be used in various sequences. Within the scope of this chapter, it is impossible to discuss every pretreatment technique that may be used in a pharmaceutical water purification system. Twelve different pretreatment techniques have been selected for discussion. The theory and application associated with each technique, design considerations, and operating and maintenance considerations are presented. The importance of proper pretreatment to the long-term successful operation of any pharmaceutical water purification system cannot be overemphasized. While there is a tremendous tendency to focus sampling and analytical monitoring on storage and distribution systems and, to a lesser extent, primary (and where applicable secondary) ion exchange unit operations, the ultimate factor that determines a system's ability to meet chemical, microbial, and, where applicable, bacterial endotoxin levels must begin with a technically sound pretreatment system and associated sampling, analytical monitoring, data trending, and preventative maintenance program.

CHEMICAL INJECTION SYSTEMS
Theory and Application
The introduction of chemicals within a pretreatment system should not be considered as "added substances," as defined within the *Official Monographs* for USP Purified Water or Water for Injection. This has been discussed in the literature (Collentro and Zoccolante, 1994). However, chemical introduction for any pharmaceutical water pretreatment system must address compendial requirements discussing "foreign substances and impurities," as presented in the *General Notices* section of the USP. In general, chemicals that may be considered for introduction within a pharmaceutical water pretreatment system can be classified as "acceptable" (with a proper raw chemical monitoring program), "marginally acceptable," and "undesirable." Prior to selecting a chemical injection application for enhancing a pretreatment operation or addressing a "problem" in a pretreatment system, it is important to recognize that any chemical introduced within the pretreatment system must, in accordance with the requirements outlined within the *General Notices* section of USP, be totally removed or clearly demonstrated by extensive product monitoring not to have an effect on product water quality or efficacy of final product. For most applications, this would appear to limit chemicals that could be introduced as part of the pretreatment system to compounds that can be readily analyzed, at trace concentrations, at final system "points of use." Furthermore, the introduction of chemicals that appear to be easy to monitor must include a structured analytical monitoring program of the "raw material," including, as a minimum, a "Certificate of Analysis" and an "internal" monitoring program.

Thus, it appears appropriate to identify specific chemicals that may be considered for introduction within a pretreatment system and demonstrate if the chemical is acceptable, marginally acceptable, or undesirable. An undesirable chemical that might be considered for a pretreatment system would include numerous classical proprietary polymers to enhance the removal of particulate matter, colloids, and naturally occurring organic material (NOM) when injected prior to a particulate removal filter. Polymers may be considered when silt density index (SDI) measurements (chap. 4) of raw feed water in the pretreatment system for a single- or double-pass RO unit are high. Unlike water purification systems used for other disciplines, such as semiconductor manufacturing, power plant applications, and numerous other industrial and commercial applications, the control of chemicals within a pharmaceutical water purification system is extremely critical. In general, a polymer, while enhancing pretreatment characteristics to reduce the fouling tendency of membranes, must be considered an undesirable material. The exact chemical composition, the percentage of chemical components, the organic structure of the components, and the difficulty in verifying the consistency of these parameters presents a serious challenge. The chemical purity of USP Purified Water and Water for Injection are determined by total organic carbon (TOC) and conductivity measurement. Trace quantities of undesirable impurities introduced by injected chemicals may not be detected by either of these two techniques at the specified maximum values. Unfortunately, if a polymer is used, the responsive pharmaceutical water system user must verify not only the chemical quality of the raw polymer being introduced, but also that the polymer, including all of its components, is not present in water used as an ingredient at point of use. This involves sophisticated analytical techniques such as gas chromatography/mass spectrometry analysis, generally not available at a pharmaceutical manufacturing facility. The complexity of chemical analysis required at point of use is much more restrictive than the indicated TOC and conductivity measurements. Furthermore, the indicated gas chromatography/mass spectrometry analysis may be incapable of identifying components within the polymer, since, in general, the chemical composition is proprietary and the specific chemical may not be in the "library" maintained by the laboratory performing the analysis. In conclusion, while polymers may exhibit excellent reduction of membrane-fouling substances, such as colloids and NOM, alternative water purification unit operations, such as synthetic organic scavengers and, in certain cases, ultrafiltration, provide a more conservative technique for removing membrane-fouling substances when the effects of a polymer on the quality of the pharmaceutical product are considered.

An example of a marginally acceptable pretreatment system chemical treatment compound includes antiscalants introduced prior to a single- or double-pass RO unit. In general, the chemical composition of several proprietary antiscalants is defined and established. Quality assurance and quality control maintained by the chemical manufacturer of the antiscalant is generally satisfactory. Obviously, limiting the selection of a particular antiscalant to a compound that has been used for many years within pharmaceutical water pretreatment systems obviously minimizes the potential negative consequences of the antiscalant on product water quality. This does not eliminate the requirement to verify that the material, on passing through the downstream membrane process, is not present in water at points of use. Depending on the application, it may be more appropriate to use conventional water softening as an alternative to injection of an antiscalant.

Examples of acceptable chemicals introduced to the pretreatment section of a pharmaceutical water purification system include a disinfecting agent, such as sodium hypochlorite, to ensure that adequate residual disinfectant is present in the raw water supply; a reducing agent, such as sodium bisulfite, which is introduced to remove residual disinfectant prior to RO systems containing membranes sensitive to residual disinfectants; and a caustic, primarily sodium hydroxide used in the feed water to an RO unit to convert gaseous carbon dioxide associated with the bicarbonate equilibrium to bicarbonate ion that will be removed by the membranes. It should be emphasized that while these materials may be classified as acceptable for introduction in a pretreatment application, the quality of material used, particularly in regard to undesirable trace impurities, must be determined. A Certificate of Analysis is critical. An internal chemical quality monitoring program (performed by an internal or contract laboratory) must be established and maintained. If trace impurities are present in the chemical compounds indicated as acceptable, that ultimately results in the presence of foreign substances and impurities, the compounds would be classified as unacceptable.

DESIGN CONSIDERATIONS
Injection of Residual Disinfectant: Raw Feed Water

The presence of residual disinfectant in the feed water supply to a pharmaceutical water purification system, particularly an USP Purified Water or Water for Injection system, is important. As indicated in chapter 2, the USP *Official Monographs* for both Purified Water and

Water for Injection, state that feed water must meet the U.S. EPA's NPDWR (or equivalent Japanese, European, or World Health Organization) "Drinking Water" criteria. A summary of the U.S. EPA's NPDWR is available in the literature (40CFR141 and 40CFR142, 2009). As discussed in chapter 2, Total coliform bacteria (with verification of the presence of *Escherichia coli* or fecal coliform) is the primary indicator of the microbial quality of U.S. EPA Drinking Water. While the U.S. EPA NPDWR does not definitively indicate a total viable bacteria limit, the *General Information* section of USP suggests that an "Action Limit" for total viable bacteria in Drinking Water be 500 cfu/mL. Within the literature, there are numerous references from regulatory personnel indicating the importance of routine residual disinfecting monitoring of feed water supplies to USP Purified Water and Water for Injection systems, establishing the fact that residual disinfecting agent is present (Avallone, 1993; Munson, 1993). Chapter 2 discussed residual disinfectant and total viable bacteria in raw feed water supplies. The absence of residual disinfectant, or extremely low residual disinfectant concentrations, in feed water supplies is not uncommon, particularly during summer months in certain parts of the United States and other countries. Subsequently, residual disinfectant, generally in the form of sodium hypochlorite, may be added to raw water in an attempt to ensure that the initial total viable bacteria levels satisfy the suggested 500 cfu/mL value and do not adversely affect system operation by increasing total viable bacteria levels within the pretreatment section of the system.

Design Parameters

If sodium hypochlorite is selected on the basis of availability and ease of handling, the material should be properly diluted, taking into account the size of the injection pump used to feed the chemical into the system. A target residual disinfectant concentration of 0.5 to 1.0 mg/L should be established. If the raw water supply contains ammonia, with associated chloramines, the target residual chloramine concentration should be 1.5 to 3.0 mg/L (Pontius, 1998). A suggested sodium hypochlorite injection system includes a storage tank with cover, a metering pump, and an injection point. An electrically actuated metering pump will generally be of positive displacement type. The pump should be equipped with provisions for adjusting the stroke of the piston within the pump, which determines the volume injected per cycle and the pulsing rate of the pump. Ideally, during normal operation, the stroke and pulse settings should be established such that the ultimate treated product water residual chlorine concentration does not vary by more than ±0.2 mg/L. In other words, it would be extremely inappropriate to inject a large volume of sodium hypochlorite, during normal operating flow rate conditions, once every five minutes. It would be more appropriate to introduce 1/10 of the volume every 30 seconds, minimizing transients in the ultimate residual chlorine concentration.

The sodium hypochlorite selected for this application should not introduce undesirable impurities, principally iron, that could interfere with downstream pretreatment components, such as the cation resin in water-softening units and RO membranes.

The injection port for introducing the chemical should be selected to ensure that potential backflow of water into the chemical injection system is not possible. Check valves should be provided in the supply line to the injection port and in the suction line from the storage tank to the pump.

For systems with relatively high flow rates (>50–100 gpm), it may be appropriate to consider pneumatically operated pumps rather than electronically actuated metering pumps. The air pressure to a pneumatically operated piston-type pump or air pulsing to a pneumatically operated diaphragm pump can be adjusted to regulate the flow of sodium hypochlorite to an injection port.

As discussed in chapter 2, the inactivation mechanisms of microorganisms are influenced by several factors, including but not limited to (Black and Veatch, 2009).

- The type and concentration of the organism being inactivated
- The disinfecting agent
- The concentration of the disinfecting agent
- Contact time

- Temperature of the water
- pH of the water
- The presence of particles and other material that "compete" for the disinfecting agent

The U.S. EPA considers a "CT" factor (concentration x time) to determine the time required for inactivation of various microorganisms (AWWA, 2009). For bacteria generally present in municipal water supplies, the time required to destroy a particular species of bacteria may be as long as 20 minutes at ambient temperature. It is highly unlikely that a 20-minute contact time between residual disinfectant, introduced as part of the chemical injection system, and bacteria will be encountered within pretreatment equipment, such as a particulate removal filter, before the residual disinfectant is removed. Subsequently, it is desirable to use a contact tank downstream of the feed water treatment point for residual disinfectant, providing a minimum contact time of about 20 minutes. Furthermore, to avoid "short circuiting" within the tank, baffles should be included.

Many pretreatment systems operate such that the feed water flow rate may vary over a period of time. For these applications, it is strongly suggested that a flow rate meter be installed in the feed water line. This meter would provide a proportional electronic signal to the residual disinfectant injection pump, increasing or decreasing the amount of residual disinfectant introduced as the flow rate increases or decreases. Some pretreatment systems use an inline residual chlorine monitor as either a "feedback" device for operating the hypochlorite injection pump, or to demonstrate that the feed water contains a minimum (or greater) residual chlorine concentration. For large systems with high flow rate capacities, particularly systems used in active pharmaceutical ingredient manufacturing operations, this technique may be appropriate. For most systems, routine monitoring (grab samples) from the effluent of the contact tank and after the sodium hypochlorite injection port should be adequate. However "grab" sampling should be collected in containers with a reducing agent, such as sodium thiosulfate (Eaton et al., 2005) which will chemically react with residual disinfecting agent to ensure that the analytical results do not reflect the reaction between residual disinfectant and bacteria present in the water sample.

A sample valve should be installed in the product piping from the injection pump to the injection port. This sample valve can be used to verify the concentration of sodium hypochlorite. Sample valves should be positioned downstream of the injection point and downstream of the contact tank.

OPERATING AND MAINTENANCE CONSIDERATIONS

The volume of disinfecting agent required to obtain a preset residual disinfecting concentration in the feed water to the pretreatment system will vary with both seasonal and climatic changes in the raw feed water supply. Obviously, this is particularly true for feed waters from a surface source or groundwater source influenced by a surface source.

It is important to note that a number of municipal systems in the United States that employ chloramines for microbial control in the distribution system change the residual disinfecting agent for a preestablished time period each year. Most of these systems operate with residual chlorine as the microbial control agent for a one-month period each year. While this technique is generally employed in a spring month such as April or May, the operating characteristics of the municipal treatment facility (long and short terms) should be verified and documented by individuals responsible for pharmaceutical water purification systems at a facility.

The analytical technique used to measure residual disinfecting agent should consider potential interference of trace concentrations of certain impurities. An overall discussion of techniques used to determine both residual chlorine and residual chloramine concentration can be found in the literature (Eaton et al., 2005, section 4500-CL; AWWA, 1973). Samples obtained downstream of the residual disinfectant injection port should be analyzed for not only residual disinfectant but also total viable bacteria and total coliform bacteria. The analytical procedures used should be consistent with the procedures used by the municipality and outlined in (Eaton et al., 2005).

The tank used for storing residual disinfectant should be equipped with provisions to allow operating personnel to measure the volume (or weight) of residual disinfectant introduced each day. This can be determined by level, or more precisely, by weight with an inexpensive "load cell." Periodic maintenance should be performed on the residual disinfectant injection pump. The "pulsing nature" associated with operation of the pumps, even with excellent selection of components and materials, results in "wear and tear" on surfaces and components.

INJECTION OF REDUCING AGENT FOR REMOVAL OF RESIDUAL DISINFECTANT AGENT
Theory and Application

Reducing agents such as sodium bisulfite may be injected to single- or double-pass RO unit feed water as part of the pretreatment system. In general, it is suggested that the use of reducing agents for residual disinfectant removal, prior to RO membranes that are intolerant to residual disinfecting agents, should be limited to feed waters from a groundwater source. Raw feed water supplies associated with a surface water source, or groundwater source influenced by a surface water source, will contain NOM. If the NOM is not removed prior to the RO system, membrane fouling will occur. Furthermore, accelerated microbial fouling of the membranes will also occur since the NOM provides a nutrient on the RO membranes for bacterial growth. Later in this chapter, chemical reactions associated with both residual chlorine and residual chloramine in water supplies is discussed. Sodium bisulfite and sodium sulfite reduces the active component (oxidizing agent) of residual chlorine and residual chloramine, as demonstrated by the following reactions:

$$Na_2HSO_3 + Cl_2 + H_2O \rightarrow Na_2HSO_4 + 2HCl$$

$$Na_2SO_3 + Cl_2 + H_2O \rightarrow Na_2SO_4 + 2HCl$$

$$Na_2HSO_3 + NH_2Cl_2 + H_2O \rightarrow Na_2HSO_4 + HCl + NH_3$$

$$Na_2SO_3 + NH_2Cl_2 + H_2O \rightarrow Na_2SO_4 + HCl + NH_3$$

The rate of each reaction is rapid. When used as pretreatment to an RO system, the slight amount of sodium introduced as part of the sodium bisulfite or sodium sulfite should not measurably affect product water quality. The rather small concentration of sodium is only a fraction of the sodium concentration in the raw feed water supplies.

Reducing agent injection may also be used to remove residual disinfectant prior to a deionization system. Again, however, it appears that this application is limited to feed water from a groundwater source, since NOM should also be removed in a system using deionization as the primary ion exchange technique. As discussed in chapter 4, NOM will organically foul anion exchange resin, resulting in increased replacement of expensive anion resin and/or maintenance-intensive, periodic hot water brining of the anion resins. For certain feed water supplies from surface sources or groundwater sources influenced by surface supplies, where TOC levels are low (approximately <1 mg/L), it may be possible to eliminate using activated carbon adsorption for residual disinfectant and NOM removal, relying on the sodium bisulfite to remove residual disinfectant prior to the ion exchange system, and acrylic-based anion resin to achieve organic removal as well as ion exchange. The use of acrylic resins is discussed in greater detail in chapter 4.

Finally, systems employing reducing agents for removal of residual disinfecting agent frequently exhibit high RO feed water total viable bacteria levels. It would appear more applicable to employ the use of 185-nm ultraviolet radiation at a high UV dose (compared with that used for UV disinfection units at 254 nm wavelength) to remove residual disinfectant for feed water from a groundwater source than inject reducing agents. Product water bacteria levels from the disinfecting agent removal ultraviolet units are significantly lower than encountered for reducing agent injection.

Design Considerations
The general design criteria for the storage tank, injection pump, injection port, and accessories, discussed earlier in this chapter for sodium hypochlorite injection, apply.

PRETREATMENT TECHNIQUES

Since sodium bisulfite is a reducing agent, it will react with dissolved oxygen in water used to prepare the sodium bisulfate solution. From a theoretical standpoint, once dissolved oxygen present in the water used to prepare the solution has been depleted, additional oxygen could be introduced from the atmosphere above the storage solution. However, this process involves gaseous diffusion, which is a very slow mechanism. Obviously, it would be inappropriate to stir continuously or otherwise agitate the prepared sodium bisulfite solution. It is recommended that the solution be stored in a "covered" container. Provisions should be included on the storage tank to drain solution, which has been stored for an excessive time period resulting in unacceptable decrease in the actual sulfite or bisulfite ion concentration, from the tank. This results in bacterial growth within the tank, delivery tubing, and RO feed water. Sampling provisions, particularly for feed water to the injection port, are strongly suggested.

"Feedback" systems can be used to ensure that appropriate amounts of reducing agent have been injected. Feedback systems can include oxidation-reduction potential (ORP) analyzers and residual disinfectant monitoring systems. It is suggested that feedback controllers, due to required maintenance, should only be considered for systems with larger flow rates (>50 gpm).

The physical location of the injection port, in relationship to the ion removal process, should be carefully considered. The injection ports provide a good method of introducing the reducing agent in a uniform concentration throughout the water stream; the kinetics of reaction are important. It would be inappropriate to position an injection port immediately prior to an RO or ion exchange unit. A few feet of piping/tubing, to provide contact time, can be extremely valuable to insure complete reduction of residual disinfectant.

Operating and Maintenance Considerations

Applicable operating and maintenance considerations presented as part of the discussion for the injection of residual disinfectant are appropriate. Potential reaction of dissolved oxygen with stored reducing agent may affect the concentration of the reducing agent solution with time. Standard Operating Procedures (SOPs) should be established, clearly indicating the "shelf life" of the reducing agent solution as well as the sampling frequency, to determine the concentration. Procedures should be established to minimize bacteria introduction from the reducing agent storage and transfer tubing. This can significantly increase microbial levels in RO system feed water. Frequent preparation of reducing agent solution (daily to biweekly) is suggested. A Certificate of Analysis for the reducing agent coupled with an "internal" chemical monitoring program should be considered.

The injection rate/volume of reducing agent should parallel seasonal and climatic changes of residual disinfectant concentration in the raw water supply. This would include adjusting the injection rate for indicated annual changes in the type of residual disinfectant (chlorine vs. chloramines) employed by some municipalities, discussed earlier.

SODIUM HYDROXIDE INJECTION IN SYSTEMS USING REVERSE OSMOSIS
Theory and Application

Injection of sodium hydroxide prior to an RO system can increase the product purity by converting gaseous carbon dioxide to bicarbonate ion, as indicated by the following reaction:

$$CO_2 + 2H_2O \leftrightarrow H_3O^+ + HCO_3^-$$

$$HCO_3^- + H_3O^+ + NaOH \leftrightarrow Na^+ + HCO_3^- + 2H_2O$$

The ionization constant associated with the first reaction establishes the concentration of reactants and product for specific water pH values (Table 2.6). As sodium hydroxide is introduced, it will react with the hydronium ion and force the equilibrium reaction to the "right," converting gaseous carbon dioxide, which will pass through RO membranes, to hydronium ion and bicarbonate ion. The bicarbonate ion is rejected by the RO membranes. Subsequently, for single-pass RO units used in pharmaceutical water purification systems (potentially units with downstream polishing unit operations), injection of sodium hydroxide obviously increases the "effective rejection" of ionic material through the RO unit. For double-pass RO units, injection of sodium hydroxide can provide product water with conductivity in the range of 0.5 to 2.0 µS/cm at 25°C.

Design Considerations

The comments associated with sodium hypochlorite injection systems also apply to sodium hydroxide injection. pH control is critical to the success of sodium hydroxide injection, specifically as it relates to RO product water purity. Quite often, small transients in raw water pH will affect the pH downstream of the caustic injection system (assuming that a feedback control system is not used). While pH is a logarithmic function (change in concentration by a factor of 10 for one pH unit), the concentration of the hydronium ion at neutral pH is minimal. Small changes in caustic can result in changes of pH beyond a "target" value, generally in the range of 8.0 to 8.5 for optimum carbon dioxide conversion to bicarbonate and hydronium ion.

It is inappropriate to inject sodium hydroxide prior to an RO system without a pretreatment step that includes water softening to remove calcium, magnesium, and other multivalent cations. Without softening as part of the pretreatment system, injection of sodium hydroxide results in the precipitation of undesirable hydroxides formed with the multivalent cations. This will rapidly result in scaling of the RO membranes.

Operating and Maintenance Considerations

The following operation and maintenance items should be considered for caustic injection systems.

- Periodic post caustic injection monitoring, by sampling and laboratory analysis, or inline analysis using a calibrated system, must be performed. Inline pH monitoring techniques require routine maintenance of the measuring electrode and periodic calibration with "buffer" solutions.
- It is extremely important to determine if trace concentrations of insoluble hydroxides are being formed. As indicated these insoluble hydroxides will affect the long-term performance of RO units. Scaling of "tail" membranes in the RO array can be determined by evaluating RO membrane precleaning data when "off-site" cleaning is performed. Scaling of RO units cleaned in place may be detected by loss in product flow verified by removal and "autopsy" of a tail array membrane.
- As indicated earlier, subtle changes in the pH of feed water can significantly affect the desired results associated with sodium hydroxide injection prior to RO units. Again, it is important to remember that pH is a logarithmic indication of the hydronium ion concentration. A change in the pH by one unit represents a change in the *measured* hydronium ion concentration by a factor of 10. More importantly, however, as discussed in chapter 2, the pH of the raw water supply is "buffered" by the presence of weakly ionized material. As a result, changes in raw water pH values as small as 0.1 to 0.3 unit can represent changes in the amount of sodium hydroxide required to obtain a desired pH in the feed water to the RO unit by a factor of 1 to 10 or greater.

THERMAL BLENDING SYSTEMS
Theory and Application

Thermal blending systems, essentially a single valve that mixes raw hot and cold water, are used for smaller capacity systems that use membrane processes, principally reverse osmosis. Colder, more viscous water encounters greater resistance to flow through a membrane. The effect of temperature on the relative capacity of both RO and ultrafiltration membranes is presented in Table 3.1. For raw water supplies from a groundwater source, where there is relatively little variation in feed water temperatures with seasonal changes ($\leq 5°C$), thermal blending systems, even for smaller capacity systems, may not be appropriate. These systems can compensate for the effects of temperature by increasing the operating pressure of the membrane process or by selecting a system with a slightly higher capacity. (RO unit capacities are generally expressed for a feed water temperature of 25°C.) However, for raw feed water

Table 3.1 Effect of Water Temperature on Reverse Osmosis and Ultrafiltration Product Water Flow Rate

Temperature (°F)	RO product water flow rate (gpm)	UF product water flow rate (gpm)
35	0.42	0.46
40	0.49	0.51
45	0.53	0.56
50	0.58	0.62
55	0.65	0.68
60	0.72	0.74
65	0.79	0.81
70	0.87	0.88
75	0.96	0.97
77	*1.00*	*1.00*
80	1.05	1.05
85	1.14	1.15
90	1.22	1.23

Note: All product water flow rate data are based on an "assigned" value of 1.00 gpm 77°F (25°C). The RO product water flow rates are for thin film-composite spiral wound polyamide membranes. The UF product water flow rates are for hollow fiber polysulfone membranes.
Source: FilmTec™ Membranes (2009) and Koch Membrane Systems (2007).

from a surface water source, particularly in northern sections of the United States, fluctuations in supply water temperature can range from 1°C to 2°C to 30°C. It is suggested that the decrease in membrane process product water flow rate, particularly for RO systems, can be addressed by system design that considers membrane area and use on a variable frequency drive motor on RO feed water pumps with controls to provide automatic system increase in RO feed water pressure. The use of thermal blending systems is discouraged since it increases the temperature of water processed by pretreatment components and subsequently increases pretreatment system total viable bacteria levels. In fact, maintaining pretreatment system feed water temperature at ≤20°C can decrease total viable bacteria levels by a factor of 10× (Collentro, 2007).

For larger pharmaceutical water systems using a membrane process, such as RO, the pretreated water is generally tempered with a heat exchanger prior to the RO. Smaller capacity systems are generally installed at facilities where steam may not be available for a heat exchanger. Domestic hot water, at a maximum temperature of approximately 120°F, is generally available for blending with cold water to achieve an operating feed water temperature in the range of 20°C to 25°C. During summer months, hot water may not be required, since the "cold" raw water supply may exhibit a temperature >25°C.

Design Considerations

The thermal blending valve should be equipped with an internal sensor that significantly minimizes or stops the flow of hot water if the blended water temperature exceeds a preset value. For applications associated with tempering feed water to systems containing an RO unit, the maximum product water temperature from the thermal blending valve should be 90°F. Figure 3.1 is a schematic drawing of a thermal blending system. The accessories are critical for most applications and discussed later in this section.

The effluent piping from the thermal blending valve should contain a direct reading temperature indicator. The temperature indicator should be selected with an appropriate range *and* degree increments so that operating personnel can periodically adjust the setting of the thermal blending valve as needed.

The selection of the thermal blending valve should consider proper pipe sizing and pressure drop for the system requirements. In general, thermal blending valves will have minimum flow rate requirements to ensure proper control function. Oversizing is undesirable. Thermal blending valves have a maximum and minimum feed water pressure rating. Proper valve function can only be achieved if the raw feed water pressure, under all operating circumstances, falls within the pressure range of the thermal blending valve supplier.

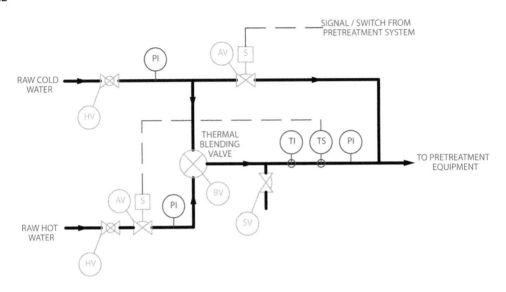

Figure 3.1 Schematic of a thermal blending system.

Thermal blending valves are generally constructed of a brass alloy. If appropriate, such as when the valves are connected to iron piping, dielectric unions should be included to eliminate galvanic corrosion. The product water piping from the thermal blending valve is the feed water to the water purification system. This feed water should meet the NPDWR as defined by the U.S. EPA. A sample valve is required downstream of the thermal blending valve to provide the required sample point. Manual isolation valves should be positioned on the raw hot and cold water supply lines to the thermal blending valve.

The thermal blending valve presents a flow restriction, as indicated by an exhibited pressure drop. For systems using thermal blending valves equipped with backflushable particulate removal filters and activated carbon units, or any processes requiring higher "backwash" flow rates, it is extremely desirable to provide a cold water bypass system around the thermal blending valve. This system achieves the desired results, providing the higher flow rates required for adequate backwash *and* more viscous cold water for the backwash operation. To provide the required bypass, considering the pressure drop through the thermal blending valve, a properly sized solenoid valve with piping can be positioned in a bypass line around the valve. The solenoid should operate in conjunction with a control system (opening the valve during the backwash operation for the individual downstream pretreatment component).

Pretreatment components positioned downstream of the thermal blending valve and RO membranes in an RO system are extremely sensitive to high temperatures. Items such as PVC (polyvinyl chloride) piping, vessel linings, controls valves, and sample valves will not withstand higher temperatures. As indicated earlier, the maximum feed water temperature to a system containing RO membranes (normal operation) should not exceed 90°F. While the thermal blending valve should be selected with an internal temperature sensor that minimizes or eliminates hot water flow on high product water temperature, a temperature switch should be positioned in the product water line from the thermal blending valve. The temperature switch should operate in conjunction with a solenoid valve installed in the raw hot water line to the thermal blending valve. The solenoid valve should fail (loss of power) in a closed position. This system would inhibit hot water flow to pretreatment components in the event of system malfunction or a loss of cold water flow.

Operating and Maintenance Considerations

Many systems are equipped with recirculation provisions in the pretreatment section prior to the primary ion removal unit operation (e.g., reverse osmosis). Thermal blending valves are generally used in smaller capacity systems. Many smaller capacity systems are operated to provide water over an eight hour period each day, five days per week. The recirculating

PRETREATMENT TECHNIQUES

provisions within the pretreatment system, as discussed later in this chapter, will introduce mechanical (Joule's) heat from the recirculation pump. This will increase the temperature of the water within pretreatment components, thus increasing the rate of bacteria proliferation. To avoid this situation, operating personnel can manually set the thermal blending valve to a colder temperature either at the end of each day or prior to a weekend, flush water through the pretreatment system, and subsequently avoid the higher pretreatment system operating temperatures with associated bacterial growth. Obviously, the intent of this operation can be achieved by slight modifications to the system design, using the "cold water" bypass valve and a "drain valve" installed in the system. This design would include a "normal" operation and "off-shift" selector switch.

The operation of the thermal blending valve should be periodically verified by limited cycling of the valve. This operation would change the position of the valve to a warmer or colder temperature and verify that the temperature, as indicated on the downstream temperature indicator, increases or decreases. Operation of the cold water bypass solenoid valve and hot water shutoff solenoid valve should also be verified periodically. Operation of the temperature switch and hot water solenoid actuation should be verified by intentionally increasing the temperature setting on the thermal blending valve and verifying that the hot water flow stops. The temperature switch and the temperature indicator should be calibrated periodically.

For the stated intended service, the thermal blending valve will require periodic adjustment to maintain the desired temperature. To achieve this objective, the operating log for the system should contain a data entry "box" for effluent temperature from the thermal blending valve. It is fully anticipated that periodic adjustment of the valve will be required about twice per month to maintain a 20°C to 25°C product water temperature value, depending on seasonal/climatic conditions.

HEAT EXCHANGERS
Theory and Application

As indicated above with thermal blending systems, membrane-based systems, primarily systems using RO with a feed water source from a surface water supply, may use tempered feed water. Smaller capacity systems may use thermal blending valves, while large capacity systems may use heat exchangers to achieve a desired feed water temperature. The following discussion is limited to applications in pretreatment systems to membrane-based processes.

Considering the nature of the application, it is *not* necessary to provide a heat exchanger of sanitary design and construction for this application. Various types of heat exchangers may be used. The most common type of heat exchangers used are plate-and-frame or shell-and-tube type. The function of a heat exchanger is to use a heating media, generally facility steam, to increase the temperature of a RO feed water stream to 20°C to 25°C. The heat load required to provide the necessary temperature increase can be calculated with the following equation:

$$\text{total heat load} = \text{feed water flow rate (gpm)} \times \frac{8.3 \, \text{lb water}}{1 \, \text{gal water}}$$
$$\times \frac{60 \, \text{min}}{1 \, \text{hr}} \times \frac{1 \, \text{BTU}}{\text{lb} \cdot °F} \times \text{desired temperature increase} \, (°F)$$

This equation provides the heat input required, expressed in British Thermal Units/hour (BTU/hr). Heat exchanger sizing should be based on the maximum anticipated flow rate and the maximum anticipated temperature increase (coldest pretreated water supply temperature). In general, heating of the pretreated water is accomplished by using latent heat from saturated facility steam. The lbs./hr. of steam required can be calculated once the steam pressure is established. Table 3.2 summarizes the latent heat (BTU/lb) of saturated steam at various pressures expressed in pounds per square inch gauge (psig). The exact latent heat can be found in the literature (Kennan and Keyes, 1963).

The above information allows calculation of the required heat for a given application and calculation of the steam flow in lbs./hr. Sizing of the heat exchanger considers the required heat transfer area for the design conditions. In addition to establishing the proper heat transfer

Table 3.2 Latent Heat of Vaporization for Water at Various Pressures

"Boiling" temperature (°F)	Gauge pressure (psig)	Latent heat (BTU/lb)
212	0	970
228	5	960
240	10	952
250	15	945
259	20	939
267	25	934
274	30	929
281	35	924
287	40	919
293	45	915
298	50	911
320	75	895
338	100	880

Note: Latent heat is calculated as the difference in the enthalpy of water (steam) less water (liquid) at the boiling temperature.
Source: From Keenan and Keyes (1963).

surface area required to obtain the desired temperature increase at the design flow rate, the surface area should not be extensive to the point where it is difficult to control "swings" in the product water temperature value. The heat transfer surface area can be calculated using the calculated required heat transfer, expressed in BTU/hr., the overall heat transfer coefficient of the specific material selected for constructing the heat exchanger, and the "average" temperature difference between the heating media and the pretreated water. This calculation is somewhat rigorous because the overall heat transfer coefficient is dependent on the materials of construction as well as the configuration of the heat exchanger surface. The temperature differential is generally expressed as a "logarithmic mean value." Rather than attempting to select (and size) a heat exchanger on the basis of the indicated rigorous equations, it is strongly suggested that a heat exchanger manufacturer calculate (with a custom-formulated, computerized program) the exact size of the heat exchanger required for specific applications. A qualified heat exchanger manufacturer will provide a detailed calculation sheet that clearly outlines the characteristics of the heat exchanger, the pretreated water flow rate, pretreated water temperature, the heating steam pressure, and the heating steam flow rate. This material is generally available without cost from heat exchanger manufacturers. It is a valuable document in determining the size of utility piping, control valves, overall heat exchanger physical size, and other accessories. The data sheet should be included as part of system documentation (e.g., Installation Qualification, chap. 13), particularly for USP Purified Water systems.

Design Considerations

The heat exchanger should be selected by considering the maximum anticipated flow rate, the minimum pretreated water temperature, and the range of facility steam pressure. The thermodynamic "quality" of the supply steam must also be considered. The literature contains specific information and procedures for determining the quality of steam (Weber and Meissner, 1959).

Considering the heat exchanger is part of the pretreatment system positioned upstream of an RO unit, a non sanitary, plate-and-frame or shell-and-tube heat exchanger may be acceptable. However, if a non sanitary heat exchanger is selected, the facility steam pressure should be regulated below the pressure of the pretreated water. This design ensures that plant steam will not be introduced into the pretreated water if a leak should occur. As an added precaution, to ensure that the suggested pressure differential is maintained, a differential pressure monitoring system with two pressure sensors, one installed in the facility steam line and the one installed in the pretreated water line, should be considered. The system should provide an audible alarm if a minimum desired differential pressure is not maintained. As an alternative, a sanitary heat exchanger with double tube sheet may be employed at a slightly higher cost.

PRETREATMENT TECHNIQUES

The location of the heat exchanger within the pretreatment system should be carefully considered. Positioning directly upstream of the RO system has several advantages. Pretreatment components, at this point in the system, should have removed particulate matter, a portion of the NOM, and, where applicable, multivalent cations that could form insoluble precipitates on the heat transfer surface areas within the heat exchanger. Table 3.3 lists some compounds that may be present in unsoftened feed water and that exhibit "inverse" solubility (i.e., the actual solubility of a compound decreases with increasing temperature). The accumulation of precipitates on heat transfer surface areas decreases the overall heat transfer coefficient, which, for a fixed heat transfer surface area, will inhibit the heat exchanger from achieving its design objectives.

Figure 3.2 presents a representative diagram of a heat exchange system with controls. To achieve the desired control of facility steam, a temperature well containing a temperature element provides a signal to a temperature controller. The temperature controller in turn provides an analog signal, generally through a processor, to a "current-to-pneumatic" converter, providing a proportional pneumatic signal to a modulating valve installed in the regulated facility steam line. Sizing of the valve should be consistent with flow and pressure

Table 3.3 Compound Exhibiting "Inverse" Solubility

Compound	Temperature (°C)	Solubility (g/100 mL)
Calcium sulfate	30	0.209
Calcium sulfate	100	0.162
Calcium sulfite	18	0.0043
Calcium sulfite	100	0.0011
Calcium hydroxide	0	0.185
Calcium hydroxide	100	0.077

Source: From Lide DR (2006–2007).

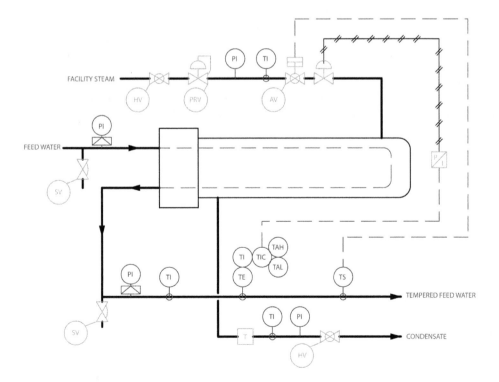

Figure 3.2 Representative heat exchange system with controls.

requirements. The temperature controller can be equipped with indicating provisions or provide a proportional output to a remotely mounted meter or processor cabinet display screen. The modulating valve in the facility steam line should be positive acting, air to open, normally closed. A properly sized steam trap should be positioned in the condensate effluent from the heat exchanger. Manual isolation valves should be positioned in the facility steam supply line and the facility steam condensate return line. A properly sized relief valve with "exhaust" piping generally piped to the roof or diverted to a location that will not expose operating personnel to steam in the event of over pressurization should also be provided. Pressure relief to the roof of a facility is preferred and often dictated by local or regional safety codes. "Drip pans" should be employed in relief lines vented to the roof to eliminate collection of rain water in the carbon steel relief lines. Calibrated temperature and pressure indicators should be provided (Fig. 3.2).

The facility steam supply line should be equipped with both a manual shutoff valve and a pneumatically actuated, "steam-tight" automatic ball valve. When steam is not used on a routine basis, the automatic ball valve should be closed eliminating long-term exposure of pressurized steam to the modulating valve. Steam pressure will gradually result in degradation of the steam regulating seat of the valve, which in turn affects the ability to control water temperature at low steam flow rates and results in low steam flow through the modulating valve when it is "closed," and ultimately valve maintenance to replace the valve seal/seat.

A final item that should be considered as part of the control system relates to a "backup" heat exchanger temperature product water monitoring and control system. The cost of RO membranes, particularly for high flow rate systems, is significant. Subsequently, it is technically easy to justify a backup heat exchanger product water temperature monitoring and control system that includes a temperature switch with the set point established at approximately 90°F. If the product water from the heat exchanger exceeds this value, a signal will be provided from the temperature switch to the pneumatically actuated ball valve, discussed above, installed in series with the modulating valve in the facility steam supply line to the heat exchanger, completely eliminating steam flow. The pneumatically actuated ball valve should be air to open, spring to close.

Sizing of the heat exchanger should not provide excessive heat transfer surface area. Excessive heat transfer surface area will produce a condition where the temperature controller and modulating valve, considering the fluctuations in pretreated water temperature, produce a cyclic situation in which the product water temperature from the heat exchanger exceeds the set point, or is below the set point, as demonstrated by the curve in Figure 3.3. In reviewing the theory and application for pretreatment heat exchangers, however, the temperature of the water may vary considerably with seasonal fluctuations. If precise temperature control is

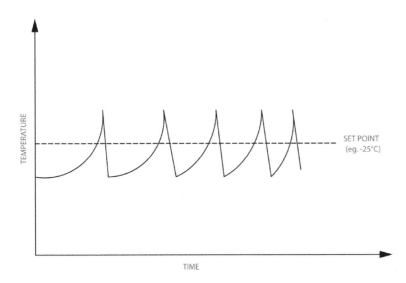

Figure 3.3 Cyclic temperature control from improper design.

desired throughout the entire range of pretreated water temperatures with seasonal fluctuations, it may be necessary to consider using two modulating valves installed in parallel within the facility steam line. The larger valve would be associated with applications during winter months, while the smaller valve would be associated with "trim" heating applications, primarily during spring and fall applications.

Selecting the right type of heat exchanger for a given application, especially if expansion of the water purification system is contemplated, is important. The area within a shell-and-tube heat exchanger is fixed. On the other hand, the heat transfer surface area for a plate-and-frame unit can be expanded by adding additional plates, assuming that potential expansion requirements are factored into the initial design and selection of the heat exchanger.

An extremely important design feature that is often overlooked relates to "residual heat" in the heat exchanger when flow is terminated. The heat exchanger generally provides tempered pretreated water to a downstream RO system. While "recirculating RO system design" is strongly suggested (chap. 4), operation of some RO units may be controlled on the basis of operating level in a storage tank positioned downstream of the RO unit. The storage tank level control system provides a signal to the RO unit, stopping the unit and terminating the pretreated water flow. The heat exchanger and associated pretreated water and product water tubing, as well as the supply steam and condensate piping, contain residual heat that will be introduced into the water. If recirculating provisions for the pretreatment system include the heat exchanger, this "residual heat" will be dissipated within the RO and polishing components. However, if the heat exchanger is *not* included in a recirculating loop, provisions must be included to remove heat, ensuring that potential overheating of the downstream RO membranes will not occur. (Many RO units include a temperature switch in the pretreated water feed line.) While this is added protection for the membranes, the condition inhibits operation of the RO unit and produces an alarm condition. Provisions to remove residual heat include, but are not limited to, a storage tank multipoint level control system that provides a signal to the heat exchanger, allowing a time period between termination of facility steam flow to the heat exchanger and termination of flow through the RO system, as well as pre or post heat exchanger "divert-to-waste" cycle (prior to or subsequent to RO system operation) to remove higher than desired pretreated feed water temperature.

For certain applications, it may be appropriate to consider a heat exchanger that has "dual function" capability. While this section has emphasized the importance of establishing appropriate heat transfer surface area in selecting the heat exchanger, it may be possible to achieve additional desirable pretreatment functions with the heat exchanger. As an example, for a recirculating hot water sanitizable RO unit and polishing components, the heat exchanger may be used for heat input during the sanitization cycle.

An additional function for heat exchangers is "trim cooling" applications. This could apply to systems using feed water supplies from a surface source, where the temperature increases to 85°F (or greater) during hot summer months. This would also apply to heat exchangers that are included within pretreatment recirculation loops. The Joule's heat introduced by the recirculating pump may increase the temperature to bacteria "incubation values," particularly when the temperature of the feed water is relatively warm. Figure 3.4 contains a flow diagram demonstrating the use of a heat exchanger for trim cooling applications. Considering the fact that this application involves removal of sensible heat by input of sensible heat to the cooling water and the differential temperatures required for effective heat exchanger selection, "chill water," in general employed for this application.

It is possible to a heat exchanger for heating recirculating water as part of a periodic sanitization program for an activated carbon and/or water-softening units. This operation is discussed further in other sections of this chapter.

Materials of construction, particularly for surfaces in contact with pretreated water, should be considered. In general, particularly for shell-and-tube heat exchangers, copper or brass alloys exhibit relatively high overall heat transfer coefficients when compared with stainless steel. However, on the basis of corrosion considerations, long-term operation with minimal maintenance and the aggressive nature of softened water, it is suggested that either 316L stainless steel tubes with 304L stainless steel shell be considered for material of

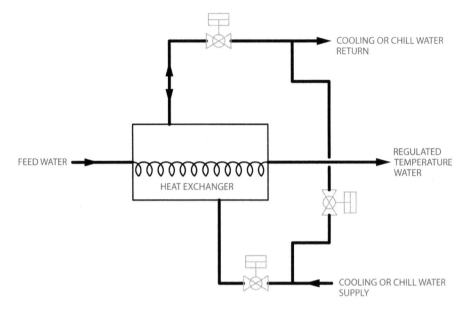

Figure 3.4 Process flow diagram: "trim cooling" heat exchange.

construction. Since it is quite possible that utility piping to and from the heat exchanger may consist of materials that are different from the heat exchanger, such as carbon steel, copper, and/or brass, the use of dielectric unions for protection against galvanic corrosion must be considered.

Many pretreatment system components, including upstream water-softening units or downstream pretreated feed water piping to the RO unit, often contain PVC or CPVC piping. It is important to consider the conduction of heat from the heat exchanger to ensure that the integrity of plastic piping, such as PVC, is not compromised. Furthermore, metallic fittings to and from a heat exchanger, such as those constructed of 316L or 304L stainless steel, should be equipped with a flange connection, mating to a plastic flange connection ("solvent" or heat welded) for any transition in the feed water and product water from the heat exchanger. It is inappropriate, particularly for this application, to thread plastic materials into metallic fittings. The physically soft PVC or CPVC will eventually "slip" from the physically "harder" metal fittings. This is true for both hot and ambient applications.

A design issue with potential significant safety applications relates to the installation of isolation valves in the feed water and product water piping/tubing of a heat exchanger and ASME (American Society of Mechanical Engineers) code requirements for the heat exchanger. If isolation valves are positioned directly in the feed water and product water lines for a heat exchanger, it should be remembered that a pressure (or vacuum) condition may occur within the heat exchanger and sections of piping "inside" the boundary of the valves if the valves are closed. Pressure within this "hydraulic system" would increase if residual heat input was available to water "trapped" in the heat exchanger when the valves are closed. It is also possible that cooling of water could occur for cooling application creating a vacuum. Since this is entirely a hydraulic system, provisions must be included for relieving the vacuum or pressure, depending on the pressure/vacuum rating of the heat exchanger and associated piping/tubing downstream of the feed water isolation valve and upstream of the product water isolation valve.

Operating and Maintenance Considerations

The function of the facility steam supply modulating valve to the heat exchanger is extremely important. Calibration of the temperature control system and associated output (current or amperage) should be determined periodically (6–12 months). The pneumatic output from the current-to-pneumatic converter should also be verified.

As indicated earlier in this section, the thermodynamic quality of the steam is extremely important since it affects the overall characteristics of the heat exchange system. The quality of the steam should be determined periodically. While the classical "drip leg" (bypass with strainer, trap, and isolation valve) should always be employed in supply steam lines, for heat exchangers installed in "branch" sections of steam piping where condensation will occur, the use of a "bypass" steam line with a valve and steam trap, should be considered. The requirement for this bypass system can be readily verified by periodic determination of steam quality.

Maintenance of the steam trap, including inspection of proper operation, is important. The steam trap should not allow liquid to accumulate in the heat exchanger, since the design of the heat exchanger assumes that the entire heat transfer surface area is capable of transferring latent heat, not sensible heat, from steam condensate.

All instrumentation associated with the heat exchanger should be periodically calibrated. The use of direct reading temperature gauges in addition to pressure gauges should be considered.

It is suggested that periodic (one to three years) inspection of heat transfer surface area be conducted. Any scale accumulating on heat transfer surface areas should be removed by an appropriate chemical cleaning program. Furthermore, a hydrostatic test, at the original test pressure for the unit, generally 50% in excess of the design pressure, should be conducted to verify that leakage, as a result of corrosion, does not occur. The presence of siliceous deposits on heat transfer surfaces will affect the overall heat transfer coefficient. Siliceous deposits can be removed with appropriate cleaning agents consistent with the recommendations of heat exchanger manufacturers. The cleaning requirements for shell-and-tube units are different from those for plate-and-frame units. For plate-and-frame units, it may be more appropriate to replace plates that are corroded or that contain deposits with new plates than attempt cleaning.

Larger systems will collect steam condensate from the heat exchange system and return the condensate to the facility steam condensate receiver, where it will eventually be used for makeup water to the boiler. The condensate return system should include a "receiver" tank and a pump. Periodic maintenance of components for the condensate return system are required to ensure proper operation of the upstream steam trap and eliminate potential flooding of heat transfer surface areas within the heat exchanger. This is a concern for systems that are infrequently used and systems with low condensate flow rates.

PARTICULATE REMOVAL FILTERS
Theory and Application

Particulate matter, defined in chapter 2, can be removed by numerous pretreatment techniques. Systems with a relatively low flow rate capacity and/or low particulate levels may use disposable cartridge filters (discussed later in this chap.). This section discusses the use of backflushable particulate removal filters, primarily classical sand filtration (surface filter) and depth filters, dual media and multimedia filters. Sand filtration, while an effective method for removing small-sized particulate matter (5–10 μm) provides filtration on a layer of sand. In general, a sand filtration unit consists of graduated levels of supporting gravel (with gravel size decreasing from the bottom of the column up through the column) and a layer of sand, generally about 0.5 mm in diameter, with a depth of 10 to 20 in. Filtration occurs at the sand-water interface in a downward filtration mode. There is limited penetration of particulate matter from the surface of the sand-water interface downward through the sand. Subsequently, the capacity of a low-velocity sand filter, expressed in the weight or volume of material removed per gallons of water treated prior to backwash, is relatively small when compared with dual media or multimedia filtration units. On the other hand, dual media and multimedia filtration units are classified as "depth" filters, since the removal of particulate matter occurs not only at the top surface of the water-media interface but also downward through the filtration media. Dual media filters generally use a supporting level of gravel positioned below two layers of filtration media. Unlike the supporting gravel where the size decreases from the bottom to the top of the column, the coarser filtration media in a dual media filter is positioned above the finer filtration media. Anthracite is a typical example of coarse filtration media, while sand, or in certain applications, manganese greensand, is used as the finer media. Larger particulate matter is removed within the coarser filtration media layer. Subsequently, the finer material is removed in the second layer of filtration media (sand) with excellent efficiency. Again, filtration occurs throughout the depth of the coarse and finer media, which results in good particulate removal capability (expressed in weight or volume) between required backwash operations. A multimedia filtration unit uses the same principles as a dual media filtration unit; however, in lieu of two layers of filtration media, it may use as many as four to six different layers of filtration media.

While classical sand filtration offers an extremely conservative method of removing particulate matter, depth filtration offers technically attractive advantages, particularly for raw water supplies containing appreciable amounts of particulate matter, colloidal matter, and heavy–molecular weight naturally occurring organic matter. To comprehend the advantages associated with depth filtration, it is necessary to understand the proper operating and, more importantly, backwashing criteria for depth filters. A common misconception associated with the operation and backwash of depth filters within a pretreatment system is that a frequently backwashed unit (e.g., daily) with "clean" media provides the most effective method of particulate removal. In actual practice, the contrary is true, assuming that particulate breakthrough has not occurred. The most important word to remember when discussing the operation and backwash frequency of a particulate removal filter is "ripening." Several articles have been written over several years discussing this phenomenon (Cleasby et al., 1984; Miltner et al., 1995; Ongerth and Pecoraro, 1995). The accumulation of operating data demonstrating this phenomenon has increased over the last few years as the "Surface Water Treatment Rule," discussed in chapter 2, was implemented, enhanced, and enforced by the U.S. EPA. The performance of a particulate removal filter can be determined by comparing product water and feed water data as a function of time since backwash (Fig. 3.5). From data presented in Figure 3.5, it is obvious that filtration ability actually increases with time (to an obvious limited extent when particulate breakthrough occurs) and is extremely low for a relatively clean unit immediately after backwash. During the ripening process, particulate matter removed by the depth filtration unit will accumulate within the filter media matrix. Over a period of time, the filtered material actually serves as the filtration media, "tightening" the bed, encouraging both the efficiency of particulate removal (in terms of volume/weight) and reducing the size of particles that pass through the filter into the final product water. In fact, both colloidal material and, to a lesser extent, naturally occurring heavy–molecular weight material will be removed in a "ripened" depth filter, even though they have a particulate size less than 10 μm, considered in this text as the size definition for particulate matter. It is

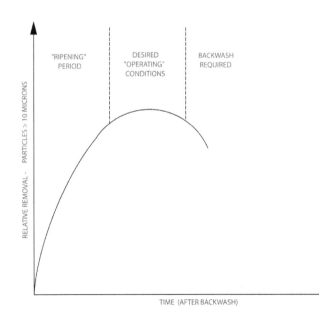

Figure 3.5 "Ripening" of depth filtration unit.

desirable to operate a depth filtration unit so that a maximum degree of ripening occurs without breakthrough of particulate matter through the finer filtration media layer directly above the supporting filter media. Subsequently, depth filtration units should *not* be backwashed on the basis of operating time; they should be backwashed on the basis of differential pressure through the filtration bed. The actual pressure drop is determined by the characteristics of the specific depth filter. In general, however, breakthrough of particulate matter will not occur if the pressure drop is limited to approximately 7 to 10 psid above the postbackwash "clean" pressure drop.

A particulate removal filter is generally positioned prior to other pretreatment unit operations, such as activated carbon adsorption and water softening. Particulate removal filters provide a valuable function within the pretreatment system. This function cannot be "supplemented" as a secondary function of other pretreatment components, such as activated carbon and water softening. Particulate matter, if not removed by a dedicated particulate removal filter, will interfere with the proper operation of an activated carbon unit, water-softening unit, and other support components, such as inline ultraviolet units, heat exchangers, and prefilters prior to a RO system.

Design Considerations

(The design consideration presented below for particulate removal vessel design are applicable to other pretreatment unit operations employing vertical cylindrical columns with media such as activated carbon and water-softening units.)

Proper sizing of dual media or multimedia filtration units is extremely critical to the successful operation of the unit. Vertical cylindrical vessels should be selected with a diameter such that at the design flow rate for the unit, the "face velocity" is approximately 6 gpm/ft^2 of the cross-sectional bed area. It is important to emphasize that higher face velocities, while often recommended by equipment manufacturers, do not allow effective filtration and, in addition to encouraging premature breakthrough of particulate matter, can result in channeling (tunneling of water through a limited section of the cross-sectional bed area), which totally eliminates the possibility of filter ripening.

The feed water piping and other face piping for the unit should be properly selected for a backwash flow rate of 12 to 15 gpm/ft^2 over the cross-sectional bed area when the feed water temperature is 60°F to 80°F. The flow rate associated with this face velocity is required for adequate bed expansion and removal of entrapped particulate matter, specifically finer (smaller) particles removed within the sand filter media (lowest filter media in the column). For some applications, the raw feed water piping maximum capacity may support the normal unit operating flow rate, while being incapable of providing adequate flow for backwash. Multiple units, operating in parallel, may be employed for this application. Using two or three smaller diameter units in an attempt to satisfy the backwash requirements may be appropriate on the basis of the raw water flow rate to the system. As an alternative to using multiple "smaller" diameter units, air can be injected with backwash water in a "scouring"-type application. The air, introduced within the support media through a dedicated distributor or custom fabricated lower distributor, will provide effective removal of entrapped particulate matter during the backwash operation at face velocities as low as 8 to 10 gpm/ft^2 over the cross-sectional bed area.

In general, it is suggested that the vertical, cylindrical filter vessel should be designed, fabricated, and tested in accordance with the ASME Code for Unfired Pressure Vessels, section VIII, division 1. While some states in the United States may not require compliance to the ASME Code for Pressure Vessels, local (city, town, etc.) and regional (county) restriction may specify code requirements. Subsequently, it would be inappropriate to provide a summary of "code" states since it does not indicate the actual requirements at a facility. In addition to state and local pressure vessel requirements, insurance underwriters at a facility may require all pressure vessels to be code stamped, even when state and local "boiler and pressure vessel" authorities may not require ASME code-stamped vessels. It is suggested that all pressure vessels be designed, fabricated and tested in accordance with the ASME code. The design pressure rating for the vertical cylindrical vessel is a function of the specific application. Considering the maximum supply pressure of most municipalities, the potential pressure

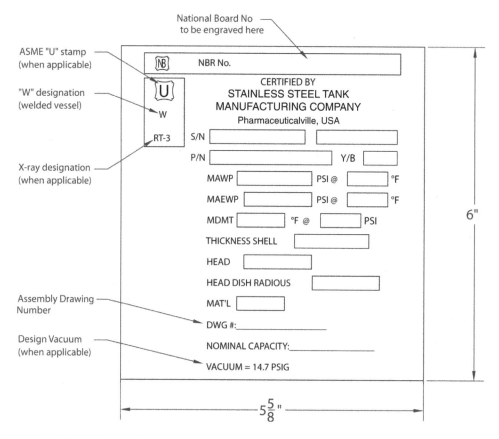

Figure 3.6 ASME data plate for vessels.

increase associated with the recirculation of pretreatment components, and the pressure drop through subsequent pretreatment components prior to the primary ion removal process, a design pressure of 100 psig is generally adequate. For an estimated 30% of all pharmaceutical water purification system applications, the design pressure may be greater, such as 150 psig. Vessels designed in accordance with the ASME code will have a visible stamp containing information shown in Figure 3.6. Furthermore, vessel design parameters, nozzle connection information, and other appropriate vessel information should be provided from the designer, fabricator, and code shop testing the vessel. Hydraulic testing of the vessel is generally performed at a pressure 50% higher than the design pressure.

The straight side height of the column should be adequate to allow backwash of the unit without impinging the filter media on the domed top of the vessel. In general, the straight side height of a column can be increased by 12 to 24 in., with minimal impact on the cost of the code vessel. Obviously, the physical area for installing the unit should have adequate height to allow any increase in the straight side height and, subsequently, the overall height of the unit. While the depth of support gravel will vary depending on the design of the lower distributor system for the unit, it is suggested that the *minimum* filter media bed depth (dual or multimedia filtration unit) be 24 in. It is further suggested that the minimum freeboard space above the filter media should be 24 in., resulting in a suggested minimum vessel straight side, considering the depth of support media, of 5 to 6 ft. The filtration unit vessel should be equipped with column interior access provisions. A side- or top-mounted access manway with minimum dimensions of 12 × 18 in. (elliptical) or, for larger columns, 18 in. in diameter (circular) should be considered. In addition, smaller "hand holes" with a minimum suggested size of 5 × 7 in. (elliptical) should be considered for accessing distributors at the base of the column. If elliptical manways or "hand holes" are used, they should be designed so that increased pressure within the column increases the

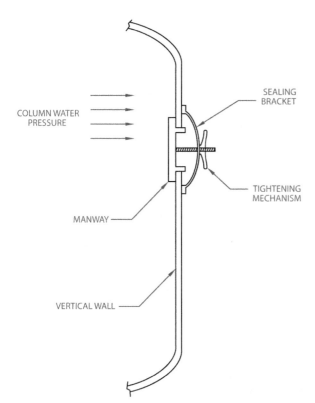

Figure 3.7 Preferred manway for vertical cylindrical column.

force on the access port, as shown in Figure 3.7. For side-mounted, circular manways (larger-diameter vessels), consider using a daveted-type system to provide access without the external rigging provisions that are required to remove the rather heavy manway.

Selection of the internal column lining or coating should consider the abrasive nature of the filter media against the walls of the vessel, particularly during the backwash operation. Filter media such as anthracite and sand, expanded in an upward direction within the vessel during the backwash operation, will tend to abrade the column surfaces in the area of the media. Generally, the top layer of supporting gravel will consist of small diameter granite, which is also very abrasive. Numerous interior column linings are available from water purification equipment manufacturers. Smaller diameter units, for relatively low flow rate systems, may employ fiberglass columns, eliminating the concerns associated with the abrasion of steel surfaces associated with larger-diameter vessels. However, the fiberglass columns are manufactured to "standard" dimensions with a relatively short straight side height. The limited ability to backwash the units properly, based on minimal freeboard space within the fiberglass columns, significantly limits the use of the vessels to applications with flow rates ≤ 45 gpm. Steel vessels should contain appropriate interior lining on the basis of the application. Protective "sprayed-on" linings of various materials, with a thickness in "mils" (thousands of 1 in.), will erode with time. The erosion generally occurs at the filter media interface at the interior walls of the column. As a result of this erosion/corrosion, some sections of the column have a protective coating, while other sections have no protective coating. This produces a "galvanic-type" corrosion situation within the column, resulting in accelerated corrosion at specific areas where the continuity of the sprayed on coating has been lost because of erosion. Eventually, this situation will result in a loss of integrity for the filter vessel. In addition during the extended operating time period before vessel failure, iron corrosion products are introduced from the walls of the filter vessel to the water, potentially increasing the concentration of iron—a highly undesirable impurity in feed water, particularly for water-softening units and RO systems. Alternatively, many acceptable sprayed and baked on linings are available including PVC. The ideal material is sheet rubber vulcanized in place. This material provides a highly effective coating of approximately 1/8 to 3/16 in. thick and has a long life. Further, if localized small failures of a rubber lining are experienced after an extended operating period, they can be repaired without replacing the entire vessel using an in-place "patching" technique. The rubber lining is relatively "soft," thus minimizing the degradation of filter media associated with the impingement of material on filter walls and the upper domed head of the filter vessel during the backwash operation. Any material used for filter lining should not introduce "foreign substances and impurities" as defined in the USP *General Notices* section.

Inlet and outlet distributors are required. It is strongly recommended that distributors for lined steel or stainless steel columns be constructed of stainless steel, since the physical forces associated with normal operation and backwash are significant. For fiberglass columns distributors of PVC construction may be considered because of the smaller diameter and resulting length of the distributors. Distribution is achieved through the filter media by back pressure exerted by the bottom distributor. To achieve effective distribution, associated with a uniform velocity of water over the entire cross-sectional bed area of the filter, a minimum back pressure of approximately 5 psid is required. Filter units operating with a pressure drop less than approximately 5 psid will not produce effective distribution, resulting in "channeling" through the unit, which decreases the performance of the unit. Assuming that adequate freeboard space is available above the filter media for the backwash operation, the inlet distributor design can be extremely simple. For example, the inlet distributor could consist of a double elbow system that directs water to the domed top of the unit, thus providing a fairly "rough" distribution of feed water, which is adequate on the basis of the back pressure exerted by the lower distributor. The use of plastic distributors (polypropylene, PVC, etc.) is not recommended for steel or stainless steel particulate removal filters columns with a diameter >36 in. While the distributors will provide the required back pressure, long-term, successful operation can be effected by a small crack in the plastic surfaces associated with the distributor. The "added cost" of providing a stainless steel distribution system will be most likely offset by the labor cost associated with removing filter media and support gravel to replace a cracked plastic distributor. Obviously, the initial loading of support gravel is difficult for units equipped with plastic distributors, since media are generally added through a top access port or manway. A section of plastic pipe may be used to deliver the support gravel to the bottom of the column until the distributor is "covered" for these applications.

Several different configurations of lower distributors can be used. Figures 3.8 and 3.9 are representative types of commonly used distribution systems. Unit design should consider that the distributors must be positioned within the vessels in a manner that minimizes or eliminates stagnant area below the lower distributor and the lower inverted "dish" of a lined steel column. For fiberglass columns, "pea" gravel may be added to cover the lower distributor.

As indicated above, proper distribution through the filter media is critical to the successful operation of a particulate removal filter. Backwash should also introduce water, upward through the filter media, at a uniform velocity over the entire cross-sectional bed area. Since the backwash flow rate is 2 to 2.5 times the normal operating flow rate, a dedicated backwash distribution system may be appropriate, particularly when the column diameter increases beyond 120 in. To ensure proper operation and backwash, it is suggested that a single (or multiple) sight glass be positioned on the straight side of the unit at the filter media–water interface and, where applicable, above and below the interface. This will allow operating personnel to determine if proper filter media bed expansion is achieved during the backwash operation and verify that the top of the filter media, after the backwash and subsequent rinse operation is completed, is in a totally horizontal position. A "nonhorizontal" filter media position after backwash will result in poor unit operation due to the decrease in depth of the filtration media and associated channeling through the unit (Fig. 3.10).

The domed top of the vertical cylindrical vessel should contain a relief valve. This valve should be selected to relieve the hydraulic pressure at the maximum design pressure rating of the column, considering the maximum feed water (or backwash water) flow to the unit. It is important to emphasize that selection of the relief valve "lift" pressure, in relationship to design pressure, is a function of the ASME code (or lack of code rating) for the column. If the

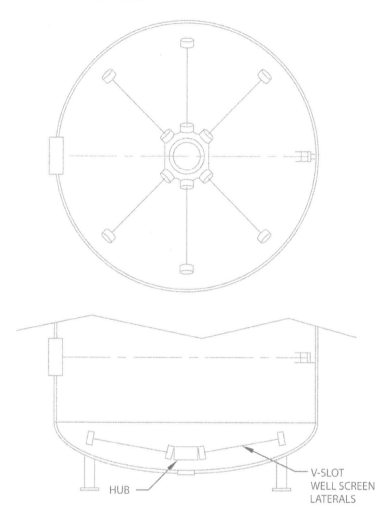

Figure 3.8 Lower distributor design of the hub radial with V-slot well screen laterals in an ion exchange vessel. (This design collects product water from the cross-section of the resin bed, eliminates stagnant areas, and delivers product to a central hub outlet.) *Source*: From Siemens Water Technologies Corporation.

vessel is designed, manufactured, and tested in accordance with the ASME Code for Unfired Pressure Vessels, the relief valve setting can be the maximum design pressure rating stated on the data plate positioned on the vertical cylindrical vessel. However, if the vessel is not ASME code stamped with a pressure rating designated by the equipment manufacturer, it is strongly suggested that the lift pressure of the relief valve be a conservative fraction of the manufacturer's recommended maximum operating pressure (e.g., 75–85% of the actual specified value). The discharge side of the relief valve should be hard piped to a physical area at the base of the column, preferably not readily accessible by operating personnel.

The backwash piping from the unit should be equipped with a transparent section of piping or sight glass that will allow operating personnel to observe gross particulate matter in the backwash water. Since particulate matter <40 μm is not visible to the human eye, the sight glass will not allow operating personnel to determine the presence of finer particulate matter in the backwash water. However, the sight glass can be used, on the basis of the observation of larger particles, to establish a conservative backwash time duration for the unit. Backwash piping should not contain any obstructions that will inhibit the flow and should be directed to a depressurized drain with an "air break" between the drain and the backwash piping to eliminate the possibility of potential microbial introduction through system piping. Face piping for the unit should be provided with properly sized manual or automatic valves. It is suggested that individual diaphragm valves or, for units with valve size >2 to 3 in., butterfly valves, be used. Pneumatically actuated valves are preferred. For diaphragm valves, selection of the pneumatic actuator should be such that the valves are positive acting (air to open, spring

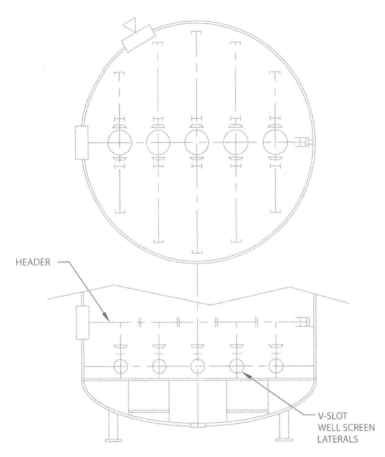

Figure 3.9 Ion exchange vessel lower distributor design showing the header with V-slot well screen laterals. (This design produces the same effect as the radial system, but it delivers the product to a pipe header, often a side discharge from the column.) *Source*: From Siemens Water Technologies Corporation.

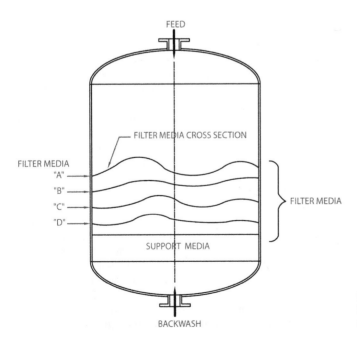

Figure 3.10 Example of an uneven filter bed as a result of poor distribution and backwash.

to close) at the maximum design pressure for the unit and 0% ΔP. Solenoid valves that control the airflow to the individual automatic valves on the basis of a signal from a local or central control panel should be mounted in the general area of the unit. Manual override provisions should be considered for the solenoid valves. If solenoid valves with manual override provisions are provided in an "uncontrolled personnel access area," the solenoid valves should be positioned in a "limited access" enclosure.

Feed water and product water pressure gauges should be provided. The gauges should be of stainless steel construction, liquid filled, with diaphragm isolators and be provided and maintained in a calibrated state. Most gauges provided with a particulate removal filter are totally inappropriate. In general, the accuracy of the gauges does not allow operating personnel to determine the pressure drop across the unit. It is not uncommon to observe a unit with a product water pressure gauge reading greater than the feed water pressure gauge reading. Since the successful operation of the particulate removal filter is associated with the ripening process, the use pressure gauges in the feed water and product water lines, supplemented by differential pressure measurement with diaphragm isolators may be appropriate. The differential pressure sensing and indicating system should be capable, as a minimum, of indicating a pressure drop through the unit as low as 3 to 5 psid.

The face piping for the unit should contain a manual regulating valve in the backwash line to allow operating personnel to adjust the backwash flow rate. The temperature of the feed water to the unit, principally for raw water from a surface water source, will exhibit significant fluctuations in temperature with seasonal changes. A direct reading temperature indicator of stainless steel construction should be installed in the feed water piping. Colder, more viscous water can achieve the same backwash efficiency as warmer, less viscous water at a lower flow rate. Temperature monitoring coupled with the observation of particulate matter in the backwash piping to drain and visual observation of bed expansion using a sight glasses on the sidewall of the column will allow operating personnel to determine the appropriate backwash flow rate as a function of water temperature.

A feed water flow rate meter should be provided, and its range should be adequate to measure the normal operating flow rate and the backwash flow rate. In general, a variable area type meter is adequate for this application; if properly selected, it will provide the required accuracy. The suggested material of construction is stainless steel.

Sample valves should be provided in the feed water and product water line of the unit. Sample valves may be of needle type or diaphragm type, with a material of construction consistent with other accessories for the unit. Ball-type valves should not be employed for sample valves.

For certain facilities, the backwash flow of a particulate filter may represent the highest flow rate to drain. The capacity of the drain system in the physical area of installation may not be adequate for the backwash flow. It may be appropriate to consider an appropriately sized backwash collection tank. The tank collects the backwash water, provides volume for storage, and discharges the water to drain at a controlled rate considering the flow rate capacity, avoiding "flooding" conditions during particulate removal filter backwash.

Operating and Maintenance Considerations

The backwash frequency should be established on the basis of the differential pressure drop through a unit. The frequency of the backwash operation, particularly for raw feed water from a surface source, will vary with seasonal and climatic conditions. The backwash flow rate should be periodically adjusted, as required, to compensate for changes in raw water temperature. The duration of the backwash operation should be adequate to remove entrapped particulate matter. This duration may also change with seasonal and climatic conditions.

Operating personnel should establish an analytical monitoring program to determine the effectiveness of the particulate removal filter. An excellent analytical indication of filter performance is the concentration of total suspended solids (TSS). The TSS concentration should be determined periodically for both feed water and product water samples. The monitoring program should include periodic multiple samples (feed water and product water) throughout the operating "cycle" of the unit (between backwash cycles).

The efficiency of the backwash operation should be verified periodically by visually monitoring bed expansion using sight glasses installed on the straight side of the particulate removal vessel, if available, and/or by observing particulate matter or water color in the transparent section of tubing in the waste line. Sight glasses in the product water piping from a particulate removal unit can also be used to verify that filter media are not passing "through" the column because of a failure of the lower distributor.

An annual inspection of the interior of the unit should be conducted. A visual inspection of "interior-coated" steel columns should primarily verify the integrity of the coating. A dielectric testing device, such as teslacoil, can be used to verify the integrity of the steel columns with PVC sprayed and baked, or rubber linings. The inspection should include visual review of distributor integrity (feed water). Replacement of manway and handhole gaskets should be performed subsequent to column access.

If air scouring is used to enhance the backwash operation, the purity of the air should be periodically determined by appropriate analytical monitoring procedures. For example, any analysis of the scouring air should verify that it does not contain oil or any other substances that would be construed as "foreign substances and impurity" as defined in the USP *General Notices* section.

While a physically challenging task, filter and support media should be replaced once every three to five years for fiberglass column units and about every five years for steel column units. The replacement frequency may be adjusted on the basis of the quality of the feed water and product water. It is strongly suggested that filter lower distributors be replaced each time media is replaced since access to the lower distributors requires removal or filter and support media that will generally require system shutdown. While residual disinfectant is generally present in the feed water to a particulate removal filter, slime and/or algae develops within the unit, particularly on supporting gravel at the base of the unit. Filter performance, in terms of the TSS measurement in the feed and product water may not be the only criteria for replacing the filter media and support gravel. For example, long-term increases in total viable bacteria levels in the product water from the unit, which cannot be reduced by backwash or sanitization, may indicate that excessive proliferation of microorganisms has occurred within the unit.

Daily logging of operating parameters, such as feed water flow rate, feed water and product water pressure, pressure drop through the unit, and backwash frequency are required. All instrumentation should be periodically calibrated. Preventative valve maintenance should be performed. For fiberglass column units employing multiport-type valves, it is suggested that an entire spare valve assembly be retained as "spare parts." For units with manual or automatic diaphragm valves, a one- to two-year diaphragm replacement frequency should be considered. For larger units with manual or automatic butterfly valves, complete rebuild of valve seals and seats should be performed every one to two years.

ACTIVATED CARBON UNITS
Theory and Application
Activated carbon, as a unit operation, can remove residual disinfectant and reduce the concentration of NOM in filtered feed water. This dual function of activated carbon is important as a pretreatment technique for feed water supplies from a surface water source. For certain raw water supplies from a groundwater source, where insignificant levels of NOM are observed, the removal of residual disinfectant can be achieved by injecting a reducing agent, or use of inline ultraviolet radiation at a wavelength of 185 nm and high intensity. Over the past few years, there has been a tendency to eliminate activated carbon units from water purification systems (raw feed water from a surface source) in an attempt to reduce system bacteria levels, which are generally highest after an activated carbon unit. While microbial control is a significant concern, it is suggested that proper design of an activated carbon unit, including the ability to periodically hot water sanitize the unit, can reduce concerns associated with bacteria levels in the product water. It should be pointed out that hot water sanitization is vastly superior to steam sanitization as indicated by data presented in Table 3.4.

The removal of residual disinfectant is desired in systems using ion exchange as the primary ion removal technique to eliminate oxidation of the cation resin and reduce organic

Table 3.4 Activated Carbon Unit—Post Hot Water Sanitization Total Viable Bacteria Levels—Steam and Hot Water

Week number	Sanitization method	Total viable bacteria (cfu/mL)
1	Steam	~1100
2	Steam	~1400
3	Steam	~2100
4	Steam	>5700
5	Steam	>5700
6	Steam	>5700
7	Steam	>5700
8	Steam	>5700
9	Steam	>5700
10	Steam	>5700
11	Hot water	180
12	Hot water	240
13	Hot water	56
14	Hot water	94
15	Hot water	40
16	Hot water	12
17	Hot water	8
18	Hot water	1
19	Hot water	2
20	Hot water	<1
24	Hot water	<1
26	Hot water	<1

Note: All total viable bacteria results by heterotrophic plate count of 1-mL sample, PCA Agar, 25°C to 30°C incubation temperature, and 48-hour incubation time period. Chemical sanitization of activated carbon unit "face piping" performed between week numbers 10 and 11 samples. Activated carbon media replaced prior to hot water sanitization on week number 13. Hot water sanitization conducted at 90°C for four hours. Steam sanitization performed at 30 psig for eight hours.
Source: From Collentro (2010).

fouling of the anion resin (chap. 4). For systems using reverse osmosis, the removal of residual disinfectant and NOM is also critical. Currently, most of the RO membranes used in pharmaceutical water purification systems are not chlorine tolerant. Furthermore, while chlorine tolerant membranes can be used, the process of removing residual disinfectant after chlorine tolerant membranes will affect the purity of the product water from the RO unit by increasing the concentration of inorganic impurities and/or increasing microbial levels. This is undesirable, since the position of this residual disinfectant scavenging/reducing operation is located closer to the ultimate "end user" and downstream of a bacteria removal unit operation, reverse osmosis. The removal of NOM as a pretreatment step to reverse osmosis is very critical because NOM will foul the RO membranes, resulting in the need for periodic cleaning. Of even greater importance, however, is the fact that the NOM will serve as a nutrient for bacterial growth on membrane surfaces. Chemical cleaning (sanitization) of RO membranes, where removal of NOM has not been provided (raw surface feed water), can result in the frequent cleaning of RO membranes for microbial control.

The removal of residual chlorine by activated carbon can be represented by the following equations:

$$C^* + 2Cl_2 + 2H_2O \rightarrow 4HCl + CO_2$$

or

$$C^* + H_2O + HOCl \rightarrow CO^* + H_3O^+ + Cl^-$$

Activated carbon removes residual chlorine through the formation of surface oxides (C^* represents the activated carbon surface and CO^* represents a surface oxide on the activated carbon). The kinetics of this reaction are extremely rapid, resulting in complete removal of residual chlorine in the first few inches of an activated carbon bed. On the other hand, the reaction of activated carbon with chloramines, principally monochloramine, while similar in nature, is not as rapid and requires increased contact time and has a much lower capacity for removal when compared with that for chlorine. Obviously, this affects the design parameters for an activated carbon unit, principally the face velocity through the bed (gpm/ft^2 over the cross-sectional bed area) and the volumetric flow rate (gpm/ft^3 of activated carbon media). While 1 g of activated carbon will remove approximately 1 g of residual chlorine, the capacity for the complete removal of chloramines is significantly less. While a properly designed unit with raw surface feed water will generally require activated carbon media replacement based on breakthrough of NOM (see below), media replacement is *often* required on the basis of chloramine breakthrough, particularly for improperly designed units.

NOM is removed by activated carbon by a physical adsorption process, which is associated with relatively weak physical forces (van der Waal's forces). As discussed in chapter 2, raw surface water supplies will contain NOM that has a significant variation in molecular weight, chemical composition, and chemical structure. The vast differences in these parameters significantly affect the adsorption coefficient for each organic compound present in the feed water to the activated carbon unit. This is a very important factor because the adsorption coefficient has a direct bearing on the ability of activated carbon to remove an organic compound. Research has indicated that the adsorption process by activated carbon is a dynamic situation, of multicomponent organic compounds on activated carbon, where organic compounds are being simultaneously adsorbed and eluted from activated carbon "sites" (Collentro, 1968). This process results in the complete removal of certain organic compounds, at a particular time in the "life" of the activated carbon media, and eventual breakthrough of the organic components at a later stage in the life of the media, when the activated carbon bed is more saturated with organic material (Fig. 3.11). Because of the elution process, it should be noted that this multicomponent adsorption phenomena may result in the presence of a particular organic fraction of the naturally occurring material at a higher level in the product water from the activated carbon unit than in the feed water. In general, activated carbon media replacement, for a properly designed unit, should be considered about once every six months. Units operating with media beyond the six-month period (surface water supplies) may exhibit the indicated breakthrough phenomena. It is important to note that this replacement frequency, while conservative for NOM, is required for both chloramine removal and physical removal of bacteria laden activated carbon media even if hot water sanitization is employed.

The adsorption of NOM by activated carbon is enhanced by the presence of cations, principally multivalent ions such as calcium and magnesium (Ong and Bisque, 1988; Ghash

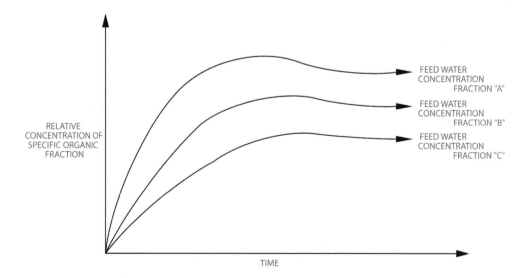

Figure 3.11 Organic "breakthrough:" multicomponent system (simulated three-fraction feed water situation). Absorption potential: fraction C > fraction B > fraction A. *Source*: From Collentro (1968).

Table 3.5 Effect of Salt Concentration on the Adsorptive Capacity of Activated Carbon—"Prepared" Fulvic Acid Solution

Adsorptive capacity (mg TOC/g of activated carbon)	Salt	Concentration (mol/L)
20	NaCl	2
20	NaCl	4
21	NaCl	6
45	$MgCl_2$	2
46	$MgCl_2$	4
47	$MgCl_2$	6
47	$CaCl_2$	1
48	$CaCl_2$	2
49	$CaCl_2$	3
50	$CaCl_2$	4

Note: Solution pH = 7.0; base solution contained 2.0 mg/L of Na^+; feed water (base) TOC level = 5.31 mg/L.
Abbreviation: TOC, total organic carbon.
Source: From Randtke and Jepsen (1982).

Table 3.6 Effect of the Addition of Calcium Chloride on Activated Carbon Removal of Fulvic Acid (Note 1)

Time (days)	Feed water TOC (mg/L)	Product water TOC (mg/L)
5 (note 2)	5.37	<0.1
10 (note 2)	5.37	4.0
40 (note 2)	5.37	4.6
80 (note 2)	5.37	–
80 (note 3)	5.37	<0.1
90 (note 3)	5.37	1.0
100 (note 3)	5.37	1.8
120 (note 3)	5.37	2.8
140 (note 3)	5.37	3.5
145 (note 3)	5.37	–
145 (note 2)	5.37	11.5
150 (note 2)	5.37	9.0
160 (note 2)	5.37	7.0
170 (note 2)	5.37	6.3
180 (note 2)	5.37	6.0

Note: Feed water pH = 8.3, dynamic "column."
(1) Feed water = 5.37 mg/L of "peat" fulvic acid at pH = 8.3 with 1.0 mol/L sodium bicarbonate.
(2) Feed water = 5.37 mg/L of "peat" fulvic acid at pH = 8.3 with 1.0 mol/L sodium bicarbonate and 100 mg/L of calcium carbonate.
Abbreviation: TOC, total organic carbon.
Source: From Randtke and Jepsen (1981).

and Schnitzner, 1979; Weber et al., 1983; Randtke and Jepsen, 1981). Tables 3.5 and 3.6 describe the effect of calcium and magnesium on the adsorption of NOM by activated carbon. Many pharmaceutical water purification systems using reverse osmosis will employ pretreatment unit operations that include both activated carbon and water softening. The sequence of pretreatment unit operations must be carefully evaluated on the basis of both feed water quality and component selection with focus on control of bacteria levels. Higher organic levels will be observed in the product water of an activated carbon unit positioned downstream of a water softener than a unit positioned upstream of a water softener (Table 3.7). This may be critical when microbial control is considered. NOM, removed at a reduced efficiency by an activated carbon unit positioned downstream of a water-softening unit, will accumulate on RO

Table 3.7 TOC Reduction by Activated Carbon—Unsoftened and Softened Feed Water

Day number	Feed water TOC (mg/L)	Product water TOC (mg/L)
1	3.89	0.76
5	3.80	0.89
10	3.81	0.79
15	3.90	0.82
20	4.16	0.91
25	3.70	1.02
30	3.68	0.96
31	3.42	3.44
35	3.23	2.88
40	3.36	3.22
44	3.35	3.20

Note: Feed water is from a surface source. The activated carbon unit was positioned prior to the water softener from the beginning of the study through day 30. Subsequent to day 30, the water softener was positioned prior to the activated carbon unit.
Abbreviation: TOC, total organic carbon.
Source: From Collentro and Collentro (1997).

Table 3.8 Elution of Organic Material from Activated Carbon by Softened Water

Average feed water TOC = 3.40 mg/L
Activated carbon unit product water TOC during operation (no recirculation) = 0.80–1.10 mg/L
Activated carbon unit product water TOC during operation (with pretreatment recirculation including water softener) = 1.0–2.2 mg/L

Note: The approximate TDS of surface feed water = 230 mg/L. The average total hardness of feed water = 100 mg/L as $CaCO_3$. The average feed water pH = 9.0. The estimated average recirculation time = 20 minutes. The maximum estimated recirculation time = 40 minutes. All data are from "grab" samples.
Abbreviation: TOC, total organic carbon.
Source: From Collentro (1995b).

membrane surfaces, providing a nutrient for microbial growth. Again, this will affect the required cleaning or sanitization of the RO unit. However, if periodic hot water sanitization of the RO unit (and downstream polishing components) is performed, the balance of water purification system-wide total viable bacteria levels must be evaluated. The evaluation will determine if allowing residual disinfectant to be present through the water-softening system positioned prior to activated carbon units ultimately results in lower water purification system-wide bacteria levels.

The final issue associated with the positioning of activated carbon and water-softening units relates to the effects of recirculation of pretreatment equipment. For units with proper design, there will be a significant reduction of NOM through the activated carbon unit. However, if water being recirculated through the activated carbon unit does not contain the same level of organic-removing cations, elution of organic material will occur (Table 3.8). While recirculation of pretreatment components, particularly activated carbon units and water softeners, is desirable, for RO-based systems, or systems where deionization units and upstream activated carbon units are recirculated, separate recirculating loops may be provided (Fig. 3.12). If separate recirculation is not provided, as supported by the information in Table 3.8, the overall efficiency of the activated carbon unit for removing NOM is reduced.

It would be inappropriate to conclude a discussion associated with the theory of activated carbon units without discussing microbial control. Since an activated carbon unit removes residual disinfectant and significant reduces NOM, one would fully anticipate that significant bacterial proliferation would occur within the activated carbon unit, particularly in the lower portion of the activated carbon bed (a warm, dark, wet area with abundant carbonaceous material). In fact, this unit operation is a bacteria generator. There are several viable methods for controlling bacterial levels in the product water from an activated carbon unit. Periodic

PRETREATMENT TECHNIQUES

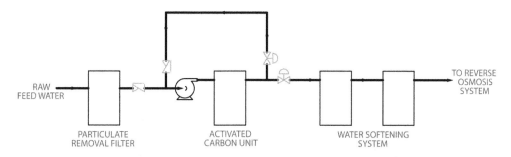

Figure 3.12 Proper activated carbon recirculation for system with feed water from surface source. (Hot water sanitation can be performed by adding a heat exchanger. An inline ultraviolet unit can be positioned directly downstream of the activated carbon unit. Dedicated recirculation can be provided for the water softening system. Hot water sanitation provisions can include both the activated carbon unit and water softening system with proper valve and piping configuration.)

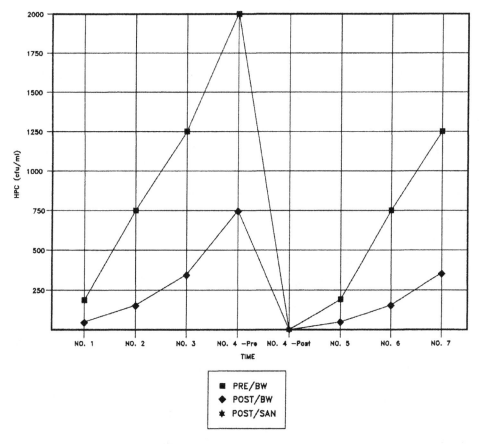

Figure 3.13 Activated carbon unit–effluent microbial levels: heterotrophic plate count versus sample time.

backwash with ambient temperature water will provide a degree of bacteria reduction. Periodic sanitization of an activated carbon unit with hot water (90°C for approximately two hours) will provide total destruction of bacteria within the bed (Fig. 3.13). It should be emphasized that the 90°C temperature, the two-hour sanitization time period, and to a lesser extent flow in the backwash direction are all critical to the effectiveness of this sanitization process. The most

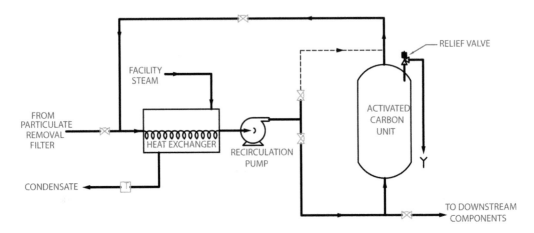

Figure 3.14 Hot Water sanitation of an activated carbon unit in a recirculating mode. [A limited number of valves are shown to demonstrate the recirculation concept without making it difficult to follow the flow path. System piping and valves may be arranged to allow the pump (and heat exchanger) to be positioned from the product water piping, if desired. Cooldown, waste piping, valves, instrumentation, and controls have been intentionally omitted.]

effective method of performing this sanitization operation is shown in Figure 3.14. The system uses a dedicated sanitary tank, recirculation/sanitization pump and heat exchanger as part of the sanitization system. This eliminates continuous heating of the backwash water and discharge to drain. Cool down can be achieved, subsequent to the sanitization operation, by displacing the hot sanitization water with ambient feed water. Using this sanitization technique, establishing activated carbon unit product water total viable bacterial limits (suggested at 500 cfu/mL) with associated trending of intracomponent microbial data will significantly reduce or eliminate microbial contamination and proliferation on downstream unit operations. For systems using water softeners downstream of the activated carbon unit, water softeners, as discussed later in this chapter, can be included in this sanitization cycle. The small amount (volume) of softened water introduced through the activated carbon unit during the sanitization operation will have minimal effect on desorption of previously adsorbed NOM.

Design Considerations
Systems using feed water from a groundwater source *may* use a reducing agent, such as sodium bisulfite or sodium sulfite, to remove residual disinfectant agent or, as indicated earlier inline ultraviolet units operating at 185 nm and elevated intensity. The feed water for these applications must contain minimal organic impurities (natural or pollutants). The system design should include the issues presented earlier in this chapter for chemical injection of reducing agents.

The use of "cartridge"-type activated carbon filters for the removal of residual disinfectant should be limited to systems with feed water from a groundwater source since "bed depth" is required for the effective removal of NOM. Furthermore, these filters have a very limited capacity for removing residual disinfectant, particularly residual chloramines.

The diameter and height of the backflushable activated carbon unit vessel is extremely critical. For systems with feed water containing residual chlorine, the diameter should be such that the face velocity through the activated carbon media is approximately 3 gpm/ft^2 over the cross-sectional bed area of the column. The volume of activated carbon should be adequate to provide a volumetric flow of about 1.0 gpm/ft^3 of activated carbon media. For systems with feed water containing residual chloramines, and/or where NOM levels are >3 to 5 ppm the diameter should be such that the face velocity through the activated carbon media is ≤3 gpm/ft^2 over the cross-sectional bed area of the column. The volume of activated carbon should be adequate to provide a suggested volumetric flow of about 0.5 to 0.75 gpm/ft^3 of activated carbon media.

As indicated, the design conditions for residual chloramine removal are much more restrictive than those for residual chlorine. As indicated in chapter 2, an increasing number of municipal water treatment facilities employ chloramines for microbial control in distribution systems. The primary chloramine compound of interest is monochloramine on the basis of its disinfection properties. Monochloramine is the predominate chloramine at pH values ≥ 7. Removal of monochloramine is enhanced by use of custom activated carbon material referred to as "catalytic carbon" (Baker and Byrne, 2004).

The removal of monochloramine by activated carbon is represented by the following equation:

$$NH_2Cl + C^* + 2H_2O \rightarrow NH_3 + H_3O^+ + Cl^- + CO^*$$

The literature (Fairley et al., 2007) suggests that a subsequent reaction may occur, as follows:

$$2NH_2Cl + CO^* + H_2O \rightarrow N_2 + 2H_3O^+ + 2Cl^- + C^*$$

Activated carbon removes monochloramine through the formation of surface oxides (C^* represents the activated carbon surface and CO^* represents a surface oxide on the activated carbon). The kinetics of this reaction are slow and influenced by several parameters. The literature indicates that the presence of NOM reduces the effectiveness of catalytic activated carbon for monochloramine removal (Fairley et al., 2006). It is suggested that NOM adsorbed to the activated carbon surface removes activated carbon "sites" for monochloramine removal. Further, the literature also indicates that monochloramine removal efficiency increase with decrease in pH (Fairley et al., 2007). The author suggests that this may be associated with the kinetics of the first reaction indicated above considering that ammonia will react with water. This reaction removes ammonia producing the ammonium ion and hydroxide ion production. At lower pH values there is a greater concentration of hydronium ion to react with the hydroxide ion. The net result is improved kinetics for the first reaction indicated above. The following summarizes suggested activated carbon unit sizing/design for monochloramine removal by a specific catalytic activated carbon:

$$\text{catalytic activated carbon volume required} = (EBCT) \times \text{flow rate}/7.48$$

where EBCT is the "empty bed contact time" expressed in minutes, flow rate is expressed in gallons per minute, and the catalytic activated carbon volume is expressed in cubic feet.

For removal of monochloramine by catalytic activated carbon at a feed water concentration of 3.0 mg/L, a good design value for many municipal distribution systems, and target product water monochloramine concentration of ≤ 0.1 mg/L, the required EBCT = 8 minutes at pH = 8.0 and 5 minutes at pH = 7.0 (Fairley et al., 2007). On the basis of this information, the design parameters for activated carbon units with 3.0 mg/L monochloramine can be calculated as follows:

Flow rate (gpm)	pH	Column diameter (in.)	Bed depth (in.)	Media volume (ft³)	EBCT (min)
12.0	7.0	24	36	9.4	5.8
25.0	7.0	36	36	21.2	6.3
45.0	7.0	42	48	38.5	6.4
75.0	7.0	48	48	50.2	5.0
100	7.0	60	48	78.5	5.9
9.0	8.0	24	36	9.4	7.8
20.0	8.0	36	36	21.2	7.9
35.0	8.0	42	48	38.5	8.2
45.0	8.0	48	48	50.2	8.3
75.0	8.0	60	48	78.5	7.8

Notes: Shaded EBCT vales are slightly below the recommended time.
Activated carbon media is 12 × 40 mesh, "Centaur," Calgon Carbon Corporation (1998), Pittsburgh, Pennsylvania, U.S.
Total organic carbon of surface feed water = 2.5 to 3.5 mg/L.
EBCT limits by Fairley et al., 2007.

In addition to the significant operating flow reduction for monochloramine removal, the catalytic activated carbon media should be replaced every six months. Media replacement frequency may be extended if feed water flow TOC levels are lower than indicated. However, frequent free and total chlorine measurements should be performed to determine the monochloramine "breakthrough" for each application.

A suggested minimum activated carbon bed depth is 36 in. As indicated activated carbon bed depths >36 in. may be required for applications where high levels of NOM or chloramines are present in the feed water.

The suggested backwash flow rate, depending on the activated carbon bed depth, should be about 4.25 to 4.5 gpm/ft^2 over the cross-sectional bed area when the water temperature is approximately 20°C to 25°C. The backwash flow rate may be reduced if the water temperature is lower (colder water has a higher viscosity than warmer water). The backwash frequency for an activated carbon unit positioned downstream of a particulate removal filter will generally be based on elapsed time rather than differential pressure. Unlike a particulate removal filter, an activated carbon unit does not benefit from the ripening process. Generally, the backwash frequency is established on the basis of product water bacterial control considerations and the performance characteristics of the unit. Excessive backwashing is undesirable because the density of the activated carbon media increases as it removes NOM. Frequent backwashing of the activated carbon unit could result in premature breakthrough of organic material, which by density is transferred to the bottom of the column during the backwash operation. Excessive backwashing can also affect the life cycle of the activated carbon unit for the removal of residual chloramines.

System control should be designed to execute ambient temperature backwash and hot water sanitization backwash, if applicable, during "off-shift" or nonpeak time periods, when the manufacturing demand for water is minimal. The system design may include provisions for periodic hot water sanitization in a "recirculating" mode, with water at a temperature approximately 90°C for approximately two hours. Sanitization in the backwash direction can provide excellent contact between the activated carbon media and the sanitizing hot water assuming that adequate column freeboard is available. If hot water sanitization is performed in the backwash flow direction, it must be preceded by ambient backwash. This sequence of operations minimizes potential hot water flow restrictions (with resulting "channeling") from an improperly expanded activated carbon bed.

Backflushable activated carbon units should employ vertical cylindrical vessels. While lower flow rate systems may use fiberglass-reinforced vinylester or polyester columns, it is more appropriate to consider steel column for higher flow rates ($\geq$40 gpm). The interior of steel columns should be lined with a "food-grade," high-temperature, sheet rubber material, vulcanized in place. Continuity of the lining should be verified by a dielectric test at a prespecified voltage. It is strongly suggested that the column be designed, fabricated, and tested in accordance with the ASME Code for Unfired Pressure Vessels, which is a requirement in most states (see discussion earlier in this chap.). Unfortunately, many units for pharmaceutical water systems are purchased each year that do not meet applicable code requirements. It is impossible to code stamp a column "retrospectively." While this particular item is not necessarily a regulatory issue, it is an issue that could become a major problem with local or state "boiler and pressure vessel" inspectors or during an insurance audit.

The ASME code vessel should be designed for operation at a suggested pressure of about 70% of the maximum design pressure. For most applications, the column pressure rating can be 100, 125, or 150 psig. Testing in accordance with the ASME code is performed at a pressure 50% higher than the rated pressure. The column should also be designed for operation at "full vacuum," if periodic hot water sanitization will be performed. It would be inappropriate to consider periodic hot water sanitization, at a suggested temperature of approximately 90°C, without providing a vessel capable of withstanding full vacuum. It is possible during the hot water sanitization process that a "water vapor space" will exist in the upper portion of the column. During subsequent column cool down with ambient temperature water, the water vapor will rapidly condense, resulting in a vacuum.

Hot water sanitization requirements for the vessel dictate the use of stainless steel piping/tubing and valves. The use of individual valves, specifically positive acting pneumatically operated diaphragm valves, should be considered as compared with a single multiport valve or ball-type valves that will increase the sanitization frequency for the unit by providing a location for bacteria to accumulate and replicate.

The freeboard space in the column (empty volume above the activated carbon media) should be adequate to allow expansion of the activated carbon bed during both ambient temperature backwash and hot water sanitizing backwash operations. It is undesirable to minimize freeboard space since impingement of activated carbon media on the top of the column during the backwash operation will result in the production of highly undesirable activated carbon "fines." Generally, the amount of freeboard space is limited by physical height restrictions at a facility. The added cost of increasing the column straight side height, assuming that there is adequate physical height at a facility, is relatively small.

Access to the interior of the column is important since media is changed more frequently than both particulate removal filters and water softeners. As a minimum, a manway should be provided for periodic inspection of column internals, specifically the lining and distributors and to allow removal and replacement of activated carbon media. As appropriate, "hand holds" may be provided, particularly in the area of the lower distributor for the unit.

The straight side of the column may be equipped with one or more vertically mounted sight glasses, positioned at and around the activated carbon–water interface. This will allow operating personnel to determine if backwash is being performed at adequate flow rates to provide desired bed expansion. The sight glasses also allow operating personnel to monitor the level of the activated carbon media periodically and verify that the media are not depleting with time. Depletion of media is associated with excessive backwash flow rates, the production of activated carbon fines, or, most importantly, failure of the lower distribution system.

The domed top of the column should have an ASME-approved relief valve. The setting for the relief valve should be consistent with the design pressure rating for the column. The exhaust side of the relief valve should be "hard piped" to a physical location at the base of the column, preferably inaccessible to operating personnel. If possible, a relief valve that can be manually operated to serve as a vent valve should be considered.

The backwash piping to waste from the unit should contain a transparent section of material to allow observation of activated carbon particles and carbon fines in the backwash water. For units equipped with periodic hot water sanitization provisions, the transparent material must be capable of withstanding hot water at a temperature of 90°C.

Waste piping from the unit to drain may require a dedicated depressurized ambient temperature mixing "trough." Many federal, state, and local restrictions limit the temperature for direct discharge of a waste line to <140°F. Obviously, the initial temperature of the postsanitization "displacement" water will be greater than 140°F. All waste connections from the unit should be to an unpressurized drain with an "air break."

The inlet and outlet distributors, positioned in the interior of the column, should be constructed of 316 or 316L stainless steel particularly for steel column units. As discussed earlier in this chapter, distribution through the activated carbon media is achieved by back pressure exerted by the lower distributor. A minimum back pressure must be achieved to ensure proper distribution through the column and eliminate the potential for channeling. Generally, a pressure drop of approximately 5 to 7 psid (after backwash) indicates that channeling will not occur.

The activated carbon unit should be equipped with a feed water flow rate indicator with a range adequate for determining the normal operating flow rate, the normal ambient temperature backwash flow rate, and the sanitizing hot water backwash flow rate.

Units equipped with automatic controls (as discussed earlier) should be provided with individual diaphragm valves. For larger units, butterfly valves may be considered. If diaphragm valves are used, they should be positive acting, air to open, spring to close. They should be designed to close at the design pressure rating for the unit and 0% ΔP. For all pneumatically operated valves, a solenoid panel may be provided in the immediate physical vicinity of the unit. Solenoid valves should be equipped with manual override provisions. It is suggested that the valves be positioned in a cabinet that provides minimum access to avoid inadvertent manual activation of an automatic valve.

The unit should be equipped with feed water, product water, and hot sanitizing water pressure gauges. It is suggested that the gauges be liquid filled, of stainless steel construction, with either diaphragm or sanitary type isolators, for direct connection to the stainless steel face piping of the unit. The accuracy of the feed water and product water pressure gauges should allow determination of pressure drops as low as 2 to 4 psid through the activated carbon unit.

For hot water sanitizable units, thermal wells and temperature indicators should be provided to allow operating personnel to monitor temperature during normal operation, ambient backwash, and hot water sanitization cycle. Temperature monitoring should be provided in the feed water to the column and, for sanitization cycles, at the "coldest" location in the system (e.g., the upper straight side of the vertical cylindrical column).

Feed water and product water sample valves, of needle or diaphragm type, should be provided for routine intracomponent sampling for appropriate chemical and bacterial analysis.

Many activated carbon units operating in pharmaceutical water purification systems employ a cartridge filtration system immediately downstream of the activated carbon unit to retain activated carbon fines. The highest water purification system microbial levels will generally occur in the product water from the activated carbon unit. Cartridge filtration to remove activated carbon fines merely provides a location with significant surface area for bacteria to accumulate and replicate in a nutrient rich environment. Selection of the "mesh size" for the lower distributor in the activated carbon unit as well as proper system design to avoid the production of activated carbon fines is critical. If possible, cartridge filtration should *not* be considered as an activated carbon "fine" trap downstream of the unit.

The use of an inline ultraviolet unit positioned in the product water from the activated carbon unit is strongly suggested for microbial control. It is suggested that "oversizing" of the unit be considered to insure that an adequate ultraviolet radiation dose is provided to bacteria present in the activated carbon unit product water, thus completely eliminating the potential for sublethal destruction of Gram-negative bacteria. Experience indicates that activated carbon unit total viable bacteria levels may be reduced by one to two orders of magnitude by the inline ultraviolet sanitization unit (Collentro, 2007).

The heat exchanger used for optional hot water sanitization may be either a plate-and-frame unit or shell-and-tube unit. A shell-and-tube unit is preferable. Heat exchanger design and operating parameters are discussed earlier in this chapter.

Replacement of the activated carbon media, specifically for applications with raw feed water from a surface source or raw feed water containing chloramines, should be considered every six months. A steel column activated carbon vessel should be equipped with a media "sluice" port consisting of a weld neck and flange positioned on the lower straight side of the vertical cylindrical column. The connection should be at least 3 to 4 in. in diameter to allow removal of the "interlocking" granular-activated carbon media.

Experience indicates that steam is a very poor sanitizing media for activated carbon units (Table 3.4). Furthermore, plant steam will generally contain volatile amines that control corrosion in the condensate and feed water systems to the facility boiler. The volatile amines would be considered as USP *General Notices* section "foreign substances and impurities." If plant steam is used for sanitization, an analytical monitoring program would be required to verify the absence of these contaminants in accordance with the requirements outlined in USP. The alternative of using USP "Pure Steam" for periodic hot water sanitization is strongly discouraged. Once again, the activated carbon unit represents the area in a system where bacterial levels will be the highest. It appears extremely inappropriate to connect a USP Pure Steam line to this bacteria-rich unit operation.

For units that include hot water sanitization provisions, the exterior of the column and "face valves and piping" should be thermally insulated to eliminate operator exposure to hot surfaces during the sanitization operation.

Operating and Maintenance Considerations

Periodic backwash (at ambient temperature) of the activated carbon unit should be performed. This backwash operation should be based on controlling effluent microbial levels. However, if a particulate removal filter is not positioned prior to the activated carbon unit, accumulation of particulate matter with associated increase in pressure drop may require an increase in the backwash frequency for the unit. Yet, excessive backwash is undesirable because it can result in premature breakthrough of residual disinfectant (assuming that the residual disinfectant agent is chloramines) and organic material, as denser activated carbon media containing organic material is literally "transferred" to the lower portion of the unit (at the effluent) because it is heavier.

Periodic hot water sanitization should be performed. The requirements for sanitization should be based on trending of product water bacterial levels from the activated carbon unit, particularly samples obtained after ambient backwash of the unit.

It will be necessary to modify the ambient temperature backwash and hot water sanitization frequency with seasonal and climatic changes. It is fully anticipated that increased backwash frequency will be required as the temperature of the raw feed water to the system increases.

In an attempt to minimize the production of activated carbon fines, assuming that there is adequate freeboard space in the activated carbon vessel, the backwash flow rate should be adjusted with changes in the raw feed water temperature. The effectiveness of the backwash operation and, subsequently, the expansion of the activated carbon bed will be enhanced by colder, more viscous water.

The duration of the backwash operation should be adequate to provide the desired bed expansion, specifically to remove activated carbon fines and reduce postbackwash product water bacterial levels. Generally, the duration is a function of the specific conditions for a unit and should be adjusted on the basis of operating experience.

An analytical monitoring program should be established to determine the concentration of residual disinfectant in product water, as well as the TOC and total viable bacteria levels. It is further suggested that periodic monitoring of the feed water be performed to determine the effectiveness of the unit, particularly with regard to reducing the TOC level. Finally, analytical monitoring should include feed and product water measurement of TSS to verify that activated carbon fines are not being generated or passing through a defective lower distribution system.

The level of activated carbon media in the column should be periodically observed and logged to ensure that media are not being removed from the column. This logging procedure will determine if excessive amounts of activated carbon fines are being produced during ambient temperature backwash, hot water sanitization, or normal operation (because of inappropriate design of the lower distributor).

As indicated, activated carbon media should be replaced about once every six months. The activated carbon sluice port on the side of the column can be used to remove spent activated carbon. It is extremely important to ensure that replacement activated carbon media are fully hydrated, acid washed, and rinsed. To adequately hydrate activated carbon (i.e., ensure that water has fully entered activated carbon "pores"), approximately 24 to 36 hours of "soak time" is required. This operation can be avoided, with its associated "downtime," by purchasing activated carbon that has been prehydrated. The use of acid washed and prebackwashed (rinsed) activated carbon is also important. Acid washing removes trace concentrations of multivalent cations such as barium, aluminum, and strontium—common natural contaminants present in commercially available activated carbon. These ions will adversely affect the performance of RO units positioned downstream of the activated carbon unit because of the formation of highly insoluble precipitates. These heavy–molecular weight multivalent cations are also undesirable in deionization systems, since they are tightly held to the cation resin (chap. 4). Finally, many sources of activated carbon will contain sodium carbonate, potassium carbonate, and silica. The presence of sodium carbonate and potassium carbonate results in product water with an elevated pH because they are "basic salts." Acid washing and pre rinsing will remove the impurities and eliminate potential problems with downstream components.

Subsequent to removal of spent activated carbon media and prior to replacement with new media, it is suggested that the interior of the column be rinsed with water and chemical disinfecting agent. A 1% solution of Peracidic Acid and Hydrogen Peroxide provides excellent removal of bacteria in a "biofilm" on the interior wall of the column.

Channeling through the activated carbon unit can be encountered if the normal operating flow rate is too high *or* too low. Operating flow rates through the unit should be provided, even during recirculating conditions, to produce the desired face velocity of 3 gpm/ft² over the cross-sectional bed area.

WATER-SOFTENING UNITS
Theory and Application
Water softening in pharmaceutical water purification systems is generally limited to three specific applications.

1. Pretreatment of feed water to a single- or double-pass RO unit to remove cations that will form insoluble precipitates within the RO membranes.
2. Treatment of regenerant water for systems using mixed deionizers as the primary ion removal technique.
3. Active pharmaceutical ingredient or consumer product applications where deionization is not required, even though product water quality is improved by using softened water.

The ion exchange process is discussed in greater detail in chapter 4. Typical reactions demonstrating the removal of multivalent cations, such as calcium and magnesium, are as follows:

$$R-Na^+ + Ca^{++} \leftrightarrow R-Ca^{++} + Na^+$$
$$R-Na^+ + Mg^{++} \leftrightarrow R-Mg^{++} + Na^+$$

Calcium and magnesium ions are removed by the ion exchange process because of their greater affinity for ion exchange sites than monovalent, light–molecular weight sodium ions. The resulting product water is free of common multivalent cations in raw water as well as trace concentrations of highly undesirable, heavy–molecular weight multivalent cations, such as aluminum, barium, and strontium.

During the regeneration process, a concentrated solution of sodium chloride passes through the ion exchange resin. The high concentration of the sodium ion reverses the equilibrium ion exchange reaction:

$$Na^+ + Cl^- + R-Ca^{++} \leftrightarrow Ca^{++} + Cl^- + R-Na^+$$
$$Na^+ + Cl^- + R-Mg^{++} \leftrightarrow Mg^{++} + Cl^- + R-Na^+$$

The regeneration process is highly effective. Regenerant sodium chloride levels are a function of the concentration of multivalent cation in the raw water supply and the ratio of the multivalent cations to monovalent ions, principally sodium. This process can provide pretreated feed water to a single- or double-pass RO system, free of potential membrane-scaling cations. For certain active pharmaceutical ingredient applications, where USP Purified Water quality is not required, water softening may be used to enhance the quality of water used during the manufacturing operation.

While not extremely common, water softeners are occasionally used to provide regenerant water for mixed-bed deionization units in the primary (single) ion exchange step in a water purification system. This application is limited to systems with extremely low volumetric demand or systems with extremely low (<15–25 mg/L) TDS levels. The mixed-bed units cannot be regenerated with water containing calcium and magnesium since these ions, during the regeneration process, would react with the hydroxide ion and produce insoluble precipitates. Not only will this significantly decrease the effectiveness of the regeneration operation, it will also result in the physical appearance of calcium hydroxide and magnesium hydroxide precipitates within the resin bed.

Design Considerations
A water-softening unit should consist of a vertical cylindrical column. For units with a relatively low flow rate, fiberglass-reinforced polyester or vinylester columns may be used. However, steel columns with appropriate interior linings are preferred. The column should be designed to contain a minimum cation bed depth of 36 in. and a minimum freeboard space of

50% to 60%. The interior of a steel-column water-softening unit should be lined with a corrosion resistant material. It is suggested that the column lining be a 3/16-in.-thick, food-grade, high-temperature sheet rubber, and vulcanized in place. Continuity of the rubber lining should be verified by a dielectric test at a specified voltage.

As indicated earlier in this chapter, most states require that the vessels be designed, constructed, and tested in accordance with the ASME Code for Unfired Pressure Vessels. The specified pressure should be consistent with the application, suggested as 100, 125, or 150 psig (maximum) and full vacuum. The full vacuum specification will meet the criteria associated with hot water sanitization, similar to that discussed for activated carbon units. The top of the column should be equipped with an ASME-approved relief valve with installation and selection similar to activated carbon columns.

The water-softening system may be designed for periodic sanitization with hot water (90°C for two hours) or periodic chemical sanitization. Face piping for hot water sanitizable units should be constructed of 316 or 316L stainless steel. While it appears inappropriate to expose austenitic stainless steel surfaces to hot water containing the chloride ion, the percent of time that the stainless steel surfaces will be exposed to the hot water and, subsequently, the degree of chloride stress corrosion or chloride pitting corrosion will not impact the operating life of the water-softening unit.

The lower straight side height of the steel column units should be equipped with a resin removal port consisting of a weld neck and flange. The minimum size for the piping connection using the weld neck and flange should be 3 in. Any connections from upstream or downstream components mating stainless steel to PVC, CPVC (copolymer of PVC), or other plastic material should be executed using flange-to-flange connections as discussed earlier for activated carbon units.

The feed water, product water, and waste lines for the water-softening system should be equipped with stainless steel pressure gauges, preferably liquid filled. The pressure gauges should have diaphragm isolators to minimize microbial introduction to the system. The range and incremental calibration for the pressure gauges should be small enough to determine the pressure drop through the water-softening unit.

The feed water piping to the unit should be provided with a flow rate meter used with a manual diaphragm valve for establishing flow during normal operation, backwash, and, where appropriate, hot water or chemical sanitization. Since it is highly desirable to establish a responsive regeneration frequency for the water-softening system, it is suggested that the flow rate meter also contain provisions for indicating the total volume of water processed between regeneration cycles.

A steel column water-softening unit may be equipped with a rectangular sight glass positioned on the straight side of the column, mounted vertically at the resin-water interface. The sight glass should be approximately 2 to 3 in. wide × 12 in. high. Multiple sight glasses may be used. Sight glasses allow operating personnel to verify that backwash of the resin is adequate, and that resin is not lost from the column because of the production of resin fines (backwashed to drain) or depleted through improper design or failure of the lower distributor. This latter condition would result in the presence of both resin fines and whole resin beads in the product water from the unit.

The product water piping from the water-softening system should be equipped with a resin fine trap. This trap should be constructed with 316 or 316L "well screen," with provisions for "backflushing" (Fig. 3.15).

An inlet distribution system to the column should be provided. The primary purpose of the distribution system is to provide "rough" delivery of feed water to the vertical cylindrical column. The inlet distributor may consist of a "double-elbow" system to deliver feed water to the domed top of the unit. Distribution through the resin bed is achieved using back pressure exerted from the lower distributor. The back pressure must be adequate to provide uniform flow velocity over the entire cross-sectional area of the cation bed. To achieve good distribution, a minimum pressure drop of approximately 5 to 10 psid (for a new or freshly backwashed and regenerated unit) is required.

The upper and lower distributors for steel column units should be fabricated from 316 or 316L stainless steel. The configuration and design of the lower distributor should ensure that

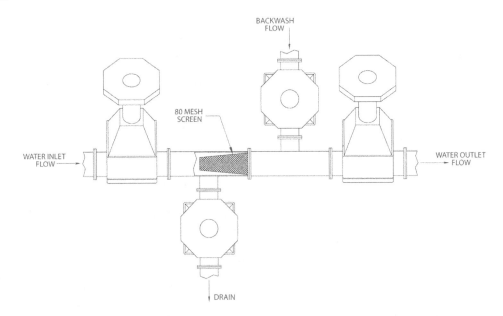

Figure 3.15 Resin trap with backwash provisions.

there is no resin "below" the distributor that is not exposed to regenerant chemicals (i.e., in a "dead area") as discussed for activated carbon unit lower distributors.

For units equipped with automatic controls, pneumatically operated diaphragm valves or, for larger units, butterfly valves should be employed. It is suggested that the valves be positive acting, air to open, spring to close, at the rated design pressure for the unit and 0% ΔP.

For units equipped with automatic controls and pneumatically operated valves, a local solenoid panel may be provided. The solenoid valves should be equipped with manual override provisions and positioned in an enclosure with limited personnel access. The system should be provided with feed water, product water, and regenerant salt water sample valves of either the diaphragm or needle type.

The waste line from the unit should be directed to a depressurized drain with an "air break" with a transparent section of material for observing the presence of resin fines. The material should be capable of operating at temperatures up to 90°C, for hot water sanitizable steel column units, during the initial phases of the displacement rinse operation, subsequent to periodic hot water sanitization of the water softener (when applicable).

A manual diaphragm valve should be provided in the ambient backwash line for regulating and adjusting the backwash flow rate due with seasonal and climatic changes in water temperature. Figure 3.16 graphically illustrates bed expansion as a function of temperature, clearly demonstrating the effect of more viscous colder water on the backwash operation.

Most water-softening systems that are employed as pretreatment to RO units will contain two or more individual units. As discussed previously, it may be desirable to position the activated carbon unit(s) prior to the water-softening units for the effective removal of NOM for certain feed waters. If activated carbon units are positioned upstream of the water-softening units, the feed water will not contain residual disinfectant. To minimize microbial growth within the carbonaceous cation beds (operating in the sodium form at a neutral pH) and to minimize the effect of resin bed stagnation on product water quality (see discussion in chap. 4), it is highly desirable to maintain continuous flow through each water-softening unit in a system. This can be achieved by one of two techniques. Progressive piping, demonstrated in Figure 3.17, can be employed when there are two units. This technique allows operation of the units in series, parallel, or individually. However, the piping arrangement will result in dead legs. As a preferable alternative for dual water-softening systems, one unit can be

PRETREATMENT TECHNIQUES

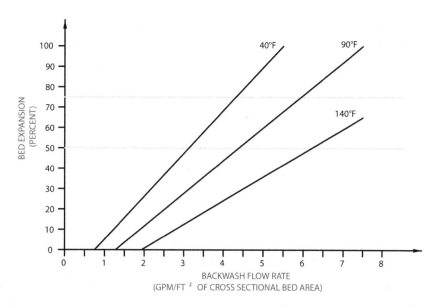

Figure 3.16 Cation resin bed expansion (during backwash) percent versus flow rate at various temperatures. (A strong acid cation resin has been used. The desired bed expansion is 50% to 75% depending on the percent free board space in the column.) *Source*: From Resintech.

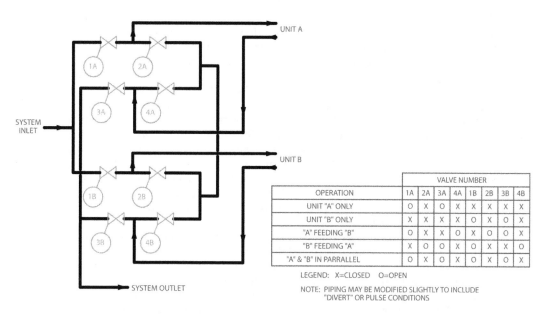

Figure 3.17 Progressive piping for dual treatment units.

designated as the "working unit," with the second unit, designated as a "polishing unit," operating in series after the first unit. In this particular case, the lead unit is regenerated on the basis of volumetric throughput or breakthrough of hardness, while the second unit is regenerated on an elapsed operating time basis. To ensure that multivalent cations are not present, even in trace concentrations, the polishing unit can be regenerated with an elevated salt concentration, which increases the conversion of ion exchange sites to the sodium form. This arrangement, with continuous flow, minimizes bacterial proliferation associated with a stagnant "standby" unit.

In general, the size of the unit (i.e., the diameter of the ion exchange column) should be based on a water flow rate associated with a face velocity of approximately 7 gpm/ft^2 over the cross-sectional bed area. It is suggested that face velocities should not be <4 or >10 gpm/ft^2 over the cross-sectional bed area to achieve the desired removal of multivalent cations, particularly as a pretreatment component to a single- or double-pass RO system. It is strongly suggested that recirculation, if provided, be performed at the suggested operating face velocity of 7 gpm/ft^2 over the cross-sectional bed area.

The selection of a regenerant salt storage system is critical to microbial control for the water-softening system. The use of "wet" salt storage systems, where the salt storage (brine) tank contains water above the salt level, is strongly discouraged. As an alternative, a "dry" salt storage system that employ a brine tank with a lower support plate, allowing regenerant brine solution to be "freshly" prepared prior to the regeneration operation, should be considered. The use of porous "fiberboard" or wood-based material as a material of construction for this application is strongly discouraged. Polypropylene or polyethylene are acceptable materials of construction.

In general, most regeneration systems for water-softening units use a simple eduction system to introduce the saturated salt solution into the ion exchange vessel. The volume of regenerant salt educted into the vessel over a set period of time during the regeneration cycle is a direct function of the feed water pressure to the eductor. To ensure proper regeneration, pressure regulation should be established prior to the water-softening system. To verify the salt concentration, a sample valve should be positioned in the regenerant salt feed water line to the column, downstream of the eductor. Operating personnel should measure the level of salt in the storage tank prior to and subsequent to the regeneration operation to determine the volume/weight of salt introduced during regeneration. Proper regeneration of a water-softening unit should be performed using 15 lb of pure (no additive for iron removal) sodium chloride per cubic foot of cation resin. The dissolved salt should be introduced at a concentration of about 10% sodium chloride and flow rate of about 0.5 gpm/ft^3 of cation resin.

If the feed water supply to the water-softening system contains measurable concentration of dissolved iron, the potential for iron fouling of the cation must be considered. Quite often, regenerant salt advertized to contain iron removal provisions is employed. Use of this salt is discouraged since the chemical "additives" in the salt are unknown. Subsequently, the additives are "foreign substances and impurities" as defined by the USP *General Notices* section. It is strongly suggested that regenerant salt be free of not only iron removal additives but debris and other material that will clog transfer piping, brine tank level control systems and, most importantly, the orifice in the educator. Further, if the dissolved iron concentration exceeds the U.S. EPA National Secondary Water Regulation concentration (0.3 mg/L), it is suggested that 12- to 24-month cation resin replacement be considered on the basis of product water monitoring for iron. The cost of cation is relatively inexpensive when compared with the consequences of iron breakthrough to downstream components particularly RO membranes were it provides a nutrient for pathogens.

For a very limited number of applications, where calcium and magnesium concentrations are significant or a high percentage of the total cations, differential conductivity may be used to determine the operating cycle of the water softener. This monitoring system compares the feed water conductivity with the product water conductivity. There is a slight difference in the equivalent conductance of the sodium ion when compared with the calcium and magnesium ions (Table 3.9). By measuring the feed water and product water conductivity and determining

Table 3.9 Equivalent Conductance—Monovalent and Divalent Cations

Ion	Relative conductance
Sodium (Na$^+$)	50.08
Calcium (1/2 Ca^{+2})	59.47
Magnesium (1/2 Mg^{+2})	53.0
Barium (1/2 Ba^{+2})	63.6
Manganese (1/2 Mn^{+2})	53.5
Iron (1/2 Fe^{+2})	54

Source: From *Handbook of Chemistry and Physics* (2006–2007) (Lide, DR, 2006–2007).

the ratio of the values internally, within a single conductivity monitor, it is possible to determine when breakthrough of hardness occurs. A much more common alternative applicable to any feed water hardness concentration is an inline "Total Hardness Monitor," which collects "grab" product water samples, conducts an automated chemical analysis, and displays or transmits the result.

Operating and Maintenance Considerations

Microbial control within water-softening systems is important. Unlike deionization systems that use bacteria-destroying acid and caustic regenerant chemicals and operate in the hydronium and hydroxyl form, a water softener is regenerated with a non-bacteria-destroying salt solution and operates at a neutral pH. During the regeneration cycle of a water softener, the backwash operation may remove some of the bacteria present in the carbonaceous cation bed. Unfortunately, however, product water bacterial levels will increase with time. Furthermore, bacteria are introduced with regenerant solutions, particularly for systems using wet brine storage. Effective control of bacteria can be achieved by periodic hot water sanitization at 90°C for two hours or, more effectively, by chemical sanitization with a 1% solution of hydrogen peroxide and peracidic acid. The later has the ability to remove biofilm with appropriate sanitization procedure.

If an existing water-softening system contains a carbon steel vessel with internal lining (or coating) incapable of withstanding hot water sanitization temperatures, it is possible to achieve effective sanitization using a 1% solution of peracidic acid and hydrogen peroxide. The chemical sanitizing solution may be introduced in a concurrent direction (downward through the cation resin bed). While discussed in chapter 9 of this text, effective chemical sanitization requires both dynamic and stagnant exposure conditions.

If the feed water to the water softener contains residual disinfectant, oxidation of the cation resin will occur. This oxidation process will remove a portion of the residual disinfectant from the feed water supply, depending on the concentration and the type of residual disinfectant. Chlorine, for example, will oxidize the cation resin more rapidly than monochloramine. Oxidation of the cation resin by residual disinfectant will result in decrosslinking of the cation resin, associated gradual loss in capacity, and elution of resin decomposition compounds. While the effect of the oxidation process will generally result in a small gradual decrease in the capacity of the resin, concerns associated with elution of organic material are more important since the presence and the composition of the organic compounds are not monitored.

For water-softening units positioned as a pretreatment to single- or double-pass RO units, the presence of trace concentrations of multivalent cations, such as aluminum, barium, and strontium, in the feed water to the softening system is extremely important. An analytical monitoring program, using "supplemental" analysis (beyond that specified in the USP Purified Water *Official Monograph*) should be established. If *trace* concentrations of these impurities are detected in the effluent from the water-softening unit, the salt dosing should be increased. This increases the sodium concentration and, subsequently, the effectiveness of the regeneration process (the sodium ion replaces the heavier multivalent cation).

As discussed earlier, the purity of regenerant salt is extremely important, particularly when the water-softening unit is used as a pretreatment component to a membrane process, such as reverse osmosis. A Certificate of Analysis should be supplied with the salt. Periodic analysis of the salt should be performed by the pharmaceutical facility ("end user"). The analysis should verify that trace concentrations of highly undesirable impurities are not present.

Certain municipalities may add proprietary chemicals during the treatment process in an attempt to keep iron in a soluble form. The appearance of particulate iron, which stains domestic sinks, basins, and so on is undesirable. However, these proprietary chemicals generally produce a colloidal form of iron. This material will generally pass through the water softener resulting in scaling of downstream RO membranes. A detailed analysis of the raw water supply, as well as a "tour" of the municipal water treatment system, is suggested. Furthermore, a contact should be established at the municipal treatment facility. It is important to conduct an annual tour with this contact to verify that treatment techniques have not changed that will affect the performance of the pharmaceutical water purification system.

The presence of resin fines in stainless steel resin traps positioned in the product water of the water-softening unit should be monitored. If resin traps are not employed, operating personnel should note the presence of resin and/or resin fines on RO prefilters. Excess production of resin fines indicates improper system operation (or design). This may be associated with an operating variable (e.g., excessive backwash rate) or a design parameter (e.g., inadequate freeboard space). The observation of resin fines will probably be associated with the presence of resin fines in the backwash water, resulting in a decrease in the volume of cation resin over a period of time. Thus, it may be necessary to periodically add cation resin to an operating unit to compensate for this situation.

Cation resin life for systems operating with iron in the feed water has been discussed previously. However, for systems where iron is not present in low concentrations in the feed water, it is important to obtain annual "core" samples from the cation bed. These samples should be analyzed for critical parameters to determine the condition of the resin. Oxidation by residual disinfectant and the percent of exchange sites converted from the calcium and magnesium to sodium sites during regeneration are important parameters.

For water-softening systems using eductors for introducing regenerant brine, it is important that the dilution water pressure be established and maintained at a fixed value throughout the regeneration process. If a noticeable decrease in the intake of salt solution is noted, the eductor should be inspected. It is possible that a small accumulation of particulate matter can significantly affect the operation of the eductor.

Periodic feed water and product water samples should be obtained and analyzed. The analysis should include total hardness, calcium concentration, and magnesium concentration, TSS, total iron, and trace impurities of multivalent cations (aluminum, barium, strontium, etc.). When two units are operated in series, the final product water should exhibit calcium and magnesium concentrations <1 mg/L as calcium carbonate. Feed water and product water data should be plotted as a function of time, as part of data trending for the system. In addition, analysis should be conducted for total viable bacteria and bacterial endotoxins (USP Water for Injection systems and low-endotoxin USP Purified Water systems). Periodic sanitization (semiannual or annual) with hot water (where appropriate) or chemical sanitization agent should be conducted to control bacteria within the carbonaceous cation bed. As discussed previously, a water-softening unit suggested total viable bacteria product water Action Limit is 500 cfu/mL.

For water-softening units used to provide regeneration for primary mixed-bed deionization units, it is important that a preoperational "rinse-to-drain" cycle be performed. This operation removes bacteria from the stagnant bed and ensures that calcium and magnesium, which may have migrated to the lower portion of the resin bed, are removed prior to the regeneration process. The multivalent cations would produce insoluble magnesium and calcium hydroxide precipitates during regeneration of the mixed-bed deionization unit.

Periodic (annual or biannual) chemical sanitization of the brine tank should be performed. An effective sanitization agent is sodium hypochlorite. Sodium hypochlorite may also be periodically added to the salt storage tank in an attempt to minimize microbial growth within the tank.

The volume of water processed through the water-softening system between regeneration cycles, as well as the results of periodic feed water analysis for total hardness, calcium, and magnesium, should be evaluated to determine if the resin capacity is decreasing with time.

ORGANIC SCAVENGING RESINS
Theory and Application
Using organic scavenging resin as a pretreatment technique has historically been limited to raw feed water supplies from a surface source with high TOC levels. Ironically, many of the applications where organic scavenging resins have been used as a pretreatment are associated with ion exchange–based systems. The resin is also used in the semiconductor industry to reduce TOC levels to extremely low values (<5 μg/L). It would be inappropriate to discuss pretreatment systems for pharmaceutical water purification systems without discussing this unit operation, which may be employed as a pretreatment operation for single- and double-pass RO units or as part of a rechargeable or regenerable ion exchange

system producing USP Purified Water or providing feed water to a distillation unit in a USP Water for Injection system.

As discussed earlier, and later in chapter 4, the presence of NOM in the feed water to a RO system can produce two highly undesirable conditions: gradual organic fouling of the RO membranes, resulting in the requirement for periodic cleaning to remove the organic material, and, more importantly, microbial proliferation on RO membrane surfaces. Feed water organic material provides a nutrient for bacterial growth on the RO membranes. Frequent cleaning (once a month or more) of RO membranes for microbial control has been directly attributed to feed water bacterial levels. However, it is suggested that while feed water bacterial levels are obviously an important factor in the microbial fouling of RO membranes, the presence of organic material, a nutrient for bacteria, is an extremely important contributor to the rate of bacteria proliferation. Organic scavenging resins can remove a significant portion of NOM from the feed water supply to an RO unit, particularly the highly undesirable heavy–molecular weight humic acid fraction. These resins can, in certain cases, be coupled with reducing agent injection to provide both effective removal of NOM and residual disinfectant agent, the function of an activated carbon unit. Microbial control within the anion organic scavenging column can be achieved by periodic regeneration, which employs a combination of sodium chloride and sodium hydroxide.

For USP Purified Water system employing either rechargeable ion exchange canisters or deionization units regenerated in place, organic scavenging resin may be used solely for organic removal or as the anion resin in a two-bed deionization system to insure TOC reduction to the 0.50 mg/L value set forth in USP *Physical Tests* section <643>.

From a theoretical standpoint, the mechanism by which organic scavenging resins remove organic material has been historically related to an adsorption process. It was suggested that the organic material is physically attracted to the organic scavenging resin by relatively weak physical forces (van der Waal's forces). Several technical papers have been presented during the 1990s that clearly demonstrate that organic material is, in fact, not adsorbed on the anion organic scavenger resin but rather removed by an ion exchange process (Gottleib, 1996; Symons et al., 1992). Research has provided valuable information demonstrating this ion exchange process and the importance of anion resin selection for organic scavenging applications. Gelular-type resins, specifically acrylic-based resins versus styrenic-based resins, are extremely effective at removing organic material, particularly membrane-fouling heavy–molecular weight material. This is clearly indicated in the data presented in Table 3.10. While macroporous resins will also provide excellent removal of NOM, the removal of the material during the regeneration cycle is complicated by the fact that the pore structure requires diffusion of exchanged organic material to the outer surface of the resin bead. While perhaps initially more effective than gelular resin, the macroporous anion resins require longer regeneration times (higher contact time) because exchanged organic material within the resin pores must diffuse to the resin surface, a process with relatively slow kinetics.

The anion organic scavenging resins are operated in the chloride form. Researchers have demonstrated that the chloride ion is 11 to 25 times more attracted to an ion exchange site than hydroxyl ions (Gottleib, 1996). This is an important item when regeneration is considered. Obviously, the exchanged organic molecule represents a very heavy complex structure with

Table 3.10 TOC Reduction by Various Anion Resins

Resin "skeleton"	Pore structure	Mean pore radius (nm)	Surface area (m^2/g)	% TOC reduction
Styrene	Gel	N/A	0.1	15
Styrene	Porous	10	409	50
Styrene	Porous	35	60	60
Acrylic	Gel	N/A	0.1	65
Acrylic	Porous	100	<5	80
Styrene	Porous	3500	7	88

Note: TOC reduction calculated for organic fraction is >10,000 d.
Source: From Symons et al. (1992).
Abbreviation: TOC, total organic carbon.

"multiple" valance that may be difficult to remove by regenerant chemicals. If the principal regenerant is sodium chloride, there is a much greater tendency to displace the exchanged organic complex with chloride ion as compared with hydroxide ions.

Design Considerations

The regeneration time required for the total removal of exchanged organic material has been estimated at approximately 16 hours. Approximately 70% to 80% of the exchanged organic material is removed in four to six hours (Gottleib, 1996). By elevating the temperature and including sodium hydroxide within the regeneration process it is possible to reduce the regeneration time to about two to three hours.

Much like an activated carbon unit, contact time is critical. The column diameter and bed depth must be carefully selected to achieve desired removal of organic material. A face velocity of 3 gpm/ft^2 over the cross-sectional bed area and volumetric flow of 1 gpm/ft^3 of resin are required. Inorganic ions, such as chloride and sulfate, affect the anion scavenging resin's ability to remove organic material. This is particularly true for sulfate. As sulfate ion breakthrough occurs through the anion column, organic material is eluted from the column. Fortunately, the elution of organic material at sulfate breakthrough is associated with the lighter–molecular weight faction of organic material, which is less likely to foul RO membranes, than the heavy–molecular weight factions of NOM. However, proper system design should include provisions for determining breakthrough of the sulfate ion, since it is extremely critical to long-term successful operation of the organic scavenging unit.

The suggested regeneration operation should be performed using a warm solution of 10% sodium chloride *and* 2% sodium hydroxide. The ideal temperature of the regenerant solution can vary from 100°F to 120°F, depending on the anion resin selected as the organic scavenger. Acrylic resins should be regenerated at the lower end of the temperature range, while styrenic resins may be regenerated at 120°F. Sodium hydroxide is used with sodium chloride to provide solubility of the organic material, increasing the effectiveness of the regeneration process. The design of the system should include provisions for the simultaneous introduction of both regenerant chemicals.

Column design, distributor design, and accessories for the ion exchange column are similar to those previously discussed for water-softening units and activated carbon units.

While the selection of the specific anion resin for each application is a function of the characteristics of the feed water supply, in general, it is suggested that gelular-type acrylic resins be considered. Other gelular resins, such as styrenic-based resins, will not provide the same degree of organic removal. Furthermore, macroporous resins, both acrylic and styrenic, require a longer regeneration time period to achieve equivalent removal of exchanged organic material, offsetting their initial increased ability to reduce organic material.

The size of the resin bead is also important. A smaller resin bead size, with associated increased surface area, will encourage ion exchange at the surface of the resin, increasing the effectiveness of the regeneration process.

Operating and Maintenance Considerations

The generation of resin fines and associated resin attrition, particularly for macroporous anion resins, is a concern. Irreversible organic resin fouling will occur with time. It is important to analyze core samples of the resin periodically (once every six months) to determine the physical condition of the resin.

The concentration of regenerant sodium chloride and sodium hydroxide is critical to the successful regeneration of the anion organic scavenging resins. Temperature is also critical. The volume of regenerant introduced, as well as the temperature, should be verified periodically.

The regeneration cycle is primarily determined by sulfate breakthrough through the anion exchange resin. A method for determining breakthrough, provided as part of the system design, should clarify the frequency of the regeneration step. This may be determined by chemical analysis or inline monitoring techniques specific for the sulfate ion. The anticipated reduction in TOC will be 50% to 80%, depending on the resin selected for the application and the "life cycle" of the anion organic scavenging resin.

If macroporous resins are used and the flow rate is terminated through the anion bed, it is quite possible, after flow through the unit is reinitiated, that a noticeable decrease in product water TOC level may be observed. While this would appear to be beneficial, it represents the diffusion of exchanged organic material into the center of the anion resin beads. This significantly decreases the potential for removing the exchanged organic material during the regeneration process. Subsequently, particularly for systems using a macroporous anion organic scavenging resin, it is suggested that continuous flows be maintained.

As indicated the required regeneration time period for anion organic scavenging resins is much longer than that required to remove the inorganic ions. Periodically, the anion bed should be "fully regenerated" by introducing regenerant salt and sodium hydroxide over a time period of 10 to 16 hours.

Feed water and product water TOC values should be determined periodically. The TOC elution from the anion organic scavenging resin should be approximately 0.04 to 0.05 mg of TOC/mL of resin/min. If calculations indicate that TOC elution is significantly higher than the indicated values, resin replacement and/or extended regeneration should be considered.

In addition to periodic measurement of TOC values from samples obtained from the unit, it is strongly suggested that periodic SDI measurements be performed to verify the effectiveness of the unit. SDI measurements are an excellent indicator of the level of RO membrane-fouling impurities present in the feed water (chap. 4). The TOC, SDI, and product water sulfate values should be used to determine the operating and regeneration cycles for the unit.

Since elution of organic material may occur, particularly if the inorganic anion concentration changes, it may be appropriate, particularly for higher flow rate systems, to consider online TOC measurement from the product water of the unit. Considering the fact that this is an ion exchange process, there will a series of "breakthrough curves" generated from the myriad of organic compounds (NOM) passing through the unit. While these curves would appear to favor the initial passage of lighter–molecular weight organic material, followed by the passage of inorganic material and finally heavier–molecular weight organic material, the actual characteristics of dynamic anion organic scavenging resin unit performance must be established for specific water supplies.

Considering the obvious maintenance-intensive requirements for organic scavengers, the use of the technology as a pretreatment unit operation is very limited, primarily to small capacity rechargeable canister ion exchange–based applications. The ever increasing performance characteristics of RO membranes coupled with decrease in membrane cost of replacement limits the use of this technology for pharmaceutical water system where the TOC product water specification is orders of magnitude greater than that for other applications such as those in the semiconductor industry.

INLINE ULTRAVIOLET UNITS
Theory and Application

Inline ultraviolet radiation is a form of energy (photon) with a wavelength longer than X-rays but shorter than the sun tanning range and visible light. This wavelength is considered to be 100 to 400 nm, generally provided by mercury vapor lamps at a "peak" value of 253.7 nm. Historically, low-pressure mercury vapor lamps have been used to provide the desired sanitizing ultraviolet radiation. These lamps use an electrical current to elevate mercury atoms to a higher energy state. As the mercury atoms return to their original state they emit ultraviolet radiation. About 82% of the ultraviolet radiation from a low-pressure mercury vapor lamp is at a wavelength of 253.7 nm, while about 6.6% is at a wavelength of 184.9 nm (Bolton et al., 2008). As ultraviolet light at 253.7 nm, passes through a water solution containing bacteria, the protein and nucleic acid contained in a microorganism absorb the energy, destroying the DNA (deoxyribonucleic acid) and, subsequently, the "viable nature" of the microorganism. As a result, microorganisms are inactivated although technically they remain "metabolically" alive. Different microorganisms require significantly different ultraviolet radiation dose rates (energy and exposure time) for *inactivation*. The required dose for the complete inactivation of an organism can be determined by multiplying the ultraviolet radiation intensity, expressed as $\mu W/cm^2$ by the contact time, expressed in seconds. Thus, the ultraviolet radiation dose is expressed in $\mu W\text{-}sec/cm^2$.

One of the primary concerns associated with the use of inline ultraviolet units is the actual ultraviolet intensity within a sanitizing chamber. This intensity is a function of the geometry associated with the unit and other critical factors, such as the number of lamps, flow rate, cleanliness of quartz sleeves, chamber baffles, and so on. The intensity can be expressed by the following equation:

$$I = \frac{S}{4\pi r^2}$$

Where I is the intensity in $\mu W/cm^2$, S is the intensity from the ultraviolet source expressed in μW, and r is the distance from the source expressed in cm. On the basis of the above equation, the ultraviolet intensity and, subsequently, the ultraviolet dose, decrease significantly as distance from the source increases. Multiple ultraviolet lamp units, used historically for pharmaceutical water purification systems, must be carefully designed to ensure that all microorganisms in the flowing water stream through the unit are exposed to adequate ultraviolet radiation to obtain complete inactivation. It should also be obvious that the contact time for properly exposing the microorganism to ultraviolet radiation is also important. One of the primary concerns expressed by regulatory personnel is that bacteria may be exposed to a "sublethal" dose of ultraviolet radiation (Munson, 1985). While it would be inappropriate within this text to imply that inline ultraviolet radiation is not a viable unit operation in a pharmaceutical water purification system, it should clearly be stated that the use of inline ultraviolet radiation, as a unit operation, requires careful unit selection, proper design, and sound preventative maintained to avoid counterproductive results as a microorganism inactivation technique. There is a debate regarding the ability of inline ultraviolet units to inactive bacteria. Published documents by FDA regulatory personnel indicate that inline ultraviolet units should not be considered to inactivate greater than 90% of the bacteria in a water stream (FDA, 1993). One inline ultraviolet unit manufacturer's literature implies that 99.9% to 99.99% bacteria inactivation can be achieved by the unit. This discussion of inline ultraviolet units has been included in this chapter because, in the author's opinion, the primary use of inline ultraviolet units can be justified as a pretreatment component. The use of inline ultraviolet units (emitting radiation at a wavelength of 253.3 nm) downstream of a bacteria specific process, such as reverse osmosis, ultrafiltration, or membrane filtration (<0.1 μm) should be carefully evaluated.

Many pharmaceutical water purification systems, particularly units equipped with single- or double-pass reverse osmosis and "polishing" ion removal techniques, provide product water with extremely low inorganic and organic levels resulting in a low-nutrient environment for microorganisms. On the basis of discussions presented in chapter 2, bacteria present in this environment will enter a defense mechanism. The physical shape and size of bacteria, responding to the low-nutrient environment, will change. Gram-negative bacteria, in particular, are extremely good at adapting to the low-nutrient environment via this defense mechanism. The literature contains very little, if any, data regarding the effects of ultraviolet radiation on bacteria that have entered a defense mechanism for survival. However, it is fully anticipated that bacteria in a defense mechanism state would, in fact, not be fully inactivated by inline ultraviolet radiation; they may be sublethally inactivated, further adapting to an inadequate dose of ultraviolet radiation for complete inactivation. This phenomenon has been observed and documented for several operating pharmaceutical-grade water purification systems (Collentro, 1995a). Regulatory personnel have also indicated that they can detect bacteria (using a 5- to 15-day culture time, R2A culture media, and a 20–22°C incubation temperature) when samples are collected from certain "high-purity" pharmaceutical water systems when the pharmaceutical facility has not detected bacteria at a 1 cfu/100 mL level using PCA culture media, 48 to 72 hours of incubation, and a 30°C to 35°C incubation temperature (Avallone, 1994).

Related to the above, some researchers have indicated that a specific Gram-negative organism, *Burkholderia cepacia*, if not totally inactivated within an inline ultraviolet unit, produces a sub lethally destroyed species that is photoreactive, replicating significantly quicker than the original organism (Carson and Peterson, 1975). In fact, the literature further states that that photoreactivation of inactivated microorganism is possible when exposed to light with a wavelength of about 300 to 400 nm (Harm, 1980).

The role of ultraviolet units should be to "control/reduce" microbial levels. It should not be considered as a unit operation, which incorrectly is designated by the word *sterilization*. Inline ultraviolet units may provide a method of bacteria control/reduction after certain unit operations, such as activated carbon, water softening, or "primary" deionization, where bacteria levels are high and downstream processes, such as ultrafiltration, reverse osmosis, continuous electrodeionization or 0.05 to 0.1 membrane filtration, will be provided to remove potentially sublethal destroyed forms of organisms.

Design Considerations

Most inline ultraviolet units are purchased as "standard" products. While it appears that the number of design options is limited, there are certain critical items that should be considered. Many of these items are available from some inline ultraviolet unit manufacturers as options.

The most desirable inline ultraviolet sanitization unit would employ a single lamp. This type of system minimizes the effect of "geometry" within the sanitization chamber associated with multiple lamp units. This enhances the capability of determining the ultraviolet radiation intensity in a "plane" of the sanitizing chamber by measuring the radiation at one point. Unfortunately, a single lamp unit does not necessarily ensure that "short circuiting" will not occur. The flow path and characteristics through the sanitization chamber are important. While the literature implies that laminar flow is undesirable, proper flow to achieve a contact time adequate to expose microorganisms to the ultraviolet radiation is desired. Turbulent flow is desired in a "plane" through the sanitizing chamber, perpendicular to the direction of flow.

The selection of an inline ultraviolet unit should include the manufacturer's information indicating the "contact time" through the unit. Internal baffles within the sanitizing chamber should be used to enhance contact time. One of the advantages of longer, single or multiple lamp units, as compared with shorter multiple lamp units, is the increase in contact time through the unit.

The presence of particulate matter, such as activated carbon or resin fines, in the feed water to the unit is highly undesirable. Prefiltration, or the addition of a manual or automatic "wiper" system on the exterior of quartz sleeves containing the mercury vapor lamps, should be considered if particulate matter is present in the feed water. Particulate matter would absorb ultraviolet radiation, thus decreasing the available radiation at a particular point in the system for inactivating bacteria. Iron the ferric form is highly undesirable since it will absorb ultraviolet radiation about 50 times greater than other material (Bolton et al., 2001).

Certain groundwater supplies may contain species of bacteria associated with iron, such as *Leptothrix* and *Crenothrix* (AWWA, 2008). These bacteria, while inactivated by appropriate amounts of inline ultraviolet radiation, tend to precipitate on the quartz sleeves of conventional inline ultraviolet units. Again, a manual or automated wiper is suggested for applications where iron bacteria are present in the feed water to a unit.

The proper selection of an inline ultraviolet unit for any application should be based on the fact that a greater than sublethal dose of ultraviolet radiation, for the full spectrum of microorganisms anticipated in the feed water to the unit, should be available.

The inline ultraviolet unit should be equipped with a radiation intensity monitor. The intensity monitor, as positioned on the sanitizing chamber, should provide an indication of the lowest anticipated ultraviolet dose within the chamber on the basis of the geometry of the chamber.

Inlet and outlet connections should be selected to minimize microbial growth. The use of threaded connections should be minimized or eliminated. Flanges or sanitary ferrule connections are more appropriate. For units with feed water or product water piping of plastic inline ultraviolet radiation sensitive plastic material such as PVC, CPVC, and/or polypropylene, "light traps" should be employed. The light traps eliminate the "reflection" of ultraviolet radiation from the sanitizing chamber to the plastic piping material.

Most ultraviolet unit manufacturers will provide a sanitizing chamber drain connection with a threaded fitting, which is undesirable. The required piping for the drain connection may provide a sizable dead leg from the sanitizing chamber, which could allow bacteria to accumulate and replicate not only in the dead leg but also on the walls of the sanitizing chamber, thus defeating the intention of the ultraviolet unit. To complicate this situation, many units installed at operating pharmaceutical facilities are provided with a threaded ball valve on the sanitizing chamber drain connection. It is suggested that the drain connection could consist of a welded sanitary ferrule with a mating cap. It is further suggested, as discussed below, that the ballast providing power to the mercury vapor lamps can be located external to the sanitizing chamber, at a position that will not be exposed to water when the sanitizing chamber is drained or when quartz sleeves are changed. When maintenance is required, the sanitary cap can be loosened, and water can be collected in a bucket (or simply directed to a drain below the unit), thus eliminating problems associated with conventional drain connections that are installed on units.

Many ultraviolet units are equipped with sanitization chambers with flanges at one or both ends. The gasket between the end flanges and the sanitary chamber flanges must be replaced frequently or else leaks will occur. In addition, the quartz sleeves are generally secured by O rings. The sealing mechanisms for both the quartz sleeves and the sanitizing chamber must be carefully evaluated to ensure that the unit, if properly maintained and will not leak.

As indicated earlier, it is preferred to select a unit with electrical supply and monitoring capability mounted at a physical location above the water containing sanitizing chamber. Many inline ultraviolet units position the electronics directly below the sanitizing chamber. While esthetically attractive and operator convenient, this arrangement, from an operating and maintenance standpoint, is undesirable because water may be introduced into the electrical enclosure.

In addition to the ultraviolet intensity monitor, the unit should be equipped with a running time meter (nonresettable display) and an individual ultraviolet lamp status indicator. Generally, two ultraviolet lamps are powered from by a single ballast.

At a minimum, even for pretreatment applications, materials of construction for the sanitizing chamber should be 316 or 316L stainless steel. The internal finish should be consistent with the application, but not less than about 15 to 20 Ra (chap. 9). Welding quality for the chamber, including inlet and outlet fittings, should be such that crevices and/or other locations for bacteria to accumulate are eliminated.

Finally, medium pressure lamp inline ultraviolet units are used at some facilities. These units provide ultraviolet radiation exhibiting multiple peaks in the 200 to 400 nm wavelength. The affect of multiple wavelength peaks of greater radiation intensity than the sanitization peak at 253.7 nm is unknown. However, the units provide similar microorganism inactivation capability when compared with low-pressure mercury lamp units.

Operating and Maintenance Considerations
Researchers, evaluating the affect of lamp life on ultraviolet dose, suggest that the decrease in ultraviolet radiation dose within the sanitizing chamber is approximately 10% for each 1000 hours of operation (Zinnbauer, 1985). Subsequently, lamp replacement should be performed when effluent bacteria levels from the unit exceed a preset value. This situation may occur before the 8000- to 9000-hour estimated operating life recommended by the inline ultraviolet unit manufacturer has expired. It is strongly suggested that the consequences of incomplete microorganism inactivation are such that conservative inline ultraviolet unit maintenance should be considered. Considering the cost of lamps and sleeves, it is suggested that lamps and sleeves be replaced every six months as part of a preventative maintenance program. It is important to avoid direct eye contact with ultraviolet radiation during lamp and sleeve replacement since this will result in cataracts (Wieringa, 2006).

Feed water and product water total viable bacteria levels should be determined as part of a microbial monitoring program for an inline ultraviolet sanitization unit. While the enumeration method employed for routine samples should be similar to that for samples collected from other pretreatment components, inactivation of bacteria is a concern. Subsequently, it is suggested that feed water and product water samples be collected prior to the recommended six-month replacement frequency for lamps and sleeves and measured with R2A media, 20°C to 22°C incubation temperature, and for 10- to 15-day incubation time period. Product water microbial levels should be plotted as a function of time. Again, lamp replacement, sleeve replacement, and/or an investigation of unit performance should be conducted if microbial levels increase above a preset value.

One of the most important operating parameter for an inline ultraviolet unit is an indication of ultraviolet radiation intensity. The intensity meter should be calibrated with a known source of ultraviolet radiation. Theoretically, this should be performed using a source calibrated against a National Institute for Standards and Technology (NIST) standard. Unfortunately, this is seldom done since the source must mate to the monitor and is generally not available from inline ultraviolet unit manufacturers or available at a high cost. As indicated earlier, the ability of an inline ultraviolet sanitization to inactivate microorganisms must be verified using feed water and product water total viable bacteria measurements. Generally, calibration information provided with equipment manufacturers' operating procedures are based on establishing a "100% transmittance" value for a unit equipped with new sleeves and lamps.

The performance of an inline ultraviolet unit is highly dependent on a supply voltage at the equipment manufacturer's specification. A small drop in the supply voltage, such as during periods of "brownout" in large metropolitan areas in summer months, can significantly decrease the actual ultraviolet intensity.

The performance of an inline ultraviolet unit is also significantly affected by an increase in temperature within the sanitization chamber. Many units can be purchased with an internal alarm or "cutoff" switch, inhibiting power to the unit if the temperature increases above a preset value. High temperatures will decrease the effectiveness of the unit for inactivation of bacteria and also degrade the ballast(s). If the unit is not equipped with a high-temperature sensor or alarm and shutoff during conditions of high water temperature, an external temperature switch positioned in the feed water line to the unit should be provided. The switch should be electrically connected to the power supply of the inline ultraviolet unit, inhibiting power to the unit when high temperatures occur and providing an audible and/or visual alarm indication.

For units equipped with access to the sanitizing chamber, it is suggested that inspection of the interior of the sanitizing chamber be performed by removing the flange and gasket about once every two to three years depending on the application. A new gasket should be provided for the access flange after the inspection. This interior inspection should also include the cleaning of surfaces with appropriate material. If extensive rouging of the stainless steel is noted, passivation or derouging of the sanitizing chamber should be considered.

The sampling procedure for monitoring unit performance should include periodic feed water particulate matter determination. Generally, TSS and turbidity measurements will provide appropriate information to determine if ultraviolet radiation will be absorbed.

Any visible leaks from either the quartz sleeves or the sanitizing chamber flange connections (or other penetrations to the sanitizing chamber) should be repaired. Leaks may introduce bacteria. In addition, the leaks could "aspirate" air into the sanitizing chamber, creating an "air bubble" and affecting the performance of the unit.

RECIRCULATION AND REPRESSURIZATION PUMPS
Theory and Application
Pumps are commonly used in the pretreatment section of pharmaceutical water purification systems. The pumps may be used to provide repressurization of water, required as a result of the pressure drop through pretreatment components and relatively low raw feed water pressure, or for recirculation purposes associated with microbial control and enhancement of pretreatment component unit operations. Most raw feed water supplies will exhibit a pressure that is adequate enough to allow water to flow through the pretreatment components to the downstream primary deionization unit operation. For applications where the raw feed water pressure is low (<30–50 psig), it may be appropriate to increase the pressure of raw water prior to the pretreatment section. Recirculation provisions within the pretreatment section can also provide an increase in pressure, if desired. Considering pretreatment backwash and water softener regeneration demands, a pump motor with variable frequency drive may be appropriate.

As indicated, recirculation pumps will enhance the operating characteristics of certain pretreatment unit operations *and* provide microbial control. For example, recirculation of an activated carbon unit will reduce bacterial proliferation within the unit with time by eliminating stagnant conditions within the carbonaceous bed. The recirculation pump for the activated carbon unit also provides a method of recirculating the water through a heat exchanger associated with sanitization provisions for the unit, as discussed earlier in this chapter. Finally, the recirculation pump for an activated carbon unit will, by maintaining a continuous flow, minimize the "diffusion" of adsorbed NOM (raw surface water supplies) from the upper portion of the bed to the lower portion of the bed. This diffusion mechanism is promoted by stagnant conditions and the concentration difference, on the activated carbon media surface, between the upper and lower portions of the activated carbon bed.

A similar multipurpose function is provided for a water softener. Recirculation of the carbonaceous ion exchange bed provides microbial control. By eliminating stagnant conditions, multivalent cations at the top of the cation bed will not migrate to exchange sites at the lower portion of the bed in the sodium form. The concentration difference of ion exchange sights, primarily in the calcium and magnesium form at the top of the resin bed is greater than the concentration at the bottom of the bed. During stagnant conditions ions will diffuse to "equalize" the sodium and multivalent cation concentration throughout the bed. Recirculation will minimize this process.

Selecting the pump for the recirculation process is extremely important. Generally, recirculation pumps are used in systems where the downstream primary deionization process does not operate continuously. It is extremely important to consider Joule's heat input from recirculating pumps. It would be inappropriate to provide recirculation of pretreatment components, with one of the primary objectives to control bacterial proliferation within the unit operation, and to introduce heat that, in fact, actually results in a situation where greater bacterial proliferation occurs as compared with a non recirculated system. If necessary, a heat exchanger can be added to the recirculating loop to remove mechanical heat introduced as part of the recirculation process. Other design alternatives may also be considered. For example, a temperature sensor with controller can be installed within the recirculating loop to monitor the temperature of the recirculating water. When the temperature reaches a preset value, a "divert-to-waste" valve can be opened, allowing water to drain. Raw, ambient temperature water would enter the system as water flows to drain, decreasing the temperature of the recirculating water. The effectiveness of this technique is directly related to the temperature of the raw water. Seasonal and climatic changes in the raw feed water temperature must be considered if this technique is used to remove pump mechanical heat from the recirculating water.

Pump selection should also consider other important parameters, such as minimum pump motor size and high pump efficiency. In general, significantly higher pump efficiencies, with resulting lower mechanical heat input, are achieved by using multistage centrifugal pumps as compared with single-stage centrifugal pumps. In selecting the pump, the pressure and flow rate must be adequate to ensure proper flow through the unit operation(s) undergoing recirculation without producing channeling, which would be counterproductive to the recirculation operation. As discussed previously, activated carbon units used in pretreatment systems with feed water from a surface source may be recirculated with softened water although slight elution of organic material from the activated carbon may occur. It is suggested that the microbial control benefits of recirculation outweigh the potential elution of small amounts of organic material.

Design Considerations

Figures 3.18 and 3.19 are performance curves for single-stage and multistage centrifugal pumps, respectively. The curves clearly demonstrate that a desired flow rate and discharge pressure can be achieved with a lower horsepower motor and higher efficiency with a multistage centrifugal pump, as compared with the single-stage centrifugal pump. The recirculation criteria must be carefully evaluated to determine the particular pump for the application. The materials of construction (surfaces in contact with pretreated water), along with the slope of the individual pump curve for the specific application, must be considered. Another item to consider is the ultimate discharge flow rate for a given horsepower motor/pump combination. Generally, for a centrifugal pump (single stage), a large flow rate range can be achieved for a given motor horsepower with little change in discharge pressure.

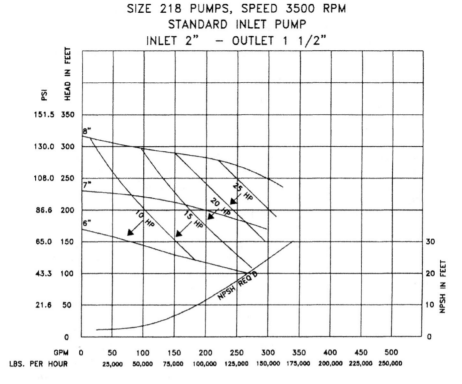

Figure 3.18 Conventional centrifugal pump curve head versus capacity curve chart. (Pump speed is 3500 rpm. Values selected from "middle" of family of pump curves in terms of impeller size.) *Source*: From Tri-Clover.

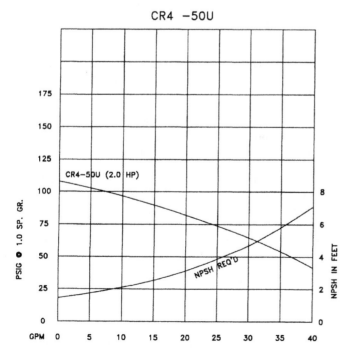

Figure 3.19 Multistage centrifugal pump curve psig versus capacity curve chart. (Pump speed is 3450 rpm. Values selected from "middle" of family of pump curves considering the number of stages.) *Source*: From Grundfos.

If a single-stage centrifugal pump is selected for the recirculation application, several important design parameters must be specified. The pump impeller size, operating speed, and motor horsepower must be determined for the desired flow rate and increase in pressure. It is highly desirable to avoid a pump equipped with a minimum or maximum impeller size. Ideally, the selected impeller size for the pump should fall in the "middle" of the impeller size range for the selected pump. This will allow potential modification to the system, specifically the recirculation provisions, for the particular pretreatment component(s) without changing the entire pump. Only the impeller would need to be changed. The pump motor size should also be selected such that changes in the impeller size, specifically increases in the impeller size to achieve higher discharge pressures for a given flow rate, can be achieved using the same pump motor. For some applications, the pump speed may be used as a variable to adjust the discharge pressure for a given flow rate. This can be easily achieved by using a variable frequency drive for the pump motor. While there is significant "flexibility" associated with the proper selection of a single-stage centrifugal pump for recirculating applications, the most important factor to consider may be the efficiency of the pump for the specific application in light of the desire to minimize mechanical heat input to the recirculating water.

In selecting a multistage centrifugal pump for a specific application, the number of stages specified is very critical. It is generally desirable to select a pump with a minimum of six to eight individual stages. Generally, at least two stages may be removed, if the application changes. This would allow potential changes to the pump performance, by removing the stages, decreasing the discharge pressure for a specific flow rate. The converse situation may also be applied. Specifically, a pump can be selected with ultimate capability for eight stages, with only six supplied, allowing potential system recirculating modification with increased pressure for a given flow rate. The pump motor horsepower should be given for the selected pump to provide the flexibility of using the maximum number of stages. Again, the of a variable frequency drive for the pump motor eliminates the addition/removal of stages for a change in performance.

The pump seal mechanism, materials, and "type" should be consistent with the pretreatment recirculation application. While not as critical as the pumps used in system product water storage and distribution systems, the proper pump seal should be such that "foreign substances and impurities" are not introduced.

The generally recommended material of construction for the pump is 316 or 316L stainless steel. While the nature of the pretreated water is not as aggressive as final product water, and the use of stainless steel would appear to be expensive for the pretreatment recirculation application, the actual cost associated with using stainless steel as opposed to alternate materials, such as brass, carbon steel, or plastic, is relatively small in light of the total system cost. If stainless steel pump selection is considered, provisions for galvanic isolation from non stainless steel materials, such as brass, copper, or steel, should be provided.

Feed water and product water connections to and from the pump should include manual isolation valves and, where appropriate, vibration isolators. Piping should be provided with a check valve capable of inhibiting flow through the pump in the reverse direction.

While the check valve mentioned above should inhibit flows to the pump in the wrong direction, it is suggested that check valves should not be used as a positive means of ensuring reverse flow through a pump, which, in a recirculating mode, would result in bypassing of the particular pretreatment unit operation being recirculated. This condition would occur if the recirculation pump was deenergized. Obviously, the pressure drop through the recirculating line, acting as a bypass, would be less than the pressure drop through the pretreatment unit operation(s). While this particular situation may seem remote, it is frequently encountered but often undetected. To eliminate this situation, an automatic valve should be installed in the recirculating piping. It is suggested that this valve be a positive acting type, closed in a deenergized mode and open in an energized mode. When the recirculating pump is non operational, the valve should be deenergized in a closed position, inhibiting potential bypass.

The pump should be provided with feed water and product water pressure gauges. As indicated earlier in this chapter, liquid-filled gauges of stainless steel construction and with diaphragm isolators should be considered. If the feed water pressure to the recirculating pump is low (<10–20 psig), it may be appropriate to consider a "compound type" pressure gauge that would clearly indicate if a positive pressure is available to the suction side of the pump. For pumps positioned directly downstream of tanks, the compound gauge would also verify that the required pump "net positive suction head" (NPSH) is met.

Considering that the pumps provide a method of enhancing the operation of pretreatment equipment, it is suggested that a single installed pump is adequate for recirculation. It is further suggested that a spare pump and motor be purchased and kept as a "warehouse spare" unit, with operator access to the pump in the event of operating pump failure.

Motor starters supplied for the pump should be of the "hands-off-auto" (HOA) type. The feed water line to the pump should contain a pressure switch set at a predetermined value. If the pump feed water pressure decreases below this preset value, the automatic provisions within the motor starter, wired locally, would inhibit operation of the pump and energize an indicating light and/or audible alarm. This provision is extremely important because it protects the pump from conditions where there is insufficient feed water flow or pressure.

As appropriate, feed water and product water sample valves should be positioned in piping to and from the pump. For pumps positioned directly downstream of an atmospherically vented tank, the feed water piping will probably not exhibit adequate pressure to obtain a sample and may be eliminated.

A visual temperature indicator, or temperature sensor to local or remote indicator, should be provided to allow operating personnel to determine if mechanical heat is being introduced to the recirculating water. If design calculations indicate that mechanical heat will be introduced, accelerating microbial levels within the recirculating pretreatment component(s), temperature indication and control will be necessary. This can be accomplished by using a temperature sensing element with temperature indicating controller.

It is very desirable to verify the performance of the recirculating pump by not only observing the discharge pressure but also the discharge flow rate. The individual pretreatment unit operations should be equipped with a feed water flow rate meter, thus eliminating the need for a dedicated meter in the discharge line from the pump.

Finally, for pretreatment systems piping may be PVC, CPVC, polypropylene or other plastic material. Pump selection should not consider the use of either male or female inlet and/or discharge connections. Pump connections should be flanges, allowing the use of mating plastic material flanges. This arrangement eliminates leaks from plastic material threaded to a "hard" pump casing material such as copper, steel, or stainless steel.

Operating and Maintenance Considerations

Operating logs should include entries for pump discharge pressure and flow rate. A log entry should be included for pump feed water pressure and any "unusual" noise generated from the pump and/or pump motor.

Pump seals should be replaced as a preventive maintenance item, with a suggested annual frequency. The interior of the pump casing should be inspected on an annual basis. Further, during seal replacement shaft "alignment" should be checked.

Every attempt should be made to run a pump continuously, where practical and consistent with system design. A pump will operate for a longer period of time with minimal maintenance when it is operated continuously as compared with operation in a cyclic mode. The use of two pumps with weekly cycling from one pump to the other is strongly discouraged.

Operating personnel should verify the rotation direction for the pump motor when it is initially installed, during start-up and anytime that the electrical supply to the pump motor is changed.

Operation of the pump feed water pressure switch, if employed, should be verified annually by deliberately restricting flow to the suction side of the pump, with associated decrease in pressure. This will verify whether operation of the pump terminates at, or around, the preset minimum feed water pressure.

Operational logs should include an entry for the temperature of the recirculating water for the specific pretreatment component(s) undergoing recirculation. This value should be plotted as a function of time, verifying that mechanical heat input from the pump is not increasing the temperature of the recirculating water and, subsequently, the water in the pretreatment component(s), which increases the rate of bacterial proliferation within the component(s).

CARTRIDGE FILTRATION
Theory and Application
Cartridge filtration, by definition in this pretreatment section, is classified as a technique for removing particulate matter by filtering the particles through synthetic or natural filter media. Cartridge filtration, as a pretreatment technique, can be used for several techniques.

For very low flow rate systems, cartridge filters may be used in lieu of backwashable particulate removal filters, which were discussed earlier in this chapter. However, backflushable units are preferred since they will "ripen," increasing their particulate removal capability and potentially removing organic and colloidal material that generally cannot be removed by cartridge filters. There maintenance requirements, cost, and bacterial proliferation issues associated with the use of cartridge filters. These factors make the use of backflushable particulate removal filters more desirable.

Particulate removal filters may be used in the product water lines from backflushable activated carbon units to remove "activated carbon fines." For systems using deionization as the primary ion removal technique, the activated carbon fines, present in small quantities from a properly designed backflushable activated carbon unit, would simply pass into the cation deionization unit of a primary two-bed unit. It is suggested that the activated carbon fines would readily be removed during backwash of the cation column, conducted as the initial step in the regeneration cycle. It is highly unlikely that the activated carbon fines could penetrate the entire cation bed, considering the small and relatively uniform size of the cation resin beads. For systems using reverse osmosis as a primary ion removal technique, the process downstream of the activated carbon filter would be a water softener (the majority of applications). The water softener provides a cation resin bed that will behave similarly to the cation resin bed in a two-bed deionization system, backwashed as the initial regeneration step. It is suggested that positioning cartridge filtration downstream of a properly designed activated carbon unit, with the appropriate stainless steel lower distributor exhibiting uniform "slot" size, is inappropriate. The cartridge filtration system provides an extended area for the small amount of activated carbon fines to accumulate. Furthermore, bacteria present in the product water from an activated carbon unit will also accumulate on the particulate removal filter media and use the activated carbon filter as a nutrient for replication. For systems equipped with poorly designed, lower activated carbon unit distributors, it is suggested that alternate methods of activated carbon fine removal, such as the use of sintered stainless steel filters, offer an effective technique for removing activated carbon fines without providing the "depth" associated with a particulate removal filter and resulting proliferation of bacteria on the filter.

Many pharmaceutical water purification systems use cartridge filtration as a resin fine "trap" downstream of water softeners or primary deionization units. Again, a properly designed lower ion exchanger vessel distributor, coupled with adequate freeboard space within the column and a regulated backwash flow rate should minimize the physical volume of resin fines in the product water from the unit. Synthetic organic scavengers, discussed earlier in this chapter as a pretreatment technique, employ anion resin, which exhibits poorer physical stability than cation resin. It is suggested that sintered stainless steel filters can be used to trap resin fines effectively, without providing a location for bacteria to accumulate and replicate within a cartridge filtration system.

Particulate removal is also used in many pharmaceutical water systems prior to inline ultraviolet units. As discussed earlier in this chapter, particulate matter will absorb ultraviolet radiation, decreasing the effectiveness of inline ultraviolet units to provide bacterial control. However, it is suggested that final product water microbial levels after inline ultraviolet units equipped with upstream particulate removal filters will not be significantly less without prefilters assuming that significant amounts of particulate matter are not present and that the activated carbon media is replaced once every 6 to 12 months. In fact, bacteria levels may be greater than levels prior to the cartridge filters upstream of the inline ultraviolet units.

In general, it is suggested that cartridge filtration as a pretreatment unit operation be limited to gross particulate removal from raw water supplies for small flow rate systems, or as a very important particulate removal step prior to primary RO units. In general, pretreatment systems equipped with multiple cartridge filtration units downstream of individual pretreatment components generally exhibit significantly higher bacterial levels than similar systems without particulate removal filters. If, as emphasized throughout this chapter, individual pretreatment unit operations, such as activated carbon, water softeners, and particulate removal filters, are designed correctly, with lower distributors of stainless steel construction and appropriate freeboard space above the media, particulate matter control should be minimal. This situation emphasizes the importance of selecting properly specified and designed pretreatment components.

Design Considerations
Before considering the use of particulate removal cartridges for specific intra-pretreatment component applications, the design of the major pretreatment components and alternative methods of filtration should be carefully evaluated.

Filter housings for cartridge filters can be provided in single or multiple filter cartridge configurations, with varying diameters and heights. Generally, 10-, 20-, 30-, or 40-in.-long (approximately) filter cartridges are employed in single or multiple cartridge filter housings. The filter housing cartridge filter element capacity is designated as "round." The length of the filter cartridges in the housing (as a multiple of 10-in. cartridges) is designated as "high." As an example, a filter housing containing three individual 20-in.-long filter cartridges is a 3 "round" by 2 "high." Filter housings should be selected to contain "popular" size cartridge, generally a 10-, 20-, or 30-in.-long element. Longer cartridge filters elements (40 in.) are available but are not as popular and are generally much harder to obtain. The filter cartridge manufacturer's recommendation for flow through "10-in. equivalent" cartridges should be carefully considered when selecting the number of 10-in. filter equivalents and, subsequently, the size of the filter housing for a particular application.

Filter housing prices, similar to any pressure vessels, increase with increases in diameter (round) and less significantly with increases in height (high). A 3 round by 2 high housing, for example, should be selected, as compared with a 6-round by 1-high housing, minimizing cost. This also minimizes the number of filter cartridges and, subsequently, the number of filter cartridge-to-filter housing "seals" per unit.

If inlet and outlet connections are properly selected and sized, it may be possible to change the "bell" of a filter housing to increase its flow rate capacity, by using longer filters, for an installed filter housing base plate.

The seal mechanism between the filter cartridge and filter housing is extremely important. While O-ring seals are preferred, they are not generally available for cartridge-type (non-bacteria-retentive) filter cartridges. If available, the O-ring seal mechanism should be strongly considered. Flat gasket seal mechanisms are less desirable. "Knife edge" seal mechanisms accomplished by exerting pressure from an inverted "V," circular raised area on the filter housing bowl and/or head to a filter cartridge provide a highly undesirable seal mechanism.

Filter cartridges may be purchased as single open end or double open end. If single open end cartridge filters are available, they should be used since there is only one filter cartridge-to-filter housing seal for each cartridge, as compared with the two required for a double open end cartridge.

Feed water and product water pressure gauges of stainless steel construction, liquid filled, with diaphragm isolators, as discussed throughout this chapter, should be provided.

Materials of construction for single cartridge units may be either unpigmented or pigmented polypropylene. The pigmentation for housings may introduce extremely low levels of eluted organic material into the water. Multiple cartridge filter housings should be of 304, 304L, 316, or preferably 316L stainless steel construction.

The housing should be equipped with high point vent and valve to allow removal of air after cartridge filter replacement. Obviously, the valve should be positioned on the "dirty side" of the unit. Valve design should consider microbial control within the unit and should be of the diaphragm or needle type. A drain valve should be provided on the filter housing. The filter housing should be designed such that the drain location is adequately below the filter cartridge-to-filter housing physical seal position. If "unfiltered water" is not fully drained from the housing, it could easily come in contact with the "clean" side of a new filter cartridge as it is being installed into the seal. This is particularly important for multiple "round" housing with 20- to 40-in.-long filter elements where the seal location may not be directly visible. From a microbial control standpoint, this is unacceptable. The criteria for selecting the drain valve attached to the filter housing drain line should be similar to those for the vent valve.

If the particulate removal filters exhibit the presence of any impurities when first operated (e.g., color, odor, etc.), a "divert-to-drain" valve should be provided in the product water piping for preoperational "rinse-down" of the filter cartridges. In general, most filters provided for pharmaceutical applications will not exhibit the properties indicated above, since they are in direct conflict with USP requirements associated with "foreign substances and impurities."

Operating and Maintenance Considerations

Considering all of the factors associated with the use of cartridge filters, frequent feed water and product water sampling and associated analysis to determine microbial levels should be established.

If cartridge filters are used for "traps" (activated carbon fines or ion exchange resin), they should be replaced on the basis of elapsed installation time rather than pressure drop. Generally, a large pressure drop would not be anticipated as a result of accumulated activated carbon fines or resin fines. However, to minimize microbial buildup within the units, periodic replacement should be performed.

For cartridge filters used to filter raw water, it is important to establish a maximum pressure drop through the filtration system. The maximum pressure drop should be based on the fact that breakthrough of material has not occurred. Subsequently, cartridge filters that have been removed during "change-out" should be periodically "autopsied." The autopsy should be performed by "slicing" the filter cartridge perpendicular to its height, verifying that particulate matter has not penetrated to the "clean" side of the cartridge. It may be necessary, on the basis of seasonal and climatic changes in the raw water supply, to change the ΔP value established for cartridge filter replacement, since the type and nature of particulate matter present in the raw water supply will change.

As filter cartridges are removed, they should be inspected for the visible growth of microorganisms, specifically slime, algae, and certain mold species. This inspection should be conducted on both the clean and dirty sides of the cartridge. Further, if odor and/or color are noted it is suggested that "swaps" or "swipes" of the filter cartridges be obtained such that the microorganisms can be identified. The presence of pathogens is an important factor in determining the sanitization frequency and program for the pretreatment system.

Feed water and product water samples should be obtained periodically for the TSS measurement. This measurement should be conducted using filter disks with a micrometer rating equal to the rated pore size of the cartridge filter element. This measurement should also be used in determining the replacement frequency of the cartridge filter elements.

As indicated, microbial growth on the cartridge filters is a concern. To minimize the effect of this situation, the filter housing should be periodically sanitized with a liquid sanitizing agent. This will minimize the potential input of bacteria from a cartridge filtration system.

The filter housing "base"-to-housing "bell" gasket should be replaced annually. Spare gaskets should be retained for non routine maintenance situations. The gasket and physical area beneath the gasket should be periodically removed and cleaned with isopropyl alcohol (IPA) since it can provide an area for bacteria proliferation.

Filter cartridges should not be stored in an area where they are exposed to fumes, dust, or other airborne contaminants. It is suggested that new replacement cartridges be positioned in airtight bags after receipt from the cartridge filter supplier.

To provide a record of cartridge filter replacement and other information, it is suggested that a label containing the cartridge filter lot number, serial number, manufacturer, and date of installation be attached to the filter housing. This information should also be recorded in the logbook for the system.

REFERENCES

40CFR141 and 40CFR142. National Primary Drinking Water Regulations and Implementation. U.S. EPA, 816-F-09-004, May 2009.

Avallone H. The Gold Sheet. Chevy Chase, MD: F-D-C Reports Inc., 1993; 26(12):1–12.

Avallone H. Paper presented at the Pharm. Tech. Conference '94, 19–21 September in Atlantic City, U.S.A., 1994.

AWWA. Water chlorination principles and practices. In: Manual of Water Supply Practices. Denver: American Water Works Association, 1973:40–43.

AWWA (American Water Works Association). Problem Organisms in Water: Identification and Treatment, Manual M7. 3rd ed. Denver, 2008.

AWWA (American Water Works Association). Water Chlorination/Chloramination Practices and Principles, Manual M20. 2nd ed. Denver, CO, April 2009.

Baker FS, Byrne JF. Methods for Removal of Chloramines with Activated Carbon. U.S. Patent 6,669,393, U.S. Patent and trademark Office, Alexandria, VA, 2004.

Black & Veatch Corporation. White's Handbook of Chlorination and Alternative Disinfectants. 5th ed. Hoboken, NJ: John Wiley & Sons, Inc., 2009.

Bolton JR. Ultraviolet Applications Handbook. 2nd ed. Edmonton, Canada: Bolton Photosciences, Inc., 2001.

Bolton JR, Colton CA. The Ultraviolet Disinfection Handbook. J Am Water Works Assoc Denver, CO, 2008.

Calgon Carbon Corporation. Centaur 12 × 40 Granular Activated carbon. Product Bulletin LC-765-02/98, 1998.

Carson LA, Petersen NJ. Photoreactivation of *Pseudomonas cepacia* after ultraviolet exposure: a potential source of contamination in ultraviolet treated waters. J Clin Microbiol 1975; 1(5):462–464.

Cleasby JL, Hilmoe DL, Dimitracopoulos J. Slow sand and direct inline filtration of a surface water. J Am Water Works Assoc 1984; 76(12):44–56.

Collentro WV. Multicomponent Adsorption in Fixed Beds. Master's Thesis.Worcester: Worcester Polytechnic Institute, 1968.

Collentro WV. Microbial control in purified water systems—case histories. Ultrapure Water 1995a; 12(3): 30–38.

Collentro WV. Unpublished data. Results from pilot study. 1995b.

Collentro WV. One Hundred Pitfalls Associated with the Design, Operation, and Maintenance of Pharmaceutical Water Systems, Part 1, Feed Water Supplies and Pretreatment Equipment. Pharmaceutical Processing, Volume 24, No. 13, Reed Business Information, Highland Ranch, CO, Dec 2007.

Collentro WV. Pharmaceutical Water System Expansion/Upgrades. Ultrapure Water Pharma 2010 Conference. Ultrapure Water, New Brunswick, NJ, May 20–21, 2010.

Collentro WV, Collentro AW. Qualifying the use of activated carbon in high purity water systems. Ultrapure Water 1997; 14(4):43–54.

Collentro WV, Zoccolante G. Defining an added substance in pharmaceutical water. Ultrapure Water 1994; 11(2):34–39.

Eaton AD, Clesceri LS, Greenberg AE. Standard Methods for the Examination of Water and Wastewater. 21st ed. American Public Health Association, American Water Works Association, and Water Environment Federation. Washington, D.C.: American Public Health Association, 2005.

Fairey JL, Speitel GE, Katz LE. Impact of natural organic material on monochloramine reduction by granular activated carbon; the role of porosity and electrostatic surface properties. Environ Sci Technol 2006; 40(13):4268.

Fairey JL, Speitel GE, Katz LE. Monochloramine destruction by GAC – effect of activated carbon type and source water characteristics. J Am Water Works Assoc 2007; 99(7).

FDA. Guide to Inspection of High Purity Water Systems. Rockville: Food and Drug Administration, Office of Regulatory Affairs, Office of Regional Operations, Division of Field Investigations, 1993.

FilmTec^{TM} Membranes, "Technical Manual Excerpt, Table 9.6, Temperature Correction Factor," The DOW Chemical Company, Form No. 609-02129-804, 2009.

Ghash K, Schnitzner M. UV and visible adsorption spectroscopic investigations in relation to macromolecular characteristics of humic substances. J Soil Sci 1979; 30:735.

Gottlieb MC. The reversible removal of naturally occurring organics using resins regenerated with sodium chloride. Ultrapure Water 1996; 13(8):53–57.

Lide DR, ed. Handbook of Chemistry and Physics. 87th ed. Boca Raton: CRC Press, Inc., 2006–2007.
Harm W. Biological Effects of Ultraviolet Radiation. Canbridge: Cambridge University Press, 1980.
Keenan JE, Keyes FG. Thermodynamic Properties of Steam. 1st ed. 35th printing. New York: John Wiley & Sons, Inc., 1963.
Koch Membrane Systems, Technical Bulletin: temperature Effect; ROGA® and TFC® Membrane Elements, August, 2007.
Miltner RJ, Summers RS, Wang JZ. Biofiltration performance. Part 2: effect of backwashing. J Am Water Works Assoc 1995; 87(12):64–70.
Munson TE. FDA Views on Water System Validation. Proceedings of the Pharm Tech Conference '85. Cherry Hill: Aster Publishing Corporation, 1985:287–289.
Munson T. The Gold Sheet. Chevy Chase, MD: F-D-C Reports, Inc., 1993; 26(12):1–12.
Ong HL, Bisque RE. Coagulation of humic colloids by metal ions. Soil Sci 1988; 106(3):220.
Ongerth JE, Pecoraro JP. Removing cryptosporidium using multimedia filters. J Am Water Works Assoc 1995; 87(12):83–89.
Pontius FW. New horizons in federal regulations. J Am Water Works Assoc 1998; 90(3):38–50.
Randtke SJ, Jepsen CP. Chemical pretreatment for activated carbon adsorption. J Am Water Works Assoc 1981; 73(8):411.
Randtke SJ, Jepsen CP. Effects of salts on activated carbon adsorption of fulvic acids. J Am Water Works Assoc 1982; 74(2):84–93.
Symons JM, Fu PL-K, Kim PH-S. The Use of Anion Resins for the Removal of Natural Organic Matter from Municipal Water. Paper presented at the 53rd Annual Meeting—International Water Conference, 19–21 October in Pittsburgh, PA. Paper No. IWC-92-12, 1992.
Weber HC, Meissner HP. Thermodynamics for Chemical Engineers. 2nd ed. New York: John Wiley & Sons, Inc., 1959:47.
Weber J, Voice TC, Jodellah A. Adsorption of humic substances: the effect of heterogeneity and system characteristics. J Am Water Works Assoc 1983; 75(12):612–619.
Wieringa FP. Five frequently asked questions about UV safety. Int Ultraviolet Assoc 2006; 8(2):28–32.
Zinnbauer FE. Ultraviolet water disinfection comes of age. Ultrapure Water 1985; 2(1):27–29.

4A | Ion removal techniques—reverse osmosis

INTRODUCTION

Ion removal is discussed in Chapters 4A, 4B and 4C. This chapter (4A) discusses primary ion removal by reverse osmosis. Chapter 4B discusses primary ion removal by ion exchange. Chapter 4C discusses post–reverse osmosis polishing ion removal techniques.

Throughout this text, an emphasis has been placed on design, operating, and maintenance considerations for individual unit operations employed in a pharmaceutical water purification system. As a unit operation in water purification systems, reverse osmosis has and will continue to displace all other ion removal technologies. With the development of RO membranes capable of rejecting in excess of 99% of ions in the feedwater, the development of the composite type membranes, and increasing use of hot water sanitizable membranes, it is suggested the majority of pharmaceutical water systems employ reverse osmosis as the primary method of ionic removal. As discussed in Chapter 4B, systems requiring infrequent or small daily volumes of water, as well as certain active pharmaceutical ingredient applications requiring significant amounts of water in a very short period of time, may continue to use ion exchange (rechargeable canisters for smaller applications and regenerative units for high volume applications).

Every attempt has been made to be highly selective in the theory and applications section of this chapter limiting discussion of items such as membrane composition and development. Unfortunately, many of the technical expressions, basic operating theory, and RO "jargon" require an understanding of the RO process by the individuals who interface with pharmaceutical water purification systems. Most of these individuals have a basic understanding of the ion exchange process but consider reverse osmosis as a "black box." This is easily demonstrated by the number of technical courses available each year that discuss reverse osmosis, when compared with the absence of courses that are dedicated to ion exchange.

As a preface to this chapter, it is appropriate to indicate that a text double or triple the size of this text could be prepared to discuss the use of reverse osmosis in pharmaceutical water purification systems. The objective of this chapter is to provide essential information that is required by individuals interfacing with pharmaceutical water purification systems. Numerous references are provided throughout this chapter. For individuals seeking greater in-depth understanding of a particular item that is only briefly discussed within this chapter, it is suggested that the reference articles as well as text and additional articles be reviewed as appropriate.

THEORY AND APPLICATION
Basic Theory—Osmotic Pressure

The literature contains excellent examples of basic RO theory (Applegate, 1984; Kronmiller, 1993; Harfst, 1995; McClellan, 1995; Amjad et al., 1996a; Collentro and Barnett, 1996; Singh, 1997). To understand the principle of reverse osmosis, the following example is often used (Fig. 4A.1). A water solution with high salt concentration is placed in an apparatus on one side of a semipermeable membrane. Pure water is placed in the same apparatus in the adjacent compartment on the other side of the semipermeable membrane. The initial water levels are established such that the volume in each of the two compartments is equal. Water passes from the dilute solution through the semipermeable membrane into the concentrated solution. This process is called "osmosis." The driving force for this natural phenomenon represents the osmotic pressure. The flow of pure water will continue until the osmotic pressure reaches an equilibrium value. At this point, the water level in the compartment initially containing the high salt concentration (horizontal position) is greater than the level in the adjacent compartment containing the pure water. If pressure is exerted on the compartment containing

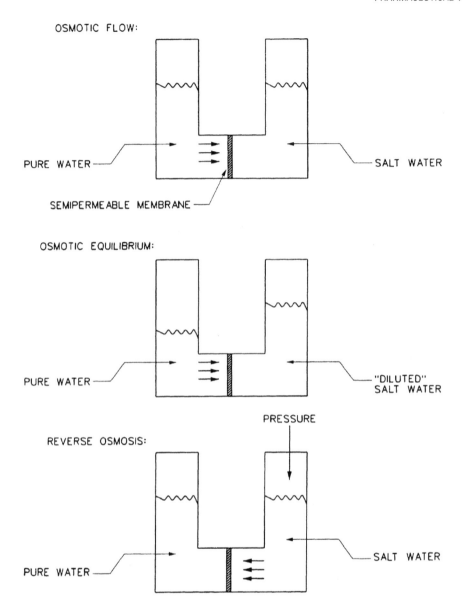

Figure 4A.1 Examples of osmosis and reverse osmosis processes. (The pure water level can be increased further by increasing the pressure, which further concentrates the salt solution.)

the "diluted" salt solution, the osmosis process can be "reversed." This pressure forces water through the semipermeable membrane, without salt; with adequate pressure, it can equilibrate the levels within each compartment to their initial position. The use of pressure and the resulting flow of water through the semipermeable membrane demonstrates the "reverse osmosis" process.

The ideal semipermeable membrane would allow only pure water to pass through it without *any* ions. Unfortunately, this is not possible. All membranes will allow small amounts of "salts," particularly those with smaller ionic radii, to pass with the pure water. The ability of a semipermeable membrane to remove ionic material is referred to as the "percent rejection of ions," which is discussed later in this chapter. The greater the concentration of ionic material passing through the semipermeable membrane in proportion to the ionic concentration of the feedwater, the lower the percent rejection of ions for a particular membrane. For a dynamic RO system, as opposed to a laboratory-type situation, a feedwater stream, a product water stream (often referred to as permeate), and a waste stream (referred to as a "reject") are provided, which increases the ability of the semipermeable membrane, in a dynamic mode, to remove ions from the feedwater stream containing "salts."

The RO process was identified by researchers more than a century ago. However, because of low product water flow rates per square foot of semipermeable membrane area, the technology was not developed. This situation changed dramatically with the development of the asymmetric cellulose acetate membrane by Loeb and Sourirajan in 1963 (Loeb and Sourirajan, 1963).

The osmotic pressure is proportional to the salt concentration and temperature of water. Since ions have different radii, the osmotic pressure is also a function of the types of ions present in the water. It has been suggested that a sodium chloride solution at ambient temperature will exhibit an osmotic pressure of 10 psig/1000 mg of TDS.

Microfiltration, Ultrafiltration, Nanofiltration, Reverse Osmosis, Double-Pass Reverse Osmosis, and Membrane Contactors

Figure 4A.2 illustrates the size of material associated with microfiltration, ultrafiltration, nanofiltration, and reverse osmosis. While microfiltration and ultrafiltration are membrane processes, they do not directly remove ionic material and, subsequently, are not discussed in this chapter.

Nanofiltration is an RO process often referred to as a "softening membrane process" because the pore size is greater than conventional reverse osmosis. Nanofiltration membranes will remove only a fraction of monovalent ions (smaller ionic radii) from feedwater and reject the majority of larger ionic radius multivalent ions. The literature suggests that the pore diameter of nanofiltration membranes is 7 to 20 Å, which is between the pore sizes for ultrafiltration and conventional reverse osmosis (McClellan, 1995; Singh, 1997). The literature also suggests that the monovalent salt passage through a nanofiltration membrane is 15% to 40% at 100 psig with feedwater containing 2000 ppm TDS (Schneider, 1994a). The same source suggests that the divalent salt passage through the membrane is 2% to 3% under the same conditions. As indicated in the preceding text, the pore diameter of nanofiltration membranes is approximately 1×10^{-9} m (i.e., 10 Å or 1 nm)—the basis for referring to this particular membrane operation as "nanofiltration." While discussed further in this chapter,

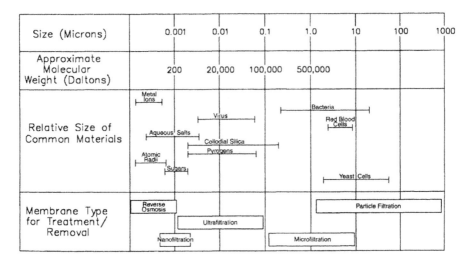

Figure 4A.2 Size of material associated with microfiltration, ultrafiltration, nanofiltration, and reverse osmosis. *Source*: McClellan (1995).

nanofiltration membranes are generally all "composite type" with a negatively charged selective skin layer.

Reverse osmosis is a process that removes virtually all ions (multivalent and monovalent) from water. Ongoing development and manufacture of RO elements has resulted in membranes capable of rejecting in excess of 99%of the dissolved ionic material present in a feedwater stream. Conventional RO membranes will remove organic material with a molecular weight greater than 150 Da and nearly all reactive silica. The literature states that in a survey of 200 RO units conducted in 1994, the average recovery of water (percentage of feedwater recovered as product water) was 71% (Harfst, 1995). Most RO membrane systems operating in pharmaceutical water system application are designed to operate at about 75% recovery of feedwater. The literature further states that "brackish water RO membranes," which employ membrane material *similar* to membranes used in pharmaceutical water systems, exhibit a water recovery of about 15% to 20% per membrane element (Parise, 1996b). Obviously, the higher operating percent feedwater recovery indicated earlier (75%) requires arrangement of membranes in pressure vessels (referred to as an "array"), which concentrates the waste stream by providing it as feedwater to subsequent RO membranes operated in series. The array is "tapered" to maintain critical flow through each membrane, considering the "removal" of product water.

Double-pass reverse osmosis is a process where two individual RO systems are operated in series. Product water from one unit is used as feedwater to the second unit. Subsequent to the change in USP Purified Water and Water for Injection chemical test methods in 1996 (USP 23 Fifth Supplement conductivity specification of 1.3 μS/cm at 25°C for Stage 1), the number of pharmaceutical water system employing double-pass reverse osmosis as a primary ion removal technique increased. However, within a few years, the introduction of "second-generation" continuous electrodeionization (CEDI) as a polishing technique for RO product water resulted in a rapid decline in the use of double-pass reverse osmosis for pharmaceutical water primary removal applications.

Finally, it would be inappropriate to conclude a discussion of membranes without briefly discussing membrane contactors. Membrane contactors are employed to remove gases from a liquid stream. They may be used to remove "reactive gases," such as carbon dioxide and ammonia (chap. 2). Gases are not removed by conventional RO membranes. Subsequently, reactive gases will pass through conventional RO membranes, reestablish equilibrium in the product water, and increase the conductivity of product water. Unfortunately, the use of membrane contactors for pharmaceutical applications appears to be limited. The units must be physically positioned downstream of an RO unit. Further, the inability to perform effective hot water sanitization is a concern. Since membrane contactors can remove "nonreactive" gases, such as oxygen, the technology may be considered for pretreatment of water to a distillation unit, where the presence of oxygen will contribute to general corrosion, chloride pitting corrosion, and chloride stress corrosion.

Reverse Osmosis Membrane Flow Characteristics: "Equations for Flux"

The flow of water through an RO membrane, referred to as membrane "flux," is the volume of water passing through the membrane per unit area per unit time. The water and salt permeability constants are membrane specific and are affected by solution temperature, pH, and ionic strength of the solution. Water permeability is affected by the presence of suspended solids, organic material, bacteria, inorganic material, and anything else that will be removed by the membrane.

A simple definition of flux is as follows (Amjad et al., 1996a):

$$J_w = \frac{\alpha_w(P_f - P_p) - \sigma(\pi_f - \pi_p)}{t}$$

where J_w is the water permeation rate (volume/time × area), α_w is the membrane constant for water, P is the pressure (P_f is the feedwater pressure and P_p is the product water pressure), σ is the reflection coefficient, π is the osmotic pressure (π_f is the feedwater pressure and π_p is the product water pressure), and t is the membrane thickness.

Assuming that the membrane thickness is fixed and that the feedwater temperature is constant, a somewhat simpler equation for flux is as follows (Parise, 1996a):

$$J_w = A[(P_f - P_p) - (\pi_f - \pi_p)]$$

where A is the water permeability coefficient, P is the pressure (P_f is the average of the feedwater and waste pressures and P_p is the product water pressure), and π is the osmotic pressure (π_f is the feedwater pressure and π_p is the product water pressure). The term ($\pi_f - \pi_p$) can be neglected for feedwater meeting the NPDWR of the U.S. EPA because the change in osmotic pressure is estimated at 1 psig/100 mg/L of TDS. Thus, it should be obvious that flux is directly proportional to a constant times "the differential pressure value."

Unfortunately, membrane flux, an excellent indicator of the performance of an RO membrane, is affected by changes in temperature. Subsequently, most RO flux calculations are conducted to determine a "normalized" permeate flux rate. This can be determined with the following equation (Parise, 1996a):

$$(J_w)_N = J_W \times \left[\frac{(P_f - P_p)_o}{(P_e - P_p)_i}\right] \times \frac{\mu_o}{\mu_i}$$

where $(P_f - P_p)_o$ is the observed (measured) feedwater and product water pressures, $(P_f - P_p)_i$ is the initial value for feed and product water pressure, μ_o is the water viscosity at the observed temperature, and μ_i is the initial water viscosity. This equation is extremely useful for calculating the normalized flux from the observed flux for a unit, considering changes in pressure and temperature. It should be pointed out that the flux can also be corrected for temperature by assuming that it changes 3% to 4%/°C (decreasing as temperature decreases and increasing as temperature increases as shown in Table 3.1, chap. 3).

The normalized flux value is extremely important. Small changes in the flux rate from an RO unit can be associated with membrane fouling (chemical and/or microbial), membrane scaling, or other situations associated with degradation of a unit operation. A significant change in flux associated with temperature must be determined to distinguish a change in flux associated with a system problem versus a change in flux associated with a temperature variation.

Equations for Salt Passage (Percentage Rejection of Ions)

The salt permeation rate can be calculated using the following equation (Amjad et al., 1996a):

$$J_s = \frac{\alpha_s \times (C_f - C_p)}{t}$$

where J_s is the salt permeation rate (volume/time × area), α_s is the membrane constant for a particular salt, C is the concentration of the salt, considering concentration polarization (C_f is the feedwater concentration and C_p is the product concentration), and t is the thickness of the membrane.

From a theoretical standpoint, the salt permeation rate can also be calculated as follows (Parise, 1996a):

$$J_s = B(C_f - C_p)$$

where B is the salt permeability coefficient for a fixed membrane thickness and C is the concentration, considering concentration polarization (C_f is the feedwater concentration and C_p is the product concentration).

From a practical standpoint, for an operating system, the percent salt rejection can be calculated from the following equation:

$$\% \text{ Salt rejection} = \left[\frac{C_f - C_p}{C_f}\right] \times 100$$

Further, the percentage of salt passage can be calculated from the following equation:

$$\% \text{ Salt passage} = \left[\frac{C_p}{C_f}\right] \times 100$$

It is suggested that the theoretical calculations for calculating the salt permeation rate, while useful in determining projections for performance from a computer-generated program generally supplied by RO membrane suppliers, are inappropriate for actual operating systems. The salt rejection can be readily calculated and, in many cases, is indicated on a direct reading meter that determines the concentration of ionic material in the feedwater and permeate piping, using in-line conductivity cells. Since, as discussed in chapter 2, the equivalent conductance (mobility) of all ions encountered in water (with the exception of the hydroxyl and hydronium ions) is fairly similar, conductivity measurements can provide an excellent indication of the percent salt rejection, employing in-line cells and a meter to determine the rejection percentage internally.

Membrane Type and Composition
General
RO membranes can be configured in a tubular, plate and frame, hollow fiber, and spiral wound configuration. For pharmaceutical applications, essentially all RO-based systems employ the spiral wound configuration. For most configurations, the flow path of product water is cross-flow through the porous membrane surface at a 90° angle to the feedwater flow direction.

As indicated earlier, all gases will pass through RO membranes, including reactive gases, such as carbon dioxide and ammonia, as well as "nonreactive gases," such as oxygen and nitrogen.

Membrane polymers contain ion exchange groups on the surface and in the pores (Schneider, 1994b). Water molecules adhere to the sides of the membrane pores, reducing the effective pore diameter and resulting in a lower "molecular weight cutoff" than suggested by historical estimates of 100 to 200 Da (McClellan, 1995).

An RO system uses a membrane array to maximize the recovery of feedwater as product water while maintaining adequate feedwater velocity to avoid precipitation of compounds in the concentrating stages of membranes in the final ("tail" array). This membrane configuration is discussed in greater detail under section "Design Considerations." In general, hollow fiber RO membranes, unlike many ultrafiltration membranes, employ flow from the outside to the inside of the membrane. Each hollow fiber membrane is about 100 to 200 µm in diameter (1–2 human hairs in thickness) with an inner flow diameter of about 1/2 of the overall fiber diameter (Paul, 1997).

Membrane Configuration—Spiral Wound
As indicated, spiral wound membranes are essentially exclusively used for RO systems employed for pharmaceutical applications. Since other membrane configurations are seldom used, this subsection limits the discussion to the assembly, flow path, and other items associated with spiral wound RO membranes.

Membrane manufacture begins with a sheet of span-bounded fibers of polyester or polypropylene. A porous polysulfone material is used to support the composite material, such as polyamide, in a thin film (skin) (Schneider, 1994b; Harfst, 1995). It should be pointed out that the porous polysulfone material is similar in nature to material commonly used for ultrafiltration membranes. Spiral wound elements have a narrow feedwater channel that is susceptible to fouling. The typical cross-flow velocity is 0.1 to 0.5 m/sec (Dudley and Fazel, 1997).

A spiral wound membrane consists of an envelope of two membranes, positioned back-to-back. The membrane sheets are separated by a permeate carrier "channel." The flat membrane sheets are glued on three sides, with the fourth side glued to a permeate collector (Fig. 4A.3). A feedwater spacer, of mesh type, is positioned between the membrane envelopes, which are "rolled" in a spiral configuration around the hollow permeate tube collector. An outer shell or restraining material provides the mechanism for encapsulating the "jelly roll" configuration. The sealed membranes are positioned in pressure vessels equipped with feedwater, product water, and wastewater connections. Historically, membranes have been provided with brine seals on the "feedwater" end of the membrane. Figure 4A.4 demonstrates

ION REMOVAL TECHNIQUES—REVERSE OSMOSIS

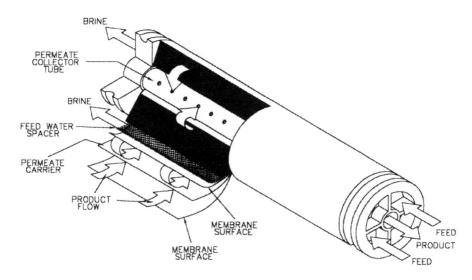

Figure 4A.3 Spiral wound reverse osmosis envelope connected to a permeate collector. *Source*: Applegate (1984).

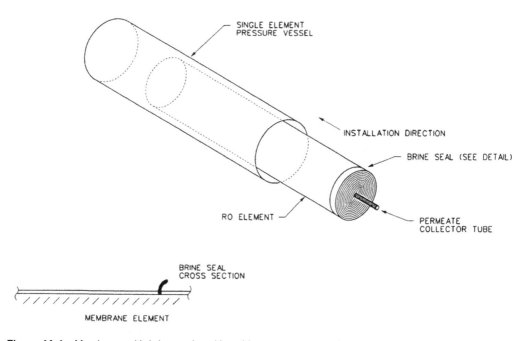

Figure 4A.4 Membrane with brine seal positioned in a pressure vessel.

a membrane positioned in a pressure vessel with a brine seal. These spiral wound membranes are referred to as "brackish water" membranes. They are not recommended for pharmaceutical systems. A "loose wrapped"/"full fit" membrane is suggested for pharmaceutical applications to eliminate the stagnant area associated with the brine seal. This item is discussed in greater detail later in this chapter.

RO membranes are positioned in single- or multiple-element pressure vessels. The feed (and concentrate) flow is in a straight axial path, parallel to the direction of the permeate collector, through the feed side channel. Product water flow through the membranes "spirals" to the center permeate collector. Concentrate flow from one element is either directed to waste or commonly fed to the next element in series.

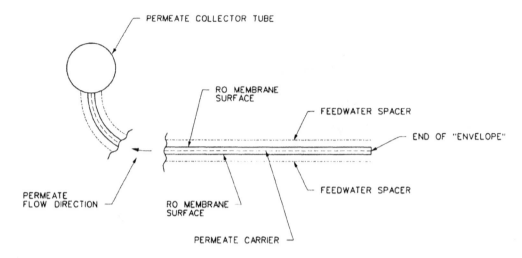

Figure 4A.5 Cross-section of a flat membrane sheet with a feedwater spacer and permeate carrier. (The feedwater flow direction is "into" the page. The permeate flow direction is "out from" the page.)

Figure 4A.5 is a cross section of a flat sheet membrane showing the membrane, feedwater spacer, and permeate carrier. Again, it is important to remember that membrane sheets, with a separating permeate carrier, are glued on three sides to establish an "envelope" that allows product water flow to a permeate collection tube in the center of the jelly roll configuration.

Feedwater spacers are generally 28 to 30 mils thick (Paul, 1997). As discussed later in this chapter, feedwater spacers may present membrane-fouling problems, but they are required to provide spacing for feedwater flow between adjacent membrane envelopes.

Cellulose Acetate Spiral Wound Membranes
While very infrequently employed for pharmaceutical water systems, cellulose acetate membranes should be discussed. Cellulose acetate membranes (including cellulose triacetate) are attacked by bacteria. To minimize microbial attack of cellulose acetate and triacetate membranes, a suggested feedwater residual chlorine concentration of 0.3 to 1.0 mg/L should be considered (Bates and Steir, 1994; Harfst, 1994). *Pseudomonas* sp. favor the carbohydrate substrate associated with cellulose acetate and triacetate membranes.

Cellulostic membranes are sensitive to pH, undergoing a process called "hydrolysis." The hydrolysis process, which gradually "dissolves" the membrane material, is minimized by maintaining a feedwater pH in the range of 4.5 to 7.5. While minimizing hydrolysis, this pH range does not completely inhibit the hydrolysis process. The literature suggests that cellulostic membranes are attacked by bacteria through enzymatic activity or the production of acidic and/or alkaline metabolites (Ridgway, 1987). While the theoretical basis of the bacterial attack of cellulostic membranes can be postulated by several different mechanisms, the important factor to consider, particularly for pharmaceutical water purification systems, is the fact that bacterial attack occurs and requires a residual disinfectant to control. Unfortunately, the residual disinfectant, present in the feed and product water, must be removed downstream of the RO unit, by a technique such as activated carbon adsorption, the introduction of a reducing agent, or reduction by ultraviolet radiation, since residual disinfectant cannot be present in USP Purified Water. Further, it would be undesirable for chloride containing disinfecting agents to be present in the feedwater to a distillation unit (vapor compression or multiple effect). Depending on the application, residual disinfectant may also have an adverse effect on pharmaceutical production or processing applications.

The literature suggests that cellulostic membranes may also be attacked by certain metallic salts, such as those of iron, as well as phenolics (Murphy and Moody, 1997). The literature further suggests that cellulostic membranes are attacked by phthalates (Nickerson et al., 1994). The many potential items that can affect the integrity of cellulostic membranes further limits the use of the membrane for pharmaceutical applications, since membrane integrity is critical.

While not a major issue, compaction of cellulostic membranes will occur over a period of time, resulting in a requirement for higher feedwater pressures and a loss in product flux (Allen et al., 1995). The same effect is not observed, to a significant extent, for thin-film composite membranes.

Spiral Wound Thin-Film Composite Membranes
Thin-film composite polyamide membranes consist of a polyamide membrane layer, approximately 3 μm thick, on a porous polysulfone support media. The "thin film" refers to the fact that the polyamide material is actually a thin layer of membrane supported on the porous, ultrafiltration-like polysulfone structure. The "composite" nomenclature refers to the fact that the actual membrane consists of the polyamide layer on a second material, physically two distinct different materials. Researchers have clearly demonstrated that thin-film composite membranes, with polyamide on polysulfone, yield higher flux rates, exhibit excellent selectivity for ions, and provide excellent mechanical/chemical properties when compared with cellulostic membranes (Hamilton and Drummonds, 1994; Bartels, 1997).

In general, thin-film composite polyamide membranes are not attacked by bacteria, unless there is a significant growth of bacteria on the membranes, *and* the membranes are allowed to "sit" in a stagnant condition for a number of days. This is a factor that is critical to the long-term successful operation of the membranes for pharmaceutical water systems.

Thin-film composite membranes are not tolerant to residual disinfecting agents such as chlorine and chloramines, requiring removal of residual disinfectant from the feedwater to avoid membrane damage. The literature suggests that the maximum calculated residual chlorine concentration for successful long-term operation of thin-film composite polyamide membranes is 0.038 mg/L (Bates and Steir, 1994). Thin-film composite membrane manufacturer's information indicates that degradation of membranes will occur after approximately 200 to 1000 hours of exposure to a 1.0 mg/L concentration of free chlorine (DOW, 2010a). Degradation may be expressed in chlorine hours, the product of the fee chlorine concentration and hours of exposure. Higher pH values, higher feedwater temperature, and the presence of iron or transition metals on the membrane surface will increase oxidative attack and subsequently membrane failure. Degradation is generally associated with initial loss in flux followed by an increase in flux and decrease in ion rejection.

Unlike cellulostic membranes, where hydrolysis restricts the pH operating range (and subsequently the pH of cleaning solutions), thin-film composite membranes may be operated over a very broad range of pH values, generally from 4.0 to 11.0. This is an extremely important property, since many municipalities, particularly in the Northeast United States, as discussed in chapter 2, increase pH to control corrosion of lead piping and lead-soldered joints as well as to increase the effectiveness of monochloramine as a disinfecting agent.

Thin-film composite membranes exhibit much higher rejection for silica than cellulostic membranes. This is an important factor when reverse osmosis is used as a pretreatment technique to unit operations such as multiple-effect distillation, Pure Steam generation and heat exchangers operating at elevated temperatures. As discussed in chapter 5, the presence of silica can adversely affect the long-term performance of vapor-liquid disengaging sections of distillation units. The literature also suggests that thin-film composite membranes exhibit a much higher rejection for both nitrates and naturally occurring organic material, when compared to cellulose acetate membranes (Singh, 1997).

Spiral wound thin-film composite RO membranes are manufactured for different industrial applications. Membranes for purification for seawater, commercial, residential, light industrial, brackish water, and pharmaceutical applications are available. It is suggested that any RO membrane used in a pharmaceutical applications be "full fit/loose wrapped" design. The full fit membranes are available in two standard diameters, 4 and 8 in., with a standard length of about 40 in. "Tape wrapped" membranes, with an order tape wrap surrounding the spiral membrane core, should not be used for pharmaceutical applications. Further, although frequently employed, brackish water membranes, using a hard fiberglass-like structure around the spiral wound membrane core should not be employed because of a required "brine seal" that provides a stagnant are of water for microbial proliferation. Membrane selection should be limited to 4 and 8 in. diameter full fit/loose wrapped non–hot water sanitizable or hot water sanitizable membranes depending on the application. The full fit/loose wrap membranes have an outer membrane wrap with mesh-type cover. These membranes do not have a brine seal but physical expand under pressure to form a seal against the walls of the RO pressure vessel. The rated flow rate capacity for an 8 in. diameter (40 in. long) hot water sanitizable pharmaceutical long grade thin-film composite polyamide membrane, at a feedwater pressure of 250 to 300 psig and a feedwater temperature of 25°C, is about 9000 gal/day (DOW, 2010b).

Concentration Polarization

It would be inappropriate to conclude a discussion associated with the characteristics of RO membranes without discussing a phenomenon referred to as "concentration polarization" (Fig. 4A.6). Concentration polarization is important since it affects the operating characteristics of an RO unit and projections for membrane performance. As feedwater/wastewater flow parallel to the permeate collector through an RO element, water is flowing radially (cross-flow) through the RO membrane from the "free stream." As feedwater, containing salts, approaches the membrane surface, the ions are restricted from passing through the membrane, while water passes through the membrane. As a result, the concentration of ions on the feedwater side of the membrane increases as water is removed. This results in a higher concentration of salts near the membrane surface than the concentration in the feedwater/concentrate stream.

Concentration polarization is associated with the formation of an ion-rich layer of water near the surface of the membrane. This layer of water is a result of the concentrating effects of the RO process as well as the fluid flow dynamics near the surface of the membrane. The concentration of salts in the water within this boundary layer is significantly higher than that of the bulk feedwater/concentrate. While there is a tendency for the more concentrated salt solution to allow diffusion back into the bulk water, the rate of water permeation through the membrane, with resulting concentration, creates concentration polarization. It should be pointed out that this situation can be improved only by creating conditions where salt is allowed to flow (diffuse) back into the bulk feedwater/concentrate stream. This diffusive process can be restricted by the low cross-flow velocity near the surface of the membrane.

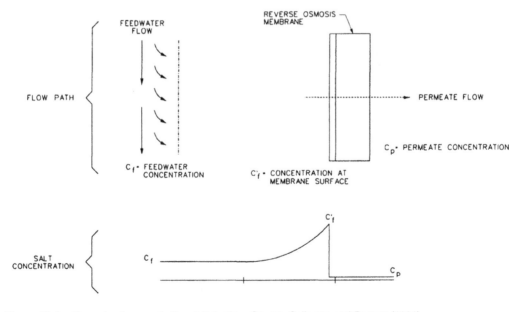

Figure 4A.6 Example of concentration polarization. *Source*: Collentro and Barnett (1996).

Fouling and Scaling

General Discussion

All RO membranes installed in a system with "natural" feedwater will eventually foul and/or scale. The degree and rate of fouling and scaling is directly related to pretreatment unit operations prior to the RO system.

Foulants can be classified as particulate matter, organic matter, colloidal material, and microbial material. Scales are insoluble chemical compounds produced by the reaction of multivalent cations with anions, principally carbonate and sulfate.

An important item to consider is that fouling and scaling of RO membranes, as suggested in the literature, occur with distinctive layering morphology (Kronmiller, 1993; Kaakinen et al., 1994). In other words, there are generally distinct layers of both membrane-fouling material and membrane-scaling material. It is suggested that calcium carbonate forms in a loose layer above a calcium sulfate layer, in an area on the "outside" of the foulant/sealant layer on the membrane surface. It is further suggested that silicates form a tightly adherent layer just on top of the membrane surface. Silicates may be in a layer with organic and colloidal material. This layer is slightly above the bacteria layer that is directly on the surface of the membrane boundary layer. This is an extremely important item to consider, particularly with regard to membrane cleaning.

A membrane "autopsy" is the definitive method of determining the degree of scaling and/or fouling; it is also an excellent method of characterizing the nature of the scalants and foulants. This autopsy, a destructive process, should be performed for RO membranes selected from various locations in a membrane array prior to membrane replacement.

Symptoms of fouling include an increase in the pressure drop through the membrane and an increase in product water conductivity, which is associated with the formation of a layer at the membrane surface, which is associated with concentration polarization a result of fouling.

Scaling/fouling will result in an increase in the required feedwater pressure over time, to obtain a preestablished permeate flow rate. There will be a decrease in the normalized permeate flow rate over time, particularly if the feedwater pressure is not increased to compensate for fouling/scaling. In addition, the transmembrane pressure drop across an RO element will increase over time. Feedwater spacers in flow channels will often provide a location for scalants and foulants to accumulate. A rapid decrease in the permeate flow rate is generally associated with the rapid fouling of membrane surfaces as opposed to membrane scaling. Particulate matter membrane fouling will generally occur in "lead" membrane elements.

As previously indicated, silica scaling can present problems. However, higher feedwater pH values increase the solubility of silica, improving the rejection of silica by thin-film composite membranes. The literature suggests that silica "passage" is reduced by 90% by increasing the feedwater pH from 8.0 to 10.0 (McBride and Mukhopadhyay, 1997).

Systems injecting antiscalants or polymers upstream of the thin-film composite polyamide membranes may experience irreversible membrane fouling, as documented by Parise (1996b). Certain surface-active chemicals can also cause irreversible membrane fouling at very low concentrations (e.g., 1 mg/L). If used, the absence of antiscalant or polymer in RO product water must be verified by chemical testing to verify the absence of foreign substances and impurities as indicated in the *General Notices* Section of USP.

Membrane Scaling

Concentration polarization can significantly contribute to membrane scaling. As previously indicated, it can be minimized by decreasing the percent recovery of feedwater through a membrane/membrane array, by increasing the feedwater and concentrate flow rate.

The most commonly encountered scale on RO membranes is calcium carbonate. In general, the carbonate ion is not present in the feedwater, while the bicarbonate ion is generally present. The literature suggests that the chemistry involved during concentration of ionic material within the membranes may result in the production of calcium carbonate as opposed to calcium bicarbonate (Amjad et al., 1996a,b; Tracey, 1996; Parise, 1996b). Table 4A.1

Table 4A.1 Rate of Salt Concentration Increase as a Function of Percent Recovery

Percent recovery (%)	Waste ion concentration/Feedwater ion concentration
10	1.06
20	1.30
50	1.95
75	3.90
90	9.75

Source: Amjad et al. (1996b).

demonstrates the rate of salt concentration increase as a function of percent recovery. This is an important item to consider with regard to production of membrane scaling compounds.

Obviously, membrane scaling can be reduced by removal of multivalent cations in the feedwater. The use of two water-softening units in series or use of countercurrent-regenerated water softener can reduce calcium ion concentrations to very low levels, eliminating the need to implement additional sale control operational changes. However, membrane scale control can also be achieved by reducing the percentage recovery, decreasing the feedwater pH, and injecting a scale inhibitor. Again, properly designed and operated water-softening units is the most attractive technique for minimizing scale formation on RO membranes used for pharmaceutical water system applications.

Most salts will exhibit increased solubility with increasing temperature. Some salts of calcium, as discussed in chapter 3, will exhibit "inverse solubility," decreasing with increasing temperature.

Without a scale inhibitor, the maximum "bulk" concentration (concentration of a particular compound in the waste stream) should not exceed 85% to 90% of the solubility constant, as discussed in chapter 2. Concentration polarization will increase the concentration of the specific material to a value greater than its solubility product in the area of the membrane surface. A scaling problem *may* be observed by an increase in the pressure drop across the *final* membrane array, since it is exposed to the highest concentration of potentially insoluble compounds.

Membrane/Colloidal Fouling
As discussed in chapter 2, colloids may exist "individually," without the presence of organic material. However, particularly for surface water supplies or groundwater supplies influenced by surface supplies, colloidal material and organic material generally exist in a complex. Colloidal fouling of RO membranes generally occurs on lead membranes and is commonly referred to as "front-end pluggage." The literature contains information associated with this process (Paul, 1996; Finan and Tracey, 1995).

It has been noted that the degree of organic and colloidal fouling can be reduced when the pH of the feedwater is elevated. Higher feedwater pH is also associated with enhanced rejection of silica. pH adjustment, as appropriate, to the feedwater of RO units will be discussed later in this chapter under section "Design Considerations."

The literature also suggests that residual chlorine may convert naturally occurring organic material to assimilable organic carbon, a nutrient for bacterial growth (Paul, 1996). As discussed in chapter 3, the proper design, positioning, and operation of an activated carbon unit is critical to the successful operation of a water purification system employing reverse osmosis. In addition to removing residual disinfectant, the activated carbon unit can effectively remove a significant portion of the naturally occurring organic material present in the feedwater. This will reduce organic and colloidal fouling of RO membranes and also reduce the concentration of organic material, a nutrient for bacteria that is also associated with membrane fouling.

Colloidal iron, silica, and aluminum, complexed with organic material, can result in RO membrane fouling. The literature suggests that ferric oxide/hydroxides, silica gels, and humic acid have been noted as foulants on membrane surfaces (Ning and Stith, 1997). Many of the observations were noted by conducting detailed analysis of operating membranes after they were removed from service (i.e., autopsied).

Microbial Fouling of Reverse Osmosis Membranes
Microbial fouling of RO membranes, in general, is the greatest reason for cleaning. It is suggested that the frequency of RO system chemical cleaning for microbial fouling can be significantly decreased by the use of proper pretreatment components, which is not only associated with the removal of bacteria but also with the removal of organic material. Further, continuous flow through a reverse osmosis unit, the ability to periodically hot water sanitize recirculating systems, and a periodic RO membrane "rotating" cleaning program can significantly extend required chemical sanitization.

The biofilm associated with microbial fouling forms in a very thin layer, directly on the membrane surface. This thin, evenly dispersed layer can result in decreased salt rejection and increased RO product water total viable bacteria levels. As the film thickness increases, salt rejection will decrease further while product water total viable bacteria levels will increase. Decreased salt rejection can be attributed to the influence of biofilm on concentration polarization.

The slight negative charge associated with bacteria as well as cell hydrophobicity contributes to the formation of the biofilm directly on the membrane surface, which is extremely difficult to remove. A sequential cleaning program is required to remove the biofilm on RO membranes, due to the "layering" effect associated with scalants and organic/colloidal fouling "above" the biofilm on the membranes. While hot water sanitizable units provide periodic destruction of bacteria on the membrane surface, the biofilm is not significantly removed. Chemical sanitization with a 1% hydrogen peroxide/peracetic acid provides and effective method of biofilm on the membrane surface and on feedwater and product water piping/tubing and support component surfaces.

Observation by some individuals indicates that microbial fouling will generally begin in the lead elements of an RO system (Amjad, 1996b). While it is suggested that there may be justification for projecting that bacterial growth starts in the lead elements, it is further suggested that microbial fouling, once initiated, will quickly proceed to all membranes in the system and, as indicated upstream and downstream surfaces.

Some pharmaceutical water purification systems employ sodium bisulfite (or another reducing agent) to remove residual disinfectant as part of the pretreatment process to an RO system (see chap. 3). It is strongly suggested that the use of sodium bisulfite (or other reducing agents) be limited to source water from a groundwater source that is not influenced by a surface water source. As discussed earlier, organic material must be removed to inhibit not only organic/colloidal fouling of the RO membranes but also to decrease the proliferation of microorganisms on membrane surfaces. Problems associated with sodium bisulfite storage and injection are discussed in chapter 3. Obviously, microbial control within a chemical injection system containing a reducing agent presents challenges that must be addressed by system design, operation, and maintenance. Bacteria introduced with the reducing agent will result in increased microbial fouling of the membranes.

Measurement of Reverse Osmosis Scaling and Fouling Indices
The Langelier Saturation Index (LSI) provides an excellent indication of the potential for feedwater to produce scaling in RO membranes. Table 4A.2 is a representative calculation for determining the LSI value. The literature contains nomographs that can be used to directly determine the index (Kemmer, 1988). Unfortunately, the index does not provide useful information for determining potential membrane fouling conditions. Fouling conditions are determined by performing an analytical procedure, referred to as a Silt Density Index (SDI) measurement on a "side stream" of feedwater to the RO system. Figure 4A.7 presents the equipment used during execution of an SDI measurement. Table 4A.3 is a representative SDI calculation for simulated data obtained during the filtration operation associated with an SDI determination.

Table 4A.2 Representative Langelier Saturation Index Calculation

$LSI = pH + pH_s$

where pH is the solution pH (7.80) and pH_s is the pH at which the solution of concern is saturated with calcium carbonate.

$pH_s = (9.30 + A + B) - (C - D)$

where 9.30 is a constant factor for specific tables, A is a factor associated with solution TDS in mg/L, B is a factor associated with temperature in °F, C is a factor associated with the calcium hardness expressed as mg/L as calcium carbonate, and D is a factor associated with alkalinity expressed as mg/L as calcium carbonate

Assumptions and "factors"

TDS = 400 mg/L (A = 0.16)
Temperature = 77°F (B = 1.98)
Calcium Hardness = 240 mg/L as calcium carbonate (C = 1.98)
Alkalinity = 196 mg/L as calcium carbonate (D = 2.29)

Substituting and calculating

$pH_s = (9.30 + 0.16 + 1.98) - (1.98 + 2.29) = 7.17$
$LSI = 7.80 - 7.17 = +0.63$

Notes: If LSI > 0, water is saturated and tends to allow formation of precipitates; if LSI = 0, water is saturated (in equilibrium) with calcium carbonate; a scale layer of calcium carbonate will not form but calcium carbonate will not dissolve; If LSI < 0, calcium carbonate will dissolve in the solution.
Sources: Permutit Company (1986), Kemmer (1988), Pontius (1990).

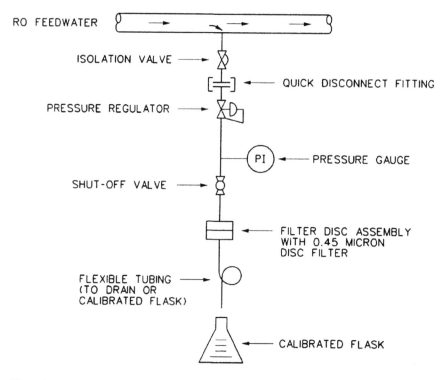

Figure 4A.7 Silt Density Index measurement apparatus. (The filter disk assembly to contain the sealing mechanism is a "disposable" 0.45 μm filter disk. To obtain accurate results, the filter disk should be "wetted"—water should cover the entire area of the disk. The filter disk assembly should employ an O-ring sealing mechanism. Once the procedure is started, flow should continue through the filter disk, either to drain to the calibrated flask. An accurate stopwatch is required for determining the time to filter a preset volume, which is suggested at 500 mL.

ION REMOVAL TECHNIQUES—REVERSE OSMOSIS

Table 4A.3 Representative SDI Calculation

Operating conditions:
- Filter pressure = 30 psig
- Filter disc type = 0.45 μm rating
- Filter volume = 500 mL

"Simulated" test observations:
- Initial time period to filter 500 mL = t_0 = 18 seconds
  Time period to filter 500 mL after 15 min of continuous filtering = t_f = 27 sec

SDI calculation:
- SDI = $100/T \times (1 - t_0/t_f)$
- SDI = $100/15 \times 18/27$
- SDI = 2.2

Note: Equation may also be expressed as SDI = $(t_f - t_0)/t_f \times 100/T$. Criteria in ASTM D4189 satisfied.
Abbreviation: SDI, Silt Density Index.
Sources: Kemmer (1988) and Kaakinen et al. (1994).

The SDI procedure is defined in ASTM D-4189. While the literature suggests that there are some execution problems (Harfst, 1994; Coulter, 1996), it is the most popular method of determining the potential for membrane fouling. The procedure involves filtration of feedwater through a 0.45 μm filter disk at a pressure of 30 psig. The time required for filtration of an established volume is recorded. The measurement is repeated sequentially, continuing the flow of water through the 0.45 μm filter, at preset periods of time, based on the level of foulants in the feedwater. It should be indicated that all major RO membrane manufacturers, within their technical and catalog information, indicate the maximum SDI value of the feedwater required to obtain proper membrane performance as a warranty for the membrane.

Subsequently, equipment manufacturers will also require measurement of RO feedwater parameters, such as free chlorine concentration, including periodic SDI measurements, to maintain the validity of the warranty for the system.

Unfortunately, execution of the SDI measurement technique, unlike "grab" sample collection, requires "field" execution. Since it is an extremely important parameter for establishing proper operation of an RO unit, by indicating that the pretreatment equipment is adequately designed and defining transient feedwater conditions associated with seasonal and climatic changes, execution is required. Automated SDI measuring devices are available. The literature suggests that the automated SDI measurement devices may present a better indication of RO system performance than plotting normalized product flux from the unit (Kaakinen et al., 1994). This justification is based on the fact that a maximum "acceptable" flux decline of 5%/month is only associated with a 0.2%/day change in the flux, a parameter very difficult to determine. The literature also suggests that it may be appropriate, for certain applications, to perform SDI measurements on the waste stream from an RO unit (Amjad et al., 1996b). This is particularly true when the feedwater SDI values are relatively low.

Figure 4A.8 is a plot of first-stage differential pressure increase as a function of time for contamination by microorganisms. Figure 4A.9 demonstrates the effect of feedwater microbial levels on SDI values, less than 3 to 5, the maximum range generally specified by RO membrane manufacturers. These data are provided to indicate that while SDI is a valuable measurement technique for determining membrane fouling, a feedwater analytical program must include periodic microbial monitoring.

Any color appearing on the 0.45 μm filter disc, used during the filtration procedure for determining SDI values, can provide an indication of the nature of the foulants. For example, organic material may exhibit a light yellow to brown color on the filter disc. Unfortunately, iron oxides may also exhibit a similar color. While it may be difficult to differentiate the foulant contributing to the color, the intensity of the color, which is associated with an increase in the calculated SDI value, will also indicate an excursion in feedwater parameters, potentially associated with transient conditions in the pretreatment equipment or a change in the raw feedwater quality to the system.

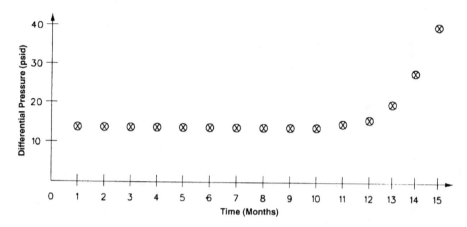

Figure 4A.8 Representative first stage differential pressure increase as a function of time. *Source*: Webb and Paul (1994).

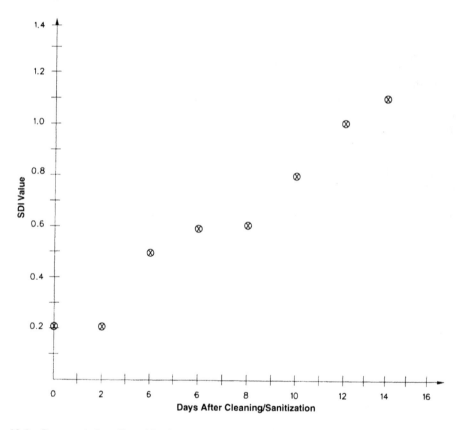

Figure 4A.9 Representative effect of feedwater microbial levels on SDI values. *Source*: Webb and Paul (1994).

As an alternative to SDI measurements, the literature contains material associated with a "cross-flow index" (CFI) (Harfst, 1994; Coulter, 1996). This technique offers a measuring procedure that truly reflects the actual flow conditions experienced within an RO element. It is suggested that the CFI measurement may be used to supplement periodic SDI measurements.

When activated carbon is used as a pretreatment technique for systems with feedwater supplies from a surface source, or groundwater influenced by a surface source, specification of

Table 4A.4 Recommended List of Raw Water Analysis for an Effective Monitoring Program

pH	Methyl orange alkalinity	Total chlorine
Total Suspended Solids	Nitrate (as nitrogen)	TOC
Chloride	Sulfate	Nitrite (as nitrogen)
Aluminum	Barium	Total viable bacteria
Copper	Iron	Calcium
Manganese	Potassium	Magnesium
Zinc	Color	Sodium
Total phosphorous	Silica	Turbidity
Chromium	Lead	Cadmium
Bromide	Coliform bacteria	Fluoride
Conductivity	Free chlorine	Phenolphthalein alkalinity

the parameters for replacing the activated carbon must be precisely defined. For any activated carbon unit installed prior to an RO unit, the activated carbon media should be acid washed, fully hydrated, and prerinsed. Nonacid washed activated carbon will elute aluminum and other multivalent ions. It may also elute both sodium hydroxide and potassium hydroxide, resulting in an elevated pH in the product water from the unit. Finally, the hydration of activated carbon coupled with an appropriate rinse will eliminate or significantly reduce the presence of activated carbon fines in the product water from the unit.

It would be inappropriate to conclude this section of potential RO element foulants and sealants without indicating that a responsive analytical monitoring program will include periodic LSI determination, periodic SDI measurement, periodic microbial determination, *and* periodic monitoring of inorganic chemical parameters that may affect the performance of the RO unit. Table 4A.4 presents a recommended list of inorganic parameters that may be monitored to assist in ensuring long-term successful operation of an RO system. This list may be altered for specific applications, based on the nature of the raw water supply and pretreatment components.

Scale Inhibitors, Polymers, and Other Feedwater
Additives to reverse osmosis units. In general, the use of antiscalants, polymers, and other RO membrane scale and/or foulant inhibitors is strongly discouraged for pharmaceutical applications. While these agents may have adverse effects on RO membranes, as indicated below, the primary objection is related to the fact that the chemicals can be considered as foreign substances and impurities per the *General Notices* section of USP.

Antiscalants may include a number of proprietary compounds. Historically, potassium pyrophosphate, trisodium phosphate, and sodium hexametaphosphate have been used as nonproprietary agents. While these antiscalants were somewhat effective, phosphate discharge presented an environmental problem. Furthermore, the phosphates formed insoluble calcium phosphate precipitates. As the use of antiscalants evolved, phosphonates exhibited better scale control than phosphates but still presented discharge problems. In addition, precipitation of complexes with calcium remained a problem. More recently, antiscalants have included polymeric compounds. Generally, these are used in a multifunctional program combined with dispersants in a polyacrylate-based pretreatment program.

For certain applications, it has been observed that the use of specific antiscalants is associated with the frequent replacement requirement for RO unit prefilters coupled with a noticeable fouling of RO membranes. The fouling is visible as a brownish gel-like slime. Table 4A.5 presents a representative analysis for this observed material. The literature indicates, as discussed earlier, that significant levels of bacteria may exist in chemical feed tanks containing polymers, antiscalants, and antifoulants (Bates and Stier, 1994). Certain antiscalants, polymers, and antifoulants may "blind" membrane surfaces with a gel coat, significantly decreasing product water flux. In some cases, the gel coat cannot be removed, requiring membrane replacement. Any antiscalant that contains phosphate will also enhance the precipitation of silica. Precipitation of silica will contribute to scale formation in an RO element and decreases the potential rejection of silica, since the solubility is decreased.

Table 4A.5 Representative Chemical Analysis—Fouling Slime from Feedwater Treatment Chemicals

Element	Weight %
Gravimetric elemental analysis	
Carbon	11.85
Hydrogen	2.05
Nitrogen	1.94
Iron	7.00
Silicon	20.45
Phosphorous	0.80
"Others"	55.91

Element	Relative Weight %
Energy dispersive X-ray analysis—cations	
Silicon	52
Iron	25
Aluminum	7.0
Potassium	6.0
Phosphorous	4.2
Calcium	2.6
Sulfur	2.1
Magnesium	0.8
Titanium	0.7

Source: Ning and Stith (1997).

The literature indicates that some municipalities employ treatment chemicals containing aluminum and/or iron (Amjad et al., 1996b; Bates and Stier, 1994; Coulter, 1996). Alum is often used as a treatment chemical by municipalities. This aluminum-containing compound will have a significant negative effect on the performance of an RO unit if a properly designed, operated, and maintained pretreatment system is not employed. Finally, potassium permanganate can present operating problems for RO systems using thin-film composite membranes. Potassium permanganate may be used as an "additive" when greensand filters, for removal of iron and manganese, are included as part of the pretreatment system.

Cleaning of reverse osmosis membranes. *It is suggested that cleaning of reverse osmosis membranes for the majority of pharmaceutical applications should be performed using a "contract" service and "off-site" execution. A recommended program is presented in the Design, Operating, and Maintenance sections of this chapter. The information presented below outlines RO membrane general guidelines that should be considered.*

The initial indication of membrane fouling is a gradual continuous decrease in the normalized product water flow rate (flux). As membrane fouling progresses, a reduction in salt rejection will probably occur, associated with increased concentration polarization within the membranes. An additional indication of membrane fouling is an increase in the feed-to-reject pressure drop.

It is suggested that an RO system, with proper pretreatment, should not require chemical cleaning with a frequency greater than once every six months. The literature suggests that excessive cleaning, primarily associated with inadequate pretreatment, occurs when the frequency is greater than once every two months (Parise, 1996b).

The specific cleaning properties of membranes, the cleaning formulation, the frequency of cleaning, and the sequence of the cleaning steps are extremely important. The cleaning program must be custom designed and consider feedwater parameters. Noncustom-selected cleaning programs can result in ineffective cleaning and potential damage to membranes.

Considering the custom nature of a cleaning program, larger flow rate systems or systems experiencing significant scaling or fouling not responsive to "classical" cleaning regimes may consider on-site testing of multiple cleaning agents. However, a responsive membrane-cleaning program can be determined by off-site analysis of membranes, prior to cleaning. This may include both lead and tail membranes. Further, the program could include destructive testing of one or two membranes. The individual membranes concentrate impurities. As such, they provide months of "accumulated" contaminant information that can be used to develop an excellent cleaning program.

The literature contains numerous references indicating that RO membrane chemical cleaning should be performed when the normalized permeate flux decreases by 10% to 15% (Amjad et al., 1996a; Paul, 1994; Coulter, 1996; Parise, 1996b). A proper chemical cleaning flow rate is critical to the successful restoration of normalized permeate flux. The literature suggests that the flow rate should be 1.0 to 1.25 times the normal operating flow rate (Parise, 1996b).

The chemical cleaning temperature is also critical. If on-site chemical is performed, the cleaning tank should be equipped with heating provisions. While not recommended, if a "contract" service with nonfacility-owned equipment is used for on-site chemical cleaning, heating provisions should be reviewed. Cleaning performed by a contract service, in general, or cleaning performed using ambient temperature water may *not* produce the desired results, particularly with regard to the removal of scale-forming compounds and, to a lesser extent, organic and colloidal material when compared to off-site membrane-cleaning programs discussed later in this chapter.

It is suggested that the lead elements in a membrane array may be cleaned in a reverse flow direction. This provides advantages, particularly with regard to the removal of colloidal and organic foulants.

As discussed earlier, the chemical cleaning regime for a specific RO system may be custom generated. However, any chemical cleaning sequence should consider layering of scalants and foulants. This would suggest that a representative program would include an initial cleaning step with a "low-pH" solution for the removal of scalants, followed by a "high-pH" cleaning solution for the removal of colloidal and organic material. Finally, a sanitizing agent would be used to destroy bacteria and remove the biofilm on the membrane surface. The indicated cleaning operation must be performed in the sequence indicated.

Circulation of the cleaning solution and rinse water should be performed at a feedwater pressure less than 50 psig or a pressure adequate to overcome the system (array) pressure drop. The suggested total cleaning volume for thin-film composite polyamide membranes should be adequate to ensure effective cleaning. In general, the volume of cleaning tanks provided for on-site chemical cleaning, or the cleaning tank volume provided by "contracted" on-site cleaning, is inadequate. For example, the literature suggests that a single 8 in. diameter × 40 in. long thin-film composite polyamide membrane will require 40 L of cleaning solution per cleaning step (Dudley and Fazel, 1997).

The chemical sanitizing agent used for bacteria destruction and biofilm removal must be capable of penetrating the membrane surface to the product water piping/tubing as well as downstream components. A suggested contact time for effective chemical sanitization (dynamic and stagnant) may range from 4 to 12 hours. In general, the greater the sanitization time period, the greater the destruction of bacteria and removal of the biofilm. However, the long-term effect of repeated lengthy chemical sanitization operations on membrane life must also be considered. The sanitizing agent must be capable of effective destruction of bacteria *and* biofilm removal. The literature indicates that a 1% solution of peracetic acid and hydrogen peroxide is extremely effective in meeting both of the indicated objectives (Mazzola et al., 2002 and Mazzola et al., 2006). This is particularly true when chemical sanitization is coupled with more frequent hot water sanitization (Collentro, 2010a).

As discussed previously, the removal of viable bacteria from the membrane surface is the most frequent condition requiring chemical cleaning/sanitization of RO units. Considering the layering of sealants and foulants, it is suggested that a three-step cleaning procedure be considered. The effectiveness of any sanitizing agent, such as a 1% mixture of hydrogen peroxide and peracetic acid, will *not* be effective if the layers of scalant and organic/colloidal foulants have *not* been removed, since the sanitizing agent will not penetrate the biofilm on the surface of the membrane.

While calcium carbonate is the most prevalent compound in scale found on RO membranes, it is generally removed quite easily with an effective low-pH cleaning solution.

The literature suggests that the reaction produces both the bicarbonate ion and carbon dioxide (Amjad et al., 1996a; Paul, 1994). During this cleaning operation, it is not uncommon to observe "foaming" in the cleaning tank. An important item to be considered with regard to the removal of calcium carbonate is that the pH in the cleaning tank must be maintained below 4.5. It may be necessary to add additional low-pH cleaner, as the cleaning progresses and bicarbonate ion and carbon dioxide are produced, to maintain the pH less than 4.5. Calcium sulfate may also be present as a scaling compound on the membranes. The removal of calcium sulfate generally requires low-pH solution and a chelating agent, such as EDTA.

While proper pretreatment to an RO unit, verified by a thorough analysis of the feedwater supply, particularly for highly undesirable trace compounds, should include water softening, certain systems may experience scalants formed by strontium, barium, and aluminum. In general, scales formed by these higher molecular multivalent cations are very difficult to remove. This is verified by information contained in the literature (Amjad et al., 1996a; Parise, 1996b). The use of series-operated softeners and/or fixed bed countercurrent-regenerated softeners essentially eliminates the presence of these heavy molecular weight cations.

An article appearing in the literature (Husted, 1998) indicates that continuous injection of a biocide, an organic compound of the isothiazolin group, at a concentration of 6 mg/L, will provide excellent microbial control during normal operation. However, the article indicates that some of the sanitizing agent will be present in the product water. Again, as discussed on several occasions, the presence of a microbial control agent of unknown chemical composition (due to the proprietary nature of the product) is a USP "Foreign Substance or Impurity."

DESIGN CONSIDERATIONS
General

While a manufacturer's "standard" RO unit may be provided for pharmaceutical applications, it is suggested that the unit considers the specific application. This may require modification of a "standard" design or custom fabrication for product flow rate considering the number of membrane diameters, and pressure vessel array developed around a computer-generated program provided by RO membrane manufacturers. It is important to understand the design considerations associated with arrangement of RO membranes in an RO system to optimize recovery of feedwater and meet physical space requirements. The product water recovery from each RO element should not exceed 15%. In fact, a conservative design figure employs a recovery of about 8% to 10% per RO element. To achieve the desired system capacity *and* recovery, individual membranes should be configured in a parallel *and* series arrangement. This arrangement directs the waste flow from an individual element as feedwater flow to a second element in series. By configuring the RO elements (within pressure vessels) in an array of decreasing number of pressure vessels ("tapering"), it is possible to maintain the required brine flow rate through membranes recognizing that the flow rate of water through the array is continuously decreasing as water (as product) is removed through each membrane. As the number of pressure vessels is reduced through the array to maintain the waste/feedwater flow rate above a minimum value specified by RO membrane manufacturers, the number of membranes is also reduced. A minimum waste/feedwater velocity is required to reduce ion buildup on membrane surfaces that will result in concentration polarization, particularly within "tail end" membranes in an RO system array.

RO membrane manufacturers provide computerized programs for establishing system design, specifically the number of membranes, membrane array, and configuration. These computerized projections consider the following operating design variables:

- Maximum feedwater flow rate to each RO membrane
- Maximum brine flow rate from each RO membrane
- Maximum recovery rate for each RO membrane
- Maximum product water flow rate through each RO membrane
- Maximum average product water flow rate for the RO system
- Maximum feedwater pressure

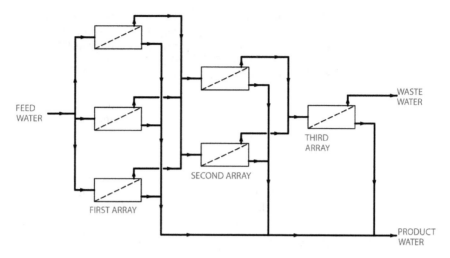

Figure 4A.10 321 RO membrane array.

The recovery rate in each membrane is established by the concentration of rejected species, particularly potential scalants. A typical membrane array is demonstrated in Figure 4A.10. The figure depicts a 3:2:1 array, which could also be referred to as a 6:4:2 array if the individual pressure vessels each contained two RO membranes (in series).

To provide appropriate information associated with preparing an RO projection using a computerized format, certain items are required. Information required for the projection includes, but is not limited to, the following:

- A comprehensive feedwater analysis is required. The feedwater analysis should present the "worst-case" conditions using historical information for seasonal and climatic changes as applicable. It is important to conduct a calcium carbonate equivalent "balance" of anions and cations to ensure that the data provided for the projection will be accurate. If there is an imbalance in the cation/anion calcium carbonate equivalent values, the program will automatically increase the sodium or chloride ion concentration for electronic neutrality.
- The computerized projection will clearly state that feedwater SDI values should be less than 3 to 5. Pretreatment techniques should be selected to ensure that SDI values meet this requirement.
- The water temperature must be specified. For feedwater from a groundwater source the temperature may vary by $\pm 5°F$ with seasonal fluctuations. For raw water from a surface water source or ground water influenced by surface water source, temperature can vary from as low as slightly above freezing to values as high as 80 to 85°F. As indicated earlier, heating of RO feedwater is discouraged since it will result in accelerated proliferation of bacteria. For geographical areas where feedwater temperatures are consistently greater than 77°F, it may be appropriate to consider feedwater cooling to control bacteria levels. It is strongly suggested that it is appropriate to address microbial control within the feedwater/pretreatment system to ensure that frequent RO membrane sanitization is not required as a result of the elevated raw water temperature. Use of a variable frequency drive RO system feedwater pump motor can be used to provide the flow rate and pressure required at various feedwater temperatures.
- Feedwater ionic concentration data for the RO projection must be expressed as the "ion" rather than the calcium carbonate equivalent. It is important to consider this fact, since analytical results for calcium, magnesium, and bicarbonate ions are often expressed as calcium carbonate. As indicated, the computerized projections automatically adjust the feedwater for "electronic balance" from an ion standpoint.

Subsequently, if calcium carbonate equivalent concentrations are used for specific ions, such as calcium, magnesium, or bicarbonate, the computer program will automatically increase the chloride (or sodium) value, incorrectly computing the projection.

- If chemicals, such as sodium hydroxide, are injected upstream of the RO system, the effect of the pH adjustment on the carbon dioxide/bicarbonate equilibrium must be calculated. The increase in the bicarbonate ion value, associated with conversion of carbon dioxide to bicarbonate ion, must be calculated in determining the input data for the RO feedwater stream. Further, it is important to remember that softened water must be "corrected" to provide an appropriate conversion of magnesium and calcium (as well as any trace cation impurities) to sodium.
- The desired product water flow rate from the RO unit should be based on the specific application. Often, this is a difficult design challenge since many facilities operate a limited number of shifts each day and only five days per week. While continuous RO recirculating system will be discussed later in this chapter, "batching operation" may require a large makeup flow rate over a short period of time. Quite often, larger than technically desired storage tank volume may be employed to reduce the product water capacity of an RO unit. Ideally, it is suggested that RO system design considers a 10 to 16 hour per day "makeup" period. However, a properly designed recirculating RO system provides desired product water flow rate for facility applications without affecting operation and maintenance of the system.
- If waste or product recycle is used, it should be specified, and the feedwater, waste, and product water flow rates recalculated, as appropriate.

Once these data are provided, the computer projection is generated. The computer program will automatically determine if the specified design (array, number of membranes per pressure vessel, total number of membranes, recycle flow rates, etc.) has violated any RO membrane manufacturer's criteria, such as maximum element recovery, minimum brine flow rate, and maximum transmembrane pressure. The designer can, as appropriate, evaluate alternative membrane pressure vessel arrays to "custom" design the RO unit or to verify that an existing standard design will meet the desired criteria, specifically the guidelines of RO membrane manufacturers.

There are some general issues that should be addressed with regard to RO system computerized projections and resulting design criteria. These issues include, but are not limited to, the following:

- The computer projection should be prepared not only for initial design but also operation with three-year-old RO membranes. In general, most computer projections will provide a "fouling factor," allowing the designer to generate projections in this manner. The "third-year" data present operating data, such as the required system feedwater pressure, critical to establishing proper system design.
- The computerized projection will specify feedwater, product water, and waste flow rates; feedwater, product water, and waste pressures; and ionic concentration of the feedwater, reject (waste), and product water.
- As discussed previously, the program does not consider gases. For example, the projected product water bicarbonate concentration does not consider the fact that *all* of the carbon dioxide present in the feedwater will reestablish equilibrium with the hydronium and bicarbonate ions in product water. Subsequently, the projected product water purity, associated with simply interpreting the product water data from the computerized projection, is not necessarily correct. The concentration of ions, such as the bicarbonate or ammonium ion, will generally be higher, resulting in a higher conductivity than that indicated by the RO computerized projection.
- Quite often, there are physical space limitations for installation of the RO unit. Further, pressure vessel size should be considered to minimize, or eliminate, waste recycle that will contain not only ionic impurities but undesirable bacteria. This may affect the number of pressure vessels, the length of the pressure vessels, and,

subsequently, the membrane array. Further, it may also require larger diameter membranes. Another issue to consider in determining the physical space available for an RO unit is the installation and removal of individual RO membranes. At least 40 in. is required on one side of the unit, with total unrestricted access. The other side of the unit should also be provided with at least 40 in. of "generally" accessible area to allow proper installation and removal of the membranes. These are *minimum* requirements, since most systems use 40 in. long elements. Unfortunately, in an attempt to standardize, most RO unit manufacturers use pressure vessels capable of containing 4 to 5 each, 40 in. long membranes. Technically, this design results in elevated RO feedwater total viable bacteria levels with associated elevated product water total viable bacteria levels, increased RO unit sanitization requirements, and increased cleaning frequency.

- Some RO systems, particularly smaller flow rate units, will conserve space by vertically mounting pressure vessels. Unless space considerations warrant the selection of a unit with vertically positioned units, it is strongly suggested that this design be avoided. The design requires the removal of pressure vessels as well as the membranes when membrane replacement is required. The new (or off-site cleaned) membranes are placed in the pressure vessels, which in turn are remounted on the unit. Experience indicates that multiple fittings must be disconnected and subsequently reconnected. Further, positioning the pressure vessels back into their original location, using mounting brackets, is tedious. Depending on the access around the unit, horizontally mounted pressure vessels capable of containing a single 4 in. diameter × 40 in. long membrane are preferred. If this factor is not considered during selection of the RO unit, it is possible that the pressure vessel with membrane will require removal "directly up" (vertically) from the system. Obviously, this is difficult to achieve and could also present problems with regard to available "head room" above the unit.
- Most RO membrane manufacturers provide double-pass RO projections as part of their standard computerized program. It is suggested that data collection from a double-pass RO projection, particularly considering the numerous recycle streams associated with the unit, be performed by two different techniques. The first technique uses the manufacturer's double-pass RO projection, while the second technique considers the units, with appropriate flow and adjusted ionic concentrations, as two single-pass units operating in series, product staged.

Microbial Control in Feedwater System

RO systems used for pharmaceutical water applications must be designed for bacteria control. There are numerous items that can be used to minimize bacteria growth within pretreatment components. These items, which were discussed primarily in chapter 3, include, but are not limited to, the following:

- Elimination of ball valves, particularly plastic ball valves from system components. Bacteria will grow not only within the "bore" of a closed ball valve but also, due to the tolerances associated with manufacturing ball valves, in the physical area between the ball and body of the valve.
- Sample valves should not be ball valves and should be positioned as close as possible to a flowing stream. Sample valves should be either diaphragm or needle type.
- Pressure gauges provide an excellent dead leg for bacterial growth. It is strongly suggested that all pressure gauges be equipped with diaphragm isolators. Diaphragm isolators should be connected as close as possible to the flowing stream.
- Where appropriate, recirculation of pretreatment components should be considered. If the recirculation time period is a significant fraction of a day (>30%), it is suggested that provisions be included to remove the mechanical (Joule's) heat associated with the recirculating pump.

- Recirculation systems should include activated carbon units and water softeners. Hot water sanitizable activated carbon units and water softeners may be considered for microbial control.
- While the use of in-line ultraviolet units within a storage and recirculation system has questionable technical benefits, their use to assist in microbial *control* within the pretreatment section of a system is appropriate.

Finally, it should be pointed out that the "six-pipe diameter rule" does *not* apply to ambient temperature systems. Dead legs should be eliminated or kept to a minimum possible length. Considering the "porous" nature of PVC piping, with related rapid bacterial proliferation, alternative piping material may be considered, such as stainless steel or unpigmented polypropylene. Threaded fittings should be eliminated if possible. Chemical sanitization must consider potential dead legs, employing both a dynamic and extended time period "stagnant" cycle.

Reverse Osmosis and Polishing System Design

It is strongly suggested that RO units be operated continuously. Further, downstream polishing components, such as CEDI discussed later in the chapter, should be included in the continuous recirculation system. As discussed in chapter 3, recirculation of pretreatment components, or the entire pretreatment system is appropriate to minimize bacteria proliferation. To facilitate the design, operation, and maintenance of the suggested "RO/polishing component loop," an RO break tank should be considered. This tank provides several functions, as follows:

- The tank provides a depressurized location for flow of pretreated makeup water for the RO unit.
- The tank provides an "air break" between the pretreated makeup water and the stored water in the tank, eliminating bacteria introduction associated with biofilm from the pretreated water piping/tubing to the tank.
- The tank provides a point for recirculation of RO/polishing component water when product water is not being delivered to a downstream tank (Purified Water) or providing makeup to a distillation unit (Water for Injection).
- For chemically sanitized systems, the tank provides a location for introduction of chemical sanitizing agent. Sanitizing agent may flow through the entire RO/polishing component, returning back to the tank.
- For hot water sanitizable RO/polishing component systems, a 316L Stainless Steel tank may be employed with heating jacket or downstream heat exchanger for periodic hot water sanitization.

Figure 4A.11 presents a typical chemically sanitized RO/polishing component system. The system includes an RO "break tank," repressurization pump, RO unit, in-line ultraviolet sanitization unit, CEDI unit, in-line ultraviolet sanitization unit, and final filtration system. When the system is in a "makeup" mode, the RO waste flow is per the discussed computerized projection, generally about 25% of the feedwater flow. When the system is in a "recirculation" mode, the RO waste flow, using a three-valve system shown in Figure 4A.12, is 5% to 10% of the feedwater flow rate (high recovery mode). The CEDI waste stream may be recirculated back to the RO break tank or diverted to waste depending on the concentration of reactive gases in the RO feedwater.

Figure 4A.13 presents a hot water sanitizable recirculating RO/polishing component system. The system uses a heating jacket around the RO break tank for heating during the sanitization process. Post hot water sanitization cooldown is performed by displacement of the hot water pretreated water. The hot water sanitization operation is generally fully automated.

The recirculating design feature with sanitization provisions provides excellent operating characteristics. Typical chemical and total viable bacteria sample results for operating systems with either design are presented in Table 4A.6.

Additional design considerations for RO units are presented below.

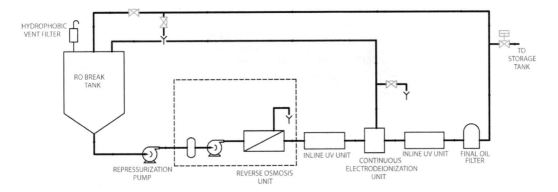

Figure 4A.11 Typical chemically sanitized RO.

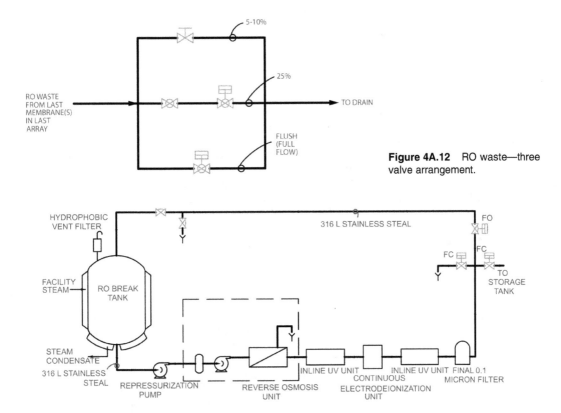

Figure 4A.12 RO waste—three valve arrangement.

Figure 4A.13 Typical hot water sanitized RO.

The Removal of Residual Chlorine or Residual Chloramines for Thin-Film Composite Membrane Applications

As indicated previously, an extremely high percentage of RO membranes used for pharmaceutical water systems are thin-film composite polyamide types. While the advantages of these membranes (when compared with cellulostic membranes) are significant, they are not tolerant to residual disinfectant agent. Considering a three-year membrane life, conventional thin-film composite membranes will only tolerate about 200 to 1000 ppm-hours of chlorine, which translates to a maximum residual chlorine concentration of 0.038 mg/L as discussed earlier. The tolerance of RO to chloramines is generally greater when compared with chlorine. This is offset by the fact that chloramine levels in feedwater are two to three times the level of residual

Table 4A.6 Typical Total Viable Bacteria Data for RO/CEDI Designs and Sanitization Provisions

Sample location	RO/CEDI system without recirculation[a]	RO/CEDI system with recirculation and hot water sanitization provisions[b]	RO/CEDI system with recirculation and chemical sanitization provisions[c]
RO Feedwater[d]	325 to >5700 cfu/mL	<1–28 cfu/mL	10–320 cfu/mL
RO product water[e]	6–~3,000 cfu/100 mL	<1–50 cfu/100 mL	<1–180 cfu/100 mL
CEDI Product Water[e]	31–218 cfu/100 mL	<1–2 cfu/100 mL	<1–12 cfu/100 mL
Post-CEDI UV product water[e]	30–228 cfu/100 mL	<1–1 cfu/100 mL	<1–3 cfu/100 mL
Final 0.1 μm filter product water[e]	3–212 cfu/100 mL	<1 cfu/100 mL	<1 cfu/100 mL

[a]Membrane cleaning and sanitization frequency biannually.
[b]Hot water sanitization twice/month, chemical sanitization every six months.
[c]Chemical sanitization monthly.
[d]Total viable bacteria by heterotrophic plate count, 1 mL, PCA media, 48- to 72-hour incubation time period, 30 to 35°C incubation temperature.
[e]Total viable bacteria by membrane filtration, 100 mL, PCA and/or R2A culture media, 72- to 120-hour incubation time period, 30 to 35°C incubation temperature.
Source: Collentro (2010b).

chlorine. Chloramine removal is an important part of the pretreatment process to RO units using thin-film composite membranes. As discussed in chapter 3, complete residual chlorine removal can be achieved with an activated carbon unit operating at a face velocity of 3 gpm/ft^2 of cross-sectional bed area and a volumetric flow rate of about 1.0 gpm/ft^3 of activated carbon media. However, to ensure the removal of chloramines, contact time is important requiring a suggested volumetric flow of about 0.5 gpm/ft^3 of activated carbon media. Type of activated carbon, pH, and TOC of the feedwater should also be considered as discussed in chapter 3. Media should be replaced at least once every six months. Exposure of thin-film composite membranes to residual chlorine and chloramines will result in membrane failure.

Prefilters to Reverse Osmosis Units
The selection of proper prefilter housings, housing installation design, and prefilters is critical to successful RO operation. Important items that should be considered for the prefiltration system would include, but not be limited to, the following:

- It is strongly suggested that all prefilter systems to an RO unit include dual filter housings. Each filter housing should be capable of supporting the full feedwater flow required to the RO unit. This allows operating personnel to isolate one of the prefilter units, replace the filter elements, rinse the filter element to drain for a preset period of time, and place the unit back into operation. Further, it is not uncommon for a prefilter housing to exhibit microbial growth on the interior surfaces. Using two filter housings, each with full feedwater flow capability, allows operating personnel adequate time to sanitize the interior of a filter housing chemically to control bacteria. Experience indicates that a design using a single filter housing often results in a "delay" of filter cartridge replacement due to the fact that operating personnel are "available" only during time periods when the RO unit is generally operational. To accomplish the stated objectives, each filter housing should be equipped with feedwater and product water isolation valves, a filter housing drain valve, a preoperational "rinse-to-drain" valve, feedwater and product water sample valves, and feedwater and product water pressure gauges with diaphragm isolators as shown on Figure 4A.14. Further, the filter housing should be equipped with a vent to ensure that all air is removed from the housing prior to placing the filter system back in service. This is an extremely important design consideration since "two-phase" flow through RO elements could result in "water hammer" and associated rapid loss in membrane integrity.

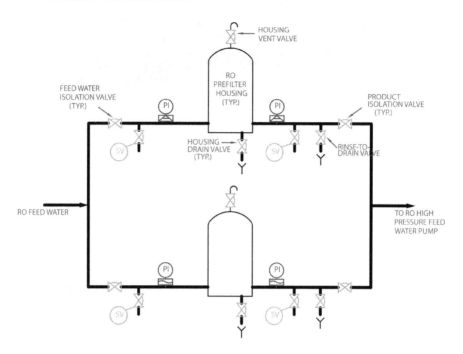

Figure 4A.14 Dual RO prefilters.

- It is suggested that the prefilter cartridges be equipped with an O-ring sealing mechanism, as opposed to a "flat gasket" sealing mechanism. Experience indicates that flat gasket sealing mechanisms may allow some particulate matter to pass to the downstream RO unit, while a single or double O-ring prefilter cartridge seal mechanism provides an absolute mechanism for eliminating particulate matter.
- The selection of filter cartridges is also important and should *not* be based solely on price. Conversely, some more expensive prefilters may exhibit undesirable technical attributes. Ideally, the filters should remove particulate matter to the micron rating specified, without releasing trace amounts of filter cartridge media to the feedwater of the RO unit. For certain applications, particularly where pretreatment is a concern (in light of meeting the maximum specified SDI value of 3–5), the use of a "charged" filter media may be considered, but only after an evaluation of a *specific* charged membrane for a *specific* application. In general, charged membranes may exhibit the removal of particulate matter, organic material, colloidal material, and, to some extent, bacteria from the feedwater.
- As previously indicated, prefiltration systems should be included with either chemical or hot water sanitization provisions. Subsequently, for most applications, it is suggested that stainless steel filter housings be considered, rather than plastic housings that limit the sanitization alternatives to chemical agents.
- For recirculation RO/polishing component applications, nonbacteria retentive RO prefilters may be positioned within the pretreatment system. In a recirculation mode, the chemical and microbial quality of the RO break tank water improves with multiple "passes" through the system. If prefilters are installed prior to the RO unit for a hot water sanitizable RO/polishing component application, the filter material, support material, and seal material must all be capable of exposure to sanitization temperature as well as the thermal cycling process.

Reverse Osmosis Unit Sampling Provisions

The sampling provisions for an RO unit should allow operating personnel to determine that the unit is operating properly; avoid frequent unscheduled maintenance operations, such as

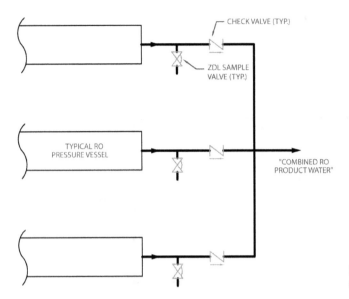

Figure 4A.15 RO pressure vessel sample requirements.

chemical sanitization for removing microbial fouling; and evaluate system parameters during excursions. Sampling provisions may include the following:

- Conductivity, TOC, pH, bacteria, and bacterial endotoxin monitoring for feedwater, product water, and wastewater.
- Provisions for SDI measurement positioned downstream of the prefilters to the RO system, providing a representative sample of RO unit feedwater. If permeate and/or waste recycle is used, the sample location for SDI determination should be positioned downstream of the recycle lines.
- For systems with multiple pressure vessels, sampling provisions should be included for each pressure vessel as shown in Figure 4A.15. For pressure vessels containing multiple RO membranes, provision may be included for "probing," a technique that allows insertion of a sample hose/tube "down" the permeate collector in an attempt to identify one or more membranes that have lost integrity or interconnector O-ring failure.
- The sampling provisions in the product water line from an RO unit should consider the potential "breach" of system integrity, specifically the introduction of bacteria to the clean side of the RO membranes. This is of particular concern since bacteria will replicate on the clean side of the membranes. While probing provisions are useful, their benefit may be offset by introduction of bacteria. Post probing chemical or hot water sanitization, if available, is strongly recommended.
- The RO system should include provisions for chemical sanitization. As indicated earlier. It is suggested that periodic membrane "rotation" be coupled with a contracted off-site RO membrane-cleaning program in lieu of performing a three-step cleaning/sanitization program in-place. Off-site cleaning of RO membranes can be customized for an application, offers "individual" cleaning as opposed to cleaning with waste from an upstream membrane, and minimizes downtime. While chemical or hot water sanitization should be performed subsequent to installation of "rotated" cleaned RO membranes, the low and high labor-intensive cleaning steps with rinse are not required.
- If on-site chemical cleaning is performed, suggested cleaning accessories, as discussed later, should include a tank, heater, pump, and filter. The strength of the cleaning agent is important. Subsequently, sampling provisions should be included on the sampling tank. This is particularly important for removing calcium carbonate

ION REMOVAL TECHNIQUES—REVERSE OSMOSIS

scalants with acid, where pH control is vital to ensuring that the scale is removed from the membranes.

High-Pressure Reverse Osmosis Feedwater Pump

The selection of the RO high-pressure feedwater pump is extremely important. The accessories required for the pump are also important to the successful operation of the system. Items that should be considered in selecting the pump, motor, and accessories include, but are not limited to, the following:

- In general, the vast majority of feedwater pumps employed for pharmaceutical water RO applications are multistage centrifugal type. The use of alternative positive displacement pumps, with the exception of extremely low flow applications, is discouraged. A positive displacement pump cannot be throttled. Subsequently, there is a great deal of inflexibility with regard to the feedwater pressure regulation required for operating the RO membranes throughout their expected life. In other words, since the pump discharge flow and pressure are fixed, the higher pressure required for an established permeate flow rate merely results in a decrease in the permeate flow rate for the fixed discharge pressure. In addition, experience indicates that maintenance requirements for smaller positive displacement pumps are extensive. The pumps will only operate without water for a very limited period of time, generally less than 5 to 10 minutes. Further, the impellers are often constructed of a plastic material that can release small fragments of impeller material to the lead RO element. If a positive displacement pump *must* be used, it is suggested that the unit be provided with a stainless steel housing and impeller. A bronze housing pump with a plastic or bronze impeller should be avoided.
- Operation of the RO unit should be initiated by starting the pump. For suggested recirculating RO/polishing component systems, this process is minimized. System design *must* ensure that any valves installed in the permeate line are either in an open mode before the RO feedwater pump begins operation. The back pressure exerted by a supply valve to a storage tank installed in the permeate piping/tubing, if not open before the pump begins operation, will irreversibly damage the RO membranes in a matter of minutes.
- It is suggested that pump selection/sizing be established on the basis of the feedwater flow rate and pressure requirements for the three-year computerized RO projection. The use of a variable frequency drive motor is strongly suggested since it provides a wide range of flow/pressure conditions and minimizes discharge throttling valve adjustments with chances in water temperature, feedwater conductivity, recirculation, etc. The pump discharge pressure (at the feedwater flow rate for the three-year period) should be at least 20% higher than the projected three-year value. Excess pressure, specifically from a multistage centrifugal pump, can be offset by throttling the discharge of the pump or preferably by automatically decreasing the electrical cycles and subsequent rotating speed of the pump. Table 4A.7 provides

Table 4A.7 Multistage Centrifugal Performance Data at Various Pump Motor Speeds

Flow rate (gpm)	Pressure (psig)	Pump speed (rpm)
160	260	3500
150	228	3281
140	199	3062
130	172	2844
120	147	2625
110	123	2406
100	102	2188

Note: Calculated values.

operating data (flow rate and pressure) for a multistage centrifugal pump at various speeds.
- Multistage RO feedwater pumps are available in vertical configuration and horizontal physical configuration, or as submersible units. The literature contains information discussing the advantages and disadvantages of each pump design (Gessner and Wolff, 1997). Vertical cylindrical multistage centrifugal pumps are highly reliable. The sole limitation is the ability to prime the pump upon loss of flow. Horizontal pumps are also reliable. However, as the number of stages increases, the "barrel" of the pump lengthens potentially placing pressure on the pump seal mechanism associated with the weight of the pump and water. Some commercially available horizontal multistage centrifugal pumps use plastic impellers, as opposed to stainless steel impellers, that can release particulate fragments to the lead RO membrane. The advantage of using multistage centrifugal pumps versus single stage centrifugal pumps relates to the required motor size to obtain the desired flow rate and pressure.
- The pump should be provided with a feedwater pressure switch that will inhibit pump operation on low feedwater pressure. This condition should also activate an audible alarm and an indicating light. Low feedwater pressure can cause pump cavitation and eventual pump failure. Further, under worst-case conditions, air may be present in the feedwater piping during low feedwater pressure conditions, resulting in two-phase flow with water hammer, which may damage the RO membranes. The control system associated with the low pump feedwater pressure switch should include a "time delay" that will inhibit rapid, repetitive cycling of the pump. This condition will also damage the downstream RO membranes. For recirculating RO/polishing component systems, a booster pump, positioned immediately downstream of the RO break tank, provides adequate pressure to avoid the low feedwater condition.
- Many booster pumps are equipped with a "hands-off-auto" switch on the face of the control panel. In the "auto" position, the pump is protected by the low feedwater pressure switch, generally through a central control panel. Unfortunately, operating personnel may try to defeat the purpose of the pressure switch, particularly during transient conditions. This generally results in rapid cycling of the pump, similar to the situation described in the preceding text. Again, the time delay, indicated in the preceding text, should be included within the software for the central control panel processor when the pump is in the "manual" position.
- For multistage centrifugal pumps without variable frequency motor drives controlled through a central panel processor, the pump curve may be such that the pressure will vary considerably with flow rate. Multistage centrifugal pump curves are relatively "flat." To protect the RO membranes and pressure vessels, it is recommended that a high pump discharge pressure sensor/switch be included with an automatic shutoff and alarm. A sensor is required for feedwater pumps with variable frequency drive motors to control pump discharge pressure/flow rate. A high-pressure switch should be included for RO units with feedwater pump motors without variable frequency drive. The sensor/switch would also eliminate potential over pressurization associated with decreased permeate and/or waste flow during normal operation.
- The feedwater line to the RO feedwater pump should include a temperature switch or temperature sensor with transmitter that inhibits operation of the pump at a high feedwater temperature to avoid membrane damage. In general, most membranes available or installed will not tolerate temperatures higher than about 113°F (DOW, 2010b). However, feedwater total viable bacteria levels increase with increasing temperature. RO feedwater heating provisions may malfunction resulting in elevated feedwater temperature. While feedwater heating is discouraged, all RO systems should be equipped with temperature-monitoring provisions and/or high temperature switch. As indicated previously, any situation that trips off the RO feedwater pump should be provided with controls including a time delay.

- The RO feedwater pump discharge piping/tubing should be equipped with a throttling valve. The throttling valve may be of modified needle type, radial diaphragm-type, ball-type, or other suitable configuration. Generally, a stainless steel ball valve may be used as a throttling valve, particularly when the upstream RO feedwater pump motor is equipped with variable frequency drive provisions since "fine" flow rate adjustment is not required by the valve.
- The feedwater and product water lines to and from the RO feedwater pump should be equipped with pressure gauges, with a suitable face diameter (e.g., 3–5 in.), manufactured of stainless steel, liquid filled, with stainless steel diaphragm isolators. Each pressure gauge should be accurate enough to allow operating personnel to determine pressure within ± 2 to 5 psig.
- The piping/tubing to the RO feedwater pump, in addition to including the indicated temperature sensor/switch, should also include a direct reading temperature gauge. If the temperature sensor is equipped with indicating provisions it is preferred in place of the gauge. If a temperature switch is used, the temperature gauge can also be used to verify proper function of the switch.
- The piping/tubing to the RO feedwater pump, downstream of any recycle lines, should be equipped with a conductivity cell. A second conductivity cell should be installed in the RO permeate tubing. The cells should be connected to a single conductivity indicator on the face of the central control panel. An analog signal may be transferred from the conductivity meter to the central control panel processor. It is suggested that the temperature of the water at each cell location can also be transferred as an analog signal. The conductivity values at each point as well as the percent rejection of ions can be displayed on a graphic display of the RO unit.
- Waste and/or permeate recycle may be provided upstream of the RO feedwater pump. Design provisions should ensure that the monitoring features, described in the preceding text are considered to measure conditions created by waste or product recycle.

Reverse Osmosis Membranes

There are a number of RO membrane element manufacturers in the United States. However, it is estimated that 90% of the RO membranes used for pharmaceutical applications are provided by about five membrane manufacturers. Most of these membrane manufacturers have highly aggressive research programs and, more importantly, provide computer programs for establishing the design of RO systems (number of membranes, configuration, and array) for a given application. Recent advances associated with the design and manufacturing of RO membranes, as discussed previously, have produced membranes with outstanding rejection properties. Over the past 10 years, significant research and commercialization has been developed to improve cleaning, particularly for microbial fouling. Hot water sanitizable RO membranes and the availability of pharmaceutical grade full fit RO membranes have greatly improved the ability to provide excellent microbial control. The parameters associated with RO membrane construction, including limitations associated with the materials used to manufacture the membranes, were presented in section "Theory and Application." Subsequently, an RO membrane purchased from one of the companies providing the majority of membranes for pharmaceutical applications would be adequate for a given application. However, a person designing a system may want to review the recent advances associated with membrane cleaning and hot water sanitization, which is discussed in manufacturer's information. While ion rejection rates are generally stated as more than 99%, the presence of carbon dioxide or ammonia, reactive gases, will result in lower actual ionic rejection for most operating systems. For RO/polishing component recirculating systems, the observed ionic rejection after recirculation for an extended time period (>1–2 hours) may actually decrease even though product RO water conductivity continues to decrease. The presence of the small ionic radius sodium ion will ultimately limit the ability of a conventional single-pass RO unit to produce theoretically "ion-free" product water.

Reverse Osmosis Pressure Vessels

The RO pressure vessel may be constructed of a fiberglass (filament wound epoxy/glass composite with high gloss polyurethane paint on the exterior) material or stainless steel. The selection of material, in general, will have little effect on the performance of the system. Some pharmaceutical manufacturers prefer stainless steel pressure vessels since they enhance the appearance of the unit.

The pressure vessels selected for an application should meet the maximum anticipated pressure requirements, dictated by the three-year RO system projection, with a safety margin factor of at least 2 to 3. Pressure vessels design and construction should be in accordance with the ASME Code for "Unfired Pressure Vessels" and may be code stamped. Vessels can be supplied with a range of pressure ratings: 300, 450, 600, 1000, and 1200 psig. It is suggested that 450 or 600 psig pressure-rated vessels are appropriate for pharmaceutical applications. Pressure vessel manufacturer's literature should contain adequate information for preparation of the Installation Qualification (IQ) for the RO system.

The pressure vessel selected for an application should be capable of using membranes from the "leading" RO membrane manufacturers. This may require the use of adapters and/or special interconnectors. However, the vessels should not be unique to a particular RO membrane manufacturer's element, limiting the potential selection of membranes to a single manufacturer. As RO membrane research continues, it is quite possible that a "different" manufacturer may commercialize a membrane with unique properties. If this membrane does not physically fit into a "specialized" pressure vessel, the pharmaceutical manufacturing firm cannot take full advantage of the technological advancements.

Pressure vessel closure systems should be selected such that a positive seal is obtained. Generally, a stainless steel grove at each end of a pressure vessel and "spiral-type" retaining ring are used to achieve the desired high-pressure seal. The use of a "split-type" ring is discouraged because of safety concerns. For hot water sanitizable RO units, interconnector, end adapter, and seal materials of construction must be compatible with the high temperature application.

Reverse Osmosis Unit Piping/Tubing

Many RO units are provided with stainless steel feedwater and product water piping/tubing, as well as a high-pressure section of stainless steel waste piping. It is suggested that a properly designed RO unit use stainless steel piping or tubing for all feedwater, product water, and waste piping/tubing. This is particularly important for recirculation RO/polishing component systems where feedwater will be in contact with high-purity water during recirculation. It is further suggested that the feedwater and product water lines be constructed of 316L sanitary stainless steel tubing. This eliminates threaded connection and nonsanitary components that will generally be dead legs for a thermally sanitized system and will be difficult to effectively chemically sanitize. The pressure rating for the piping/tubing should provide a safety margin factor of at least 2 above the maximum anticipated operating pressure of the system. For systems employing 316L stainless steel tubing using sanitary ferrules, the use of high-pressure clamps should be considered downstream of the RO feedwater pump through high-pressure RO pressure vessel connections. Unfortunately, many low product water flow rate RO units contain PVC piping and flexible plastic tubing. On the basis of experience, perhaps the only reason the use of this material has not resulted in personal injury to operating personnel is that hydraulic pressure is relieved very quickly with limited flow of water. It is strongly suggested that units using PVC piping and various types of plastic tubing be operated at a pressure less than about 60% of the maximum pressure. Further, high-pressure lines using plastic components should contain a nonisolating positive means of mechanical relief for pressure.

Many standard units use plastic tubing for high-pressure connections on feedwater, product water, and even waste lines. It is suggested that the "additional cost" to purchase or convert the plastic piping to compression type stainless steel tubing on feedwater and high-pressure waste piping is extremely small when one considers the potential consequences to personnel safety. It is further suggested that the use of plastic tubing on the *product* water line from a low flow rate unit will inevitably result in microbial contamination of the "clean side" of an RO membrane. Alternatively, sanitary polypropylene piping or tubing should be considered for these applications.

Reverse Osmosis Unit Integral Cleaning System

As previously discussed, a responsive controlled off-site membrane chemical cleaning is encouraged. However, if three-step on-site cleaning is desired several design factors should be considered. Specific items associated with the design of the cleaning system may include the following:

- The cleaning system piping/tubing should be designed to allow operating personnel to clean each array separately, increasing the effectiveness of the cleaning operation.
- The cleaning pump capacity (pressure and flow rate) should be specified for cleaning the first array of the system, which contains the greatest number of membranes.
- The cleaning piping/tubing should be arranged, as indicated, to allow cleaning of separate arrays. For example, assume that an RO unit is arranged in a 2:1 array, the cleaning system, should be capable of cleaning half of the first-stage membranes at one time.
- The cleaning tank should be properly sized for the cleaning application and include heating provisions, as discussed previously. While it is suggested that electrical heaters in plastic tanks are questionable from a safety standpoint, many RO unit-cleaning systems include this type of heating provision. It is strongly suggested that composite type tanks, using fiberglass-reinforced polyester or vinylester, may provide an alternative to polyethylene tanks for the cleaning operation. This would eliminate concerns associated with the use and physical support of electrical heating elements through the relatively thin wall of a polyethylene storage tank.
- The cleaning system piping/tubing should be configured to allow cleaning of the lead elements in the reverse flow direction. This is important, since the flow path will allow removal of material beneath feedwater spacers in the lead elements. Historically, this area has been referred to as the "shadow side" of the feedwater spacers. During normal operation and cleaning in the direction of normal operation, the "shadow" area will never be effectively exposed to the cleaning agent. However, by reversing the direction of flow, the "shadowed" area is exposed to the cleaning agents. Scalants or foulants beneath the feedwater spacer, in the area of concern, will be removed.

Reverse Osmosis Monitoring

The RO unit, depending on capacity (product water flow rate) and application, should contain critical instrumentation for monitoring functionality and performance. Suggested monitoring functions should include, but are not limited to, the following:

- Feedwater pressure to RO Pump
- Discharge pressure from RO pump
- Feedwater pressure to first array
- Intra pressure vessel (array) pressure
- Reject pressure
- Permeate pressure
- Feedwater flow rate
- Permeate flow rate (sanitary or by difference of feedwater and waste/reject flow rate)
- Reject water flow rate
- Feedwater pH (optional)
- Feedwater oxidation-reduction potential (ORP) (optional, generally used with system employing injection of reducing agent for removal of residual disinfectant)
- Feedwater temperature
- Product water temperature
- Feedwater conductivity

- Product water conductivity
- Percent rejection of ions
- Waste recycle flow rate
- Operating speed of RO pump with variable frequency drive
- RO pump status (on/off)

Accessories

Many accessories can be considered during the design of RO systems, including the following:

- For RO system that do not contain continuous flow provisions (recirculating RO/polishing component systems) product water conductivity monitoring system should operate in conjunction with a "divert-to-waste" system. This system should automatically divert water with conductivity greater than a preset value to waste, eliminating potential contamination of the downstream storage tank with below quality product water. For continuous recirculating systems an alarm should sound and be displayed if the conductivity increases above a preset value.
- It is imperative that the routine waste piping/tubing from the RO unit be physically separated from the product water rinse-to-drain piping/tubing. Each line should be directed to a depressurized drain. Segregation of the two lines is required to eliminate microbial back contamination of the rinse-to-drain line from the product water piping/tubing.
- If the product water piping/tubing to downstream components are equipped with automatic shutoff valve, the system should be designed such that either the rinse-to-drain valve or the product water valve to the downstream components is open whenever the RO unit pump is energized. Failure to provide unrestricted product water flow will result in rapid irreversible damage to the RO membranes.
- In an attempt to minimize the potential effects of back pressure on RO membranes, the product water line from each pressure vessel should be equipped with a check valve. The check valve should be of stainless steel construction. A positive acting check valve, such as a spring-loaded ball check valve, may be considered. As discussed earlier, the check valve, coupled with a sample valve, provides a means of sampling the product water conductivity from each pressure vessel.
- The waste line should be equipped with a regulating valve to allow operating personnel to adjust the waste flow rate to a preset value that is based on the computerized projection for estimated recovery of water. For continuously operating systems, three valves configured in parallel are employed as shown in Figure 4.12.
- RO cleaning system, if employed, should include a cleaning tank, a centrifugal pump, heaters, a cartridge filtration system, appropriate piping/tubing, and fittings. The cleaning tank volume should be sized on the basis of the volume of the maximum number of membranes and pressure vessels that will be cleaned. The cleaning temperature should be approximately 30°C.
- In certain systems, single pretreatment components are provided, such as a single multimedia filter and/or an activated carbon unit. At least two water softening units are generally included as part of the pretreatment system, considering the length of the regeneration cycle. Controls for these systems may contain provisions to "lock out" operation of the RO unit when single pretreatment components are in a backwash mode. It is inappropriate to rely on an "emergency/transient" monitor, the pressure switch (or pressure sensing element) installed in the feedwater line to the RO feedwater pump, to provide this function. For continuous operating RO/polishing component systems, the makeup water flow from the system may be terminated during backwash with the system in a recirculating condition (RO unit high recovery mode). The RO break tank volume should be adequate to allow uninterrupted recirculation.
- A sanitary type pressure sensor should be installed in the product water tubing from the RO unit. This pressure sensor with transmitter (through the central control panel)

will detect high RO product water pressure, indicating gross failure of the RO membranes. High product water pressure should activate an audible alarm and displayed indication.
- The product water rinse-to-drain cycle should be initiated upon shutdown and start-up of the RO unit that is *not* operated in a continuous mode. Provisions may also be included to start the RO unit automatically, in a rinse-to-drain mode, if the unit has not been operational for a preestablished time period. It is suggested that an RO system should operate for at least 15 minutes every 2 hours when continuous operation is not used.
- The cleaning system for an RO unit should be designed to use permeate for mixing cleaning chemicals.
- The product water line from the RO unit, for noncontinuous operation, should be designed to avoid a "solid" water condition subsequent to RO unit shutdown. The solid water condition, particularly when increase in water temperature will occur, will increase pressure on the product side of the RO membranes.

OPERATING AND MAINTENANCE CONSIDERATIONS
General

To reduce microbial fouling of RO membranes, microbial control in the pretreatment is required. The microbial control program for the pretreatment system should include periodic sanitization of piping and components, as appropriate.

Activated carbon media used for any application preceding RO membranes should be acid washed and prerinsed. The chemical constituents present in the prewashed media should be determined. Subsequent to acid washing of activated carbon to remove naturally present undesirable multivalent cations such as aluminum, suppliers generally conduct a rinse with neutralizing basic solution. Rinsing of the acid washed material must be performed. Experience indicates that pH excursions may occur if the activated carbon product water pH is not verified prior to placing a unit into operation. The specified rinse pH requirement for product from acid-washed, neutralized, and rinsed activated carbon is suggested as more than 6 but less than 8.

Prefilter cartridges to the RO system should be changed on the basis of differential pressure and/or elapsed time. The established differential pressure value for replacement is based on particulate loading. However, even if the maximum particulate loading has not been reached, the filters should be changed periodically (e.g., once every other week maximum) to avoid microbial accumulation on the filters, with subsequent bacterial introduction to the downstream RO system.

System design should be such that the maximum operating temperature for the unit is less than 95 to 113°F. Data should be maintained to demonstrate that the system is not operating at the higher temperatures, which are in the incubation range for bacteria, resulting in high product water total viable bacteria levels and increased hot water and/or chemical sanitization frequency.

An established preventive maintenance program should include the following:

- Periodic membrane replacement with rotated off-site cleaned membranes (every 6 months)
- Periodic membrane sanitization (every 2–4 weeks for hot water sanitizable membranes and every 3–6 months for chemically sanitized membranes)
- Periodic prefilter replacement (every 1–2 weeks)
- Periodic regeneration and associated maintenance for the upstream water softening units (where applicable)
- Maintenance of the chemical feed systems (where applicable)
- Calibration of critical instrumentation (every 6–12 months)
- Replacement of interconnector O-rings (every 12 months)

- Replacement of end adapter O-rings (every 12 months)
- Replacement of diaphragms in diaphragm-type valves, such as sample valves (every 1–2 years)
- Replacement of rotated RO membranes (every 3 years)

To assist in the evaluation of system performance, it is beneficial to examine the "exhausted" RO unit prefilters. The nature of the contaminants present and the "color" of the membrane surface may provide an indication of excessive contaminants. For example, a red to brown color may be associated with the presence of iron. A yellow to brown color may be associated with organic material, potentially complexed with colloidal material.

Periodic SDI analysis, using will also provide valuable information. Examination of the filter disk can identify impurities similar to those discussed in the preceding text. Finally, visual inspection of RO membranes upon removal and destructive membrane autopsy can provide additional information regarding the design, operation, and maintenance of pretreatment components.

Specific Monitoring and Trending Items

Operating personnel should perform an SDI measurement at least once each week, or as frequently as daily, particularly when raw feedwater characteristics are changing. The frequency of the SDI determination should be consistent with the RO element manufacturer's recommendation *and* the warranty provided by the RO system manufacturer.

Adequate feedwater data for the RO unit should be obtained from analytical results to determine the LSI value. It is suggested that LSI values be compiled monthly or at a frequency that allows operating personnel to determine if potential scaling of RO membranes will occur. The frequency of SDI measurement is generally greater for a system with raw water from a surface supply or groundwater supply under the influence of a surface supply while the frequency of LSI determination may be greater for feedwater from a groundwater supply.

Larger systems may be equipped with residual disinfecting agent analyzers. Systems using a reducing agent for removal of residual disinfecting agent use an in-line ORP analyzer. Generally, the analyzers are installed directly in the feed line to the RO unit (side stream), verifying that residual chlorine or chloramines are not present. As discussed in chapter 3, ORP analyzers are a critical accessory in systems using a reducing agent to remove residual disinfectant prior to an RO system containing thin-film composite polyamide membranes.

Data trending is extremely important. As a minimum, the percent rejection of ions, the normalized product water flow rate, and the feed-to-reject pressure drop should be logged and recorded daily (once each shift for larger systems) and plotted as a function of time. This may also be achieved by the use of sensors and transmission of an analog signal to the central control panel processor with input/output provisions to a centralized data collection system.

The RO feedwater pump should be properly maintained. Monitoring of booster pump feedwater and product water pressure, as well as the feedwater flow rate, should be monitored continuously and recorded daily (system operating log). Pump seals should be replaced as part of a periodic maintenance program. Air should be bled from vertical multistage centrifugal pumps whenever the feedwater piping is drained or if air accumulation is suspected. Since vertical multistage centrifugal pumps have pump seals at the high-pressure side of the multiple casing arrangement, the pump seal maintenance is important. For small-capacity systems using positive displacement pumps, pulsation dampers should be installed downstream of the pumps. The pulsation dampers should be inspected periodically, by observing the downstream pressure, to verify that they are performing appropriately, reducing or eliminating rapid, frequent changes in RO feedwater pressure.

Reverse Osmosis Cleaning Considerations

Again, while a responsive "controlled" off-site cleaning program with rotation of RO membranes is strongly recommended, on-site cleaning programs should consider several factors. A definitive three-step program must be performed. The sequence should introduce a low-pH agent to remove scale and some colloidal material, a high-pH solution to remove organics and complex organic material, and a sanitizing agent to remove bacteria and biofilm from the surface of the membrane.

Cleaning should be considered when the *normalized* product water flow rate decreases by ≥10%. Cleaning should never be delayed to the point where the normalized product water flow rate decreases by 15% or greater. Cleaning should also be considered if the feedwater pressure, corrected for temperature, increases by 10% and when the permeate quality decreases by 10% to 15% (increase in salt passage of 10–15%).

When RO element cleaning is performed, particularly to remove microorganisms, the lead elements within the RO membrane array may be cleaned in the reverse direction. Biofilm forms directly on the membrane surface. Feed channel spacers contribute to biofilm formation by "shadowing," as discussed earlier. The only effective method of ensuring that all surfaces of the membrane are exposed to sanitizing agent, particularly for the lead membranes, is to clean in a reverse direction. While this can readily be performed when off-site cleaning is employed, it is much more difficult to perform in-place.

As discussed previously, cleaning to remove scalants, specifically calcium carbonate, should be conducted with a low-pH cleaner that will release carbon dioxide to the cleaning tank. The release of carbon dioxide is associated with "foaming" in the cleaning tank and, more importantly, an increase in the pH of the cleaning solution. During the cleaning operation to remove calcium carbonate, it is important to readjust the pH value periodically, by introducing additional cleaning agent, to maintain the pH value at approximately 4.5 (or value specified for the selected low-pH cleaning agent).

Subsequent to rinsing the low-pH cleaning agent from the membranes, a high-pH cleaning should be preformed. The high-pH cleaning removes foulants such as organic material and colloidal material complexed with organic material. A light yellow to brown color may appear in the cleaning tank from removed naturally occurring organic material. If the color of the water in the cleaning tank becomes dark yellow or brown, it is suggested that the tank be drained and refilled with fresh high-pH cleaning solution. Subsequent to completion of this cleaning step, a thorough rinse should be performed.

Sanitization should be conducted, as required, using a 1% solution peracetic acid and hydrogen peroxide (Collentro, 2010a), at temperature of 25°C and RO feedwater pressure of about 40 psig. Chemical sanitization should only be performed after low-pH and high-pH cleaning *or* subsequent to installation of rotated membranes for continuously recirculating systems. The chemical sanitization operation should be performed using a multiple-step process outlined as follows:

- The 1% peracetic acid/hydrogen peroxide-sanitizing agent solution should be prepared in the cleaning tank.
- The solution should be fed to the RO unit, or feedwater piping/tubing of the RO unit, at a pressure adequate to achieve product flow but no higher than about 40 to 45 psig. If downstream polishing components are included in the sanitization sequence, flow should be established through the components.
- The 1% concentration of sanitizing agent should be verified at sampling points throughout the system using "test strips." For the RO unit, this should include feedwater, product water, wastewater, and individual pressure vessel sample valves if available.
- Once the 1% sanitizing agent concentration has been verified, allow the sanitizing agent to circulate back to the cleaning tank for a time period of 15 to 20 minutes.
- Terminate electrical power to the cleaning pump and allow the 1% sanitizing solution to remain in the RO unit for a time period of two to four hours. As discussed further in chapter 9, unlike a hot water–sanitized system where a "dead leg" can be defined as a multiple of "pipe diameters," any stagnant area is a dead leg in a chemically sanitized system. During the dynamic recirculation step of the sanitization operation, sanitizing solution passes by the dead legs without destroying bacteria or oxidizing biofilm constituents such as bacterial endotoxins. During stagnant conditions, diffusion of sanitizing agent, a process driven by concentration difference, will result in destruction of bacteria in both dead legs and biofilm and oxidize

constituents in the biofilm. Experience indicates that this lengthily but important step is required for effective sanitization including biofilm removal (Collentro, 2010a).
- After the stagnant period, the electrical power to the cleaning pump is restored. The sanitizing agent is recirculated for about 15 to 20 minutes.
- The power to the pump is terminated, the cleaning tank drained, and filled with pretreated water or RO permeate, if available. The chemical sanitizing solution is "displaced" from the RO unit until test strips and conductivity measurements verify that it has been removed.
- The RO unit is returned to normal operation.

The effectiveness of a cleaning program should be verified by analysis including total viable bacteria and quantitative bacterial endotoxin measurements.

Troubleshooting
An effective troubleshooting program can be established if a log (or data from a central facility collection system) is maintained of critical RO operating parameters. The log, used to trend data, can identify the following situations:

- *Low transmembrane pressure:* Low transmembrane pressure may be associated with an improperly set reject valve, an improperly set RO feedwater pump discharge valve, a decrease in pretreated water pressure (resulting in a decrease in the RO feedwater pump discharge pressure), and failure of the booster pump.
- *Observation of feed-to-product bypass:* This situation is associated with the failure of interconnector or end adapter O-rings used to seal the product stream from the feed stream and failure of the RO elements at the "glue" line.

This situation may also be associated with interconnector or end adapter O-ring failure. During RO membrane replacement, it is possible to "turnover" an O-ring during installation. It is suggested that USP Glycerin be applied on every O-ring seal. Membrane interconnectors and end adapters should not be physically "forced" into an engagement position. Both interconnectors and end adapters should be carefully inspected for cracks or other defects that will result in feed-to-product bypass.

- *Observation of feed-to-reject bypass (decrease in salt rejection):* This situation can be the result of improperly seated or damaged brine seals. In general, it is difficult to determine this condition because most parameters will not vary to a significant (measurable) extent. This situation may result in higher membrane rejection, with increased concentration polarization and membrane fouling. The results may be noted after an appreciable period of time. This would be associated with an increase in the cleaning frequency for the RO system. As indicated, conventional brackish water membranes, with brine seals, should *not* be used for pharmaceutical applications. The full fit nature of pharmaceutical "loose wrapped membrane eliminates this situation.
- *Element telescoping:* This condition is caused by high feed-to-reject pressure drop. It is associated with a high feedwater flow rate and/or excessive membrane fouling due to inadequate cleaning. This condition results in deformation of RO elements, allowing the membrane envelope to move physically in relation to adjacent membrane envelopes. This results in membrane surface damage and rupture of the "glue seals." Ultimately, this condition will require membrane replacement.

As indicated earlier, may RO units are standard design from an equipment supplier. To meet a specific product water flow rate requirement, each pressure vessel may not be filled with membranes. Again, while the use of standard products for applications is discouraged, pressure vessels containing "blanks" or "spacers" to adapt a standard product to a specific capacity may encourage telescoping. This is a direct function of the "blank/spacer" employed.

Experience indicates that a section of tubing with O-ring end adapters may be used for this application. The lack of an adjacent full physical surface area can result in telescoping.

- *Gradual increase in product water flow rate:* This situation may occur as a result of increased feedwater temperature, an increase in the transmembrane pressure drop, or a slight increase in the RO feedwater pump feed pressure.
- *Sudden increase in product water flow rate:* This situation is associated with catastrophic RO element failure. This failure may be a result of membrane damage, the failure of one or more O-rings, cracking of the end-cap adapters, and failure of a permeate collector "end plug."

SUMMARY

As noted earlier, there are many conditions that can result in RO unit malfunction. With the exception of cleaning requirements, most RO systems operate highly effectively, without problems. This is particularly true for units operating in a continuous flow mode. However, RO systems are similar to other unit operations in a water purification system. Maintenance is important. Data logging is also important. Perhaps this is summarized best by an article in the literature that states that most RO problems result from a failure to record data, a failure to analyze data, and a failure to respond after data analysis (Lueck, 1998).

REFERENCES

Allen V, Silvestri NJ, Fenton M. A comparison of double-pass RO treatment approaches. Ultrapure Water 1995; 12(3):22–29.
Amjad Z, Pugh J, Zuhl RW. Reverse osmosis element cleaning. Ultrapure Water 1996a; 13(7):27–32.
Amjad Z, Pugh J, Harn J. Antiscalants and dispersants in reverse osmosis systems. Ultrapure Water 1996b; 13(8):48–52.
Applegate LE. Membrane separation processes. Chem Eng 1984; 91:64–89.
Bartels CR. Designing membranes for specific needs. Ultrapure Water 1997; 14(3):43–50.
Bates W, Stier R. Recent developments in RO pretreatment technology. Ultrapure Water 1994; 11(2):20–27.
Collentro AW, Barnett SM. Predicting the performance of reverse osmosis membranes for the production of high-purity water. Ultrapure Water 1996; 13(7):40–46.
Collentro WV. Pharmaceutical Water System Fundamentals - Ion Removal - Reverse Osmosis. The Journal of Validation Technology. Duluth, MN: Institute of Validation Technology, Summer 2010a;16(3):66–75.
Collentro WV. Unpublished data 2010b.
Coulter SL. Organics, TOC, color, turbidity, and SDI pretreatment for unit operations. Ultrapure Water 1996; 13(7):54–59.
DOW Chemical Company, FilmTec^{TM} Membranes, Water Chemistry and Pretreatment: Biological Fouling Prevention, Form No. 609-02034-1004, 2010a:1–4.
DOW Chemical Company, FilmTec^{TM} Membranes, FilmTec Heat Sanitizable RO Elements, Form No. 609-00215-0408, 2010b:1–2.
Dudley LY, Fazel M. Crossflow studies to evaluate cleaning programs. Ultrapure Water 1997; 14(2):49–55.
Gessner TC, Wolff EM. Selecting a pump for membrane system. Ultrapure Water 1997; 14(4):56–60.
Finan MA, Tracey DA. Selection of a novel multifunctional antifoulant. Ultrapure Water 1995; 12(3):61–65.
Hamilton R, Drummonds D. Reverse osmosis versus ion exchange—part 1. Ultrapure Water 1994; 11(7): 22–32.
Harfst WF. Pretreatment requirements for reverse osmosis systems. Ultrapure Water 1994; 11(8):42–44.
Harfst WF. Types of water treatment membranes. Ultrapure Water 1995; 12(7):34–38.
Husted GR. Biocides for thin-film composite RO elements. Ultrapure Water 1998; 15(3):29–30.
Kaakinen JW, Moody C, Franklin J, et al. SDI instrumentation to estimate RO feedwater fouling potential. Ultrapure Water 1994; 11(5):42–54.
Kemmer FN. The Nalco Water Handbook. 2nd ed. (Nalco Chemical Company). New York: McGraw-Hill Book Company, 1988:4.15–4.17.
Kronmiller D. RO permeate water flux enhancement. Ultrapure Water 1993; 10(2):37–40.
Loeb S, Sourirajan S. Sea water demineralization by means of an osmotic membrane. Saline Water Conversion II: AdvChem 1963; 38:117–132.
Lueck S. Computerized data acquisition and reverse osmosis. Ultrapure Water 1998; 15(3):37–40.
Mazzola P, Martins A, Penna T. Identification of bacteria in drinking and purified water during the monitoring of a typical water purification system. BMC Public Health 2002; 2(13):1–11.

Mazzola P, Martins A, Penna T. Chemical resistance of the gram-negative bacteria to different sanitizers in a water purification system. BMC Infect Dis 2006; 6(131):1–11.

McBride D, Mukhopadhyay D. Higher water recovery and solute rejection through a new RO process. Ultrapure Water 1997; 14(5):24–29.

McClellan SA. Membrane process technology basics—nanofiltration. Ultrapure Water 1995; 12(7):39–46.

Murphy AP, Moody CD. Deterioration of cellulose acetate by iron salts, oxygen, and organics. Ultrapure Water 1997; 14(l):19–22.

Nickeson G, Mclain W, Bukay M. Investigation of severe flux decline in a two-pass CA/thin-film composite membrane RO system. Ultrapure Water 1994; 11(4):26–37.

Ning RY, Stith D. The iron silica and organic polymer. Ultrapure Water 1997; 14(3):30–33.

Parise PL. RO system troubleshooting: Diagnoses and remedies—part 1. Ultrapure Water 1996a; 13(8):21–24.

Parise PL. RO system troubleshooting: Diagnoses and remedies—part 2. Ultrapure Water 1996b; 13(9):54–60.

Paul DH. Obstacles to the effective chemical cleaning of a reverse osmosis unit. Ultrapure Water 1994; 11(7):33–38.

Paul DH. Biofouling of reverse osmosis units. Ultrapure Water 1996; 13(4):64–67.

Paul DH. A review of membrane water treatment technologies. Ultrapure Water 1997; 14(3):39–42.

Permutit Co. Water and Waste Treatment Data Book. 15th printing. Publication No. 24781-7M-2/86. Paramus, NJ, USA: The Permutit Company, 1986.

Pontius FW. Water quality and treatment—A handbook of community water supplies. 4th ed. (American Water Works Association). New York: McGraw-Hill Book Co., 1990:1074–1091.

Ridgway HF. Microbial fouling of reverse osmosis membranes: Genesis and control. In: Mittelman MW, Geesey GG, eds. Biological Fouling of Industrial Water Systems. San Diego, California, USA: Water Micro Associates, 1987; 138–193.

Schneider BM. Part 1—nanofiltration compared to other softening processes. Ultrapure Water 1994a; 11(7):65–74.

Schneider BM. Softening process comparisons: Degree of softening and types of ions removed. Ultrapure Water 1994b; 11(8):22–31.

Singh R. A review of membrane technologies: Reverse osmosis, nanofiltration, and ultrafiltration. Ultrapure Water 1997; 14(3):21–29.

Tracey D. Membrane fouling—What is it? Where does it come from? and What does it mean? Ultrapure Water 1996; 13(7):47–53.

Webb WG, Paul DH. Surface water reverse osmosis system biofouling. Ultrapure Water 1994; 11(8):36–40.

4B | Ion removal techniques—ion exchange

INTRODUCTION

Ion exchange, as an ion removal technique, has historically played a very important part in the production of pharmaceutical grades of water. While the use of reverse osmosis as a primary ion removal technique has increased dramatically over the last 20 years, certain applications still employ the use of ion exchange. Water softening, a pretreatment technique to reverse osmosis (see chap. 3), or applications requiring softened water for direct use in product, such as many active pharmaceutical ingredient applications, continue to rely on ion exchange technology. USP Purified Water systems with high water instantaneous flow rates (200–300 gpm and greater) for batching applications may also employ ion exchange to minimize storage volume requirements. Conversely, USP Purified Water systems that use minimal volumes of water each day (e.g., <500 gal) may rely on rechargeable ion exchange canisters, with minimal maintenance, thus eliminating the need for extensive pretreatment, periodic flushing, and membrane cleaning for a small RO-based system. This section of chapter 4 discusses ion exchange as a primary removal technique in pharmaceutical water systems.

DESCRIPTION AND CLASSIFICATION OF ION EXCHANGE RESINS

In general, ion exchange resin consists of spherical beads approximately 0.5 to 1.2 mm in diameter. While the color of the resin will vary, it is generally an opaque yellow. Resin, particularly anionic, will exhibit a characteristic amine type odor. Ion exchange resin is classified as either cation or anion. The cation resin contains functional sites capable of exchanging positive ions, while the anion resin contains functional sites capable of exchanging negative ions. Resin can be further characterized into four basic types: strong acid cation exchange resin, weak acid cation exchange resin, strong base anion exchange resin, and weak base anion exchange resin. Each type of resin may be produced in either a gelular or macroreticular (porous) form, discussed later in this chapter.

Strong Acid Cation Exchange Resin

Most strong acid cation exchange resins have a chemical structure consisting of styrene cross-linked with divinylbenzene. Sulfonic acid radicals provide the functional groups associated with ion exchange. A strong acid cation exchange resin, properly regenerated, is capable of removing all cations from a typical water supply.

Weak Acid Cation Exchange Resin

Weak acid cation exchange resin may consist of an acrylic acid–divinylbenzene matrix with a carboxylic functional group. The use of weak acid cation exchange resins is generally limited to unique situations where cation removal associated with alkalinity is important (Rohm and Haas Co., 1978; McGarvey, 1983). This may be appropriate for certain industrial applications where the feedwater source is from a groundwater supply that contains a significant percentage of alkalinity (bicarbonate ion) as the anion and hardness (calcium and magnesium) as the cations. In this situation, weak acid cation exchange resin may be used to remove the hardness associated with the alkalinity. The product water from this weak acid cation exchanger can be passed through a degasifier, removing carbon dioxide generated from the upset in the carbon dioxide-bicarbonate equilibrium reaction associated with the reduction in pH of the weak acid cation product water. The use of weak acid cation exchange in Purified Water systems should be limited to unique feedwater sources.

Strong Base Anion Exchange Resin

Classical strong base anion exchange resin has a chemical structure similar to that of strong acid cation exchange resin (styrene cross-linked with divinylbenzene). The functional group is based on the quaternary ammonium ion. In general, strong base anion exchange resins will remove all anions from a water stream when the feedwater has been processed through cation exchange resin. Strong base anion exchange resin is subdivided into two types: type 1 and type 2. A type 1 strong base anion exchange resin will remove all anions present in a water stream with the greatest efficiency. A type 2 strong base anion exchange resin will remove all anions with the exception of silica, where removal efficiency is less than that of a type 1 strong base anion exchange resin.

Weak Base Anion Exchange Resin

Weak base anion exchange resin has a chemical structure consisting of styrene and divinylbenzene, acrylic and divinylbenzene, and, in certain cases, epoxy material. A weak base anion exchange resin does not have the capability to remove ions that exist in a chemical equilibrium, such as bicarbonates and silica. These resins are effective at removing totally ionized substances, such as chlorides and sulfates. These resins also exhibit much higher exchange capacity than strong base anion resin. The use of weak base resin in Purified Water systems is determined by the raw water characteristics. Weak base resin may be used with a separate column of strong base anion resin, thus combining high capacity with the ability to remove totally ionized and partially ionized material.

Gelular and Macroreticular (Macroporous) Resins

Until the late 1960s and early 1970s, gelular resins were the only commercially available type of resins. These resins have a smooth surface with virtually no porosity (Rohm and Haas Co., 1975). If a gelular resin were sliced in half, a flat, nonporous, circular structure would be visible under a microscope (Fig. 4B.1). In an attempt to develop ion exchange resins exhibiting superior physical stability to gelular resins, ion exchange manufacturers introduced macroreticular resins, often referred to as macroporous resins. The exterior structure of macroreticular resin is porous (Rohm and Haas Co., 1979). When a macroreticular resin bead is sliced in half, the circular structure appears to contain numerous pores (holes) (Fig. 4B.2), which is associated with the method of manufacturing where agglomeration of numerous smaller particles has been achieved to obtain the porous structure. Macroreticular strong acid cation exchange resin may be considered for those USP Purified Water applications where oxidation of the cation resin is a concern. The degree of cross-linking for macroreticular cation resin is 2.5 to 3 times that of gelular strong acid cation resin. Consequently, the resin is much more resistant to oxidation by residual disinfecting agents, since it contains a much greater degree of cross-linking—the primary structural factor affected by oxidation. Many USP Purified Water systems utilize macroreticular strong base anion exchange resin. However, the advantages of

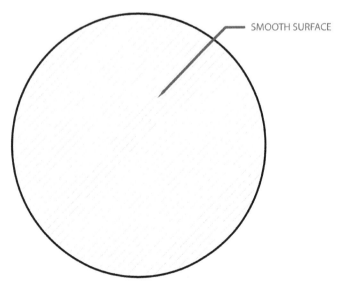

Figure 4B.1 Cross-sectional view of gelular resin bead.

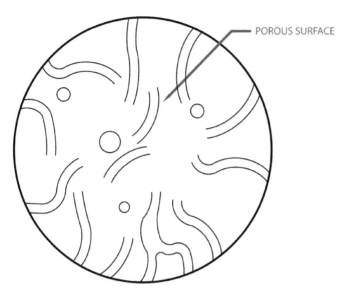

Figure 4B.2 Cross-sectional view of macroreticular resin bead.

this particular resin may also be obtained by acrylic resins (Symons et al., 1992; Myers, 1995; McCullen, 1984).

Macroreticular strong base anion resins were primarily developed to improve the ability to resist organic fouling with naturally occurring organic material. Researchers suggest that dissolved organic material fouls anion exchange resin by a combination of an electrostatic mechanism with van der Waals' interaction (Gustafson and Lirio, 1968). Strong base gelular anion resin will gradually lose capacity with time due to fouling with organic material, which continually reduces the number of accessible exchange sites. While macroreticular resins offer much more surface area to resist organic fouling, they also exhibit a reduction in ion exchange capacity over time. For USP Purified Water applications utilizing ion exchange as a primary ion removal technique, with a feedwater supply from a surface source containing moderate to significant concentrations of naturally occurring organic material (NOM), long-term routine operation with *either* gelular or macroreticular strong anion exchange resin will result in problems. The loss in capacity results in more frequent regeneration. To overcome this situation, depending on resin life, caustic brining of the resin bed may be performed to remove the organic material (Rohm and Haas Co., 1973). On the basis of direct observation, however, the elapsed time period encountered before it is necessary to perform this operation is less for gelular resins than for macroreticular resins. This would indicate that claims of increased physical stability for macroreticular resins versus styrenic (or more importantly gelular acrylic resins) may be subtle, at best. In other words, the percentage of whole resin beads present after multiple caustic brining cycles has been observed to be higher for gelular resins than macroreticular resins.

Styrenic and Acrylic Resins

Acrylic strong base anion exchange resin has a chemical structure associated with an acrylic matrix cross-linked with divinylbenzene. The functional group of the acrylic resin is the quaternary amine. Acrylic resins have been used for a number of years in Europe for deionization applications with feedwaters containing appreciable amounts of NOM (Mansfield, 1976). Unlike gelular or macroreticular strong base styrenic resins, operating experience indicates that acrylic resin tends to exhibit significantly less organic fouling (Baker et al., 1978). In fact, during periodic routine regeneration with sodium hydroxide, a faint yellow to light brown color can be observed in the regenerant waste from an anion column containing a gelular strong base acrylic material. Table 4B.1 summarizes the published data for strong base acrylic resins versus strong base styrenic gelular and strong base styrenic macroreticular resins. For USP Purified Water applications, where the feedwater supply contains a moderate

Table 4B.1 Organic Removal—Strong Base Acrylic Resin and Strong Base Styrenic Resin (Gelular and Macroreticular)

Resin type	Pore size (Å)	Initial TOC (mg/L)	Final TOC (mg/L)	TOC capacity (mg TOC/g of resin)
Gelular Styrene	N/A	27	31	19
Macroreticular styrene	350	16	27	30
Gelular acrylic	N/A	13	20	38
Macroreticular acrylic	10,000	10	18	70
Macroreticular styrenic	30,000	10	21	50

Notes: All TOC values are for organic fractions with a molecular weight >10,000 Da. The initial TOC value was 34 mg/L and pH value 7.5.
Source: Symons et al. (1992, Figure 4).

to heavy concentration of NOM, strong base gelular acrylic resins offer an excellent alternative to macroreticular resins. The use of acrylic resins in USP Purified Water systems will be discussed later in this chapter.

BASIC THEORY
Ion Exchange, Demineralization, and Deionization

It is appropriate to begin by briefly defining terms associated with ion exchange resins. Ion exchange is a process by which undesirable ions are removed from a water stream and replaced with another ion. A good example of ion exchange is the water softening process, where calcium and magnesium ions are removed and replaced with sodium ion. This reaction is demonstrated by the following simplified equation:

$$Mg^{2+} + R-Na^+ \leftrightarrow Na^+ + R-Mg^{2+}$$
$$Ca^{2+} + R-Na^+ \leftrightarrow Na^+ + R-Ca^{2+}$$

The term *demineralization* is somewhat misleading. Minerals are generally positive ions in solution. Therefore, demineralization would tend to imply a process where only positive ions are removed. However, the terms *demineralization* and *deionization* are frequently used synonymously. Deionization is a process by which both positive and negative ions are removed from solution and replaced with hydronium and hydroxyl ions, respectively, as exemplified by the following simplified equations:

$$Na^+ + R-H_3O^+ \leftrightarrow H_3O^+ + R-Na^+$$
$$Cl^- + R-OH^- \leftrightarrow OH^- + R-Cl^-$$
$$H_3O^+ + OH^- \rightarrow 2H_2O$$

Affinity of an Ion for an Exchange Site

It is important to remember that the only way the ion exchange process can work is if it is a reversible reaction. It is fortunate that the two ions necessary to produce water are monovalent and of relatively low molecular weight. This allows deionization, since the affinity of an ion for an exchange site increases with increasing charge ($+3 > +2 > +1$ and $-3 > -2 > -1$) and molecular weight. Table 4B.2 summarizes the affinity of typical cations and anions for an ion exchange site. Figure 4B.3 presents representative equations of the equilibrium occurring during normal *operation* of a deionization unit.

It is important to emphasize that ions that have a lower affinity for exchange sites will be the first to be detected in the effluent from an exchange column approaching depletion; in fact, they may always be present at minor (trace) concentrations during the operating cycle. This is particularly true for the sodium ion. "Sodium leakage" is one of the major factors affecting the ability of a conventional two-bed deionization unit to produce high quality (<1 μS/cm conductivity) water.

ION REMOVAL TECHNIQUES—ION EXCHANGE

Table 4B.2 Relative Selectivity of Common Ions for Typical Strong Acid and Strong Base Resin Sites

Cation	Selectivity coefficient vs. H_3O^+
Lithium (Li^+)	0.8
Sodium (Na^+)	2.0
Potassium (K^+)	3.0
Ammonium (NH_4^+)	3.0
Magnesium (Mg^{+2})	26
Calcium (Ca^{+2})	42

Anions	Selectivity coefficient vs. OH^-
Bicarbonate (HCO_3^-)	6
Chloride (Cl^-)	22
Nitrate (NO_3^-)	65
Sulfate (SO_4^{-2})	5.2

Sources: Cations—From Rohm and Hass (1965); Anion—From Gotlieb and DeSilva (1990).

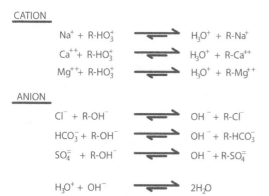

Figure 4B.3 Representative equations for a deionization process.

Ion Exchange Terms, Calculations, and Capacity Units

To individuals without a water purification background, it certainly appears that a chemist or chemical engineer has developed a system that makes it impossible to understand terms and calculations critical to ion exchange. A basic summary of terms and conversion factors used in ion exchange systems is given below.

Calcium Carbonate Equivalents

Calcium carbonate equivalents for anions and cations compensate for the molecular weight and charge of ionic species in the chemical reaction associated with the ion exchange process. Calcium carbonate was selected because it has a molecular weight of 100 Da. The calcium carbonate equivalent for any substance can be calculated by dividing the molecular weight of calcium carbonate (100) by the molecular weight of the ion and multiplying by the absolute valence divided by 2, which is summarized by the following equation:

$$\text{Conversion factor to } CaCO_3 \text{ equivalent} = \frac{100}{\text{molecular weight}} \times \left(\frac{\text{charge}}{2}\right)$$

For example, the calcium carbonate equivalent for sodium (molecular weight = 23, valence = +1) is (100/23 × 1/2) or 2.18.

Calcium Carbonate Balance

For any water analysis, the cations expressed as calcium carbonate should equal the anions expressed as calcium carbonate (excluding extremely weakly ionized species such as silica).

This value should equal the total dissolved solid (TDS) level expressed as calcium carbonate. This method quickly verifies the validity of a chemical analysis. To assist in this evaluation, the TDS level as calcium carbonate can be approximated by multiplying the conductivity (expressed in µS/cm at 25°C) by 0.5.

Conversion to Grains/Gallon

The information outlined in the preceding section provides a method for converting the concentration of a species, expressed as mg/L (or ppm) as the ion (or radical) to mg/L or ppm as calcium carbonate. The concentration in grains/gal as calcium carbonate can be obtained by dividing this value by 17.1. Multiplying by the total number of gallons treated, the number of grains as calcium carbonate can be obtained. This calculation is important since resin capacities are generally expressed in grains (or kilograins)/ft^3 of media. Thus, this conversion can be used to calculate the anticipated "throughput" (volume of water) between regeneration cycles.

Physical Factors Affecting Ion Exchange Resin

Several commonly encountered physical factors will affect ion exchange resin. These factors are briefly summarized as follows:

- The maximum continuous operating temperature for ion exchange resin is limited by the anion resin rather than cation resin. Generally, strong base anion resin should not be continuously operated at temperatures greater than 120°F.
- Ion exchange resin should not be exposed to temperatures less than 32°F, since freezing of water retained in the resin matrix will occur, which will physically fracture the resin, particularly macroreticular resin.
- Resin should remain in a moist condition. It should be stored in sealed plastic bags. Repeated wetting and drying of the resin will result in expansion and contraction and ultimately leading to fracture of the resin beads and production of resin fines.
- Resin should not be exposed to *high* concentrations of chemicals, particularly acids and caustics. This results in chemical shock.
- Oxidation of resin with associated de-cross-linking will gradual occur if the resins are exposed to an oxidizing agent, such as residual disinfecting agent.
- Cation resins will foul with iron.
- Anion resins will tend to foul with organic material.
- Resin fragmentation will occur due to mechanical/physical shock. This may be caused, for example, by excessive backwash, where resin particles are impinged on the side or top walls of an ion exchange column.

Pharmaceutical Water Systems Applications

While there are numerous ion exchange configurations and applications for systems producing pharmaceutical waters, this section of this chapter will primarily address four specific applications:

1. Water softening by cation resin in the sodium form for pretreatment of feedwater to an RO system or for providing regeneration water for a mixed-bed deionization system operated as a primary ion removal technique.
2. Two-bed ion exchange for primary ion removal in both conventional co-current regeneration mode and fixed resin bed units, regenerated in a countercurrent mode.
3. Cation polishers operated in the hydrogen form to remove trace concentrations of sodium after a two-bed deionization unit to decrease conductivity, producing water with a quality meeting the USP Purified Water conductivity specification.
4. Mixed bed ion exchange applications for primary ion removal after two-bed deionization or RO systems.

Many of the design, operating, and maintenance considerations, as well as the potential problems, associated with these ion exchange applications are similar. For example, distribution is a concern in any ion exchange vessel. Therefore, in an attempt to avoid repetition, comments made in the initial detailed discussion of ion exchange generally pertain to other applications and will not be restated. Appropriate references will be made to the initial discussion.

WATER SOFTENING
Theory and Application
An overview of the water softening process was presented in chapter 3. The theory, design, operating, and maintenance considerations associated with water softening as an ion exchange process are presented in this chapter.

Water softening is employed in many pharmaceutical water purification systems, particularly as pretreatment to reverse osmosis. It can be used as a technique to reduce the total hardness in water supplies used to produce active pharmaceutical ingredients, particularly in initial rinse applications. Water softening is generally used as a pretreatment technique to reverse osmosis in a USP Purified Water system. It is required, for a very limited number of applications, during regeneration of a mixed bed deionization unit used as the primary ion removal technique in a USP Purified Water system. Finally, water softening may be used as a (stand-alone) pretreatment technique for feeding vapor compression distillation units (chap. 5).

Water softening is an ion exchange process. In its regenerated state, the resin in a water softener exists in a sodium form. As raw water containing "hardness," primarily calcium and magnesium, passes through the water softening resin bed, the multivalent cations are removed from the water and replaced with sodium, which is represented by the following reaction:

$$Mg^{++} + R-Na^+ \leftrightarrow Na^+ + R-Mg^{++}$$
$$Ca^{++} + R-Na^+ \leftrightarrow Na^+ + R-Ca^{++}$$

During the regeneration process, the equilibrium associated with all ion exchange techniques is shifted as a brine solution, while high sodium concentration is fed through the resin bed. Magnesium, calcium, and other multivalent cations are displaced from the ion exchange resin sites and released to the wastewater from the water-softening unit. Higher concentration of sodium in the regenerant solution will result in more effective conversation of multivalent cation exchange sites to sodium sites (specifically as it relates to equilibrium conditions). In a water softener regenerated in a co-current fashion (regenerant chemical introduced in the same direction of flow as the water being processed during normal operation), higher regenerant sodium levels will result in lower concentrations of multivalent cations during normal operation (after regeneration), which is primarily attributed to displacement of multivalent cations from ion exchange resin sites on resin in the lowest portion (discharge) of the ion exchange column.

Design Considerations
Design considerations for water-softening units are presented in chapter 3.

Operating and Maintenance Considerations
Operating and maintenance considerations for water-softening units are also presented in chapter 3.

CONVENTIONAL, CO-CURRENT-REGENERATED, TWO-BED DEIONIZATION SYSTEMS
Theory and Application
A conventional two-bed deionization system consists of a separate cation exchange unit followed by an anion exchange unit. During normal operation, the flow of water is from the top to the bottom of the ion exchange columns. Regenerant chemicals, acid for the cation unit and caustic for the anion unit, are introduced in the same direction, from the top to the bottom of the columns. For a system utilizing strong acid cation exchange resin and type 1 strong base anion exchange resin, there is an imbalance associated with the capacity of equivalent volumes

of resin. In general, the capacity of the cation resin per unit volume will exceed that of anion resin by more than 20% to 30%. Anion resin is regenerated with decationized water. Subsequently, a portion of the "excess" cation capacity is depleted during anion backwash, caustic regeneration, and displacement (slow) rinse. It is uncommon for a smaller diameter or shorter cation column to be used when compared with the anion column. Most conventional two-bed deionization units utilize equal volumes of cation and anion resin positioned in similar size (diameter and straight side height) exchange columns.

Unlike the product water purity from an RO unit, which is directly related to the TDS level of the feedwater, the product water purity from a two-bed deionization unit generally remains constant for feedwater supplies with significant variations in TDS level and ionic profile. While there may be some minor effects on product water purify associated with sodium leakage for applications where a significant portion of the cation concentration is sodium (generally surface water supplies), a conventional co-current-regenerated, two-bed deionization unit will produce product water with a resistivity of 50,000 to 100,000 Ω-cm or greater (10–20 µS/cm or lower). While conventional two-bed deionization units may be considered for any USP Purified Water system application, they are economically attractive when feedwater TDS levels are in the range of 25 to 200 mg/L. However, it should be emphasized that selection of the primary ion removal technique must be carefully evaluated on the basis of the characteristics of the feedwater *and* the nature of the application. Operations requiring small volumes of water sporadically may employ rechargeable two-bed deionization canister systems.

Design Considerations

The ion exchange column diameter should be selected to produce a face velocity within the resin beds of 5 of 10 gpm/ft^2 over the cross-sectional bed area. A generally accepted face velocity is 7 gpm/ft^2 over the cross-sectional bed area.

Lined steel column exchange vessels should be designed, constructed, manufactured, and tested in accordance with the ASME Code for Unfired Pressure Vessels. Design pressures should be 100 to 150 psig, based on the specific application. The interior of lined steel ion exchange column(s) should be lined with 1/8 to 3/16 in. thick sheet rubber that has been vulcanized-in-place. Continuity of the rubber lining should be verified by a dielectric test. Fiberglass-reinforced polyester columns or vinyl ester columns with top inlet and bottom outlet fittings may also be used, principally for smaller flow rate systems. Rechargeable fiberglass reinforced polyester columns with top inlet and outlet connections are also employed for selected applications as discussed earlier.

Adequate access should be provided to the columns. Manways and hand holes should be provided for the removal/repair of distribution systems and for general access to the interior of the columns. Access should accommodate repair of the rubber lining.

A resin media removal port should be positioned on each ion exchange column, located at the lower vertical straight side of the cylindrical exchange column. Design of the removal port should not introduce an area for resin "hideout" or microbial proliferation (associated with a stagnant area).

The suggested minimum bed depth for ion exchange resin is 30 in. Generally, a bed depth of 36 to 48 in. can be effectively used for conventional two-bed deionization units.

Both the cation and anion resin columns should contain approximately 50% to 75% of freeboard space to allow adequate expansion of the resin beds during backwash, the initial regeneration step. This will significantly increase the effectiveness of the subsequent regeneration operation. Inadequate freeboard space, and subsequent insufficient backwash, will not allow proper exposure of all resin sites to regeneration chemicals.

A 3-in. wide by approximately 12-in.-long rectangular sight glass may be mounted vertically on the straight side of the column at the resin-water interface. The sight glass allows verification of bed expansion during the backwash operation and observation of a decrease in resin level, which is associated with potential system operating or maintenance problems.

The inlet to each column should be equipped with a distributor. Distributors for the anion column should be fabricated of 316L stainless steel. While many cation exchange units are equipped with 316L stainless steel distributors, alternate materials, resistant to acid attack,

ION REMOVAL TECHNIQUES—ION EXCHANGE

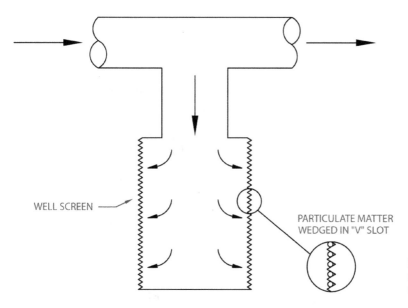

Figure 4B.4 Improper use of inlet "screen" for distribution.

may be considered. The purpose of the inlet distributor is to disperse the feedwater stream. This can generally be accomplished by using fittings that divert the water to the upper domed top of the resin column. Distribution through the resin bed is provided by back pressure exerted from the lower distributor. Using a stainless steel mesh inlet distributor will prevent problems due to the potential buildup of material in the V-shaped slot of the distributor (Fig. 4B.4).

The lower distribution system provides adequate back pressure to ensure uniform water velocity through the entire cross-sectional area of the resin bed. To achieve adequate distribution, the pressure drop through a freshly backwashed (and regenerated) two-bed deionization unit should not be less than 15 psid. The construction materials for the lower distributors should be consistent with the materials for the upper distributors. Distributor type and configuration should ensure proper flow characteristics (elimination of channeling) during normal operation, backwash, and regeneration. Examples of typical distributor configurations are shown in Figures 3.8 and 3.9. Lower distributor positioning within the column should ensure that all resin is exposed to regenerant chemicals.

For most applications, based on the design of the distribution system and ion exchange column, it is inappropriate to employ a single lower distributor for both operations and regeneration. For larger diameter columns, it may be necessary to include a separate distributor for the lower flow rates associated with the regeneration cycle compared to the operating cycle.

The two-bed deionization system should be equipped with an inlet flow rate meter to verify the operating flow as well as the backwash flow rate for both the cation and anion columns. It is highly desirable to combine flow rate monitoring with flow-totalizing capability. While regeneration of the two-bed deionization unit will be performed on the basis of an increase in conductivity (decrease in resistivity), the volume of water processed between regeneration cycles, coupled with information associated with the TDS level of the feedwater supply, can be used to evaluate the performance of the two-bed deionization unit, including the ion exchange resins.

Pressure gauges with appropriate range and accuracy should be positioned in the feedwater piping to the cation column, the product water piping from the cation column, and the product water piping from the anion column. The pressure gauges are used to identify potential operating problems, which may be indicated by an increase or decrease in the pressure drop through a resin column. Gauges should be liquid filled with diaphragm isolators.

The sizing of system piping should be based on a "classical" velocity value of 6 ft/sec. This velocity ensures that there is minimal pressure drop through the piping and fittings.

As a general guideline, cation backwash piping should be sized for a flow rate of 6 gpm/ft^2 over the cross-sectional bed area. Anion backwash piping should be sized for a flow rate of 3 gpm/ft^2 over the cross-sectional bed area. These values are general guidelines, based on a backwash temperature of 70 to 80°F. While the actual backwash flow rate for a particular resin should be establishing using resin manufacturer-suggested values, the flow rate should not vary considerably from the stated values. The backwash piping for both the anion and cation column should be provided with a manual diaphragm valve for regulation of the backwash flow rate as the feedwater temperature changes. More viscous cold water requires a lower backwash flow rate. Excessive backwash flow rate can result in impingement of resin on the side walls and domed top of the column, producing undesirable resin "fines."

The backwash piping to drain (depressurized with an air break) should be equipped with a vertical section of transparent piping to allow operating personnel to observe the presence of particulate matter, activated carbon fines, or resin fines during the backwash cycle. Excessive amounts of any of these materials could indicate system operating problems.

It is strongly suggested that resin fine traps using stainless steel "well screen" material be installed after the cation column and the anion column. The resin traps employ slotted stainless steel, similar to the material used for the lower distributor in ion exchange columns, to remove resin fines (Fig. 3.15). The resin fine traps should be periodically backwashed (approximately once every 1 to 3 months). The installation of a resin fine trap between the cation and anion columns is extremely important. If cation resin enters the anion column (as cation fines), the material will eventually work its way to the bottom of the anion bed during periodic backwash, since cation resin is denser than anion resin. During the regeneration cycle of the anion column, the cation resin will be converted to the sodium form, which results in the inability of the two-bed deionization unit to produce acceptable product water quality. Of even greater importance, however, is the fact that it is extremely difficult to identify this particular situation. Cation resin fines are not visible, and sodium "leakage" is the limiting factor to product water quality from a two-bed deionization unit.

While there are many types of valves that can be used in the assembly of a two-bed deionization system, it is strongly suggested that individual, pneumatically operated diaphragm valves are the most reliable type of valve for successful, long-term operation. Each pneumatically operated valve should be positive-acting, spring-to-close, air-to-open. The spring and pneumatic actuator sizing should be selected such that the valve will close at the maximum system design operating pressure at 0% ΔP. To provide flexibility with regard to long-term system operation, each valve should contain flanged ends, mating to flanges installed in the face piping of the two-bed deionization unit. This provides significant flexibility for valve maintenance/replacement as well as replacement of any other section of the face piping. As an alternative, valves with union ends may be used for face piping constructed of PVC or CPVC.

Operation of the two-bed deionization unit may be controlled by a remotely positioned panel. However, it is suggested that a panel containing locally mounted solenoid valves with manual override provisions be positioned in the immediate area of the two-bed unit. The panel can be provided with appropriate input/output capability to a central control panel. Pneumatic tubing from the solenoid valves to the individual valves may be polyethylene or another flexible plastic material. For aesthetic purposes, this material can be positioned in electrical conduit. The use of copper or stainless steel pneumatic tubing is not only expensive but also difficult to install and maintain in a perfectly horizontal (and vertical) position.

Pneumatically actuated diaphragm valve position indicators should be used to indicate the valve status. Further, pneumatically actuated diaphragm valves can be provided with "travel stops" to limit flow rate if appropriate operating experience indicates that positive-acting diaphragm valves will open and close properly, provided that adequate air pressure is available for opening the valve. The use of individual pressure gauges with a face diameter of approximately 1 in., mounted on a fitting connecting the air supply to each pneumatic valve, may be considered if air supply pressure fluctuation is a concern.

ION REMOVAL TECHNIQUES—ION EXCHANGE

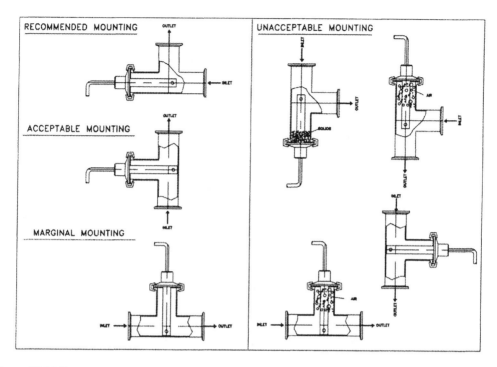

Figure 4B.5 Proper and improper positioning of conductivity cells. *Source*: Thornton Associates.

Pneumatically actuated diaphragm valves can be purchased with microswitches that verify if a valve has fully opened or closed. Generally, the microswitches operate in conjunction with a PLC (programmable logic controller) to monitor the operating and regeneration cycle of the two-bed deionization unit. Because of the limited "travel" of an automatic pneumatically actuated diaphragm valve, operating experience indicates that the switches generally present operating and maintenance problems. Occasionally, the operating or regeneration cycles are inhibited when the micro switch fails. For the vast majority of USP Purified Water applications, it is suggested that micro switches should not be considered.

Sample valves should be positioned in the feedwater piping, the cation product water piping, and the anion product water piping. Further, sample valves should be positioned in the dilute acid and dilute caustic regenerant chemical lines to allow operating personnel to verify the strength of the regenerant chemicals. Sample valves should be of a diaphragm or needle type of appropriate material of construction.

A product water conductivity (or resistivity)-monitoring system should be provided. Proper positioning of the conductivity cell is extremely critical. Figure 4B.5 demonstrates the proper installation method for a conductivity cell. The cell must be positioned so that the full flow of product water passes directly into the cell. Cells positioned in "branch" sides of piping tees may provide erroneous indication.

In addition to providing proper installation of a conductivity cell, it is important to position the cell within the face piping so that it is in a fully flooded condition. Cells should never be positioned at the top of a vertical section of pipe, since air can accumulate, resulting in an erroneously high resistivity (low conductivity) indication. If possible, conductivity cells should be positioned downstream of resin fine traps. Resin fines, if not removed, can become lodged in conductivity cells, resulting in an erroneous product purity indication.

Any two-bed deionization unit operated in a cyclic fashion, and without recirculation provisions, should be provided with a preoperational rinse-to-drain cycle. This cycle will divert a preestablished volume of water (based on time) to drain, overcoming the effects of ion "migration" discussed earlier in this chapter.

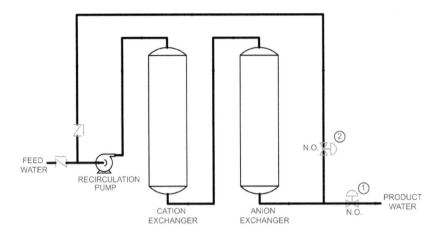

Figure 4B.6 Proper recirculation for a deionization system. (A two-bed deionizer is shown here. Valves 1 and 2 should operate from the control system such that one valve is always open when the other is closed. Recirculation piping can be enhanced for bacteria control by adding a "double block and bleed system," if desired. Pressure increase by the recirculation pump should be >1.3 times the maximum pressure drop through the deionization system.)

For two-bed deionization units that sporadically provide product water (e.g., for a single shift each day), a recirculating loop should be considered. If recirculating pumps are utilized, it is important to install a pneumatically operated diaphragm valve in the recirculating line. For most applications, the recirculating flow will stop when two-bed product water is required downstream. A check valve installed in the recirculating line is *not* a positive method of eliminating bypass of the two-bed unit associated with water flow through the idle centrifugal recirculating pump (Fig. 4B.6) in a reverse direction. If a deionization unit is allowed to remain in a stagnant condition for an extended period of time, ions may "diffuse" through the bed. This results in production of below-quality water when the unit is placed back into operation and may decrease the volumetric throughput between regeneration cycles.

The use of a conductivity/resistivity monitoring system to determine the regeneration requirements of the two-bed deionization unit, by establishing a preset maximum conductivity (minimum resistivity) value, must consider momentary operating transients, which could be associated with periodic cycling of the unit, occasional "slug" of below-quality water associated with ion exchange kinetics, and other subtle operating factors. The conductivity/resistivity monitoring system for the product water from the anion unit should provide a signal to visible display *and* audible alarm, alerting operator personnel that regeneration is required. This system should include a time delay (e.g., 1–2 minutes) to avoid alarms associated with system transients. It is preferable to initiate regeneration manually—by depressing a button on the central control panel. This action can ensure that appropriate regeneration conditions have been verified by operating personnel, such as availability of adequate acid and caustic to perform the regeneration.

Warm caustic regenerant solution should be used to remove exchanged silica during the regeneration process. The most effective way of providing the warm sodium hydroxide solution is to heat the dilution water prior to mixing with caustic. A plate-and-frame or shell-and-tube heat exchanger may be employed for this application. The temperature of the warm caustic regenerant is a function of the type of resin utilized. For styrenic-based anion resins, the regeneration temperature should be approximately 120°F, while regeneration of strong base acrylic resins should be performed at 95–100°F. To avoid potential overheating of the resin associated with a malfunction of the temperature control system, a temperature switch should be installed in the heated dilution water line, prior to caustic injection. This switch should operate in conjunction with a solenoid valve installed in the heating steam line to the exchanger, as discussed in chapter 3.

While many two-bed deionization units used in USP Purified Water systems are equipped with eductors to introduce regenerant acid and caustic, much greater reproducibility and reliability can be achieved by combining diluting eductors with pneumatically operated diaphragm pumps. The acid and caustic flow rate can be adjusted, as required, by adjusting the air supply pressure (or air pulsing frequency) to the pumps. There is a significant safety factor associated with the use of pneumatically operated diaphragm pumps for this application (as opposed to electrically operated positive displacement pumps). If pressure-regulated air is supplied to the pumps, the pressure of concentrated acid and caustic can never increase to a value greater than the supply air pressure to the diaphragm pumps. On the other hand, an electrically operated positive displacement pump (with relief valve failure) can pressurize a piping line until it bursts, releasing hazardous concentrated acid or caustic to the water purification area.

While bulk storage of regenerant acid and caustic is appropriate, "day tanks" are helpful, since the volume of acid and caustic introduced during the regeneration process can be verified by recording day tank levels at the beginning and end of the regeneration cycle. If hydrochloric acid is used to regenerate the cation column, a fume adsorber should be provided on the acid day tank.

If dual or multiple two-bed deionization systems are employed, progressive piping may be considered (Fig. 3.17), assuming that the system feedwater pressure is adequate.

Operating and Maintenance Considerations

The two-bed deionization unit equipment manufacturer should provide the minimum and maximum design flow rate for the unit to avoid channeling. Operating personnel should verify that variations in flow rate from the minimum to maximum values do not produce a considerable (>20%) increase in the product water conductivity value. This evaluation should be performed as part of the Operational Qualification (OQ) of the Purified Water System.

As discussed earlier in this chapter, migration of exchanged ions through a stagnant ion exchange bed will result in the production of below-quality product water. A recirculating pump can eliminate/minimize this situation. The use of an in-line ultraviolet sanitization unit(s) within recirculating loops may be appropriate for microbial control during recirculation where pH "swings" through the cation and ion column are decreased as water conductivity decreases.

Since sodium leakage from the cation unit is the primary factor associated with product water quality from a two-bed deionization system, occasional samples should be obtained from the sample valve positioned between the cation and anion columns. Samples should be analyzed for pH. If two-bed conductivity increases and the pH value has not changed, or slightly decreased, sodium ion concentration should be determined. Since the analysis for sodium ion will generally require the resources of an outside laboratory, it is suggested that this analysis be performed about once every six months; if excessive sodium leakage is suspected, it is generally noted by an increase in two-bed deionization system product water conductivity.

Since sodium will be present in the product water from the two-bed deionization unit, particularly for water supplies containing high percentages of sodium ion (such as surface water supplies), it can be anticipated that the sodium ion will be "balanced" with a hydroxyl ion. As a result, two-bed product water will generally have a pH greater than 7. The use of a mixed bed or, more appropriately, cation-polishing unit for Purified Water Systems using conventional two-bed deionization units is often required to meet the USP/EP Purified Water Conductivity Specification. If below-quality product water is experienced, the feedwater TDS level to the two-bed deionization unit should be determined prior to initiating further system evaluation.

Iron fouling of cation resin will occur if there is a significant concentration of iron in the feedwater to the two-bed deionization unit. This was discussed in detail earlier in the section on water-softening units.

Oxidation of the cation resin will occur if residual disinfectant is present in the feedwater to the two-bed deionization unit (as discussed earlier in this chapter). It should be noted that cation decomposition products, associated with oxidation, may adversely affect the anion

Table 4B.3 Regenerant Chemical Characteristics—Hydrochloric Acid

Percent HCl	Concentration (°Baumé)	Specific gravity	lbs. of HCl/gal
1	0.5	1.0032	0.0837
2	1.2	1.0082	0.1683
4	2.6	1.0181	0.3399
6	3.9	1.0279	0.5147
8	5.3	1.0376	0.6927
10	6.6	1.0474	0.8741
12	7.9	1.0574	1.059
14	9.2	1.0675	1.247
16	10.4	1.0776	1.439
18	11.7	1.0878	1.634
20	12.9	1.0980	1.833
22	14.2	1.1083	2.035
24	15.4	1.1187	2.241
26	16.6	1.1290	2.450
28	17.7	1.1392	2.662
30	18.8	1.1493	2.887
32	19.9	1.1593	3.096
34	21.0	1.1691	3.317
36	22.0	1.1789	3.542
38	13.0	1.1885	3.769
40	24.0	1.1980	3.999

Source: The Permutit Company (1986).

resin. Further, unless the residual disinfecting agent is chronically elevated, the acid aldehyde generated from this reaction may not be detected in final two-bed product water.

Organic fouling of the anion resin will occur if NOM is present in the feedwater to the two-bed deionization system. The degree and extent of organic fouling is a function of resin selection, pretreatment component selection (and operation), and the nature of the organic material in the feedwater. Organic fouling was discussed earlier in this chapter.

Sodium hydroxide is used for regeneration of anion resin. It is suggested that the sodium hydroxide solution be purchased in a concentrated liquid form and be of *Rayon Grade* quality. The use of inferior grades of sodium hydroxide can introduce substances that could be considered as USP "foreign substances and impurities."

For many Purified Water applications, the preferred cation regenerant solution is hydrochloric acid. Alternatively, the use of sulfuric acid will generally require a two- or three-step regeneration procedure. It is suggested that concentrated *Technical Grade* hydrochloric acid ($\sim 20°$ Baumé) be utilized, where available.

The volume and concentration of regenerant acid and caustic are *critical* to the successful regeneration of a two-bed deionization unit. Tables 4B.3 and 4B.4 provide data associated with hydrochloric acid and sodium hydroxide solutions at various concentration values. While the acid and caustic concentration and volume introduced during the regeneration cycle are a function of the cation and anion resins utilized for a particular application, there are some approximations that can be applied to ensure that over- or underdosing does not occur. For strong acid cation resin, an acid dosing level of 2 gal of 20° Baumé ($\sim 32\%$) hydrochloric acid/ft^3 of resin is appropriate. The acid should be diluted about 5:1. The resulting diluted acid should be introduced to the resin bed over a period of approximately 20 to 30 minutes. For strong base anion resin, a suggested dosing level is approximately 1 gal of 50% liquid sodium hydroxide/ft^3 of resin. The caustic should be diluted about 15–19:1 and introduced over a 30- to 60-minute period. Again, it should be emphasized that the information presented above represents guidelines. Specific volume and concentration of regenerant chemicals should be consistent with the resin manufacturer's data sheets for the specific resins in the system.

As discussed earlier, the backwash flow rate for the resin beds must be adjusted with changes in feedwater temperature. More viscous colder feedwater will require a lower

Table 4B.4 Regenerant Chemical Characteristics—Sodium Hydroxide

Percent NaOH	Specific gravity	lbs. of NaOH/gal
1	1.0095	0.0842
2	1.0207	0.1704
3	1.0318	0.2583
4	1.0428	0.3481
5	1.0583	0.4397
6	1.0648	0.5332
7	1.0758	0.6284
8	1.0869	0.7256
9	1.0979	0.8246
10	1.1089	0.9254
12	1.1309	1.133
14	1.1530	1.347
16	1.1751	1.569
18	1.1972	1.798
20	1.2191	2.035
30	1.3279	3.324
40	1.4300	4.773
50	1.5253	6.364

Source: From The Permutit Company (1986).

Table 4B.5 Viscosity of Water at Various Temperatures

°C	°F	Viscosity (μpa-sec)
0	32	1793
10	50	1307
20	68	1002
30	86	798
40	104	653
50	122	547
60	140	467
70	158	404
80	176	354
90	194	315
100	212	282

Source: From Lide DR, Handbook of Chemistry and Physics (2006–2007).

backwash flow rate to achieve the same bed expansion, as compared with warmer backwash water, as demonstrated by the data in Table 4B.5.

To evaluate the long-term performance of the anion and cation resins, it is suggested that a "core" sample of resin be obtained for analysis about once per year. This core sample is obtained by lowering a resin sample tube vertically down through the resin bed, preferably the entire depth of the bed. This provides a representative sample of resin throughout the bed.

The presence of reactive silica in the product water from a two-bed deionization unit will *not* be detected by conductivity/resistivity monitoring due to the weakly ionized nature of silica. While the USP *Official Monograph* for Purified Water does not include a specific test for silica, and, in general, manufacturing applications requiring pharmaceutical grades of water *may* not be affected by the presence of silica, its presence *may* be concern. A fair number of water purification systems produce water for feeding single- or multiple-effect distillation units and Pure Steam generators. Silica is an extremely unique substance that can exhibit volatility in operations including a phase change (liquid to steam) and also precipitate on hot surfaces. Subsequently, it may be carried with steam in distillation units (particularly multiple-effect units) and Pure Steam generators. The white colored, physically hard precipitate produced by this silica will significantly decrease the effectiveness of distillation and Pure Steam generation unit operation with time. The precipitates may deposit on vapor-liquid disengaging sections of the distillation units and Pure Steam generators. If not removed by laborious cleaning methods, these precipitates can eventually inhibit the ability of a distillation unit or Pure Steam generator to produce water meeting the USP/EP Water for Injection and/or Pure Steam bacterial endotoxin limit. For distillation on Pure Steam generation feedwater applications, it is critical to periodically monitor the product water from a two-bed deionization unit for silica. Further, it is suggested that internal specifications be established for silica.

Unlike a mixed bed deionization unit (discussed later in this chapter), which should be completely exhausted prior to regeneration (to ensure separation of the anion and cation resins), a two-bed deionization unit can be "short cycled." In other words, the unit can be regenerated at any time during the operating cycle. This can be extremely beneficial for water purification systems consisting of a single ion exchange train, where regeneration operations can be scheduled during off-shift hours (or on weekends) when there is no demand for water. "Short cycling" should also be considered for microbial control. In general, it is suggested that the maximum cycle time between regenerations not exceed 7 to 14 days.

Validation procedures, including SOPs, should include specific references to the calibration of instrumentation used in a two-bed deionization system. Of particular concern is periodic calibration of the product water resistivity/conductivity monitor. Calibration should include evaluation of meter response with cell simulators (precision resistors) and less frequent calibration with a solution of known conductivity, such as a standard potassium chloride solution. To achieve the latter calibration, it will be necessary to remove the conductivity cell from the piping. Since two-bed unit product water will not meet the conductivity criteria for Purified Water, the calibration criteria set forth in USP *Physical Tests* Section <645> are not required but may be used. In addition, pressure gauges and flow rate indicators should be periodically calibrated.

COUNTERCURRENT-REGENERATED DEIONIZATION UNITS (FIXED BEDS)
Theory and Application
Countercurrent-regenerated, two-bed deionization systems, commonly referred to as fixed beds, have historically been widely utilized in Europe. While the technology has been available for a number of years (Abrams, 1972), the number of pharmaceutical water systems employing the technology is limited. This particular technology could be attractive for USP Purified Water applications (with deionization as the primary ion removal technique), since fixed beds, or fixed beds coupled with cation polishers, can produce product water with a conductivity significantly less than 1.3 µS/cm at 25°C, the "Stage 1" specification. However, as discussed earlier in this chapter, reverse osmosis has displaced the use of regenerative deionization for the majority of USP Purified Water applications.

A fixed bed deionization system differs from a conventional two-bed deionization system in that techniques are employed to maintain the physical location of resin beads within the cation and anion beds. This situation is somewhat complicated by the fact that both anion and cation resins will undergo expansion and contraction from the regenerated form to the exhausted form. While the degree of expansion and contraction is a function of the specific resin, the degree of expansion and contraction is sufficient to present concerns, particularly with regard to larger diameter columns. Numerous techniques have been developed to maintain fixed resin beds, including internal column, inflatable (collapsible) air bags at the top of the bed; resin expansion columns extending from the ion exchange columns; multiple distribution systems to create a blocking flow; and inert material with expansion and contraction capability positioned above the resin bed. Over the past several years, there has been an increasing tendency to use "partially" fixed bed deionization units by providing ion exchange columns with significant bed depth (7–8 ft or greater). These systems have worked quite well. To understand the theory involved with this technology, it is important to understand the basic principles associated with fixed bed deionization.

In a conventional two-bed deionization unit, the cation and anion columns will generally contain a 3 to 4 foot bed depth of resin. With the requirement of 50 to 75% freeboard space, the straight side height of the ion exchange columns are in the range of 6 to perhaps 7 ft. The initial step of the regeneration process for conventional co-current regenerated units is backwash, expanding the bed to provide good contact between the regenerant chemical and the ion

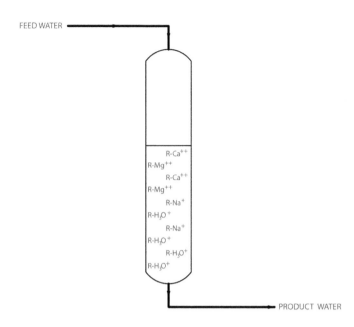

Figure 4B.7 Typical ion exchange column operating profile through a resin bed (Cation exchange is shown as an example. Trace "unconverted" exchange sites are not shown. Sodium leakage was neglected. The resin bed, as shown, is about 40 percent exhausted. The feedwater contains only Ca^{2+}, Mg^{2+}, and Na^+ as cations).

exchange sites. Regenerant chemicals are introduced downward through the resin bed. If the characteristics of the depleted resin bed were evaluated, an examination of resin exchange sites down through the bed, prior to regeneration, would indicate that the majority of resin sites at the top of the bed have been fully converted to the exhausted form. In the case of a cation exchange unit, the resin sites at the top of the bed would primarily be in the calcium and magnesium form; these have a greater affinity for ion exchange sites (McGarvey, 1983). If resin is examined down through the bed, the percentage of calcium and magnesium (as well as other heavier molecular weight multivalent cations) will decrease, while sites in the sodium form will increase. Near the bottom of the bed, exchange sites will exist in both the hydronium and sodium forms (Fig. 4B.7). During the regeneration process for the cation bed, acid is introduced downward through the column. At the top of the resin bed, pure acid, without previously exchanged ions, will convert a high percentage of the ion exchange sites back to the hydronium form. As the regenerant acid passes through the column, the concentration of previously exchanged ions (from the operating cycle), such as calcium, magnesium, and sodium will increase. Subsequently, the conversion of ion exchange sites to the hydronium form decreases as the regenerant chemical passes downward through the column. In fact, the resin with the lowest conversion to the hydronium form will be that at the bottom of the column (Fig. 4B.8). When the unit is placed back into operation after the regeneration cycle, it is this last resin, at the bottom of the column, that water will pass through. Subsequently, the limiting factor of the cation exchange unit to produce product water free of all cations (particularly sodium ion) is a function of the degree of conversion to the hydronium form during the regeneration cycle and resulting sodium leakage.

By physical "fixing" the resin bed and introducing chemicals in a direction countercurrent to the direction of normal operating flow, it is possible to improve product water quality significantly. In the case of a cation column (with downward operating flow and upward regeneration flow), fresh regenerant acid is exposed to resin at the bottom of the bed, converting a high percentage of the exchange sites to the hydronium form. Resin with the least conversion to the hydronium form is located at the top of the bed (Fig. 4B.9). Earlier, it was indicated that a conventional two-bed deionization unit will produce product water with a resistivity in the range of 50,000 to 100,000 Ω-cm (10–20 µS/cm conductivity). A properly designed fixed bed unit will produce product water with a resistivity of 1 to 10 MΩ-cm (0.1–1.0 µS/cm conductivity). Product water from a fixed bed deionization system will meet the Stage 1 conductivity specification for USP Purified Water.

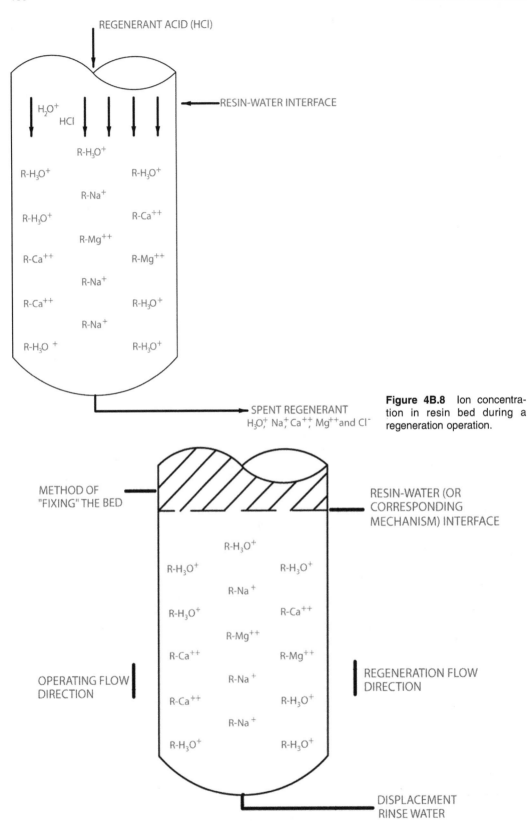

Figure 4B.8 Ion concentration in resin bed during a regeneration operation.

Figure 4B.9 Ion concentration in a resin bed during countercurrent operation (neglecting silica for simplicity).

It is also important to discuss partially fixed bed deionization systems. As indicated previously, water purification equipment manufacturers have attempted to develop simulated fixed bed units by increasing resin bed depths to 6 to 10 ft. For these applications, a limited degree of freeboard space is provided to allow expansion and contraction associated with operation and regeneration of the resin (change in density from the regenerated to exhausted state). While such systems may not physically "fix" the entire resin bed depth, a significant portion of the bed is fixed simply due to the weight of the ion exchange resin above material in the lower portion of the bed. These units are extremely attractive because they eliminate complicated techniques to fix the bed, which may provide chemical and/or microbial hideout. On the basis of the operating experience, it would appear that the quality of the product water from these units, when compared with a fully fixed bed unit, is not significantly different.

Design Considerations

The use of fixed bed deionizers provides more effective use of regenerant chemicals. While it has been suggested that the regenerant chemical dosing may be reduced by as much as 50% (Dow Chemical, 1982), it is recommended that the dosing be maintained at a level of 65% to 75% of that used for a conventional two-bed deionization unit. This ensures product water quality with the purity indicated above.

Depending on the percent freeboard space within the bed, the height of the resin bed, and the nature of the pretreatment equipment, it may be necessary to provide a separate backwash column and a method for periodically transferring the upper 15% to 20% of the resin bed to the column. Generally, such systems are unnecessary, if proper pretreatment and system design are provided.

The extensive list of design considerations presented for conventional two-bed deionization units generally apply to fixed bed units.

The distributor design for a fixed bed is extremely critical. Separate operating and regeneration distributors are generally used. A regeneration distributor positioned at the top of the column (for countercurrent regeneration) is essential to maintain back pressure on the regenerant chemical flowing upward through the resin bed.

Freeboard space within a partial fixed bed must be sufficient to allow expansion and contraction of the resin bed from the exhausted to regenerated state. If freeboard space is insufficient, expansion of the resin bed could result in physical shock to the resin and produce resin fines. Further, column integrity could be an issue, particularly for units equipped with minimal or no freeboard space.

In selecting an ion exchange resin for this application, uniformity of resin bead size is important. In a partially fixed bed, there will be a tendency for separation of different size resin beads during the operation and regeneration cycles.

Some fixed bed units are designed with upward flow during normal operation and downward regenerant chemical flow. Distribution systems for units designed to operate in this manner must be carefully selected, since the density of regenerant acid and caustic will tend to produce channeling.

Operating and Maintenance Considerations

The operational and maintenance considerations for conventional two-bed deionization units generally apply to fixed bed units.

Proper pretreatment is extremely critical to the successful operation of fixed bed units, since, in general, backwash is either eliminated or minimized. Particulate, colloidal, and NOM should be significantly reduced in the feedwater to fixed bed units.

For units equipped with an external backwash column with a resin transfer system, procedures must be followed to avoid mechanical and physical shock to the resin beads during hydraulic transfer from the ion exchange columns to the backwash column, and from the backwash column back to the resin columns.

Because of the large volume of resin in the columns (compared to a conventional two-bed unit), several operating items must be considered, such as the requirement to perform a preheat cycle (with deionized water) prior to performing warm caustic regeneration of the anion column. The mass of anion resin is simply too large to achieve desired temperature for exchanged silica removal during normal regeneration, as indicated for a conventional two-bed deionization system.

CATION POLISHERS
Theory and Application

Cation polishing can be employed to produce product water meeting the USP Purified Water Stage 1 conductivity limits when positioned downstream of a conventional two-bed deionization system. There are several conventional two-bed deionization systems in operation that provide product water that meets all USP Purified Water attributes except conductivity. The majority of these systems can meet the USP Purified Water chemical specifications by addition of a cation polisher downstream of a conventional two-bed deionization system. In fact, it is suggested that many USP Purified Water systems needlessly employ polishing mixed bed units to produce higher water quality than required (from a conductivity standpoint) and with much higher microbial levels than systems using cation polishing.

A cation polisher removes sodium ions, the primary ion (balanced with a hydroxyl ion) contributing to the resistivity/conductivity of product water from a two-bed deionization system. As a result, the cation polisher lowers the pH by removing sodium, replacing it with the hydronium ion that reacts with the hydroxyl ion to produce water. In general, product water conductivity from a system consisting of a conventional two-bed deionization unit, followed by a cation polisher, will be less 0.1 to 1 μS/cm at 25°C. Furthermore, product water quality from a system consisting of a fixed bed deionizer and cation polisher should exhibit a final conductivity of less than 0.1 to 0.25 μS/cm at 25°C.

Design, Operating, and Maintenance Considerations

The comments associated with the design, operation, and maintenance criteria for the cation portion of a conventional two-bed deionization unit applies to the cation polisher.

Because of the limited ion exchange burden on the cation polisher, operating flow rates in the range of 10 to 15 gpm/ft^2 over the cross-sectional bed area can be employed. The regeneration frequency for a cation polisher can be determined by product water conductivity, pH, elapsed time, and/or throughput. While pH is an excellent indicator of cation polisher performance, maintenance of the pH probes can present a problem. While significant advancements in the development of pH probes (requiring less frequent cleaning and maintenance) has occurred over the last 10 years, it is suggested that conductivity and/or throughput are valuable indicators for determining the regeneration frequency of a cation polisher.

A system may be equipped with a single cation polisher, even if dual upstream two-bed units are used. As a result, regeneration is usually conducted at a "planned" time to avoid interruption of system product water flow. Quite often, the regeneration frequency is such that the cation polisher is short cycled.

Since the primary ion removed by the cation polisher is sodium, and since conversion of ion exchange sites from the sodium form to the hydronium form is less than ideal, higher acid regeneration concentrations should be considered.

Because of the issues presented earlier, countercurrent regeneration of cation polishers should also be considered. For existing units regenerated in a co-current manner, the use of a postregeneration "clean," moisture and oil-free filtered air (or nitrogen) mixing operation may be considered to "distribute" ion exchange resin at the bottom of the bed (in a non-fully converted state to the hydronium ion) throughout the bed, significantly enhancing the percentage of resin with exchange sites nearly completely in the hydronium form to the bottom of the bed. Obviously, this operation will increase the ability of the cation polisher to produce water with a lower conductivity than a unit that has not been "mixed."

There are significant benefits associated with use a polishing cation polisher in lieu of a mixed bed unit. In-situ regeneration of a mixed bed unit, discussed later in this chapter, involves multiple steps including physical separation of the anion resin from the denser cation resin. Unlike two-bed regeneration, automated regeneration requires operator presence to verify critical parameters during the regeneration cycle. Further, since a cation polisher operates in the hydronium form, bacteria growth is lower than a neutral pH carbonaceous mixed bed polishing unit.

MIXED BED DEIONIZATION UNITS
Theory and Application

Mixed bed deionizers may be utilized as a primary ion removal technique when feedwater TDS levels are extremely low (<20–50 mg/L). Mixed bed deionizers may also be utilized as polishing units downstream of two-bed or RO units. A mixed bed deionization unit consists of a mixture of cation and anion resin. Generally, an attempt is made to balance the exchange capacities of the resins. A typical mixed bed deionization unit will contain about 60% anion and 40% cation.

There are two major classifications of mixed bed deionization units: regenerative units (units that are regenerated-in-place) and rechargeable units (generally canister type units). A mixed bed deionization unit is capable of producing high quality water, as compared with water produced from a conventional two-bed deionization unit. Product water from a mixed bed unit operating in a *primary* exchange mode will generally exhibit a conductivity of less than ≤ 1 µS/cm at 25°C. On the other hand, mixed bed units operating in polishing modes can produce water with a conductivity of ≤ 0.060 µS/cm at 25°C. It is suggested that systems employing two-bed deionization units installed mixed bed polishing units subsequent to the revision of the USP Purified Water and Water for Injection chemical specification established by the Fifth Supplement to USP 23. Unfortunately, addition of the polishing mixed bed unit *could* present problems due to the fact that microbial growth in a mixed bed deionization unit is much greater than that in a cation or anion column. This is of particular importance for rechargeable canister systems.

The regeneration cycle for a regenerative mixed bed unit will generally include a backwash operation to separate anion resin from cation resin, a settling step, simultaneous or (nonsimultaneous) regeneration of the separated anion and cation resins, a displacement (slow rinse) step, drain down of the water in the column to a level slightly above the anion resin, mixing of the resin with clean oil-free air or nitrogen, column refill, and final rinse. These steps are provided for information purposes only to show the complexity of the regeneration operation. The specific regeneration steps for mixed bed units provided from various water purification equipment suppliers may vary.

A regenerative mixed bed deionization unit can be successfully regenerated because the anion can be physically separated from the cation due to density. Anion resin is less dense than cation resin (Table 4B.6). The density difference can be enhanced by conducting a preregeneration "caustic kill" cycle, introducing caustic down through the entire resin bed.

Table 4B.6 Resin Manufacturer's Data—Density and Expansion

Resin type	Approximate shipping weight (lbs/ft³)	Swelling (% exhausted to regenerated form)
Strong acid cation—gelular styrenic	50 (H_3O^+)	5–9
Strong acid cation—gelular styrenic	52 (Na^+)	5–9
Strong acid cation—macroreticular styrenic	48 (H_3O^+)	4–7
Strong acid cation—macroreticular styrenic	50 (Na^+)	4–7
Weak acid cation—macroreticular acrylic	47 (H_3O^+)	27–80[a]
Weak acid cation—macroreticular acrylic	44 (Na^+)	27–80[a]
Strong base type 1 anion—gelular styrenic	44 (Cl^-)	18–20
Strong base type 1 anion—gelular styrenic	41 (OH^-)	18–20
Strong base type 1 anion—macroreticular styrenic	43 (Cl^-)	20–27
Strong base type 1 anion—macroreticular styrenic	40 (OH^-)	20–27
Strong base type 2 anion—gelular styrenic	44 (Cl^-)	10–15
Strong base type 2 anion—gelular styrenic	41 (OH^-)	10–15
Strong base type 1 anion—macroreticular acrylic	42 (Cl^-)	15–20
Strong base type 1 anion—macroreticular acrylic	40 (OH^-)	15–20
Strong base type 1 anion—gelular acrylic	45 (Cl^-)	10–15
Strong base type 1 anion—gelular acrylic	42 (OH^-)	10–15

[a]Swelling from H_3O^+ form to Na^+ or Ca^{+2} form.
Source: From Resintech, Inc. (1996).

This operation converts many of the cation exchange sites to the sodium form (greatest density) and anion exchange sites to the hydroxyl form (lowest density). Excellent separation of anion from cation resin is obtained during the subsequent backwash operation.

Design Considerations: Regenerative Mixed Bed Units

While a mixed bed column contains both anion and cation resins, many of the design considerations presented for a conventional two-bed deionization unit apply to the design of mixed bed columns with exceptions presented below.

The minimum suggested bed depth for a properly designed mixed bed unit is 48 in., with a suggested minimum freeboard space of 100% resulting in a column straight side height of about 96 in. It is strongly suggested that the straight side of a mixed bed deionization column be equipped with 1 or 3 sight glasses to observe resin levels, primarily identification of the cation-anion interface and upper anion level. These observations are critical to the proper regeneration of the unit.

Mixed bed units should be provided with an interface distributor positioned precisely at the horizontal level in the column where exhausted cation will separate from exhausted anion resin. The interface distributor will be employed during the regeneration cycle to allow acid and caustic regenerant chemicals to flow to waste without cross-contaminating the anion above the distributor with regenerant acid or the cation below the distributor with regenerant caustic. A distributor should also be positioned above the exhausted anion resin level in the column to provide proper introduction of caustic during the regeneration cycle (and alternate deionized water used as a blocking flow for nonsimultaneous regeneration operations).

The top of the column must have a vent to allow draining and refilling during the regeneration cycle. When operated in a polishing mode, face velocities as high as 15 to 30 gpm/ft^2 over the cross-sectional bed area can be considered.

The lower regenerant distributor design is critical since the relatively low flow rate of acid introduced during the regeneration cycle is much lower than the normal operating flow rate. Dual function use of a lower distributor for normal operating flow rates and acid introduction flow rates is not recommended. The use of dual velocity strainers (The Permutit Co., 1981) or a separate regenerative acid distributor should be considered.

Operating and Maintenance Considerations: Regenerative Units

Unlike a two-bed deionization unit, where regeneration is a relatively simple process, the numerous steps associated with mixed bed regeneration, including items such as observation and verification of anion and cation separation during the initial backwash operation, require operator presence, even for units equipped with automatic controls.

If product water purity after regeneration is not obtained in a reasonable time period (60 minutes maximum), it is suggested that the regeneration cycle be reinitiated at the column drain-down step. This allows remixing of anion and cation resin a potential item of concern.

To regenerate a mixed bed unit effectively, the resins must be completely exhausted (or mostly exhausted) or the cation intentionally exhausted by introduction of caustic through the bed (caustic kill), prior to initiation of the regeneration cycle. If either of these conditions is not established, it is quite possible that the anion and cation resins will not fully separate.

Depending on the types of resins selected, a phenomenon referred to as "clumping" may occur, particularly when a unit is operated with new ion exchange material (Rohm and Haas Company, 1977). This process significantly inhibits the ability to separate anion and cation resins, since resin beads will join together to form clumps. Often, operating personnel are not aware of the situation, particularly when only one (small) sight glass is provided with the unit. While declumping agents are available, proper resin selection and multiple regeneration cycles tend to eliminate this condition.

The cation-to-anion ratio is critical to balance exchange capacity as discussed earlier. During the regeneration cycle, the anion-to-cation interface must lie precisely at a horizontal position (within the vertical cylindrical column), consistent with the location of the interface distributor. The cation-to-anion ration may not be changed for a specific unit without significant column and distributor modification.

The quality of air or nitrogen utilized for mixing the resins subsequent to regeneration and decrease of water level in the mixed bed unit should be free of all foreign substances and impurities as defined in the *General Notices* Section of USP. This would include moisture potentially containing gram-negative bacteria and oil.

Resin exhaustion of a mixed bed deionization unit generally occurs quite rapidly when compared with a two-bed deionization unit. A rapid increase in conductivity over a relatively short period of time will be noted when the bed is exhausted.

Because of the complex nature of the regeneration cycle, and improper distributor design, it is possible to experience regenerant chemical hideout during the regeneration process, which will significantly affect the rinse down time of the unit (to purity). For example, rinse down times for units equipped with improper distributors can be as long as one to two days. Modification of these same mixed bed units to include proper distribution systems can reduce rinse down times to less than one hour.

In selecting ion exchange resin for a mixed bed unit, uniformity of resin bead size should be considered. This will encourage proper separation of anion and cation during the regeneration process and promote proper remixing of the resins subsequent to the regeneration process. The resin bead size distribution for acrylic resins will generally eliminate consideration for regenerative mixed bed applications.

Design, Operating, and Maintenance Considerations: Rechargeable Units

As discussed previously, several pharmaceutical water purification systems continue to employ rechargeable ion exchange units. It is estimated that >95% of all rechargeable canister ion exchange systems utilize mixed beds as opposed to two beds or alternative configuration. The majority of rechargeable canisters used for pharmaceutical water purification applications have *poor* distribution characteristics. For convenience associated with transport of the units, standardization, and cost considerations, the units are equipped with top inlet and outlet connections. Flow characteristics within the units are poor, which encourages microbial growth in the neutral pH units and results in channeling (Fig. 4B.10).

A significantly superior design for rechargeable canisters employs the use of a top inlet and bottom outlet vessels. While these units are a bit more difficult to transport than typical rechargeable canisters, they can be adapted for mobile applications. When equipped with

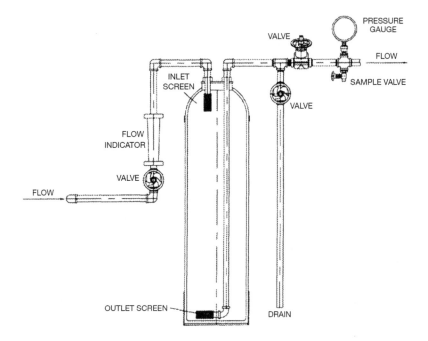

Figure 4B.10 A typical rechargeable canister.

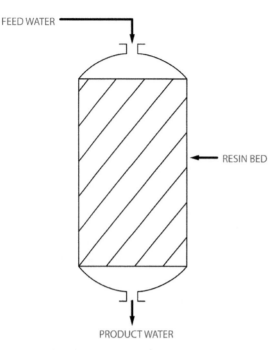

Figure 4B.11 A superior rechargeable ion exchange canister in countercurrent operation. (The resin bed is fixed. The distributors are not shown. The resin charge/removal port is not shown. The material of construction selected for this application is fiberglass-reinforced polyester, 316L stainless steel, and PVDF-lined carbon steel.)

appropriate distributors, the units will exhibit excellent ion exchange characteristics. Furthermore, microbial growth within the units is orders of magnitude lower than that of conventional rechargeable canisters. The design of these units is shown in Figure 4B.11.

To achieve desired microbial control, it is possible to operate units with a top inlet and a bottom outlet at face velocities of 20 to 30 gpm/ft^2 over cross-sectional bed area. On the other hand, the face velocity through standard rechargeable canisters is much lower and nonuniform (over the cross-sectional resin bed area).

Quite frequently, it is necessary to replace standard rechargeable canisters to achieve microbial control as opposed to resin exhaustion. Further, personnel may not routinely sample canister product water for total viable bacteria. This can result in frequent sanitization of downstream components such as storage tanks and distribution loops without identification of the source of bacteria. Again, this situation can be significantly improved by utilizing the alternative canister configuration suggested above.

To complicate microbial issues associated with standard rechargeable canisters, it is not uncommon for pharmaceutical facilities with water purification systems using rechargeable canisters to have numerous "spare" canisters available "on-site." Further, the rechargeable canister providers frequently "store" spare units at their facility. These canisters provide an ideal environment for bacterial growth because of their stagnant condition. The canisters will require extensive rinse down once they are placed online. Unfortunately, the rinse down time is generally determined by the ability to meet a preestablished conductivity specification. Quite frequently, TOC is not monitored and is *not* within specification when conductivity meets specification. Finally, there are generally no criteria for total viable bacteria. Obviously the minimum two-to three-day delay in obtaining total viable bacteria results would delay the ability to use water from freshly installed mixed bed rechargeable canisters. The conditions indicated above are significantly accelerated with ambient temperature increase.

Most standard rechargeable canisters are equipped with quick disconnects for easy removal and installation into systems. Unfortunately, most of these quick disconnects are of a nonsanitary design and provide "crevices" for bacteria to accumulate and replicate. Flat gaskets seal mechanisms, as an example, provide an excellent location for microbial proliferation.

Many rechargeable canister systems, due to a maximum flow restrictions of about 15 gpm for the largest (standard size canister), configure multiple units in a series/parallel arrangement. The vast majority of these applications do not have provisions for determining flow balancing, which is required to obtain proper distribution through the parallel canister "banks." It is impossible to guarantee that the pressure drop through each rechargeable canister will be the same. Subsequently, the use of multiple units in a series/parallel configuration will not only produce poorer quality product water, result in premature canister replacement, but will also encourage microbial growth, since lower (and higher) flow rates may be encountered through units operated in a parallel configuration.

Because of their tendency to increase microbial levels, rechargeable canister applications may be physically repositioned for USP Purified Water system. Canisters may be removed from the USP Purified Water distribution loop and repositioned in makeup water systems to USP Purified Water storage tanks. This is consistent with a trend to utilize stainless steel USP Purified Water storage tanks and distribution tubing, which *can* be hot water sanitized or compatible with the use of ozone for microbial control.

The physical relocation of rechargeable canisters from USP Purified Water distribution loops to the makeup portion of the system may result in a slight increase in the conductivity of the recirculating USP Purified Water, since "continuous" ion exchange is no longer available. This is particularly true in situations where carbon dioxide may be present in increased quantities in the gaseous space above the stored Purified Water. While not recommended, nitrogen blanketing through a hydrophobic vent filtration system can minimize the effects of atmospheric carbon dioxide adsorption associated with large volume tank drawdowns (generally for batching applications). For non-high-volume batching requirements, fluctuations in the USP Purified Water storage tank level can minimized, decreasing the concentration of carbon dioxide above the stored water.

In an attempt to remove canisters from polishing loops, some systems utilize rechargeable canisters positioned in the product water piping/tubing from a single- or double-pass RO unit. It is suggested that this is an extremely inappropriate design alternative. Bacteria will grow against the direction of flow (through a biofilm), particularly for systems with stagnant condition. Subsequently, bacteria will proliferate from the standard rechargeable canister to the clean side of the upstream RO membranes. The use of check valves will not inhibit this problem. Periodic sanitization of the RO membranes (as frequently as once every 2–4 weeks) may be required for microbial control.

The qualifications of the company performing canister regeneration and the condition at the regeneration facility are important. Some regeneration facilities do not use warm caustic for the regeneration of the anion resin. This presents an obvious problem for applications where silica concentration in the product water is a concern. Housekeeping practices as well as conformance to standard operating procedures are important. The ability of regenerated resin to produce a specified conductivity, specified TOC value, and process a specified volume of water per unit volume prior to exhaustion are all important.

Some contract regeneration facilities utilize an in-canister regeneration process. While this process may ensure that a pharmaceutical facility will have "dedicated resin" for its application, the hydraulics of in-canister regeneration *may* provide inferior regeneration and, more importantly, less effective bacterial control.

When a rechargeable canister leaves a facility, there is a loss in "control" of the unit. Validation procedures cannot ensure that the resin is being handled appropriately. Very few contract resin regeneration facilities have provisions for resin segregation. Subsequently, it is impossible to guarantee segregation of material from other resins employed for other applications (including potential waste treatment applications) due to the size of the resin beads.

Consistent with the concerns expressed above, it may be appropriate to consider use of "virgin" resin for certain applications for some facilities. This program minimizes concerns associated with the control of resin once it leaves a facility. In addition, requirements (standard operating procedures) for presanitization of the rechargeable mixed resin beds prior to resin installation can be established.

Using a system with separate cation and anion canisters and a cation-polishing canister should be considered in lieu of mixed bed canisters. The three-bed system provides product with adequate purity to meet the USP Purified Water conductivity specification and, more importantly, offers *much* better bacterial control. The separate bed units operate in acidic or basic environments, assisting in bacterial control. On the other hand, mixed beds operate at a neutral pH.

REFERENCES

Abrams IM. Counter-current ion exchange with fixed beds. Paper presented at the 10th Annual Liberty Bell Corrosion Course. Course No. 4. Philadelphia: Diamond Shamrock Chemical Company, 1972.

Baker B, Davies VR, Yarnell PA. Use of acrylic strong-base anion exchange resin in organic bearing waters. Paper presented at the 38th International Water Conference, Pittsburgh, 1978.

Dow Chemical. Countercurrent ion exchange. Idea Exchange 1982; 4(1), Technical Brochure No. 177-1287-82. Dow Chemical Co. USA.

Gotlieb MC, DeSilva F. Factors in high purity mixed-bed demineralizers. Ultrapure Water 1990; 7(2):61–70.

Gustafson RL, Lirio JA. I&EC Product research and development 1968; 7:117.

Lide DR. Handbook of Chemistry and Physics. 87th ed. Boca Raton, FL: CRC Press, Inc, 2006–2007.

Mansfield GH. The Assessment of Anion Exchange Resin Capacity with Respect to Fouling by Naturally Occurring Organic Materials. Paper presented at the International Conference on the Theory and Practice of Ion Exchange, Cambridge, UK, 1976.

McCullen WL. The Use of Acrylic Resins in Demineralization. Presented at the 22nd Annual Liberty Bell Corrosion Course, September 24–26. Philadelphia: Rohm and Haas Company, 1984.

McGarvey FX. Introduction to Industrial Ion Exchange. Technical Brochure No. NDJ. Birmingham, NJ: Sybron Chemical Company, 1983.

Myers PS. Operating Experiences with a New Organic Trap Resin. Paper presented at the 56th Annual Meeting—International Water Conference, October 31–November 2, Pittsburgh. Paper No. IWC-95-1, 1995.

The Permutit Co. Dual Velocity Strainers. Bulletin No. 6169. U.S. Patent 4,162,975. Paramus, NJ, USA: The Permutit Company, 1981.

The Permutit Co. Water and Waste Treatment Data Book. Manual 24781-7M-2186. Paramus, NJ, USA: The Permutit Company, Inc., 1986:71–77.

Resintech, Inc. Product Reference Guide (Various Resin Data Sheets). Cherry Hill, NJ, USA: Resintech, Inc., 1996.

Rohm and Haas Co. The deionization of water, part I: The hydrogen cycle operation. In: Amber-Hi-Lites, No. 86. Philadelphia: Rohm and Haas Company, 1965.

Rohm and Haas Co. Helpful hints in ion exchange technology. In: Amber-Hi-Lites, No. 134. Philadelphia: Rohm and Haas Company, 1973.

Rohm and Haas Co. The use of ion exchange resins in condensate polishing. In: Amber-Hi-Lites, No. 148. Philadelphia: Rohm and Haas Company, 1975.

Rohm and Haas Company. Helpful hints in ion exchange technology (The Application of Murphy's Law #1). In: Amber-Hi-Lites, No. 156. Philadelphia: Rohm and Haas Company, 1977.

Rohm and Haas Co. Ion exchange ... industrial water treatment. Technical Brochure IE-247. Philadelphia: Rohm and Haas Co., 1978:15.

Rohm and Haas Co. Two decades of macroreticular ion exchange resins. In: Amber-Hi-Lites, No. 161. Philadelphia: Rohm and Haas Company, 1979.

Symons JM, Fu PL, Kim PH-S. The Use of Anion Exchange Resins for the Removal of Natural Organic Matter from Municipal Water. Paper presented at the 53rd Annual Meeting—International Water Conference, October 19–21, Pittsburgh. Paper No. IWC-92-12, 1992.

4C | Additional ion removal techniques

INTRODUCTION
The final section of this chapter discusses ion removal techniques other than classical ion exchange and reverse osmosis. The primary focus of this section is continuous electrodeionization (CEDI). Two ion removal techniques employing principles leading to the development of CEDI, electrodialysis and electrodialysis reversal, will briefly be discussed. While both electrodialysis and electrodialysis reversal are infrequently employed, a historical summary of each technology is appropriate.

Subsequently, this section of the chapter has been divided into two major sections. The first section will present a historical overview for electrodialysis and electrodialysis reversal. The second section presents design, operation, and maintenance considerations for CEDI systems. As indicated, CEDI technically "evolved" from electrodialysis and, to a lesser extent, from electrodialysis reversal. Once the theory of electrodialysis and electrodialysis reversal is understood, it is technically feasible to discuss the subtle changes associated with the development of CEDI.

ELECTRODIALYSIS AND ELECTRODIALYSIS REVERSAL
Historical Overview
Electrodialysis was developed, in the research stage, around the turn of the century. However, researchers recognized that commercialization of the technology was limited by the need for ion exchange membranes with good physical properties and high capacity. As a result, further development of electrodialysis was delayed until ion exchange material for ion exchange membranes was developed.

From the beginning of the century to the early 1940s, little was done to develop electrodialysis further. Research and development was reinitiated in the early 1940s. In 1940, Meyer and Stauss developed a multicompartment electrodialysis cell with ion-selective membranes. In 1948, full commercialization of electrodialysis was initiated with the marketing of viable ion transfer membranes. Ultimately, this resulted in the first commercial unit installation in 1954. Electrodialysis reversal, an enhancement to the electrodialysis process, was commercially introduced in the early 1970s. Commercially available electrodialysis systems in the 1950s were provided for Drinking Water applications from brackish water. Electrodialysis reversal units were used for Drinking Water and boiler feed applications. In the 1980s, electrodialysis reversal units were provided for Drinking Water and boiler feed requirements and as part of the primary ion removal technique for high purity water systems, including semiconductor applications.

Theory and Application (Electrodialysis)
In electrodialysis, ions are removed from water by an electronic force that literally pulls the ions from the solution through a semipermeable membrane. This electrochemical separation process, in which ions are transferred through anion- and cation-selective membranes, results in an "ion-concentrating" section and an "ion-depleting" section. With semipermeable membranes and a direct current, the electrochemical process provides the separation required to remove ions from water. This process is unique, compared to other membrane technologies, in that electrodialysis achieves deionization by "pulling" the ions away from the water rather than moving water away from the ions. Further, the semipermeable ion exchange membranes operate at low pressure versus the high pressure associated with other membrane processes, such as reverse osmosis. The process is not classified as a dewatering process nor does it require water to be forced against and through a membrane surface.

Electrodialysis uses alternating cation and anion exchange membranes in a specific arrangement, as shown in Figure 4C.1. This arrangement, with the use of a direct current,

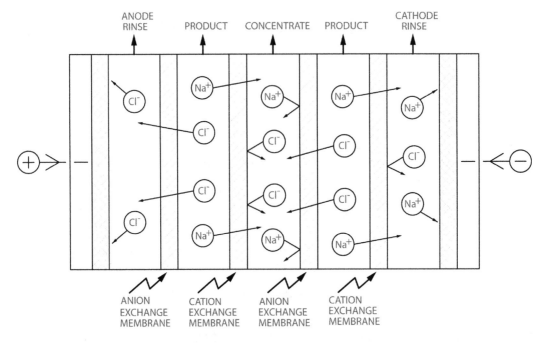

Figure 4C.1 Basic electrodialysis membrane arrangement.

provides a membrane-controlled process with ion movement. As shown in Figure 4C.1, it provides a deionized water compartment and a concentrated ion (brine) compartment. The driving force for pulling ions through the semipermeable membranes is a direct current applied to the water solutions within the membrane configuration shown in Figure 4C.2. Figure 4C.2 depicts multiple "compartments" that are typical of an electrodialysis unit. As shown by the flow path in Figure 4C.2, the process results in the removal of ions from some compartments and simultaneous concentration of ions in alternate compartments. This produces a concentrated stream, often referred to as a brine stream, and an ion-depleted stream, referred to as the deionized water stream, from the unit. Negative ions pass through the anion membranes, while positive ions are rejected by the anion membranes. Conversely, positive ions pass through the cation membranes, while negative ions are rejected by the cation membranes.

Figure 4C.3 clearly shows the two electrodes used to establish the flow of direct current through the electrodialysis compartments. The reactions occurring at each of the electrodes are as follows:

Anode:

$$H_2O - 2e^- \rightarrow 1/2 O_2 \uparrow + 2H^+$$

$$2Cl^- - 2e^- \rightarrow Cl_2 \uparrow$$

Cathode:

$$2H_2O + 2e^- \rightarrow H_2 \uparrow + 2OH^-$$

The gases produced at the anode and cathode (oxygen, chlorine, and hydrogen) are removed by a dedicated water stream that rinses the electrodes. The ion exchange membranes are impervious to water. In a typical electrodialysis unit, the numerous ion exchange membranes are arranged in a "stack." Water containing ions is pumped between cation and anion exchange membranes guided by a polyethylene spacer that establishes the flow path. It is important to understand that the water flow path is *between* membranes rather than *through* the membranes. This differentiates electrodialysis from reverse osmosis, where the flow of water is through the semipermeable type membranes.

ADDITIONAL ION REMOVAL TECHNIQUES

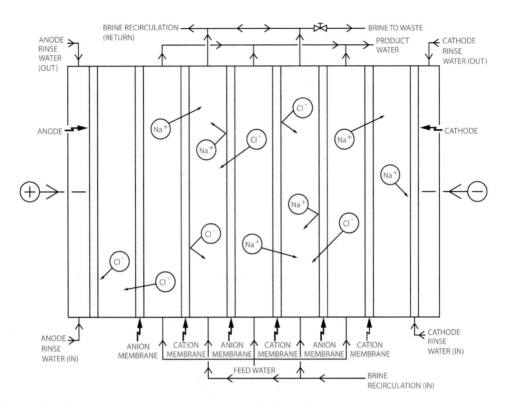

Figure 4C.2 Typical electrodialysis cell arrangement with flow paths.

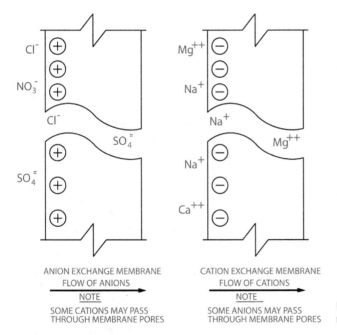

Figure 4C.3 Anion and cation exchange membranes—ion flow.

Operating Description

It is appropriate to briefly discuss some of the basic operating and unit assembly criteria for electrodialysis. The ion exchange membranes are constructed in flat sheets. Earlier electrodialysis units employed anodes constructed of stainless steel and cathodes constructed of platinum-coated tantalum or niobium. Newer electrodialysis and electrodialysis reversal units exclusively employ anodes and cathodes constructed of platinum-coated titanium.

For structural support, the ion exchange membranes are fabricated for long-term operation. This is generally achieved by applying cation- or anion-selective polymers to a fabric type material. Cation exchange membranes may use sulfonated cross-linked polystyrene-based media, while the original anion membranes used cross-linked polystyrene containing quaternary ammonium groups. Newer anion membranes may use aliphatic compositions that exhibit better properties for long-term operation with poorer quality feedwater. The use of acrylic-based anion membranes is popular based on the ability of the membranes to tolerate high concentrations of disinfecting agents. Ion exchange membranes may not be perfectly semipermeable. In other words, the membranes may not reject *all* ions of the same ionic charge. The selectivity of the ion exchange membranes should be at least 90%. Newer membranes exhibit higher selectivity.

Most ion exchange membranes are about 0.5 mm thick. Cation membranes used in the physical stack area adjacent to the electrocompartment, sometimes referred to as "end" membranes, will be both heavier and thicker than the other membranes in the stack. The membranes are tolerant to pH values, during normal operation, in the range of 1 to 10. For cleaning purposes, the "short-term" pH range may be between 0 and 13. Historically, the life of ion exchange membranes has been about 5 to 10 years. The membranes exhibit a residual chlorine tolerance of 0.3 ppm during normal operation and concentrations as high as 20 ppm during cleaning operations. Polyethylene spacers approximately 1 mm thick are positioned between the ion exchange membranes. The spacers create turbulence and allow water flow parallel to the anion and cation membrane surfaces.

As mentioned earlier, gases generated at electrodes are removed by a rinse water stream. The flow of rinse water to the anode compartment is continuous. The flow of rinse water to the cathode compartment may be intermediate, allowing the acidic environment (pH ~ 2) to dissolve any scale-forming components present in the compartment. However, periodic gas removal is performed by rinsing.

Polarization
The individual solution compartments and the ion exchange membranes are electronically positioned in series. The individual compartments conduct the same electrical current through the stack. Within the compartments where ions are being depleted, both cations and anions conduct the current. However, in the ion exchange membranes, only one type of ion (cation or anion) will conduct the current. While there are many methods of explaining polarization, it is important to remember that ions within the membrane structure must travel twice as quickly as ions in the water solution to conduct the same electric current. The net result is the rapid removal of ions as they approach the individual ion exchange membranes. Obviously, the decreasing ion concentration within the area immediately adjacent to the ion exchange membranes can control the ability to conduct an electric current through the water solution containing ions.

As discussed previously, the electrodialysis process involves the transport of ions through the membrane surfaces by an electrical driving force. For a specific ion exchange membrane, only one half of the ions approaching the membrane surface from the water solution being deionized are physically transported through the membranes due to the semipermeable nature of the membrane (anion or cation selective). To establish the electric current through the unit, the balance of ions arriving at the membrane surface from the water being deionized flow by convection and diffusion. Both of these processes are relatively slow. Again, this will result in a significant decrease in the concentration of ions in a thin layer immediately adjacent to the membrane surfaces. Polarization occurs when the water at the membrane surface is depleted of ions. This situation decreases the ability of the cells to conduct the electrical current and increases the electrical resistance through the membrane compartment.

Water, without any ions, will provide hydronium and hydroxyl ions by the following reaction:

$$2H_2O \leftrightarrow H_3O^+ + OH^-$$

During normal conditions, the ionization constant associated with this reaction is extremely small. At a neutral pH (pH 7), with no other ions present, the concentration of hydronium and hydroxyl ions is equal (1×10^{-7} moles/L). However, during polarization within the electrodialysis unit, the electric current will "use" the hydronium and hydroxyl ions to "help" transmit the electric current. This results in hydronium and hydroxyl ion passage through the appropriate ion exchange membrane. If a significant percentage of the current is carried through the ion exchange membranes by the hydroxyl ion, the pH within the "brine" compartment will increase, resulting in the precipitation of compounds such as calcium carbonate.

The polarization phenomena, undesirable for electrodialysis, results in the following:

- The current efficiency through the "cell" is decreased.
- Energy consumption through the cells is increased.
- The transfer of "other" ions into the polarization region is reduced, since the hydroxyl and hydronium ions are conducting the electric current in this region.

It would be inappropriate to discuss the polarization phenomena without mentioning "current density." This term is defined as the amount of current carried by a unit area of membrane surface. The current density can be increased until nearly all of the ions next to the membrane surface are removed. When all of the ions are removed in the area adjacent to the membrane surface and polarization occurs, the "limiting current density" is reached.

Colloidal/Organic/Microbial Fouling
There are limitations associated with the use of electrodialysis. These limitations evolved into an improved technology, electrodialysis reversal. Prior to completing a discussion associated with the theory and application of electrodialysis, it is appropriate to briefly discuss the somewhat unique nature of the fouling mechanisms associated with electrodialysis.

When describing the electrodialysis process, specifically the method of ion removal, it appears that weakly ionized material, such as *silica*, and other material present in pretreated feedwater supplies, such as organic material, colloidal material, and bacteria, are not removed by the electrodialysis process. This is based on the fact that the electric current is unable to upset the equilibrium required for silica removal and seemingly has no effect on organic material, colloidal material, and microbial contamination because they do not exhibit an *adequate* charge. However, colloidal material, organic material, complexes of colloidal and organic material, and bacteria exhibit a *slight* negative charge. Earlier in this chapter, it was indicated that the electric current associated with electrodialysis can be strong enough to affect material with extremely low ionization constant. The slight negative charge associated with the impurities mentioned in the preceding text is adequate to attract these materials to the surface of anion exchange membranes. This is an important item, since one of the advantages of electrodialysis is the ability to avoid fouling with colloidal, organic, and microbial matter. As the materials are attracted to the anion exchange membrane, they will be deposited and held on the surface of the membrane, fouling the membrane and resulting in a decline in ionic transport across the membrane surfaces for an established current density.

Parameters to avoid this fouling situation, including both turbidity and SDI values, should be determined for feedwater to the electrodialysis unit. Operating experience with electrodialysis units indicates that fouling will not be observed when the five-minute SDI value is less than 12 (Pontius, 1990). Fouling will be observed when the SDI value is more than 16. If turbidity is determined, fouling will not occur when turbidity values are less than 0.25 to 0.52 NTU (nephelometric turbidity units).

While other techniques may be used to minimize the degree of fouling by colloidal, organic, and microbial material, such as the use of a tortuous path spacer design that creates turbulence to minimize deposition, the most effective method of eliminating the situation is to consider periodic reversal of electrode polarity, via a process referred to as electrodialysis reversal.

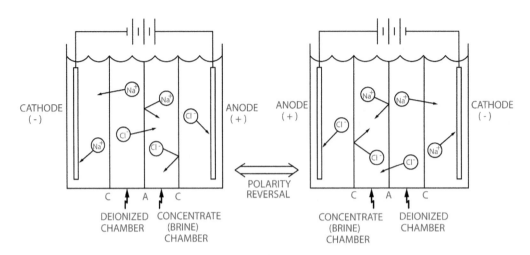

Figure 4C.4 Electrodialysis reversal—ion removal and polarity reversal.

Electrodialysis Reversal

Electrodialysis reversal is extremely similar to electrodialysis, with the exception that the polarity of the anode and cathode providing the electric current through the ion exchange membranes is periodically reversed to eliminate fouling and scaling. Figure 4C.4 shows a cell operating with one polarity and a second cell with the polarity reversed. Electrodialysis reversal technology was developed in the early 1970s and represents a significant improvement to the electrodialysis process. With a few exceptions, it has replaced the use of electrodialysis.

Figure 4C.5 depicts the beneficial effects of polarity reversal and the removal of foulants from the surface of the anion membranes as well as the removal of scale from the concentrating (brine) chambers. Upon polarity reversal, the foulants are "lifted" from the surface of the anion exchange membranes. It is important to remember that the foulants are generally on the surface of the anion exchange membranes, not "within" the membrane structure. The polarity reversal literally "pulls" the foulants, exhibiting a slight negative charge, from the surface of the anion membranes. As a result, this process is highly effective at removing the foulants. As shown in Figure 4C.5, the flow of deionized water through chambers that were concentrating (brine) chambers prior to polarity reversal, removes potential scale-forming ions, tending to make the unit "self-cleaning." Furthermore, for most applications, electrodialysis reversal can be operated without the introduction of acid and/or antiscalants to the feedwater.

Electrodialysis reversal units are capable, without the addition of acid and/or antiscalants, of operating with concentration (brine) stream calcium sulfate values of 150% above the solubility limit (von Gottberg and Siwak, 1997). In addition, the concentrate (brine) stream can operate with an LSI as high as 2.1 to 2.2. Generally, the maximum salt removal per pass is 40% to 50% (Siwak, 1993).

The literature contains information associated with the development of polyethylene spacer designs that increase the available membrane area and enhance turbulence at the ion exchange membrane surface (von Gottberg and Siwak, 1997). This spacer design allows a higher current density before polarization will occur. The stated theoretical ion removal is 50% to 75% per stage. A conservative ion removal projection is 40% to 60% per stage. This technology also uses a four-way valve in the streams to and from the electrodialysis reversal unit, as shown in Figure 4C.6. The primary advantage of the advanced spacer design is that it lowers the overall electrical resistance across the ion exchange membrane stacks. This makes it easier for ions to travel from the dilute stream to the ion exchange membrane surfaces and subsequently into the concentrating (brine) stream.

Additional spacer data and information are available in the literature (von Gottberg and Siwak, 1997). Further, additional operating data are also available from the literature

ADDITIONAL ION REMOVAL TECHNIQUES

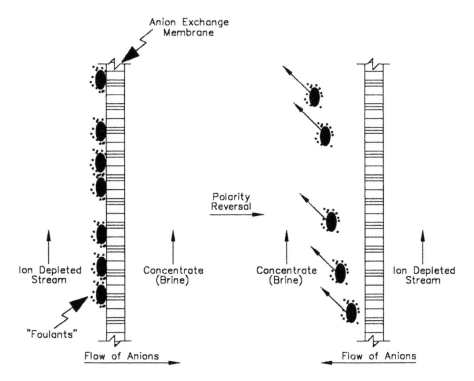

Figure 4C.5 Removal of foulants from anion exchange membranes by polarity reversal.

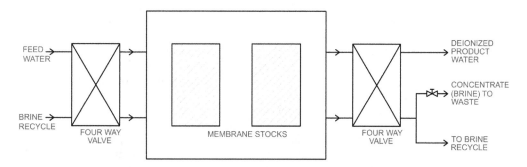

Figure 4C.6 General electrodialysis reversal flow path with four-way valves. *Source*: From von Gottberg and Siwak (1997).

(Kiernan et al., 1992). These data support the requirement to combine electrodialysis reversal with other ion removal techniques to achieve product water quality for pharmaceutical applications. Table 4C.1 presents operating data to indicate the nature and characteristics of anticipated product water.

Design Considerations

Pretreatment Requirements

While electrodialysis reversal does not involve the flow of water through a membrane, there are certain pretreatment criteria that must be considered. Pretreatment considerations include the following raw water impurities:

- *Particulate matter*. The design of an electrodialysis reversal system should incorporate cartridge filtration as the first component. The cartridge filtration system should be designed to remove particulate matter with a size greater than 10 μm.

Table 4C.1 Observed Continuous Electrodeionization Unit Product Water Quality

Parameter	Observed operating range
Conductivity at 25°C	0.055–0.85 µS/cm
TOC	15–35 µg/L
Total viable bacteria (note)	<1–2 cfu/100 mL
Bacterial endotoxins	<0.001 EU/mL

Note: Total viable bacteria measurements by membrane filtration of a 100 mL sample, R2A or PCA culture media, 30 to 35°C or 22°C incubation temperature, and 72 to 120 hour incubation time period.

- *Organic removal.* As indicated during the discussion associated with fouling of the surface of ion exchange membranes used for electrodialysis and electrodialysis reversal, the removal of organic material may be appropriate. While electrodialysis reversal with periodic changes in polarity will reduce the effects of organic fouling, a careful analysis of the raw feedwater to the facility must be considered.
- *Colloidal removal.* The presence of colloidal material, principally colloids of silica, iron, and aluminum, in the feedwater to an electrodialysis reversal unit is undesirable.
- *Iron and manganese.* Iron and manganese removal should be considered if the iron and manganese feedwater concentrations are excessive. The literature suggests a maximum feedwater total iron concentration of 0.3 ppm and a maximum manganese concentration of 0.1 ppm (AWWA, 1995).
- *Residual chlorine.* As discussed earlier, some anion exchange membranes are sensitive to oxidation by a residual disinfecting agent. The suggested maximum normal operating residual chlorine level should be ≤0.3 ppm.
- *Hydrogen sulfide.* Occasionally, trace concentrations of hydrogen sulfide may be present in raw feedwater supplies. For brackish water supplies containing hydrogen sulfide at concentrations more than 0.3 ppm, removal will be required. Removal of hydrogen sulfide is achieved by techniques similar to those employed for iron and manganese removal.
- *Reactive silica.* The electrodialysis reversal process will tolerate significant levels of reactive silica. Reactive silica removal will not be required unless the feedwater concentration exceeds about 200 ppm. It is suggested that unit operations required to remove silica at concentrations more than 200 ppm may eliminate consideration of electrodialysis reversal as a primary ion removal technique.
- *Temperature.* The maximum feedwater temperature should not exceed 45°C. High operating temperatures will affect other parameters. For example, higher operating temperatures will be associated with lower water viscosities, thus affecting polarization. Higher feedwater temperatures are also associated with accelerated microbial growth in the feedwater and on anion exchange membrane surfaces.
- *Silt Density Index.* An SDI determination of the feedwater, performed after five minutes of filtration, should indicate a value less than 12; it should not exceed 16.
- *Turbidity.* The feedwater turbidity should be less than 2.0 NTU. The normal operating turbidity of the feedwater should be less than 0.25 to 0.5 NTU to minimize the frequency of cleaning.
- *Total hardness.* For certain applications, high relative concentrations of multivalent cations (as a percentage of total cations) can affect the operation of the electrodialysis reversal unit. A reduction in the total hardness level may be necessary to avoid undesirable conditions.
- *Supply pressure.* The feedwater supply pressure to the electrodialysis unit is suggested as 20 to 30 psig. The literature suggests a supply pressure of 2 to 40 psig (AWWA, 1995).

Limitations of Electrodialysis Reversal

The electrodialysis reversal process will not "directly" remove organic material. Some organic material removal may be observed, along with colloidal removal, as a result of periodic polarity reversal. Organic and/or colloidal material attracted to the surface of the anion-selective membrane will be removed upon reversal of polarity. As a "stand-alone" process, unless the feedwater is from a groundwater source with low natural TOC levels, it would be inappropriate to use electrodialysis reversal for the direct production of USP Purified Water. A secondary ion removal technique or a process capable of removing organic material would be required. Single- or double-pass reverse osmosis would provide an excellent technique for removing all of this material. However, an RO unit with polishing CEDI system offers superior performance with less equipment for the vast majority of pharmaceutical water purification systems.

CONTINUOUS ELECTRODEIONIZATION
History, Theory, and Application

The material presented in this section has discussed the history, use, and limitations of electrodialysis and electrodialysis reversal. For pharmaceutical water purification systems, the use of either technology is limited. However, as indicated earlier, the technology associated with electrodialysis has led to the development of new technology referred to by ASTM as "continuous electrodeionization". It is strongly suggested that the use of CEDI is limited to post-RO polishing applications. Over the last several years, CEDI has evolved from a "first-generation" non–hot water sanitizable polishing technique to a standard unit operation for most USP Purified Water systems and feedwater systems to Pure Steam generators and multiple-effect distillation units.

CEDI technology was initially developed in the mid-1950s. Commercialization of the process began in the 1980s and proceeded to the marketing stage around 1987. Development and commercialization of the technology was delayed primarily due to extensive R&D efforts to define critical parameters for proper unit operation, such as the spacing between ion exchange membranes.

A CEDI unit consists of an electrodialysis unit, described earlier in this chapter, modified to include the ion-depleting and ion-concentrating compartments. The compartments generally contain mixed bed ion exchange resin. The physical arrangement of the internal compartments for a CEDI unit is demonstrated in Figure 4C.7. As indicated earlier, the literature states that successful operation of the technology requires that the process must be positioned downstream of a single- or double-pass RO unit (Hernon et al., 1994a). As RO

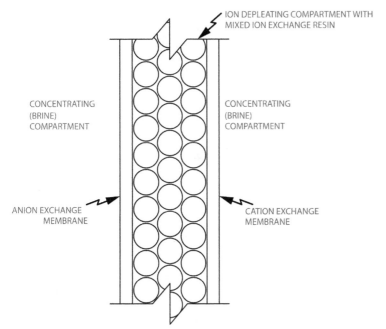

Figure 4C.7 Continuous electrodeionization chamber with resin.

product water enters the ion-depleting compartments of the CEDI unit, trace concentrations of both strongly and weakly ionized material are removed. Initially, the ion-depleted chamber, containing mixed resin, will remove the strongly ionized material. The electric current within the CEDI stack carries these ions to the ion exchange membranes or onto the ion exchange resin.

In this portion of the unit, the electric current is transmitted primarily by ions in the feedwater, with transfer of monovalent ions from the ion-depleted chamber to the concentrate (brine) chamber. However, as water and weakly ionized material proceed through the ion-depleted chamber, only weakly ionized material remains. As a result, polarization occurs, resulting in "water splitting," with the subsequent release of hydronium and hydroxyl ions into the water stream. While some of these ions will pass through the ion exchange membranes, a significant portion will be attracted to resin sites, releasing any exchanged weakly ionized material, such as bicarbonate ion, from the carbon dioxide – bicarbonate equilibrium. Silica would also be removed on the anion resin, which is rich in the hydroxyl ion. Since salt splitting is a continuous process, resin regeneration is also continuous.

Deionization occurs in a manner similar to that of a continuously regenerated mixed bed deionization unit. The function of the ion exchange resin is to enhance mass transfer (ion removal) within the unit. By removing ionic material, the resin also encourages water splitting, which is desired for this polishing ion removal technique. The ion exchange resin also reduces the resistance to electric current through the stack, since the resins are typically 2 to 3 (or greater) orders of magnitude more conductive than water in the ion-depleted chamber. In fact, in the portion of the chamber where water splitting is occurring, electric current passes through the ion exchange resin with little, if any, transfer directly through the water depleted of ions.

As previously indicated, water splitting provides continuous resin regeneration. As the resin is regenerated, exchanged ions, primarily weakly ionized material, is transferred from the ion exchange resin area through the ion exchange membranes to the concentrate (brine) stream. The literature suggests that water splitting is greatest in the area where resin beads are in contact with each other or with the ion exchange membranes (Allison, 1996). The literature further suggests that the extent of water splitting is a function of water purity, applied voltage, and water flow velocity through the ion-depleted chamber (Ganzi and Parise, 1990).

To verify the theory associated with salt splitting and the removal of both strongly ionized and weakly ionized substances, cell "autopsies" have been performed. The literature states that these studies clearly indicate that a significant portion of the resin within the ion-depleted cell is in the regenerated form, literally years after initial operation (Hernon et al., 1994b). Furthermore, the internal resin pH is significantly higher than the bulk pH of the RO feedwater to the ion-depleted chamber. The internal anion resin pH increases significantly from the point where RO water enters the unit to the region of the chamber where salt splitting is occurring. This property is extremely important because it promotes the removal of weakly ionized material common to water supplies, such as the bicarbonate ion (in equilibrium with carbon dioxide) and silica. Figure 4C.8 demonstrates resin positioning within the ion-depleted chamber, as well as the flow of ions in the ion-depleted and concentrate (brine) chambers during operation. It should be pointed out that ion transport within the ion-depleted chamber occurs by diffusion to the ion exchange resin, through the resin, and eventually through the ion exchange membranes with direct current as the driving force.

Enhancements to initial CEDI units include the addition of ion exchange resin to both the ion-depleting and ion-concentrating chambers. The literature describes the advantages of employing ion exchange resin in *both* the ion-depleting and concentrate (brine) chambers (Ganzi et al., 1997). It is suggested that this approach provides support for the ion exchange membranes, a decrease in the resistance through the relatively low ionic concentration concentrate (brine) chamber, and the potential for enhancing the attraction of ions from the ion-depleted chamber to the concentrating (brine) chamber.

The recovery rate for units will vary from 90% to 95% depending on the quality of feedwater from the upstream RO unit. The RO unit will remove particulate matter, colloidal material, and other contaminants and significantly reduce total viable bacteria, bacterial endotoxin, and organic material levels that are not effectively removed/reduced by CEDI.

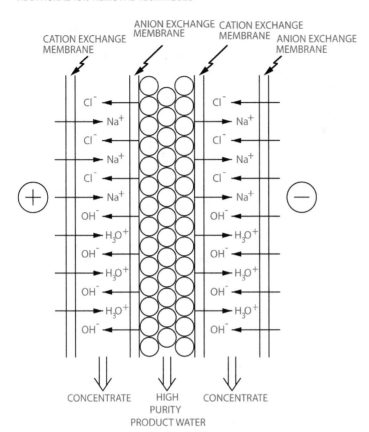

Figure 4C.8 Dynamics of operating a continuous electrodeionization system. *Source*: From Ganzi and Parise (1990), Ganzi et al. (1997).

Further, reverse osmosis decreases the ionic concentration to a point where CEDI can be operated in the water splitting mode described earlier.

The removal of weakly ionized material, such as the bicarbonate ion will be achieved in the water splitting area of the ion-depleted chamber. The literature states that the exchange rate for weakly ionized substances can be increased by improving the effective surface area of the ion exchange resin (Ganzi et al., 1997). Unfortunately, the volume of resin that can be positioned in the ion-depleting chambers is limited. However, the literature further suggests that this problem can be addressed by using tightly packed ion exchange resin (Ganzi et al., 1997). This can be achieved by using resin that exhibits 95% of the resin bead size (anion and cation) within 10% of a specified diameter.

The removal of organic material and the effect of organic material on the operation of CEDI must also be considered. In addition to selecting ion exchange resin with "uniform" bead size, it is extremely important to use resin that will not leach compounds into the ion-depleted water stream. While salt splitting provides regeneration, chemicals are not used. This can provide a significant degree of stability with regard to potential elution of organic material from the resin beads. Further, gas chromatography/mass spectrometry analysis performed on the effluent of an operating CEDI unit (2- to 3-year operation history) did not indicate the presence of any resin decomposition products (Collentro, 1997).

Another item that must be considered is the effect of organic fouling of the ion exchange material. Ideally, this should not be a concern, since pretreated feedwater has passed through an upstream RO unit. While there would appear to be limited potential for removing organic material, it is suggested that the concentrate (brine) stream for a CEDI unit may exhibit a higher TOC value than the ion-depleted stream. It is further suggested that the effect of organic material on both operation of the CEDI unit and the ability of the unit to remove the material are a function of the chemical structure of the organic material. Operating experience indicates that organic fouling of ion exchange resin does not have an effect on product water quality from a CEDI System.

The bacteria and, to a lesser extent, bacterial endotoxin levels on CEDI product water quality do not appear to be a concern for properly designed systems. The literature suggests that it *may* be possible to consider a CEDI unit as bactericidal (Ganzi et al., 1997). It is suggested, based on operating experience, that CEDI unit product water total viable bacteria levels are generally less than 1 cfu/100 mL (membrane filtration of 100 mL samples, R2A or PCA culture media, 30–35°C incubation temperature and 72–120 hour incubation time period). However, this experience is based on continuous CEDI unit operation and an inlet total viable bacteria level (RO product water level) less than 10 cfu/100 mL (same enumeration conditions). CEDI unit product water total viable bacterial levels simply reflect feedwater bacterial levels. In other words, as the feedwater bacterial levels to the CEDI unit increase, product water bacterial levels may increase. As a result, proper maintenance and cleaning of the upstream RO unit, particularly for controlling total viable bacteria, (sanitization) are critical. Furthermore, the design of feedwater and product water tubing for the CEDI unit should be of sanitary type. Because of the electric current through the unit, the sanitary feedwater and product water connections must be nonconducting, unpigmented polypropylene, or other materials, such as polyvinylidene fluoride (PVDF). Transition from the nonconducting material to 316L stainless steel is required for hot water sanitizable units. The use of PVDF for RO/CEDI systems in not suggested.

Bacterial endotoxins do not appear to affect the performance of a CEDI unit. Generally, bacterial endotoxin levels in the product water from the upstream RO unit are less than the minimum detectable. As discussed in chapter 2, it takes a considerable amount of dead gram-negative bacteria to increase the bacterial endotoxin concentration. It is not anticipated that a *significant* degree of bacterial destruction would occur within the CEDI unit, resulting in an increase in the product water bacterial endotoxin levels.

The literature indicates that silica removal by CEDI is significantly affected by water temperature (Hernon et al., 1994a). Lower water temperature is associated with a decrease in silica removal. For a specific application, the reactive silica removal at 16°C was 95%. A decrease in temperature to 10°C resulted in a decrease in silica removal from 95% to 90%. Considering the increase in silica removal associated with enhanced RO membrane performance, it is suggested that silica removal by CEDI units is not a concern.

Design Considerations

CEDI is employed for polishing application of single- and double-pass RO unit product water. Generally, product water from a single- or double-pass reverse osmosis unit is fed to a storage and distribution system. Without post-RO unit polishing, product water conductivity at point of use will not meet the USP Purified Water criteria. Further, the literature states that one particular CEDI manufacturer will not install the unit without an RO system upstream of the unit (Hernon et al., 1994a). Subsequently, system design generally includes an appropriately designed pretreatment system and single pass RO unit with feed to a CEDI unit. While continuous deionization unit stack life may be extended with the use of a double-pass reverse osmosis system, it is suggested that the capital expenditure, complexity, operating and maintenance cost do not justify the use of double-pass reverse osmosis.

While many RO/CEDI systems are designed for cyclic operation, it is strongly suggested that *continuous* flow, even when there is no demand for water, be considered. A suggested chemical sanitized RO/CEDI system is shown in Figure 4C.9. Pretreated feedwater is fed to a conical bottom RO break tank. Tank material of construction is unpigmented polypropylene or polyethylene. The tank is equipped with a tight fitting, gasketed, full-diameter cover. Tank volume is adequate to avoid "short-cycling" of the pretreated feedwater makeup valve. Tank accessories should include the following:

- Flanged and gusseted fittings for pretreated makeup water (top), discharge to the downstream repressurization pump (center bottom), RO/CEDI "loop" return (center top), level sensor (top), hydrophobic vent filter (top), pressure relief system (top),

ADDITIONAL ION REMOVAL TECHNIQUES

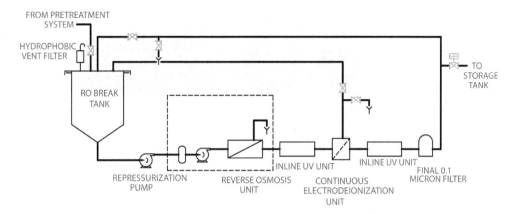

Figure 4C.9 Typical chemically sanitized RO.

concentrated liquid–sanitizing agent introduction port (top), and CEDI waste optional recovery stream (top).
- Hydrophobic vent filtration system with "inverted" unpigmented polypropylene housing and O-ring filter-to-housing seal mechanism.
- Conductance-type or equivalent-type level monitoring system with analog outlet capability.
- Relief mechanism to eliminate implosion or explosion of the tank in the event of accessory malfunction or failure.
- Steel support stand with corrosion-resistant finish.
- The pretreated feedwater piping/tubing should include a positive acting (air-to-open, spring-to-close) pneumatically actuated diaphragm valve. A sample valve and pressure indicator should be positioned directly upstream of the valve. The piping/tubing downstream of the valve should be of minimum physical length and arranged to be fully drainable.
- The bottom discharge line from the tank to the repressurization pumping system shall contain a compound-type pressure gauge, zero dead leg–type drain valve, and inline full-diameter pump feedwater isolation valve as shown in Figure 4C.10. The compound pressure gauge is required to verify that pump cavitation is not occurring. The "low" level tank set point may adjusted to satisfy pump net positive suction head (NPSH). A sampling valve is intentionally excluded at this point since it is

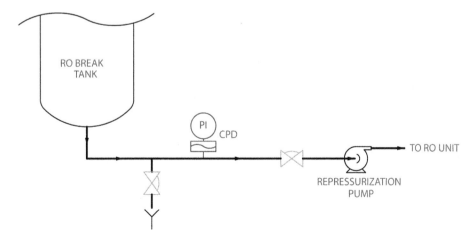

Figure 4C.10 RO Break tank discharge tubing.

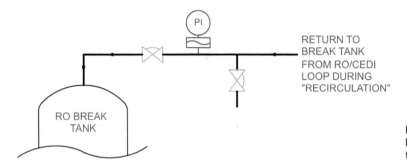

Figure 4C.11 RO CEDI loop return to RO break tank.

possible that the pressure may be close to 0 psig or at a slight vacuum during operation. Sampling at this point could create introduce undesirable impurities into the system. A sample valve should be positioned downstream of the repressurization pump. The rinse-to-drain valve is included to allow infrequent draining of the tank for maintenance operation. Of greater importance however, the valve is used during the chemical sanitization cycle for the entire RO/CEDI loop including the RO break tank. Upon completion of chemical sanitization, water with sanitizing agent in the RO break tank should be drained by closing the repressurization pump feedwater isolation valve and opening the drain valve. It is important to indicate that the discharge from the drain valve (generally equipped with pre-sanitized clean hose) should be directed to a depressurized drain *with* an air break between the discharge of the hose and the drain. It is also important to note the level in the RO break tank to avoid fully draining the tank, which would allow atmospheric bacteria and other potential aqueous contaminants to enter the tank.

- The return tubing from the RO/CEDI loop should contain an inline manual diaphragm-type isolation valve, pressure indicator, and zero dead leg–type drain valve as shown in Figure 4C.11. The inline manual valve is throttled slightly closed to exert back pressure on the RO/CEDI loop when it is in a recirculating mode. Ideally, the pressure through the RO/CEDI loop during makeup to the USP Purified Water storage tank or Water for Injection feedwater tank and during recirculation should be equal. The drain valve is employed at the end of the chemical sanitization cycle. While the majority of the chemical disinfecting agent can be removed by carefully draining nearly all of the water from the RO break tank, some residual disinfecting agent will still be present. Further, the individual unit operations in the RO/CEDI loop will contain residual disinfecting agent. After the draining operation for the RO break tank is completed, the repressurization pump feedwater valve can be opened. The RO/CEDI loop tubing inline valve can be closed and the drain valve in the return tubing opened. Note that the precautions for hose handling and direction to drain indicated earlier apply. The repressurization pump can be energized, delivering water through the RO/CEDI loop (with a small flow to RO waste). The RO/CEDI return containing disinfecting agent is diverted to waste rather than returning to the tank. This displacement operation proceeds until all sample points in the RO/CEDI loop indicate the absence of chemical disinfecting agent. RO/CEDI loop return valve positions are restored to normal operating positions upon completion of chemical disinfecting agent removal.
- The CEDI waste is high-quality water when compared with the pretreated feedwater to the RO break tank. Ideally, it is desirable to recover the waste to the RO break tank. Unfortunately, the CEDI waste stream may contain carbon dioxide or ammonia gas depending upon RO feedwater characteristics. The presence of either of these reactive gases could result in an increase in CEDI product water conductivity. Figure 4C.12 depicts a system with automatic valves that can either divert CEDI waste to a depressurized drain with air break or return the waste to the RO Break Tank based on CEDI product water conductivity.

ADDITIONAL ION REMOVAL TECHNIQUES

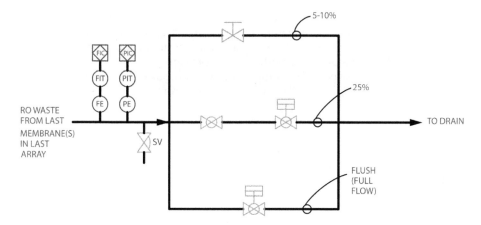

Figure 4C.12 RO waste tubing.

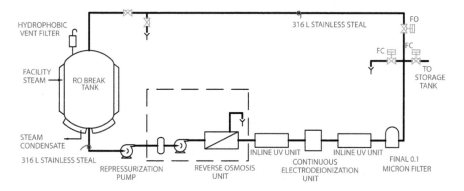

Figure 4C.13 Typical hot water sanitized RO.

The chemical sanitization provisions indicated in the preceding text should be included in RO/CEDI system with hot water sanitization capability. As discussed in chapter 9, periodic hot water sanitization will not remove biofilm. Chemical sanitization, properly executed with a 1% solution of peracetic acid and hydrogen peroxide will remove biofilm assuming that there are no dead legs or that dead legs are exposed to the peracetic acid and hydrogen peroxide solution. The suggested chemical sanitization frequency for non–hot water systems is three to six months. The suggested chemical sanitization frequency for hot water sanitizable RO/CEDI systems is annually.

A suggested hot water–sanitized RO/CEDI system is shown in Figure 4C.13. Pretreated feedwater is fed to a 316L stainless steel RO break tank. The tank should be designed, fabricated and tested in accordance with the American Society of Mechanical Engineers (ASME) Code for Unfired Pressure Vessels, Section VIII, Division 1, as discussed in chapter 6. Heating for the hot water sanitization process can be introduced to a heating jacket around the RO break tank or a shell and tube type heat exchanger positioned downstream of the RO break tank but upstream of the RO unit. Tank accessories are similar to the accessories for a chemically sanitized RO/CEDI loop with the following exceptions:

- Fittings should be sanitary ferrule type.
- The tank should be provided with a sanitary access manway, 18 to 24 in. diameter. A top-mounted manway is preferred.
- Hydrophobic vent filtration system with "inverted" unpigmented 316L stainless steel housing and O-ring filter-to-housing seal mechanism should be provided. Further, housing heating provisions should be included as discussed in chapter 6.

- The tank level sensor should be fabricated of material compatible with hot the RO/CEDI loop hot water sanitization temperature of 80 to 90°C. Differential pressure level sensing with analog output to the central system control panel is suggested.
- The tank should be provided with sanitary pressure relief provisions. A rupture disc with support accessories, including a discrete output to the central control panel if disc failure occurs, should be included. The recommended disc rupture pressure should not be greater than 50% of the design pressure rating for the tank.
- As indicated earlier, the tank may be equipped with a single- or multiple-zone heating jacket(s) positioned around the exterior of the vessel (chap. 6). If an external heat exchanger is employed in lieu of the heating jacket, 2 in. thick chloride-free insulation contained in a 304L stainless steel outer shroud should be provided.
- The pretreated feedwater piping/tubing and accessories indicated for chemically sanitized system should be included. However, material of construction should be 316L stainless steel.
- The bottom discharge line from the tank to the repressurization pumping system shall contain a compound-type pressure gauge, zero dead leg–type drain valve and inline full-diameter pump feedwater isolation valve as shown in Figure 4C.10 with surfaces in contact with water constructed of 316L stainless steel. Teflon or Viton gaskets are recommended. Diaphragms should be Teflon with EPDM backing.
- The return tubing from the RO/CEDI loop should contain an inline manual diaphragm-type isolation valve, pressure indicator, and zero dead leg–type drain valve as shown in Figure 4C.11. Material in contact with water should be of 316L stainless steel construction with Teflon or Viton gaskets. Valve diaphragms should be Teflon with EPDM backing.
- CEDI 316L stainless steel waste tubing shall be employed.
- The RO/CEDI loop hot water sanitization cycle requires temperature control. If the RO break tank is equipped with a heating jacket, the temperature-sensing element should be positioned on the lower straight side of the tank. If a dedicated shell and tube heat exchanger is used (prior to the RO unit), the temperature sensor should be positioned downstream of the heat exchanger. The temperature sensor must provide analog output to the central control panel. Further, analog signals for temperature should be provided downstream of the RO unit, downstream of the CEDI unit, and downstream of the final filtration system. Modulated plant steam is used to heat water until the tank (or RO feedwater temperature) increases to a value of 80 to 85°C. Insulation of RO/CEDI tubing and accessories is suggested. The hot water sanitization time period, suggested as two hours in duration, should not begin until the post final filter housing temperature is 80°C. It is suggested that the hot water sanitization cycle be automatically controlled after manual initiation.
- The RO break tank or heat exchanger will require valves, instrumentation and controls, as discussed in chapter 6.

Additional RO/CEDI loop hot water sanitization items to be considered include, but are not limited to, the following:

- The RO break tank water level should be established at a maximum level, such as 90% capacity, prior to initiation of the cycle. This minimizes the volume of softened water makeup to the tank during the sanitization process. If feasible, a dedicated USP Purified Water loop "drop" or distillation/Pure Steam generator feedwater drop should be considered for makeup during hot water sanitization. The inorganic and organic material in pretreated water can have an effect on both the "conductivity recovery time" for the CEDI unit and long-term successful operation of the CEDI unit. Material may be "eluted" or pass through the RO membranes at sanitizing temperature. This material will accumulate on ion exchange resin and membranes in the CEDI unit (Collentro, 2003). This is demonstrated in Table 4C.2.
- During RO/CEDI loop hot water sanitization, both the RO transmembrane pressure drop and feedwater pressure should not exceed the RO membrane manufacturer's

Table 4C.2 RO Product Water TOC Data During Hot Water Sanitization Cycle

Elapsed time after start of hot water sanitization operation (minutes)	Phase of sanitization cycle	"Grab" TOC sample results (μg/L)
0	Start	25
20	Heat-up	120
40	Heat-up	230
60	Heat-up	340
80	Heat-up	510
100	Temperature hold	605
120	Temperature hold	810
140	Temperature hold	980
160	Temperature hold	1070
180	Temperature hold	1180
200	Temperature hold	1200
220	Cooldown	1120
240	Cooldown	1080
260	Cooldown	910
280	Cooldown	680
300	Cooldown	560
320	Cooldown	100
340	Cooldown	50
600	Normal operation	20

Source: From Collentro (2003).

recommendations. Generally, this value is 40 to 45 psig. The RO break tank repressurization pump motor should be provided with a variable frequency drive. The speed of the pump can be established through the central control panel to maximize pressures without exceeding the indicated feedwater and transmembrane pressure criteria.
- It is suggested that the RO feedwater pump motor be de-energized throughout the RO/CEDI loop hot water sanitization cycle.
- Normal operating alarms for parameters such as flow rates, pressures, conductivity, and percent ion rejection alarm must be inhibited during the sanitization operation. Alarm conditions for these parameters should be established for the hot water sanitization cycle.
- The RO unit shall be designed in accordance with criteria set forth earlier in this chapter.
- The RO unit waste line "three parallel valve" arrangement discussed earlier in this chapter should be configured such that the 5% to 10% waste valve is open.
- The electrical power to inline ultraviolet units in the RO/CEDI loop should be inhibited upon initiation of the hot water sanitization cycle.
- The electrical power to the CEDI "stack(s)" should be inhibited during hot water sanitization.
- The makeup valve to the downstream USP Purified Water storage tank (or feed to a distillation unit and/or Pure Steam generator) should remain in a closed position during the hot water sanitization cycle.
- The recirculating valve to the RO break tank should remain in an open position throughout the hot water sanitization phase of the cycle. A divert-to-drain valve should open, allowing cool pretreated makeup water to the RO break tank to displace the hot sanitizing water as the final step of the RO/CEDI loop sanitization cycle.

The information presented above has focused on chemical and hot water sanitization for RO/CEDI systems. Specific design parameters for CEDI units are available from the unit

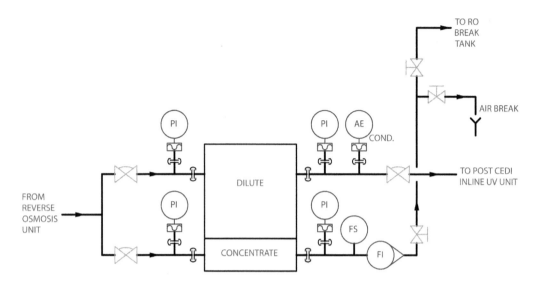

Figure 4C.14 CEDI unit P&ID.

manufactures. However, specific design considerations for CEDI units are presented as follows:

- Figure 4C.14 depicts a typical single CEDI stack with suggested instrumentation. Feedwater to a CEDI stack is split into two streams. One stream feeds the ion-depleting chambers while the second stream feeds the ion-concentrating chambers. Individual manual diaphragm valves are provided on each of the feedwater lines to the stack. Sanitary pressure gauges are positioned downstream of each manual feedwater valve. Feedwater pressure is regulated such that the pressure to the ion-depleting chambers is 5 to 10 psig greater than the pressure of the feedwater to the ion-concentrating chambers. Obviously, the pressure regulation provides a lower flow rate to the ion-concentrating chambers when compared with the ion-depleting chambers. The waste flow line from the CEDI stack should be equipped with a regulating valve, pressure gauge, direct reading flow meter (variable area type), and flow switch. While the exact waste flow rate, as a percent of feedwater flow rate, is a function of the quality of the upstream RO product water quality, the value is generally about 5% to 10%. RO product water containing low bicarbonate ion levels (low carbon dioxide concentration) and related low conductivity indication (<5 µS/cm) may only require a 5% CEDI unit waste flow. Conversely, RO product water with higher bicarbonate/carbon dioxide levels and associated higher conductivity values will generally require a waste flow rate as high as 10%. While the indicated data (Collentro, 2010) are for makeup conditions to a downstream tank (single pass), it is assumed that the CEDI unit is installed in a recirculating system, as discussed earlier in this chapter.
- It is suggested that the tubing, valves, and accessories be constructed of 316L stainless steel. The two feedwater connections, product water connection, and waste connection to the stack must transition from 316L stainless steel to unpigmented polypropylene or PVDF fittings on the stack to electrically isolate the stack from the conductive stainless steel tubing. This is accomplished by mating stainless steel sanitary ferrules to the polypropylene or PVDF fittings with sanitary ferrules. Suggested sanitary ferrule gasket material is Viton. Further, low-pressure sanitary ferrule clamps are suggested. The use of transition "hoses" is discouraged.
- 316L stainless steel tubing with compression-type fittings may be used for the waste line. The flow rate control valve may be of needle type with bellows-type seal

- mechanism. The variable area flow rate meter may be equipped with threaded end connections. Materials of construction may be transparent nonorganic eluting plastics or glass. Meters for hot water sanitizable units must be capable of operating at sanitizing temperatures of 80 to 85°C.
- The electrical supply to a CEDI stack must be terminated if there is no flow through the stack. It is suggested that a nonsanitary flow switch with discrete signal to the central control panel through the CEDI power panel can provide the required termination of stack electrical power. By positioning the switch in the waste line, it is possible to eliminate a sanitary type flow meter, or meter of nonsanitary design, in the feed or product water tubing for the CEDI stack. However, a stainless steel construction flow switch is suggested to provide long-term reliable flow rate monitoring and compatibility with hot water for hot water sanitizable units.
- The product water tubing from the CEDI unit should contain a manual diaphragm isolation valve zero dead leg sample valve, sanitary conductivity cell, and sanitary pressure gauge. For systems equipped with multiple CEDI units operating in parallel, it is suggested that each unit be equipped with the indicated conductivity cell. While conductivity monitoring of the "common" product water from the multi stack system may be routinely performed, the individual probes, perhaps with "lead wires" not permanently connected to a conductivity meter, can be used for determining individual CEDI stack product water conductivity values in the event of common CEDI system conductivity increase. Further, the cells can be used for periodic stack performance monitoring.
- While the feedwater, product water, and even wastewater from a CEDI stack contains a low ionic content, it is possible that catastrophic stack failure could result in a high electric current input to the water solution. Grounding rods should be installed in feedwater, product water, and waste lines to insure personnel safety.
- Waste recirculation to the RO break tank should be considered on the basis of product water conductivity as discussed earlier.
- The power supply for a CEDI unit will be equipped with transformers. Heat generated from the transformers will increase the temperature of the power panel that may also contain instrumentation or control wiring for communication with the central control panel. It is strongly suggested that the power panel be equipped with cooling devices. The use of vortex-type coolers using a clean, dry, oil-free air supply provides excellent cooling.
- Product water from the CEDI unit may pass through a final filter before feeding a USP Purified Water storage tank. It is suggested that 0.1 µm membrane filters be selected for this application. It is further suggested that an inline ultraviolet sanitization unit be positioned between the final membrane filtration system and the CEDI unit to minimize/eliminate microbial growth back to the CEDI unit.

Operating and Maintenance Considerations

General CEDI manufactures feedwater requirements suggested for operation are presented in Table 4C.3. It is implied that exceeding one or more of the values presented in Table 4C.3 would result in long-term operating problems. However, as the design of CEDI units and RO membranes have been enhanced, extensive operating experience indicates that feedwater conductivity values ≤10 µS/cm, TOC concentration less than 25 µg/L, and total viable bacteria less than 10 cfu/100 mL provide excellent long-term operation of CEDI units. Considering the indicated feedwater conditions, CEDI product water quality is less than 0.10 µS/cm at 25°C. Typical performance characteristics for the unit are presented in Table 4C.4.

Power requirements for the units are generally extensive. One CEDI unit manufacturer's bulletin suggests that the standard unit for a flow rate of 50 gpm will require a 50 A breaker, 480 V, 3 phase, 60 cycle electrical supply (EDI Membrane Systems, 1995).

Table 4C.3 CEDI Unit Manufacturer's Feedwater Recommendations

Parameter	Maximum recommended value
Conductivity at 25°C (including carbon dioxide and silica)	<40 µS/cm
Free chlorine as Cl_2	<0.02 mg/L
Iron	<0.01 mg/L
Manganese	<0.01 mg/L
Sulfide	<0.01 mg/L
pH	4–11
Total hardness as $CaCO_3$	<1.0 mg/L
TOC as carbon	<0.5 mg/L
Silica	<1.0 mg/L

Source: From Siemens (2009).

Table 4C.4 CEDI Unit Manufacturer's Typical Performance Data

Parameter	Projected value
Recovery	90–95%
Differential pressure (feed to product)	20–30 psid
Silica removal	90–99%

Source: From Siemens (2009).

Operation and maintenance items that should be considered include, but are not limited to, the following:

- As indicated, CEDI unit product water conductivity should be less than 0.10 µS/cm at 25°C. The CEDI unit amperage/voltage may require adjustment to achieve this value. Further, for systems with high RO unit carbon dioxide/bicarbonate concentrations, CEDI unit waste reclaim may not be possible.
- CEDI unit product water conductivity will increase during and subsequent to both hot water and chemical sanitization. During both operations power to the stack is inhibited. Prior to chemical sanitization the ion exchange capability of the CEDI stack is inhibited by salt introduction, as discussed below. Subsequent to hot water sanitization, CEDI unit product water conductivity will slowly decrease from a value as high as 1 to 5 µS/cm to a value that is generally lower than the pre–hot water sanitization value. The decrease in post-CEDI unit product water conductivity to a value less than 0.10 µS/cm subsequent to hot water sanitization generally takes two to four hours. Further CEDI product water conductivity may be noted for a period of about 12 to 24 hours. Subsequent to chemical sanitization, CEDI unit product water conductivity will slowly decrease from a value generally more than 25 to 50 µS/cm to a value that is generally lower than the pre–chemical sanitization value. The decrease in post-CEDI unit product water conductivity to a value less than 0.10 µS/cm subsequent to chemical sanitization generally takes 8 to 24 hours. Further CEDI product water conductivity may be noted for a period of about 48 to 72 hours.
- As indicated earlier, the ion exchange capacity of CEDI units must be exhausted prior to chemical sanitization. Subsequent to de-energizing the CEDI stack, sodium chloride solution is fed to the CEDI stack. The volume and concentration of sodium chloride used for this ion removal "kill" operation should be consistent with the manufacturer's literature. However, experience indicates that the recommended salt volume is excessive. This may result in extended CEDI unit product water "recovery" time subsequent to chemical sanitization. It is suggested that a salt volume of about 1 pound per 5 gpm CEDI unit product water flow rate capacity is adequate. However, use of the lower salt volume should be verified by the CEDI unit

manufacturer. The method and components employed to introduce the salt solution to the CEDI stack should not introduce contaminants.
- Subsequent to chemical sanitization or repeated hot water sanitization operations, a drop in CEDU unit product water flow rate may be noted. Generally, design product water flow rate can be recovered by a brief (5–10 minute) flow through the CEDI stack in the reverse direction (product-to-feed). However, since the characteristics of CEDI stacks from different manufacturers are unique, manufacturer confirmation should be obtained before executing the operation. If executed, the "backwash" water should be of USP Purified Water quality or greater.
- CEDI stack life will vary with manufacturer. Experience indicates that a minimum CEDI stack life is about two years. With appropriate upstream RO unit pretreatment and RO unit maintenance stack lives of three years to more than five years have been noted.
- When a CEDI unit requires replacement, it is important to note that the entire unit does not need to be replaced. The unit can be "restacked" at a cost of about 50% of that for total CEDI unit replacement. For single- or multiple-stack units, it is suggested that a one or more "spare" CEDI units, treated with "preservative" be purchased as "spare parts." This eliminates any downtime should unlikely catastrophic stack failure occur. For systems with multiple stacks at multiple locations, CEDI unit manufacturers may agree to a "rotating" spare CEDI unit inventory at their facility. This concept, negotiated during purchase of the CEDI unit but, as indicated, is limited to purchase of several CEDI units.
- CEDI units may be provided with a microbial control preservative. The material is generally removed from the CEDI unit quickly as noted by product water conductivity. However, it is important to note the CEDI product water "conductivity cycle" that may be observed upon start-up of a new or restacked unit. As preservative is removed conductivity will decrease. It is quite like that this may occur in a few minutes. Product water conductivity, regardless of the CEDI unit amperage/voltage settings, will likely approach 0.055 µS/cm at 25°C for 1 to 12 hours depending upon RO product water conductivity. An increase in CEDI unit conductivity may be noted after this initial period. During this period, ions are being removed by the resin in the stacks. Adjustment of amperage/voltage to the required settings for the specific application may be required. "Tuning" of the amperage/voltage values may take as long as one to two weeks. The tuning process is a bit difficult due to the delayed response of CEDI unit product water conductivity to changes in stack amperage/voltage adjustment.
- Initial experience with CEDI units in the late 1990s indicated stack leakage and the requirement to frequently tighten bolts on rods compressing the individual "chambers." Extensive experience with CEDI units indicates that stack leaks are simply not observed. Leaks are limited to the polypropylene fitting-to-stack adapters, primarily at the sealing gasket. Further, it is suggested that periodic bolt tightening is not required.
- Recent observations indicate that monochloramine is not removed by CEDI units (Collentro, 2010a). An increasing number of RO pretreatment systems are improperly designed or maintained. RO will not remove monochloramine (with the exception of undesirable membrane oxidation). Monochloramine in the feedwater to a CEDI stack will ultimately result in an increase in product water conductivity (Krpan and Wu, 2010). It appears that the mechanism associated with this observation is irreversible, resulting in the need for CEDI unit replacement.

REFERENCES

Allison RP. The continuous electro-deionization process. Ionics Bulletin No. TP-374. Presented at the American Desalting Association Biennial Conference and Exposition, August 4–8, Monterey, CA, 1996.

AWWA. Electrodialysis and electrodialysis reversal. In: Manual of Water Supply Practices. Manual M38. Murray P, ed. Denver: American Water Works Association, 1995.

Collentro WV. Unpublished data, Pennsylvania Pharmaceutical Manufacturing Facility, 1997.
Collentro WV. "Hot Water Sanitization of RO and Double Pass RO EDI Systems", presented at the UltraPure Water Journal Executive Forum – Expo. Plainfield, NJ: Tall Oaks Publishing, 2003.
Collentro WV. Pharmaceutical Water System Fundamentals – Ion Removal – Reverse Osmosis, The Journal of Validation Technology, Summer 2010. Duluth, MN: Institute of Validation Technology, 2010; 16(3):66–75.
Collentro WV. Pharmaceutical Water System Fundamentals – Ion Removal – Resin Based Systems, The Journal of Validation Technology, Fall 2010. Duluth, MN: Institute of Validation Technology, 2010a; 16(4).
EDI Membrane Systems. Ionics electrodeionization process. Ionics Bulletin No. 154-E, 1995.
Ganzi GC, Parise PL. The production of pharmaceutical grades of water using continuous deionization post-reverse osmosis. J Parenter Sci Technol 1990; 44(4):231–241.
Ganzi GC, Jha AD, DiMascio F, et al. Theory and practice of continuous electrodeionization. Ultrapure Water 1997; 14(6):64–69.
Hernon BP, Zanapalidou RH, Zhang LI, et al. Applications of Electrodeionization in UltraPure Water Production: Performance and Theory. Ionics Bulletin No. TP-371. Presented at the 55th Annual Meeting of the International Water Conference, 30 October–2 November in Pittsburgh, PA, 1994a.
Hernon BP, Zanapalidou RH, Zhang L, et al. Electrodeionization power plant applications. Ultrapure Water 1994b; 11(5):33–41.
Kiernan JC, Harvey WT, Burrage D. Electrodialysis reversal brine concentration for zero liquid discharge: The ocean state power story. Ionics Bulletin No. TP-356, 1992.
Krpan N, Wu L. "Chlorine Species Passage Through Polyamide Reverse Osmosis Membranes", presented at Ultrapure Water Pharma. New Brunswick, NJ, 2010.
Pontius FW, ed. Water Quality and Treatment—A Handbook of Community Water Supplies. (American Water Works Association) New York: McGraw-Hill, Inc., 1990:735.
Siemens Water Technologies Corporation. Ionpure® HWS-1 Instant Hot Water Sanitizable CEDI Module, ION-HWS.S-DS-0309, Lowell, MA, USA, March 2009:1–2.
Siwak LR. Here's how electrodialysis reverses ... and why EDR works. Ionics Incorporated Bulletin. The International Desalination & Water Reuse Quarterly 1993; 2(4).
von Gottberg AJM, Siwak LR. Re-engineering of the Electrodialysis Reversal Process. Ionics Incorporated Bulletin, The International Desalination & Water Reuse Quarterly, 1997.

5 | Distillation and pure steam generation

THEORY AND APPLICATION

Distillation is the primary process used for the production of USP Water for Injection for pharmaceutical applications. It is the only process allowed for the production of EP Water for Injection. USP essentially states that Water for Injection may be produced by distillation or any process that can be demonstrated, on a consistent basis, to provide water equal or superior to distillation. However, the number of facilities employing membrane or other processes for the production of USP Water for Injection is extremely small. While USP, EP, and JP have attempted to "harmonize" the method for production of Water for Injection, it is strongly suggested that the EP (as well as certain other pharmacopeias) will require distillation for production of Water for Injection. The EP position regarding this matter refers to the "robust" nature of the distillation process specifically as it relates to the phase change of water to steam. Papers and presentations have addressed this issue (Meltzer et al., 2009; Collentro, 2005; and European Medicines Agency, 2008).

It should also be indicated that some pharmaceutical facilities use distillation as the method for producing USP/EP Purified Water. This is particularly true when USP Purified Water is desired for small daily volume demands that require low bacteria levels. Pure Steam generators are used to provide steam that, when condensed, meets the chemical, bacterial, and bacterial endotoxin specifications outlined in the USP *Official Monograph* for Pure Steam. Distillation, as a unit operation, provides bacterial endotoxin removal and reduces dissolved ionic material, particles, colloids, and nonvolatile organic material by a phase change related to the conversion of water in the liquid state to the gaseous state (steam). Since this process occurs at a minimum temperature of 100°C, bacteria are also destroyed. During the phase change process associated with distillation (or Pure Steam generation), the steam generated must be free of any water that may contain undesirable impurities, specifically bacterial endotoxins. Single-effect distillation units, multiple-effect distillation units, vapor compression distillation units, Pure Steam generators, and condensing units are all used to achieve this goal. These units will be discussed in this chapter.

To understand the principle associated with a phase change, the basic principles of distillation and Pure Steam generation, and to emphasize the importance of producing water as a gas, steam without *any* water as liquid, it appears appropriate to discuss the thermodynamic principles related to the process. For the purposes of this discussion, it is assumed that the phase change occurs at atmospheric pressure, which is equivalent to 0 psig (pounds per square inch gauge) or 14.7 psia (pounds per square inch absolute—atmospheric pressure at sea level).

The phase change of water at atmospheric pressure occurs at 212°F (100°C). As ambient temperature water is heated to 212°F, the energy required, per pound of water, is about 1 BTU (British thermal unit, a unit of heat energy) per °F. Assume that water existed at an initial temperature of 70°F. The energy required to increase the temperature of 1 lb of water from 70 to 212°F would be 142 BTU/lb, calculated as follows:

Energy required = (mass of water in lb) × (specific heat of water at constant pressure)
                  × (temperature increase in °F)

or

$q = mc_p T$
$q = 1\,\text{lb} \times 1.0\,\text{BTU/lb}\,°\text{F} \times 142°\text{F}$
$q = 142\,\text{BTU}$

Once water reaches 212°F, the phase change will occur. The amount of energy required to change 1 lb of liquid water at 212°F to 1 lb of vapor (steam), also at 212°F, can be calculated using a thermodynamic function called "enthalpy," which represents the "total heat" or "heat content" of a substance at a particular temperature and pressure. The enthalpy of steam, without any water, at 212°F and 14.7 psia, is 1150.4 BTU/lb.

$$\text{Note}: P_{\text{absolute}}(\text{psia}) = P_{\text{gauge}}(\text{psig}) + 14.7$$

On the other hand, the enthalpy of liquid water at 212°F and 14.7 psia is only 180.1 BTU/lb. The amount of energy required to convert liquid water at 212°F and atmospheric pressure to steam can be calculated by subtracting the enthalpy of liquid water from the enthalpy of steam at 212°F. This results in a heat requirement of 970.3 BTU/lb of water. The energy required per pound of water is summarized by the following equation:

$$q = (\text{mass of water in pounds})$$
$$\times (\text{enthalpy of steam at 14.4 psia} - \text{enthalpy of water at 14.7 psia})$$

or,

$$q = m(h_{\text{steam}} - h_{\text{liquid}})$$
$$q = 1\,\text{lb}(1150.4\,\text{BTU/lb} - 180.1\,\text{BTU/lb})$$
$$q = 970.3\,\text{BTU}$$

This value is significantly greater than the 142 BTUs required to increase 1 lb of water from 70 to 212°F. The energy associated with increasing the temperature of water from 70 to 212°F is referred to as "sensible heat," while the energy required to convert 1 lb of water to 1 lb of steam is referred to as "latent heat."

It is important to point out that the temperature of the heated water will *not* increase above 212°F at atmospheric pressure no matter how much heat is added. Additional heat will simply convert more water to steam. However, heat transferred to Pure Steam, not water, could result in "superheat." This process increases the temperature of steam at a certain pressure. To simplify, steam, as a gas, is heated with the resulting heat input increasing the temperature, at a constant pressure, provided that the steam volume is not fixed. Superheat is undesirable for Pure Steam as discussed later in this chapter.

It would be inappropriate to conclude this discussion without explaining why the term *enthalpy* has been introduced. It has been indicated that one of the most important factors associated with the production of Pure Steam and Water for Injection (by distillation) relates to the fact that there is absolutely no water in Pure Steam. Various measuring devices are available to determine the amount of liquid water in steam (Cal Research Inc., 1998). These devices use the heat content, enthalpy, as a measuring device. For example, the device may indicate that the enthalpy of the steam is not 1150.4 BTU/lb but rather 1100 BTU/lb. This situation indicates that total evaporation (phase change) of water to steam has not occurred. The "percent conversion" is directly proportional to the measured change in enthalpy versus the required change in enthalpy (970.3 BTU/lb). In fact, the percent conversion is expressed as the "quality" of the steam, a ratio of the measured enthalpy change, divided by the required enthalpy change for total conversion. A quality of 1.00 would indicate that there is no water in the steam. It is the objective of any distillation unit or Pure Steam generator to produce Pure Steam (water as a gas) with a quality of 1.00, ensuring that there are no contaminants or impurities present, including bacterial endotoxins.

The quality of Pure Steam as well as the presence of noncondensable gasses is a concern. This is particularly important for sterilization applications. The European Committee for Standardization (CEN) requires that Pure Steam quality, superheat, and noncondensable gases be measured periodically (HTM 2031, 2007 and British Standards Institute, 2006). This matter is further discussed in the literature (PDA, 2010). Table 5.1 presents Pure Steam attributes (with explanation) and physical parameters with accepted testing limits. It should also be indicated that Pure Steam generator feedwater and condensed Pure Steam chemical parameters may be specified by regulatory agencies (AAMI/ANSI, 2006; British Standards Institute, 2006; and HTM 2031, 2007).

To achieve proper distillation unit or Pure Steam generator performance, a method of ensuring that water is absent in the steam must be provided. Various manufacturers of distillation units and Pure Steam generators will refer to a vapor-liquid disengaging process by

Table 5.1 Pure Steam Attributes

Parameter	Limit
Noncondensable gases	$\leq 3.5\%$
Superheat	$\leq 25°C$
Dryness value for metal loads	≥ 0.95

Sources: From HTM 2031 (2007) and British Standards Institute (2006).

patented names. This chapter will use the same terminology for this operation. Again, it is the major factor associated with appropriate unit design and operation.

Another issue associated with the overall purity of steam produced during the distillation and Pure Steam generation processes relates to certain volatile organic impurities and other inorganic impurities that may be present in the feedwater supply to the units. Trace concentrations of volatile organic compounds, or inorganic compounds that exhibit "volatile" characteristics, will compromise the chemical purity and, in certain cases, the bacterial endotoxin content of the Pure Steam and, in the case of distillation, the resulting condensed distillate product.

The numerous impurities that may be present in raw water supplies and, subsequently, in feedwater supplies to distillation units or Pure Steam generators (based on the nature and type of the water purification equipment in the feedwater treatment system) are discussed, in detail, in chapter 2. Volatile organic impurities of concern include trihalomethanes (THM) compounds, such as chloroform, and monochloramine, a disinfecting agent used by municipalities with feedwater sources from a surface water supply containing appreciable amounts of naturally occurring organic material (NOM). In addition, there are several volatile organic pollutants that could be present in trace quantities, particularly if the raw feedwater supply to the pretreatment system for the distillation unit or Pure Steam generator is from a "private" source such as a well. Further, the presence of "reactive gases" such as carbon dioxide and ammonia, in equilibrium with bicarbonate and ammonium, respectively, are also important. These gases, if not removed, will be carried over with the steam. Equilibrium will be reestablished in the condensate, degrading distilled product water quality. In general, vapor compression distillation units, which operate at lower temperatures than multiple-effect distillation units, are equipped with provisions to remove volatile impurities. However, both ammonia/ammonium and monochloramine decomposition by-products have been observed in vapor compression unit product water (Collentro, 2005–2009). While the higher operating temperatures associated with multiple-effect distillation units and, to a lesser effect, single-effect distillation units may result in some "outgassing" within the condensing section, trace quantities of impurities will be present in the distillate product.

The presence of reactive silica and, to a lesser extent, colloidal silica, in feedwater to a multiple-effect distillation unit or Pure Steam generator is undesirable. Quite often, multiple-effect distillation unit and Pure Steam generator manufacturers will establish a warranty of equipment performance based on a maximum silica concentration in the feedwater. Unfortunately, reactive silica and, to a lesser extent, colloidal silica, will be literally volatile; when present, they will be carried over with Pure Steam. While silica may affect product water quality, it is very weakly ionized. Thus, it is difficult to detect in distilled product water. However, the major concern associated with silica (reactive or colloidal) relates to the fact that it will cause deposits on heat transfer areas within the vapor-liquid disengaging section and, to a lesser extent, within the condensing sections of distillation unit. The degree of silica "carryover" with steam is a function of distillation unit or Pure Steam generator operating pressure. It is a greater concern in multiple-effect distillation units that operate at higher pressures and, subsequently, higher temperatures, when compared with vapor compression distillation units or single-effect distillation units. Obviously, it is also a concern for Pure Steam generators. The siliceous precipitates formed, primarily on vapor-liquid disengaging systems, result in a physically hard, tightly adhered material with a "sandpaper-like" surface, significantly reducing desired Pure Steam velocity and smooth flow of Pure Steam required by many vapor-liquid disengaging systems to remove any entrapped water.

The production of distilled water requires heating *and* cooling. The selection of a distillation unit for a given application is a function of the available facility steam and cooling water. The selection of a particular type of distillation unit is also a function of the volume of distilled product water required each day and, more importantly, the volumetric flow rate (e.g., gal/hr) of distilled product water required. Single-effect distillation units will require more steam per volume of distillate product water than multiple-effect distillation units. Vapor compression units will require less steam and cooling water per volume of distilled product water than either multiple-effect or single-effect distillation units. However, single-effect and multiple-effect distillation units have minimal "moving parts," while vapor compression units have multiple pumps, a vapor compression unit, and other support accessories with moving parts.

The heating requirements for a single-effect distillation unit, per volume of distillate, are greater than equivalent heating requirements for multiple-effect distillation units. The facility steam requirement decreases, per volume of distillate produced, with the number of effects in a multiple-effect unit. While there are no "rules" associated with selecting a given distillation process for a defined distillate water flow rate, it is suggested that single-effect distillation units are most appropriate for applications less than 200 gal/hr, multiple-effect distillation units for applications in the range of 200 to 2000 gal/hr, and vapor compression units for applications greater than 2000 gal/hr. However, specific facility requirements as well as "customer preference" generally dictate the type of distillation unit selected for an application.

Single-effect distillation units consist of an evaporator section, a vapor-liquid disengaging section, and a condensing section. The units may be electrically or facility steam heated. The vast majority of applications use facility steam. As discussed earlier, the latent heat required to produce Pure Steam within the evaporator section is quite high resulting in extremely high electrical requirements, depending on the capacity of the unit. Single-effect distillation units generally operate at atmospheric pressure, vented through a hydrophobic vent filtration system installed on the condensing section of the unit. As a result, product water from a single-effect distillation unit is depressurized and would flow to a storage tank (or sanitary distillate collection tank with sanitary pump) positioned as close as possible to the distillation unit condenser. The flow of product water from the condensing section of the distillation unit to the storage or collection tank is by gravity. As a result, the design and arrangement of the tubing between the condenser and the storage or collection tank is critical, discussed later in this chapter. Single-effect distillation units may use heat exchangers to recover sensible heat from the process. Generally, the heat exchangers would recover heat by passing feedwater through the "nondistillate" side of the condensing unit. Table 5.2 summarizes the supply steam flow (lb/hr) for a single-effect distillation unit as a function of distillate water flow (gal/hr).

For certain applications, specifically "small volume per day" Water for Injection requirements *and* a need for Pure Steam, it may be appropriate to consider a Pure Steam generator with a condensing unit to produce Water for Injection. Many of the requirements for this technology are small pilot plants or research and development laboratories. The Pure Steam generator provides the same function as the evaporator and vapor-liquid disengaging section of a single-effect distillation unit. The condensing unit, a separate unit operation from

Table 5.2 Required Single-Effect Distillation Unit Facility Steam Flow for Distillate Product Water Flow Rate

Facility supply steam flow (lb/hr)	Distillate product water flow rate (gal/hr)
140	15
280	30
475	50
700	75
950	100
1425	150
1900	200
2850	300

Source: From Vaponics (1984).

the Pure Steam generator, can be used, particularly during off-shift hours, to provide the required volume of Water for Injection or Purified Water used during normal operating hours. When a condensing unit and Pure Steam generator are "coupled," it is strongly suggested that the Pure Steam generator provide *either* Pure Steam for operations such as an autoclave *or* feed to the condensing unit. Simultaneous operation of the Pure Steam generator for multiple applications is *not* recommended.

Multiple-effect distillation units are more efficient than single-effect distillation units, in terms of facility steam and cooling water requirements per unit volume of distillate, since they use the latent heat of evaporation multiple times. Unlike single-effect distillation units, multiple-effect distillation unit pressure vessels operate at high pressures (e.g., 100–150 psig). The number of individual evaporator units or pressure vessels is referred to as "effects." Cooling water requirements are generally minimal, since heat can be effectively transferred to the feedwater.

Vapor compression distillation units operate at much lower temperatures than multiple-effect distillation units. While the exact flow path through vapor compression units may vary from manufacturer to manufacturer, generally, low-pressure steam from an evaporator section passes through a vapor-liquid disengaging section to a vapor compressor. This compressor increases the pressure and, subsequently, the temperature of the steam. Table 3.2 provides data showing the increase in temperature with pressure for saturated steam. The higher pressure and temperature steam from the effluent of the vapor compressor is used to heat new feedwater in the evaporator section. This technique is highly efficient with minimal facility steam and cooling requirements. Distilled product, pressurized by a pump, can be delivered to a storage tank that is not adjacent to the unit by a distillate pump on the unit. Maximum recovery of heat is achieved. However, the units are equipped with several mechanical parts and accessories, such as heat exchangers, pumps, and, perhaps most importantly, the vapor compressor. The compressor will require oil cooling for operation. Since the operating temperatures are relatively low when compared with multiple-effect distillation units, a portion of the noncondensable gases are removed by treatment of the feedwater to the unit or product water from the evaporator section of the unit.

A multiple-effect distillation unit has no moving parts. Maintenance is limited. The primary item requiring maintenance is the vapor-liquid disengaging section, which can accumulate deposits, particularly if the feedwater is not properly pretreated.

Most distillation units, particularly multiple-effect units and vapor compression units, and, to a lesser extent, single-effect units, use 316L stainless steel for surfaces in contact with water. Some units, particularly single-effect units, will limit the use of 316L stainless steel to surfaces in contact with Pure Steam or distillate. Chloride stress corrosion as well as chloride pitting corrosion is a concern. Figure 5.1 presents information associated with chloride stress corrosion of austenitic (series 300) stainless steels. This process can occur at temperatures as low as 150°F, with a chloride concentration of 0.5 ppm and a dissolved oxygen concentration of 0.14 ppm. The potential for chloride stress corrosion increases with temperature. Higher distillation unit operating temperatures are associated with greater possibility of stress corrosion for a given chloride and oxygen concentration. Some multiple-effect distillation unit manufacturers address this problem by substituting titanium tubes in the heat exchangers (effects). Obviously, chloride stress corrosion and chloride pitting attack are of less concern in lower operating temperature vapor compression units.

In chapter 2, during the discussion of impurities in raw water, thermal decomposition of NOM, which releases the chloride ion, was discussed. Many single- and multiple-effect distillation units do not use a membrane process, such as reverse osmosis or ultrafiltration, to remove the heavy molecular weight NOM that contributes to the introduction of the chloride ion during thermal decomposition. It is important to consider the total organic carbon (TOC) values in the feedwater to a distillation unit (single effect or, more importantly, multiple effect) even when feedwater is essentially "ion-free" but produced via ion exchange.

Single-effect and multiple-effect distillation units are Pure Steam generators that require deionized feedwater. It is suggested that the feedwater quality to the units exhibit a conductivity less than 1 μS/cm. Obviously, the concentrating effect associated with the evaporation process, coupled with potential thermal decomposition of NOM, is a concern.

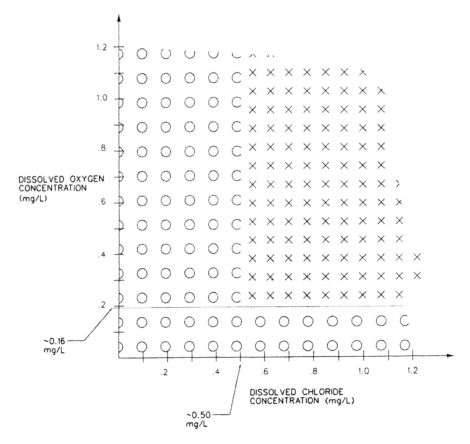

Figure 5.1 Chloride stress corrosion austenitic stainless steel. X = tube failure conditions, O = no tube failure. (The temperature is 250°F, and the austenitic stainless steel surfaces are under stress.)

While the manufacturers of vapor compression units state that the units may be operated with softened feedwater, it is suggested that deionized feedwater be considered, with a suggested conductivity in the range of 1 to 10 µS/cm. While the temperatures are lower in a vapor compression unit, periodic acid cleaning of the evaporator and heat exchanger surfaces is required. Deionization of the feedwater can reduce the cleaning frequency. It is *not* suggested that any distillation unit or Pure Steam generator feedwater be ion-free, since the water will be highly aggressive and will accelerate general corrosion within the distillation unit or Pure Steam generator. This phenomenon may be visually noted, during the periodic inspection of interior surfaces, by accelerated corrosion of surfaces in contact with the high purity feedwater.

Multiple-effect distillation units can be designed such that the first effect is capable of supplying Pure Steam. This is performed by "oversizing" the heat transfer surface area for the first effect. It is possible, based on system control provisions, to allow simultaneous production of Pure Steam *and* normal operation of the multiple-effect distillation unit. This function is *not* possible with vapor compression units because of the relatively low steam operating pressures and the nature of the vapor compression process.

When a Pure Steam generator is required for a facility with a vapor compression distillation unit with "softened" feedwater, the Pure Steam generator feed is generally Water for Injection. From an energy conservation perspective, the Pure Steam requires the latent heat of vaporization twice; once to produce Water for Injection and once to produce Pure Steam. From an engineering and regulatory perspective, the Pure Steam generator feed should *not* be directly from a zero dead leg valve in the recirculating Water for Injection loop. A dedicated Pure Steam generator feedwater tank with "air break" is suggested as shown in Figure 5.2. As indicated earlier, if the feedwater to the vapor compression distillation unit is from a

DISTILLATION AND PURE STEAM GENERATION

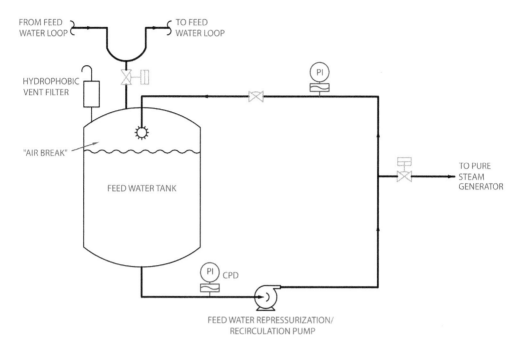

Figure 5.2 Pure Steam Generator feedwater tank and pump.

pretreated source that reduces the conductivity to a value ≤ 1 μS/cm, both the still and Pure Steam generator can be fed directly from the deionized source.

Generally, the time from unit "start-up" to availability of distilled product water from the condenser is less for a single-effect and multiple-effect unit than a vapor compression unit. However, it should be pointed out that this is a function of the operating pressure and design features of the vapor compression unit and should not be interpreted as a definitive statement for *all* units.

Since the first effect of a multiple-effect distillation unit operates at pressures as high as 150 psig, the feedwater generally requires repressurization. In general, the pressurization system should include a pump and a tank. Feedwater pressurization is generally not required for single-effect distillation units or vapor compression distillation units.

DESIGN CONSIDERATIONS

To adequately address design considerations, this section will be divided into several "subsections" as follows:

- Single-effect distillation units
- Multiple-effect distillation units
- Vapor compression distillation units
- Pure Steam generators
- Condensing units

Each subsection will discuss commercially available "standards" for the particular unit operation. For example, the multiple-effect distillation unit discussion will include units with boiling/evaporation on the inside of the tubes, boiling on the outside of the tubes, external boiler/evaporator units, and individual effect preheaters. Obviously, the entire myriad of units supplied throughout the pharmaceutical industry cannot be presented. However, the examples provided should demonstrate the basic principles of most commercially available units.

Single-Effect Distillation Units
Low-Velocity Units

Figure 5.3 depicts a low-velocity, single-effect distillation unit. This basic distillation unit represents general principles used for all distillation units. The evaporator sections of low-velocity distillation units are generally of a horizontal cylindrical type. A U-tube steam coil (or "bayonet-type" electrical heater) is positioned through one end of the evaporator. The heat source provides slow, uniform boiling in the evaporator section. The heating coil or electrical heating element(s) should be totally immersed in water.

Proper system design should include a full-diameter flange on one end of the horizontal cylindrical evaporator with a mating flange plate. The flat mating flange plate, connected to the evaporator section with a gasket, should have a bolt pattern similar to the flange on the evaporator section, allowing the plate, with attached steam coil or electrical heating element(s), to be easily removed from the evaporator section. The heating coil, or electrical heating element(s), mounted "through" the flat flange plate, is easily removed. Obviously, this will facilitate inspection, cleaning, and even replacement of the heating elements without major modification to the system. Easy access to the evaporator section is critical.

The heating coil is generally constructed of copper, while electrical heating elements may be stainless steel or Inconel. Optionally, a stainless steel heating coil may be used. However, the overall heat transfer coefficient for stainless steel, when compared with copper, is relatively low. Water in contact with the copper heating surface is used to produce Pure Steam. The Pure Steam passes to a condensing section fabricated of stainless steel.

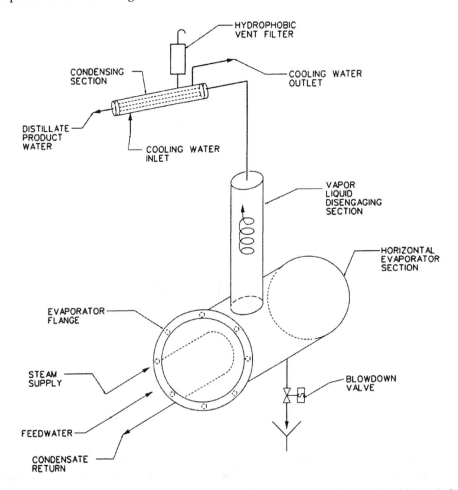

Figure 5.3 A low-velocity, single-effect distillation unit. [Electric heating may be employed instead of steam. Condenser cooling water may be used feedwater (assuming proper pretreatment). A support stand is required.]

The evaporator section of the unit may be constructed of copper. However, interior surfaces of the evaporator are generally treated with multiple "hot tin wipes" to provide a uniform tin-coated surface. Tin will not allow bacteria to proliferate on its surface. Pretreated feedwater to the evaporator should have a conductivity of about 1 µS/cm. Higher-quality water is not required and may be detrimental to the long-term successful operation of the unit, since it is ion-free and highly aggressive. The center of the upper portion of the horizontal cylindrical evaporator section should be equipped with vertical cylindrical steam discharge provisions. Again, for flexibility, it is suggested that the transition employ a flange, with a mating flange on the vertical cylindrical section. Level control within the evaporator can be achieved by a constant bleed or a simple level control system, such as a "float feeder."

A blowdown valve should be positioned at the center of the base of the horizontal evaporator section. This will provide a location for removing feedwater impurities concentrated as part of the evaporative process. It is strongly recommended that the blowdown valve be of automatic type, connected to a timer, ensuring that blowdown occurs periodically when the unit is operational. The frequency of blowdown should be adjustable, based on the purity of feedwater, the nature of the remaining impurities in the feedwater, the daily operating history for the unit, and, perhaps most importantly, scaling tendencies on the heating coil or electrical heating element(s), determined during periodic inspections of the interior of the evaporator section by an established maintenance program.

The design of the evaporator section, including the diameter of the evaporator, volume of water in the evaporator, configuration of heat transfer surfaces, and operating facility steam pressure should be such that steam is generated at atmospheric pressure and slow velocity. The steam will rise from the water in the evaporator section to the vertical cylindrical section, which contains vapor-liquid disengaging provisions positioned near the top (exit) of the vertical cylindrical section. Within the "space" between the hot water level and the vapor-liquid disengaging section, it is highly likely that the low-velocity nature of steam will result in the condensation of some of the steam, with return of water (condensate) back to the evaporator section. While this slow boiling process may appear to be "counterproductive" to ensuring that there is no water in the steam, the steam reaching the vapor-liquid disengaging section should contain relatively little water.

The vertical cylindrical section of the unit should contain a vapor-liquid disengaging section. Generally, manufacturers will provide a "trademark" name to the proprietary mechanism employed by the vapor-liquid disengaging device. Some manufacturers will position additional heat transfer surface within the lower section of the vertical cylindrical section, promoting the ability of the unit to provide "water (liquid)-free" steam to the vapor-liquid disengaging section. The heat exchangers also increase the velocity of the steam prior to entering the vapor-liquid disengaging section, a highly desirable condition. Many of the vapor-liquid disengaging sections employed rely on centrifugal devices that literally "spin" the steam, "throwing" any entrapped water to the outside of the vertical cylindrical section where it condenses and flows back to the evaporator section. The "enriched" Pure Steam, containing no liquid, passes from the centrifugal separating device to the condenser, which is positioned above the vapor-liquid disengaging section and mounted in a "sloped horizontal" fashion. Obviously, the vapor-liquid disengaging device should not be designed to provide any "holdup" of water. The use of material such as wire mesh or common industrial distillation unit "packing material" is generally unacceptable, since it provides water holdup. Within a well-designed vapor-liquid disengaging section, shortly after shutdown, there should be no liquid present on or in surfaces that are used to assemble the vapor-liquid disengaging section.

The materials of construction for the vapor-liquid disengaging should be 316L stainless steel since they are in direct contact with Pure Steam. If the entire vertical cylindrical column containing the vapor-liquid disengaging section and optional additional heat transfer surfaces is of stainless steel construction, and the evaporator section is of copper construction, the two cylindrical vessels must be galvanically isolated to avoid corrosion.

Pure Steam from the vapor-liquid disengaging section flows to a condensing section. This section of the distillation unit, as implied, condenses the Pure Steam, yielding distilled product water. If properly designed, the distilled water quality should meet the chemical, bacterial, and bacterial endotoxin attributes for USP Water for Injection. (As discussed earlier, it is assumed that the implied 10 cfu/100 mL total viable bacteria "Action Limit" for USP Water for Injection, as defined in the *General Information* section, is, in fact, a maximum acceptable value from a regulatory investigator's standpoint.) Since the distillation process occurs at a minimum temperature of 100°C and since a phase change has occurred, viable bacteria should not be present in the distilled product water. To conserve plant steam, the feedwater may be preheated in the condensing section of the distillation unit. However, it should be clearly indicated (especially for a single-effect distillation unit) that cooling water will still be required for the condensing process.

Generally, for low-velocity, single-effect distillation units, the condenser materials of construction as well as the condenser design are the single most important factors in determining the ability of the unit to address criteria presented in the current good manufacturing practices (cGMPs) (Tarry et al., 1993). As previously indicated, the condenser should be of a vertical cylindrical type, properly pitched to ensure that there is no water holdup in the heat exchanger. It is strongly suggested that the condenser be provided with a double tube sheet. It is further suggested that *all* surfaces in contact with distilled product water be of 316L stainless steel with Teflon® (gasket) material. It should be noted that Teflon "envelope" gaskets may be used, as appropriate. The condenser should also be provided with a hydrophobic vent filter, preferably with a steam-heated jacket or electrically heated "blanket." Many single-effect distillation unit manufacturers provide inappropriate vent filtration systems. Some of these vent filtration systems use a "hybrid" filter containing a 0.2 μm "disk-type" membrane. This filter generally does *not* provide a positive method of atmospheric bacteria removal, cannot be integrity tested, and does not meet guidelines set forth in the cGMPs (FR, 1976).

The condenser should be designed and positioned so that it is fully drainable. The elevation of the condenser should be higher than the Water for Injection or Purified Water storage tank. The flow of distillate from the condenser to the storage tank is by gravity. Occasionally, due to height restrictions, a "distillate collection system," with a sanitary tank, a hydrophobic vent filter, and a transfer pump, is provided. However, the use of this system is strongly discouraged because it provides multiple components that must be maintained to ensure proper operation as well as avoid potential contaminants. It is not uncommon to increase the height of the condenser, or the height of the entire distillation unit, to decrease the level of the storage tank, or to provide other techniques for ensuring that gravity flow from the condenser to the storage tank occurs.

The suggested distillate water temperature from the condenser should be consistent with the desired temperature in the Water for Injection storage and distribution system, but it should not be less than 85 to 90°C. It is suggested that any condenser be employed solely for removal of energy (heat) required to provide water (liquid) from water (Pure Steam). If additional cooling is desired, a separate "after cooling" heat exchanger is suggested.

The transfer tubing arrangement and orientation from the condenser to the storage tank is critical. This applies not only to low-velocity, single-effect distillation units but also to all distillation applications where product water is discharged at atmospheric pressure. The storage tank should be positioned as close as possible to the distillation unit. The section of tubing from the distillation unit should be equipped with sampling provisions, a sanitary conductivity sensor (probe/cell), and a divert-to-drain system.

The sampling valve should be installed in a manner that does *not* introduce bacteria-laden air into the distillate product tubing, which defeats the purpose and objective of the hydrophobic vent filtration systems installed on the condenser *and* downstream storage tank. There are many ways of achieving the desired results. A "U bend" with sanitary tubing, as shown in Figure 5.4, can be fabricated to achieve the desired results, assuming that the sample valve does not allow operating personnel to withdraw sample at a rate greater than the rate of distilled water production *and* that the sample is collected while the distillation unit is operational. Improper positioning and use of a sampling valve in the distillate tubing line from the condenser allows atmospheric air to enter the tubing, flowing to the condenser and/or vapor space above the water in the downstream storage tank. The effect of this situation is similar to removing the hydrophobic vent filtration systems installed on the condenser and storage tank. It is a clear conflict with criteria set forth in the cGMPs (FR, 1976).

DISTILLATION AND PURE STEAM GENERATION

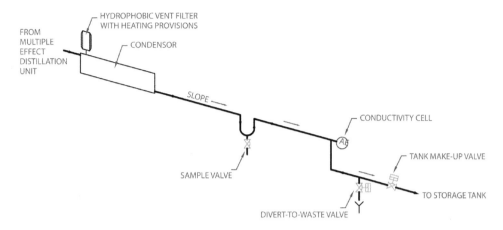

Figure 5.4 Multiple effect distillation unit—sampling and divert provisions.

The conductivity sensor for measuring the purity of the distilled product water must be positioned within the tubing from the condenser to the downstream storage tank in a manner that ensures that it is fully immersed in distilled product water when the distillation unit is operational. This is extremely important, since a conductivity probe (cell) in steam or filtered air will indicate an erroneously low conductivity value. This situation could occur when product water from the distillation unit is of unacceptable quality. In other words, with improper positioning of the conductivity cell, operating personnel may not be alerted to a transient situation resulting in below-quality distilled product water, believing that the unit is producing "in-specification," low-conductivity product water.

The divert-to-waste system should include two sanitary diaphragm valves. One valve will stop the flow of product water to the storage tank, while the second valve opens with a very slight time delay, diverting water to a depressurized drain (with an air break). It should be noted that the two sanitary diaphragm valves can be replaced with a single sanitary three-way valve, a relatively simple design for a distillation unit. However, once again, provisions must be included to avoid contamination of the storage tank, tubing between the storage tank, and the condenser with bacteria-laden atmospheric air. If a valve is simply installed on the tubing line that delvers distilled product water from the condenser to the storage tank, water will flow to waste at a rate that will not "fill" the tubing. As a result, atmospheric air with bacteria will be introduced, defeating the function of the hydrophobic vent filtration systems on the condenser and storage tank. This situation can easily be avoided by positioning a U bend–type liquid trap in the section of tubing from the divert-to-drain valve to the depressurized drain with an air break as discussed earlier.

Tubing from the condenser to the storage tank, with the exceptions noted above, should be as straight and direct as possible. Since it is highly desirable to ensure that the line is fully drainable, it is suggested that a significant slope, in a direction from the condenser to the storage tank, be considered. This may be achieved by elevating the condenser discharge and inlet to the storage tank, using a slope of 1 to 3 in./linear ft. Obviously, the positioning of the divert-to-drain valve and the feedwater connection, with valve, to the tank are also critical. The positioning of these two valves determines the potential for liquid holdup within the section of tubing between the condenser and the storage tank. The tubing should be thermally insulated to ensure that the temperature of the water remains above a recommended 85 to 90°C value. To ensure that the quality of distilled water delivered to the tank meets chemical, bacterial, and bacterial endotoxin criteria, it is suggested that the system control cycle include a distilled product water pre makeup divert-to-drain cycle. This cycle, with recommended duration of 5 to 10 minutes, would divert distilled product water to waste, "flushing" the section of tubing between the distillation unit condenser and storage tank. In addition to a predelivery time delay, distilled product water should not be delivered to the storage tank until a preset product water conductivity value is achieved for a suggested minimum time period of two to three minutes.

The heating steam pressure for single-effect distillation units is generally in the range of 30 to 60 psig. Most units will operate with facility steam at a maximum pressure of 100 psig. Some commercially available units will operate with facility steam pressures as low as 10 to 15 psig. Often, these units are essentially "derated," larger-capacity units. While it is not suggested that manufacturers' "down-rating" criteria are unacceptable, it is suggested that operation of a single-effect distillation unit with low-pressure steam (<10 psig), even with manufacturer's approval, can produce undesirable results. This is primarily associated with the fact that units designed for a higher distillate product water flow rate (containing an evaporator, a vapor-liquid disengaging section, and a condenser) are simply being downrated based on heat transfer area within the evaporator section. Some manufacturers will responsibly construct a standard "low-pressure" unit with a properly sized boiler, vapor-liquid disengaging section, and condenser.

High-Velocity Units

A typical single-effect, high-velocity distillation unit is depicted in Figure 5.5. This unit consists of the same three major components as a low-velocity unit: an evaporator, a vapor-liquid disengaging section, and a condensing section. However, the evaporator section and vapor-liquid disengaging section are generally combined as one vertical cylindrical column.

Heating facility steam flow is through the inside of U-tubes, while feedwater is delivered to the shell side of the evaporator section. The U-tubes are generally configured such that the steam inlet and outlet are at the base of the vertical cylindrical column. Steam generated within the evaporator section flows upward to the vapor-liquid disengaging section. While the units are designed to operate at atmospheric pressure, a double tube sheet is strongly suggested for applications where the distilled product water will be designated as USP Water for Injection. The units should include components constructed of 316L stainless steel, such as the evaporator shell, heating tubes, condenser, vapor-liquid disengaging section, and so on.

Steam flow is at high velocities since boiling is rapid. In fact, it is quite likely that the upper portion of the tube bundle may not be fully submerged in water. The "falling film evaporative process" (McCabe et al., 1993) provides high-velocity steam and efficient heat transfer. However, proper system design is important to ensure that high-velocity steam does not result in the entrapment of water particles. The falling film evaporative process, with alternating rapid "wetting and drying" of tube surfaces, can also result in corrosion if the feedwater is not properly pretreated. Subsequently, it is strongly suggested that the feedwater conductivity to a high-velocity, single-effect distillation unit be less than 1 µS/cm to minimize potential scaling of heat transfer surfaces. The frequency and duration of evaporator blowdown should be adequate to ensure that undesirable impurities do not precipitate on heat transfer surface areas.

A high-velocity, single-effect distillation unit is similar to an individual "effect" of certain multiple-effect distillation units (evaporator and vapor-liquid disengaging section). However, while the appearance may be similar, the single-effect unit is designed to operate at atmospheric pressure, while multiple-effect columns are designed to operate at much higher pressures (e.g., 100–150 psig).

To minimize both chloride stress corrosion and chloride pitting attack of stainless steel tubes, titanium tubes may be considered. Chloride stress corrosion and chloride pitting corrosion can occur at 212°F as discussed earlier in this chapter. While the potential for chloride stress corrosion and chloride pitting is significantly less in a single-effect unit when compared with a multiple-effect unit, the economic factors associated with employing titanium, as a substitute for stainless steel tubes, are favorable.

With the exception of the "falling film evaporative" concept, resulting in a high steam velocity, and the "combination" evaporator and vapor-liquid disengaging section in a single column, the principles associated with low-velocity and high-velocity distillation units are similar.

DISTILLATION AND PURE STEAM GENERATION

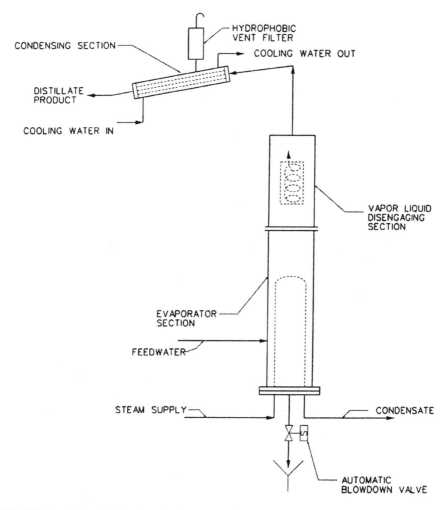

Figure 5.5 A high-velocity, single-effect distillation unit. [Electric heating may be employed instead of steam but is not recommended. Condenser cooling water may be used as feedwater (assuming proper pretreatment). A support stand is required. All sections are of cylindrical configuration.]

High-Velocity Units with an "External" Boiler

Figure 5.6 depicts a single-effect, high-velocity distillation unit with an external evaporator. These units have three major components: a separate evaporator section, a vertical cylindrical column for evaporated water recirculation (lower portion) and vapor-liquid disengaging (upper portion), and a condensing section. Water, from the base of the vertical cylindrical column, flows to the evaporator section, with natural thermal circulation as the driving force. Facility steam is delivered to the shell side of the evaporator section and inputs latent heat to water on the *inside* of the tubes, generating steam. The steam flow is subsequently directed back to the vertical cylindrical column. The steam produced from the external evaporator flows to the middle of the vertical cylindrical column. The steam is delivered in such a manner that it "spins" by centrifugal force. This spinning action, with proper steam velocity and column design, is such that any water present will accumulate on the interior walls of the vertical cylindrical vessel and flow down, by gravity, to the lower portion of the column containing feedwater to the external evaporator section. Subsequently, the high-velocity steam process, in a properly designed unit, removes water before the steam enters the vapor-liquid disengaging section.

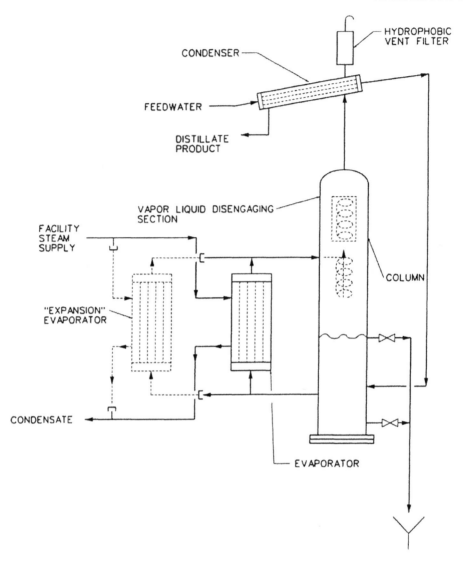

Figure 5.6 A high-velocity, single-effect distillation unit with an external boiler/evaporator.

The vapor-liquid disengaging section is positioned at the top of the vertical cylindrical column. The high velocity of steam, coupled with enhanced centrifugal separation provided by the vapor-liquid disengaging components, ensures that extremely high-quality Pure Steam passes from the vertical cylindrical column to the condenser. Any entrapped water accumulates on the inside of the vertical cylindrical column and drains, by gravity, to the base of the column.

A constant bleed device, positioned in the lower water-containing section of the vertical cylindrical column, removes impurities associated with the concentrating effect of the evaporation process. The constant bleed device should be adjustable to allow for changes in feedwater characteristics, long-term decreases in system capacity (distillate flow) resulting in scaling of the heat transfer surface area, changes in facility steam pressure, and operator evaluation of system parameters determined by data trending.

The feedwater to the system passes through a preheater in the condenser. The tempered feedwater is then introduced to the vertical cylindrical column at a "constant water level control device" that is physically positioned exterior to the column but equilibrates the water level within the vertical cylindrical column, since the unit operates at atmospheric pressure.

A manual bottom drain valve should be physically positioned to allow effective blowdown of the vertical cylindrical column. While the constant bleed device will remove impurities, it is fully anticipated that particulate matter, corrosion products, and insoluble precipitates will accumulate at the base of the vertical cylindrical column. These impurities, concentrated by the evaporative process, can only be effectively removed by blowdown from a location at the base of the column.

The distillation unit operates at atmospheric pressure, vented through a hydrophobic vent filter positioned on the condensing section.

There are two advantages of an external boiler unit that should be considered compared to a single-effect distillation unit without a boiler ("classical" unit). The first advantage is associated with the ability to remove the evaporator section (boiler) easily, inspect it, perform maintenance on it, and, if required, replace it. This is extremely simple to perform when compared with a high-velocity, single-effect distillation unit. The second advantage of an external boiler unit relates to the ability to add additional boiler capacity, increasing the ultimate distilled product water capacity. A review of manufacturer literature suggests that the capacity could be doubled by just adding another evaporator section (Paul Mueller Company, 1985). The capacity *may* be quadrupled when three additional evaporators (boilers) are added with a second condenser, operating in parallel with the original condenser (Paul Mueller Company, 1985). It is suggested, however, that any expansion be carefully evaluated. For example, while three additional boiler sections and an additional condensing section *may* increase the distillate flow rate by a factor of four, the diameter of the vertical cylindrical column has not changed. The characteristics of the vapor-liquid disengaging section have not changed. The expansion should not compromise the quality of the distilled product water by negatively impacting Pure Steam quality to the condensing section.

The materials of construction for the external evaporator, the vertical cylindrical column, and the condensing section can vary. Some manufacturers use carbon steel shell evaporators with copper or tin-coated interior tubes. The vertical cylindrical column may also be constructed of double hot–wiped, tin-coated copper. Other manufacturers provide units constructed of double tin–wiped copper, 316L stainless steel, or titanium. It is suggested that the units be constructed of 316L stainless steel. Feedwater should be properly pretreated to remove inorganic compounds, organic compounds, colloids, and silica. Since the unit operates at atmospheric pressure, it is difficult to justify the use of titanium tubes. Once again, however, the "adder" for titanium tubes in the boiler section should be evaluated in light of anticipated maintenance, including replacement of the evaporator section of the unit.

Condensate Feedback

Figure 5.7 presents a single-effect distillation unit with a "condensate feedback" feedwater system. Condensate feedback, a technique limited to single-effect distillation units (or Pure Steam generators), can be used for applications where facility steam condensate does *not* contain amines or any other volatile impurities. Amines are volatile organic compounds used in many facility steam boiler applications to control corrosion within the feedwater and condensate system of the facility boiler. If the steam condensate is free of all volatile organic impurities, such as amines, *and* the steam is of high quality (low conductivity), it can be used directly as feedwater to a single-effect distillation unit. This eliminates the requirement for distillation unit pretreatment components.

To ensure that the condensed facility steam quality is appropriate from an inorganic standpoint, it is suggested that the conductivity be less than 5 µS/cm (at 25°C) *and* total suspended solids (TSS) level should be less than 0.5 ppm. Generally, this technique is *not* applicable for distillation units with a steam supply from a large, central boiler system. Unfortunately, facility steam condensate quality, associated with the presence of dissolved and particulate iron oxides, is a concern. Further, larger systems would generally use a volatile, boiler water treatment program, including amines, to assist in corrosion control within the facility steam condensate and feedwater systems.

The use of condensate feedback increases the cooling water requirements for the condenser because facility steam condensate, which becomes feedwater, is already hot and

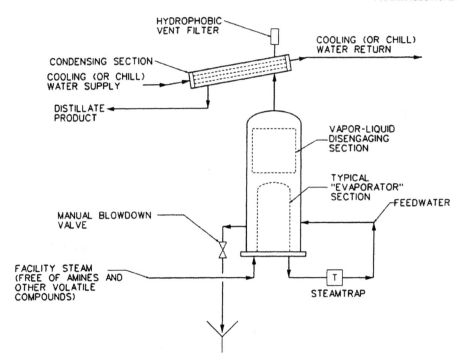

Figure 5.7 A single-effect distillation unit with a condensate feedback feedwater system.

cannot be used to remove heat during the condensing process. The use of condensate feedback is extremely limited, primarily for smaller (low daily volume requirements) applications, such as those requiring 1 to 10 gal of distilled product water per day.

Condensate Feedback Purification
Figure 5.8 presents a schematic of a single-effect distillation unit with condensate feedback purification. Condensate feedback purification is similar to condensate feedback. It uses facility steam condensate as feedwater to a single-effect distillation unit. However, before the facility steam condensate is used as feedwater, it passes through a heat exchanger (or multiple heat exchangers) and a deionization system, to improve the quality of the feedwater and, in certain cases, remove up to about 2 ppm of *neutralizing* boiler water treatment amines. The technique cannot be used when the steam condensate contains "filming" amines or neutralizing amines at a concentration greater than approximately 2 ppm, because of the volatile nature of the amines and the inability of the purification train to remove filming amines or neutralizing amines at a concentration greater than 2 ppm.

Initially, the facility steam condensate must be cooled to less than 120°F to avoid thermal degradation of the anion resin in the downstream purification system—an extremely important design consideration. In lieu of a single heat exchanger, reducing the temperature to less than 120°F, a second "regenerative" heat exchanger can be used to recover heat after the facility steam condensate has been processed through the water purification system. However, thermal efficiency can be achieved by using the relatively cooler (<120°F) water for cooling in the condensing section of the single-effect distillation unit.

The water purification train generally employs rechargeable canisters. Four rechargeable canisters are suggested: two containing mixed ion exchange resin, one containing organic removal media, and one containing amine removal resin (Vaponics, 2010). The organic removal media may be activated carbon or, preferably, activated carbon mixed with a macroporous anion resin. The amine removal canister, for most applications, would consist of strong acid cation resin.

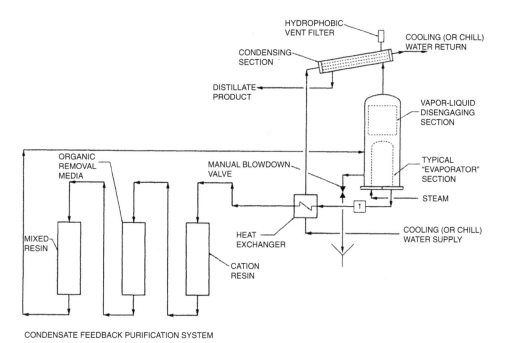

Figure 5.8 A single-effect distillation unit with a condensate feedback purification system. (Additional heat exchangers may be included to increase the thermal capacity of the unit. Sequencing of the purification units may be altered for specific applications. The heat exchanger is a tube-in-tube type. Accessories, such as a temperature sensor and a conductivity monitor, are not shown.)

Condensate feedback purification should only be considered for single-effect distillation units with very low daily volumetric. Since the condensed facility steam will contain a fraction of the ionic material present in raw feedwater, rechargeable canister replacement frequency is reduced. Further, other impurities potentially present in raw water, such as NOM, colloidal material, suspended solids, and residual disinfectant will not be present in the distillation unit feedwater. Further, the technology may be appropriate when a "dedicated," electrically heated or gas-heated boiler is used with the condensate feedback system, which was discussed in the previous section. If a dedicated boiler is used, the total dissolved solid (TDS) level of the condensate should be less than 2 to 5 ppm.

Finally, when either condensate feedback or condensate feedback purification systems are employed, it is necessary to provide a source of feedwater during initial start-up to fill the evaporator section of the single-effect distillation unit. This source of water must be available from time to time, particularly if the distillation unit is operated only sporadically.

Multiple-Effect Distillation Units
Boiling on Shell Side, No External Evaporator
Figure 5.9 demonstrates a four-effect distillation unit, designed such that boiling occurs on the shell side of the evaporator sections and where there are no external evaporators/boilers.

Commercially available, multiple-effect distillation units can be conceptually observed as multiple effects, assembled by positioning as few as three and as many as six, single-effect units in series. The individual effects would be designed, constructed, and tested per the American Society of Mechanical Engineers (ASME, Section VIII, Division 1—Unfired Pressure Vessels) since they operate at pressures as high as 100 to 150 psig. The theory associated with multiple-effect distillation units was discussed earlier. The selected design pressure should be consistent with the maximum anticipated operating pressure. The test pressure should be 50% greater than the design pressure. Historically, multiple-effect distillation units of the type discussed in this chapter are designed to operate at maximum pressure values of about

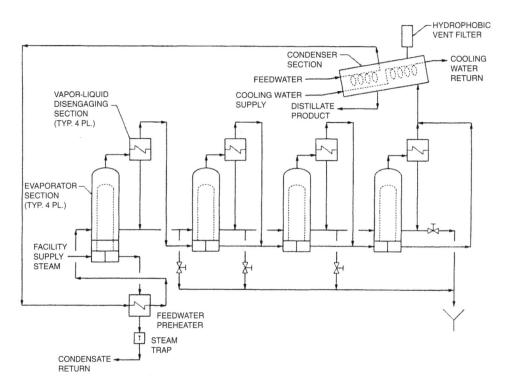

Figure 5.9 A four-effect distillation unit with boiling on the shell side and no external evaporation. (An additional heat exchanger may be added to increase the thermal efficiency of the unit. The column level control system and the condenser "reboiler" are not shown.) *Source*: From Vaponics (1989).

125 psig. All evaporator (effects) shells and tubes, as well as all surfaces in contact with Pure Steam or distillate should be constructed of 316L stainless steel with Teflon gaskets. Feedwater to each effect should initially flow through the condensing unit for preheating, recovering heat from the condensation of Pure Steam, from all but the first effect. The first effect, as a minimum, must be equipped with a double tube sheet. An option may be considered to have all effects constructed with a double tube sheet. However, this particular option is not necessary for a properly designed, instrumented and controlled unit.

Within a multiple-effect distillation unit, where boiling of feedwater occurs on the outside of the tubes, the upper portion of the U-tubes may *not* be totally immersed in water. Flash evaporation, associated with a "film" of water, particularly on the upper sections of the vertically positioned tube bundle, provides rapid and efficient heat transfer. This method of evaporation will produce steam free of water. However, this same principle, coupled with high operating temperatures (e.g., 338°F at 100 psig) *and* the stress placed on the U-tube bundle may result in chloride stress corrosion cracking of the stainless steel tubes over time. This can be controlled by treating the feedwater to reduce the chloride ion concentration to as low a level as practical, performing continuous blowdown of each effect, or using U-tubes of titanium construction tubes in lieu of 316L stainless steel tubes or other austenitic stainless steel material (Vaponics, 1997b).

An additional item that must be considered is the long-term effects of chloride pitting corrosion of 316L stainless steel tubes, specifically at the tube-to-tube sheet joints. Again, appropriate operating and maintenance considerations or use of titanium tubes can reduce or eliminate this situation.

The level control within each effect for this particular multiple-effect unit design may be difficult to achieve, particularly with transients in the facility steam pressure to the first effect. Heat distribution in each effect and feedwater flow must be balanced by orifices installed within tubing connecting the steam and feedwater lines to each effect. A multiple-effect distillation unit employing this design must be thoroughly tested prior to shipment.

DISTILLATION AND PURE STEAM GENERATION

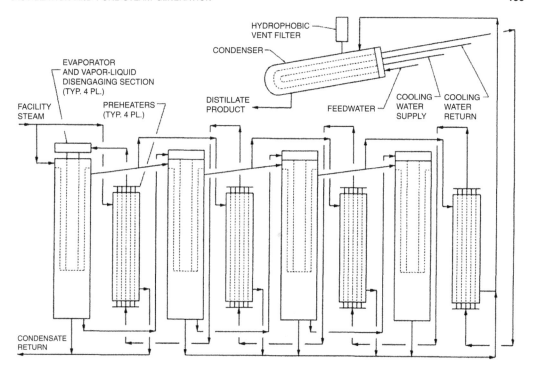

Figure 5.10 A multiple-effect distillation unit with boiling inside tubes and no external boiler (see Fig. 5.9 for details on the vapor-liquid disengaging section). *Source*: Courtesy of Finn-Aqua America, Inc. (1989b).

In addition to recovering heat from the condensing unit, it is also appropriate to recover heat from the supply steam condensate to the first effect, using a heat exchanger to transfer sensible heat to the feedwater. To ensure proper system operation, each effect should be equipped with a sanitary-type level-sensing device and a sanitary temperature sensor with a transmitter and a remote indicator. Additional control and monitoring features associated with multiple-effect distillation units will be discussed below.

Boiling Inside Tubes and No External Boiler
Figure 5.10 presents a schematic of a multiple-effect distillation unit where boiling occurs inside the tubes and no "external" evaporators are used. Commercially available units employing this technology rely heavily on the unique properties of the evaporator column. By far, the design of the evaporators (effects) is the key to the successful operation of these units.

The principle of evaporator design is associated with the flash evaporation of feedwater *and* the mechanical separation of bacterial endotoxins within droplets of water (and other impurities) using centrifugal force. The flash evaporation concept produces high-velocity vapor, which is directed to a narrow channel, where the steam begins to "rotate," thus creating centrifugal force. The Pure Steam flow rate exposes water droplets and impurities, containing bacterial endotoxins, to a centrifugal force about 500 times greater than that of gravity (Finn-Aqua America, Inc., 1986). All impurities in the vapor are forced to the outer parameter (inside of the column), where they are collected and flow, by gravity, to the base of the column. Pure Steam without the presence of impurities including water flows from the top of the column. The columns for each effect operate in series, with only the first column heated by facility steam. As discussed earlier, Pure Steam from the first column is used to heat the second column, transferring latent heat to water in the second effect. Pure Steam from the second effect is directed to the third effect. Pure Steam and Pure Steam condensate from the Pure Steam originally generated in the first effect flows to the condenser along with Pure Steam and Pure Steam condensate from all effects subsequent to the first effect. In this particular design,

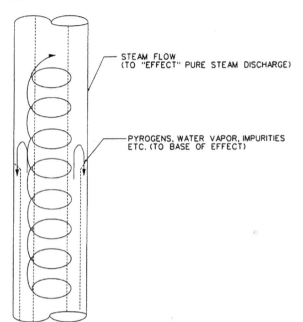

Figure 5.11 A double-jacket patented centrifugal pyrogen separation system. *Source*: Courtesy of Finn-Aqua America, Inc. (1989a).

each column consists of two pressure vessels. The inner vessels serve as a heat exchanger (steam in the shell side supplying latent heat to water in the tubes). The outer vessel is fitted with a double jacket. It allows the steam generated in the inner vessel (tubes) to rotate at a high velocity, forcing all particles and bacterial endotoxins to a channel on the inside wall of the outer pressure vessel for each effect. This design does *not* use a dedicated vapor-liquid disengaging section; it relies on the construction described above to achieve the separation of steam from all impurities, including bacterial endotoxins and water. Figure 5.11 depicts the unique nature of the double pressure vessel arrangement and the unique vapor-liquid disengaging mechanism.

As previously indicated, this design employs a falling film evaporation process. Feedwater is provided to the inside of the inner tube. Boiling occurs within the tube, producing Pure Steam that passes to the "outer tube." The heating steam contained in the shell side of the inner tube supplies the latent heat for boiling inside the tube. The heating Pure Steam is collected as condensate (possibly with some Pure Steam), and for all but the first effect of the multiple-effect unit flows to the condenser.

This particular design is thermodynamically effective, limiting the use of cooling water to the condenser. Feedwater may be heated to as high as 150°C (pressurized to allow feed to the pressurized effects. The distillate outlet temperature is approximately 90 to 95°C. Manufacturers' data for this type of multiple-effect distillation unit design (Finn-Aqua America, Inc., 1989a) suggests that distillate product water will have a conductivity of 0.2 to 0.5 µS/cm when feedwater is "deionized."

The required facility steam pressure to the first effect is a function of the unit selected, including the number of effects. The facility steam pressure value increases as the number of effects increase. In general, a steam pressure as low as 75 psig and as high as 150 psig may be required for operation, depending on the unit selected.

While all pressure vessels are designed in accordance with the ASME Code for Unfired Pressure Vessels for operation at the maximum operating pressure (e.g., 125 psig), a positive method of mechanical pressure relief should be provided for each effect to eliminate potential column accidental overpressurization. While the stated manufacturers' information references distilled product water quality with deionized water, it is suggested and implied within the literature that the conductivity of feedwater quality should be less than 5 µS/cm, with the total reactive silica less than 1 ppm and free of both amines and residual disinfecting agent.

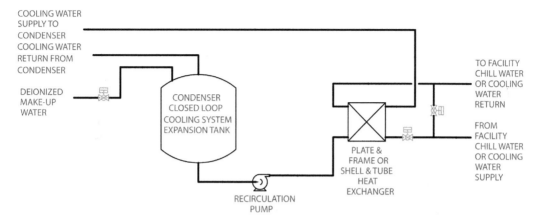

Figure 5.12 Multiple-effect distillation unit condenser—closed loop cooling system.

The individual columns (effects), preheaters, and interconnecting tubing should be insulated (chloride-free material) with outer jacket (PVC or aluminum).

The cooling (or chill water) water, used in the condensing section, can contain scale-forming compounds. Manufacturers' literature suggests that the cooling water temperature be 45 to 70°C, a function of the application. Further, the cooling water total hardness should be less than 100 ppm (Finn-Aqua America Inc., 1989a). Considering the indicated cooling water requirements, it is necessary to "blend" chill water to obtain the indicated temperature values to avoid thermal "shock." An attractive alternative to direct use of blended chill water and or other cooling source consists of a closed loop cooling system with heat exchanger, depicted in Figure 5.12. While multiple effect distillation units require minimal maintenance, experience indicates that fouling and scaling or the "cooling media side" of the condenser presents problems. The system shown in Figure 5.12 allows cooling water (or chill water) to flow to a plate and frame-type heat exchanger. A closed loop cooling system containing "RO product" water quality recirculates the actual condenser cooling water to the heat exchanger. This arrangement transfers heat exchanger maintenance from the sanitary condensing unit on the multiple-effect distillation unit (often positioned in a difficult access area) to the plate and frame heat exchanger that can be physically positioned in a convenient maintenance area. Periodic cleaning and/or replacement of the "plates" in the plate and frame heat exchanger can be performed as part of a preventative maintenance program. Further, maintenance of the plate and frame heat exchanger is not intrusive to the sanitary condenser and distillate tubing on the multiple effect distillation unit.

There are several additional features, discussed earlier for multiple-effect distillation units, that apply to this specific unit design. Specifically, cooling water is only required in the condensing section of the unit. The materials of construction for the columns and the condenser should be 316L stainless steel. The condensing section should be equipped with a double tube sheet of sanitary design.

In the previous section, which presented information for an alternate design multiple-effect distillation unit, control and monitoring functions were not discussed. General instrument (local and remote) and control requirements for multiple effect distillation units include, but are not limited to, the following:

- Facility steam pressure
- Feedwater flow rate
- Feedwater pressure
- Continuous blowdown flow rate
- Distillate outlet temperature
- Cooling water flow rate to the condenser
- Distillate product water high-conductivity alarm "set point" (normal operation)

- Distillate divert-to-drain performance characteristics, including preoperational rinse duration
- Feedwater high-conductivity alarm set point

Regulation of these parameters will require appropriate sensors, monitors, transmitters, and controllers. In addition, provisions should be included for starting and stopping the distillation unit based on a signal from the level control system on the downstream storage tank. Further, it is strongly suggested that recording (or remote data logging provisions) be provided for the conductivity and temperature (as a function of time) of the distillate product. Finally, it is suggested that direct reading, individual temperature indicators be provided for each effect and for the condenser (in addition to the temperature element, temperature transmitter, and temperature-indicating controller).

Alarm conditions for a multiple-effect distillation unit should alert operating personnel to conditions that will potentially result in product water that does not meet the chemical, bacterial, or bacterial endotoxin specifications for USP Water for Injection as well as other unit excursions such as high pressure in an effect. Alarm conditions should include, but not be limited to, the following:

- High distillate conductivity
- High water level in any effect
- High feedwater conductivity
- Low control air pressure
- Low facility steam pressure to the first effect
- High pressure in any effect
- Low distillate product water temperature
- High distillate product water temperature
- Low feedwater pressure
- Low pretreated water pressure to the feedwater pump

Distillation unit operating pressures, particularly within the first effect, dictate feedwater pressurization requirements. A feedwater system consisting of a stainless steel tank with hydrophobic vent filter and a multistage centrifugal pump capable of delivering water at pressures as high as 150 psig should be considered. Some feedwater systems do not employ a tank, although it is strongly suggested. There are many advantages, such as elimination of pretreated water system loop back contamination obtained with the air break provided with the feedwater tank.

A properly functioning distillation unit should provide at *least* a 3-log reduction in bacterial endotoxins, preferably a 5-log reduction. Bacterial endotoxin challenge test results should be provided during a suggested Factory Acceptance Test (FAT). If the bacterial endotoxin challenge test is initially performed at the pharmaceutical manufacturer's facility, servicing a potential manufacturing or design problem is much more difficult. The literature suggests that a "successful" challenge is associated with a reduction of bacterial endotoxins in feedwater, at a level of 25 endotoxin units (EU)/mL, with product water less than 0.125 EU/mL after the unit has been operating for 10 to 15 minutes (Finn-Aqua America, Inc., 1989b). It is suggested that the actual "operating" distillate product water bacterial endotoxin level of the unit should be less than 0.01 IU/mL. This is particularly true for a bacterial endotoxin challenge test performed on a "new unit" as part of the FAT.

Boiling Inside Tubes and External Boilers
Figure 5.13 depicts a commercially available, multiple-effect distillation unit with boiling occurring inside tubes and with external boiler (evaporator) sections. This particular design includes four basic components: a condensing section, external evaporator, a column/effect with a vapor-liquid disengaging section, and individual interstage heat exchangers for each effect. The evaporators are mounted externally to the individual effects and are designed to operate by natural thermal circulation from adjacent effects. Each evaporator is

DISTILLATION AND PURE STEAM GENERATION

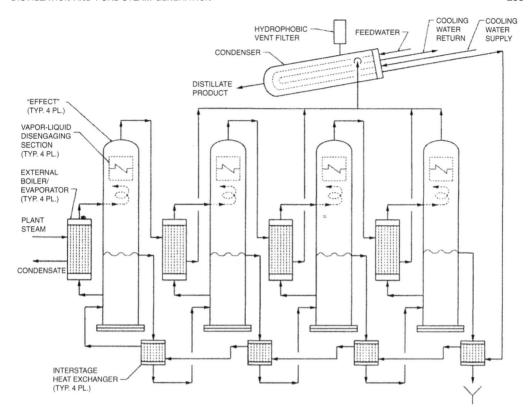

Figure 5.13 A four-effect distillation unit with boiling inside tubes and a boiler. *Source*: Courtesy of Mueller-Barnstead (1987a).

"interchangeable," with the exception of the first-effect evaporator. The externally mounted, evaporator heat exchangers can be easily removed for cleaning, inspection, maintenance, or replacement. This operation does not require space above the vertical cylindrical effects, or "rotation" of an effect to a horizontal position, to remove the shell. Boiling of steam occurs inside the evaporator tubes, and steam is provided to the shell of the evaporator. All components within the evaporator sections should be manufactured of 316L stainless steel. The individual evaporator sections should be designed in accordance with the ASME Code for operation at pressures of 125 to 150 psig. This design maximizes the overall "wetting" of surfaces, prevents "hot spots," and subsequently reduces the possibility of chloride stress corrosion and chloride pitting attack.

Steam produced within the individual evaporators is directed to a vertical cylindrical effect that provides three separate functions. The lower portion of the effect contains water that circulates to the external adjacent evaporator. The center section provides primary vapor-liquid disengaging, while the upper section is provided with a vapor-liquid disengaging device. Steam produced in the individual evaporators flows to the individual adjacent column/effect, where it is directed in a spiral motion. The centrifugal action forces water to the inside wall of the effect. Any water removed from the steam flows by gravity to the reservoir of water in the lower section of the vertical cylindrical effect. The Pure Steam produced within this primary vapor-liquid disengaging section of the effect flows upward and is directed through a "multidirectional flow path" (vapor-liquid disengaging device) that provides Pure Steam free of all impurities, including water. All surfaces within the vertical cylindrical effects are constructed of 316L stainless steel, mechanically polished and electropolished. The Pure Steam generated from an "upstream" effect passes to a "downstream" effect, where it releases its latent heat, undergoes condensation, and is discharged to a "common" condenser. The condenser is a U-tube heat exchanger with a double tube sheet. It is physically positioned such that distilled water is fully drainable from the unit. Cooling water is provided to one set of U-tubes in the condenser, while multiple-effect distillation unit feedwater is provided to a second set of U-tubes in the heat exchanger. Condensation of any remaining Pure Steam occurs in the shell side of the condensing unit. The condenser is provided with a heated and jacketed hydrophobic vent filtration system, as discussed earlier in this chapter.

This particular multiple-effect distillation unit design employs interstage heat exchangers. Each heat exchanger is equipped with "straight tubes." Feedwater progressively flows through the individual interstage heat exchangers. The feedwater flow pattern to the multiple-effect distillation unit is directed to the U-tube heat exchanger of the condenser, with effluent to the last effect interstage heat exchanger. Feedwater would then flow to the next to last effect interstage heat exchanger and, subsequently, to the first-effect interstage heat exchanger. This arrangement provides the highest temperature feedwater to the first effect, which subsequently provides feedwater to the second, third, and so on effects. The heating media (which is the "overflow" of liquid from an "upstream" vertical cylindrical effect) passes through the tube side of the individual feedwater heat exchangers. The flow of feedwater is from the last effect, through the interstage heat exchangers, to the first effect, since the pressure (and subsequently the temperature of the water) within the vertical cylindrical column is higher for the first effect than the second, third, and so on effects. The feedwater temperature is constantly increasing until it is supplied to the first effect. As overflow water from the first effect passes through the tube side of the first-effect interstage heat exchanger, releasing its heat on the shell side, it provides feedwater for the second-effect evaporator section. This process continues from the first effect to the final effect, which maximizes the recovery of heat.

For this particular design, manufacturers' literature indicates that the distilled product water temperature should be 80 to 90°C.

Individual controls are provided to regulate the inlet steam pressure, the feedwater flow rate, the distillate temperature, and supply steam pressure. The face of the control and monitoring panel for this specific unit contains indicators for the following:

- Distillate product temperature
- Facility steam pressure
- Feedwater pressure
- First-effect temperature
- Second-effect temperature
- Third-effect temperature
- Fourth-effect temperature (if applicable)
- Fifth-effect temperature (if applicable)
- Sixth-effect temperature (if applicable)
- Feedwater flow rate
- Feedwater conductivity
- Product water conductivity

Alarm conditions, with individual, properly labeled, indicating lights include the following:

- High product water conductivity
- High feedwater conductivity
- Cooling water failure (low flow rate)
- High supply steam pressure
- Low supply steam pressure
- High column pressure—first effect
- High water level—any effect
- Low water level—first effect

The product water conductivity monitor/controller operates in conjunction with a "divert-to-waste" system, similar to the system described for previous multiple-effect distillation units. This feature includes a distillate divert-to-drain cycle when the unit is started.

DISTILLATION AND PURE STEAM GENERATION

Table 5.3 Facility Steam Consumption for Various Multiple-Effect Distillation Unit Product Water Flow Rate

Distillate flow rate (gal/hr)	Facility steam flow rate—three effects (lb/hr)	Facility steam flow rate—four effects (lb/hr)	Facility steam flow rate—five effects (lb/hr)	Facility steam flow rate—six effects (lb/hr)
50	153	115	–	–
100	306	230	–	–
150	459	345	–	–
200	612	460	367	–
300	765	575	550	–
400	–	690	734	611
500	–	1150	918	764
650	–	–	1193	993
800	–	–	1468	1222
1000	–	–	1835	1528
1500	–	–	2749	2291
2000	–	–	3665	3055

Note: Facility steam supply pressure = 100 psig.
Source: From Vaponics (1997a).

Table 5.4 Multiple-Effect Distillation Unit Capacity as a Function of Facility Steam Supply Pressure

Facility steam supply pressure (psig)	Multiple-effect distillation unit product water flow rate (gal/hr)
40	370
60	507
80	634
100	739
120	793

The final effect "overflow" from the vertical cylindrical effect is directed through an interstage heat exchanger and then to waste. This basically provides continuous "blowdown" of the system.

Table 5.3 summarizes multiple-effect distillation unit capacity versus steam consumption for units with three to six effects. The data clearly demonstrates the reduction in facility steam consumption per unit volume of distillate product water as the number of effects increase.

The anticipated product water quality is 0.2 to 0.5 µS/cm at 25°C, assuming that the feedwater quality is ≤ 1 µS/cm conductivity and that the reactive silica concentration is ≤ 1 ppm.

Table 5.4 presents a summary of unit capacity at 40, 60, 80, 100, and 120 psig facility supply steam pressure. For a given unit, this demonstrates the advantage of higher pressure steam, when available.

Vapor Compression Distillation Unit
Boiling Inside Tubes: Pre-evaporation Degasification

Figure 5.14 depicts a vapor compression distillation unit where evaporative boiling occurs inside the tubes. The unit is equipped with pre-evaporation degasification provisions. This unit may be provided with softened feedwater per the manufacturer's recommendation. Boiling occurs inside the tubes of a vertical shell-and-tube heat exchanger, which is considered the "evaporator" section for the unit.

As the phase change from liquid feedwater to steam occurs within the tubes, the steam rises to the top of the vertical tube bundle, where it is released into an open portion of the vessel above the vertical shell-and-tube heat exchanger. The steam then passes through a vapor-liquid disengaging section that is physically positioned inside the vessel at the domed top of the evaporator. The purpose of the vapor-liquid disengaging section is to remove any water and other particles from the steam at this point in the system. The Pure Steam from the

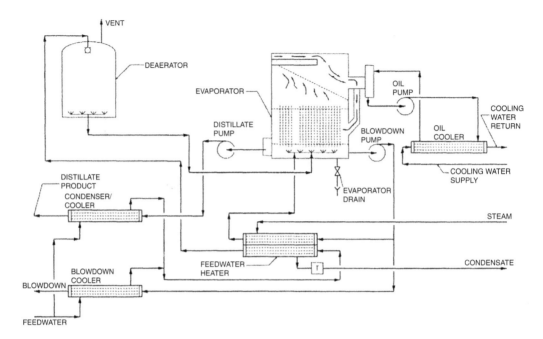

Figure 5.14 A vapor compression distillation unit with boiling inside tubes. *Source*: Courtesy of MECO (1991).

vapor-liquid disengaging section flows to the suction side of a compressor. This vapor compressor increases the pressure of the Pure Steam from about 1.0 psig (saturation temperature = 215°F) to 3.5 psig (saturation temperature = 223°F). It should be noted that the operating pressure within the tubes is extremely low, and the increase in pressure is small. These two factors are extremely important in selecting a vapor compression unit. Not only will the low operating pressure result in low operating temperatures, with a related decrease in the tendency for chloride stress corrosion or chloride pitting attack on austenitic stainless steel surfaces, the lower pressure also decreases the start-up time of the unit.

The 3.5 psig Pure Steam from the effluent of the vapor compressor flows back to the vertical cylindrical vessel on the shell side of the shell-and-tube heat exchanger. The 3.5 psig Pure Steam passes around the outside of the tubes within the heat exchanger, transferring its latent heat to feedwater inside the tubes, and subsequently condenses. The condensed Pure Steam is collected in the lower section of the shell side of the heat exchanger. The collected distillate product is transferred by a distillate pump to a distillate cooler (i.e., a condenser).

A portion of the feedwater from the tube side of the vertical cylindrical heat exchanger is removed as blowdown, since the feedwater is constantly concentrated during unit operation. When operational, the unit requires a continuous supply of feedwater.

The features for commercially available vapor compression unit using this design are as follows:

- *Evaporator*: The evaporator section of the unit consists of a vertical shell-and-tube heat exchanger with internal self-supporting "straight" 316L stainless steel tubes.
- *Distillate piping*: The distillate piping delivers condensed Pure Steam from the shell side of the vertical heat exchanger to a distillate pump. The distillate pump then delivers distillate through a distillate cooler, a double-tube-sheet heat exchanger. If desired, the distillate pump is capable of providing distilled product water at a positive discharge pressure, as compared to atmospheric pressure (which requires "gravity" for flow to a downstream storage tank). All distillate tubing should be of 316L stainless steel material.
- *Heat exchangers*: To maximize the efficiency of the vapor compression unit, in addition to the vertical shell-and-tube heat exchanger within the evaporator section,

the unit should include additional heat exchangers for blowdown cooling, feedwater preheating, and distillate cooling.
- All heat exchangers should be of a shell-and-tube design and be fabricated of 316L or 316 stainless steel. As previously indicated, the tubes in the distillate cooler (condenser) should be constructed of 316L stainless steel. The distillate cooler heat exchanger (condenser) should be designed so that the distillate water flow is through the tubes of the unit, while feedwater flow is through the shell side of the unit. The unit should be provided with a double tube sheet if a U-tube arrangement is used or two double tube sheets if a "straight-tube" arrangement is employed.
- The heat exchanger used to cool the blowdown water from the evaporator section should be designed so that the feedwater flow is in the shell side of the heat exchanger, while the blowdown flow is in the tube side. While a double tube sheet should be considered for this heat exchanger, a single tube sheet may be adequate; however, it requires increased monitoring of the feedwater to the unit to verify that it meets the U.S. EPA's National Primary Drinking Water Regulations (NPDWR).
- *Feedwater and blowdown tubing*: Feedwater and blowdown tubing may be constructed of either 304L or 316L stainless steel. The properties of 316L stainless steel, considering the relatively small cost difference (when compared with 304L stainless steel), make it the material of choice for the feedwater and blowdown tubing.

Specific information associated with operation of the vapor compression unit is presented as follows:

- Feedwater to the unit, as a minimum, should be softened. It is strongly suggested, however, that the feedwater be treated by reverse osmosis or reverse osmosis and continuous electrodeionization (CEDI) be employed. This suggestion is *not* based solely on the desire to remove all ionic material but the desire to remove impurities, such as NOM, colloidal material, particulate matter, etc. Feedwater flow to the vertical cylindrical evaporator section is controlled by a level-sensing device. Since the vapor space above the vertical cylindrical heat exchanger of the evaporator is slightly pressurized, a differential-type liquid level detection system, with controller, is required to properly regulate the feedwater flow rate. The feedwater controller provides a 4 to 20 mA signal to a "current-to-pneumatic" converter, which in turn provides a proportional pneumatic signal to a modulating valve installed in the feedwater supply line to the unit. In other words, the feedwater flow rate, during operation, should *not* be "on"/"off", but continuous, modulated as required.
- As the feedwater enters the system, as shown in Figure 5.14, a portion of the feedwater flow is directed to the shell side of the blowdown heat exchanger. The balance of the feedwater flow passes through the shell side of the distillate cooler (condenser). This technique maximizes heat recovery from the system. The two tempered streams rejoin, prior to the feedwater heater.
- After the two feedwater streams are rejoined, the entire feedwater flow may be directed through the tube side of a feedwater heater. The same heat exchanger may be used to reheat recirculated water "around" the lower portion of the evaporator section, beneath the inlet to the tubes in the vertical cylindrical heat exchanger. This recirculating stream also serves as a discharge point for blowdown from the evaporator section (a fraction of the recirculated water). The shell side of the feedwater heat exchanger, when employed, is heated by facility steam.
- The feedwater may contain dissolved inorganic gases, such as carbon dioxide and/or ammonia, as well as volatile organic compounds such as chloroform, a THM. The removal of noncondensable gases is critical to the successful operation of a vapor compression unit, since the gases could create a situation within the evaporator tubes where a gaseous bubble inhibits the evaporative process by restricting water flow to heat transfer surface area. Differential temperatures (and associated pressures) within the vertical cylindrical heat exchanger are extremely small. Any outgassing that creates a vapor blanket will significantly reduce the capacity of the distillation

unit and *may* affect the quality of the distillate product water, a critical item associated with this particular type of vapor compression unit. It is strongly suggested that provisions are included for periodic sampling of the feedwater to determine the concentration of noncondensable gases.

- A degasifier is provided to treat feedwater prior to the vertical cylindrical heat exchanger within the evaporator. Gases are removed by a number of techniques. The preferred technique is to provide a means of mechanical separation coupled with a means of heating the feedwater. Steam to heat the water may be provided from the "shell" side of the vertical cylindrical heat exchanger. The steam and noncondensable gases removed by the degasifier may be vented directly to the atmosphere. A check valve should be provided in the discharge line to the atmosphere to avoid/minimize air introduction during unit shutdown, which could result in steam condensation within the degasification unit.
- A pressure sensor monitors the Pure Steam pressure at the domed top of the vertical cylindrical evaporator section. Pressure is maintained in the range of 0.8 to 1.0 psig by regulating the facility steam flow to the feedwater heat exchanger, which is achieved by providing a modulating valve on the facility steam supply line to the shell side of the feedwater heater. The latent heat from facility steam is transferred to the feedwater and the recirculating water from the base of the evaporator section. The facility steam required during normal operation is associated with heat losses from the evaporator and downstream components. This would include heat losses in the blowdown water after the blowdown heat exchanger. A higher facility steam flow rate to the feedwater heat exchanger is required during periodic start-up of the unit.
- As previously discussed, the feedwater flow rate is modulated based on a proportional signal from a differential level sensor and controller monitoring the evaporator water level.
- If desired, an evaporator drain valve may be positioned at the bottom "inverted head" of the unit to allow draining of the unit.
- A relief valve, set at approximately 5 psig (based on the 1.0 psig operating pressure suggested throughout this section), should be positioned in the upper portion of the evaporator to prevent overpressurization.
- The vertical cylindrical heat exchanger is positioned in the lower section of the evaporator. Feedwater and recirculated water are blended in the lower section of the evaporator beneath the tubes of the heat exchanger and are fed to the inside of the heat exchanger tubes. Steam is generated within the tubes, flows upward to the vapor space above the tubes, but still within the evaporator, through a vapor-liquid disengaging section and then to the vapor compressor.
- The vapor compressor generally consists of a V-belt driver, a high-speed centrifugal-type compression section, external compression housing, a compressor drive shaft, bearings, a shaft seal, oil sump, and an electrically driven motor. A lubrication oil pump supplies oil through nozzles to lubricate and cool the bearings adequately at each end of the high-speed operating shaft. Temperature gauges should be mounted on the inlet and outlet of the oil cooler to ensure that it is functioning properly and that overheating of the compressor does not occur. Further, it is strongly suggested that a temperature switch be provided on the discharge side of the oil cooler to energize an audible alarm and visual indicator when the effluent oil temperature is too high. It is further suggested that pressure gauges be positioned in the inlet and outlet piping to and from the oil cooler and that a pressure switch be installed in the discharge line of the oil cooler. The pressure switch should also energize an audible alarm and visual indicator in the event of a low oil cooler discharge pressure.
- A separate differential level detecting system monitors the liquid level of *condensate* within the *shell* side of the vertical cylindrical heat exchanger to regulate the flow of distillate product from the shell side of the vertical cylindrical heat exchanger to the downstream distillate pump. If the unit is equipped with a distillate flowmeter, it should be of a sanitary type.

- A conductivity sensor (cell) should be provided, with a monitor, to indicate the purity of the distillate product and control the divert-to-waste system. This system would be similar to the system employed to monitor distillate from a multiple-effect distillation unit, including either a three-way valve or two individual diaphragm valves. The system should divert below-quality water to waste and in-specification water to a downstream storage tank. A "time delay" should be included to divert water to waste for a preset period after start-up of the unit.
- A recirculating pump should be provided, with feed from the lower portion of the evaporator (beneath the tube side of the vertical cylindrical heat exchanger), to constantly recirculate water. A portion of the water, as appropriate to inhibit unacceptable concentrations of impurities associated with the evaporative process, is removed and delivered to a blowdown heat exchanger and subsequently to waste.
- Operation of the distillation unit is generally controlled by level in the downstream storage tank. When the level in the storage system has decreased to a "low" set point, the distillation unit would commence operation. After start-up, the set point for pressure in the upper portion of the evaporator is generally increased by about 50% to approximately 1.5 psig. This higher pressure setting would only be maintained for about three to five minutes to decrease start-up time. The pressure setting would then be reduced to the normal operating value of 1.0 psig.
- The evaporator section of the unit, particularly when softened feedwater is used, will accumulate scale over time and require maintenance. The degree of scaling and, subsequently, the cleaning frequency is a function of the quality and characteristics of the feedwater. Scaling is detected by a reduction in the distillate flow rate (volume of water processed per given time period). When the distillate flow rate has decreased below the manufacturer's recommended value, the evaporator section of the unit, including the vertical cylindrical heat exchanger and areas of all heat exchangers where scale formation is possible, should be cleaned with a manufacturer approved acid solution. The type, concentration, contact time, and so forth for acid cleaning should be defined in the operating and maintenance manual for the distillation unit, which should be provided by the manufacturer.

Boiling Outside Tubes: Degasification After Evaporation
Figure 5.15 demonstrates a commercially available vapor compression distillation unit where boiling occurs outside the tubes and degasification is performed subsequent to evaporation. This unit operates on a different principle than the unit discussed in the previous section. Feedwater to the unit flows through two shell-and-tube heat exchangers. One of the heat exchangers, operating in parallel, is the condensing unit for the distilled product water. The second heat exchanger removes heat from blowdown water. A recirculating pump removes water from the evaporator section of the unit, discharges a portion of the recirculated water to the blowdown heat exchanger, joins with the tempered feedwater stream from the blowdown heat exchanger and condensing unit operating in parallel, and returns the water to spray nozzles positioned above steam-heated evaporator tubes within the evaporator section. A portion of the sprayed water is rapidly converted to Pure Steam, while the balance of the feedwater is collected at the base of the evaporator section and recirculated by the evaporator recirculation pump. The evaporator contains heating provisions for start-up and makeup requirements. The Pure Steam generated from the evaporator section flows through a vapor-liquid disengaging section to remove any entrapped water. The Pure Steam is subsequently directed to a gas centrifugal compressor. The Pure Steam pressure is increased, with an associated increase in the temperature of the Pure Steam. The discharge from the centrifugal compressor is fed to the inside of the evaporator tubes, releasing its latent heat to water "sprayed onto" the outside of the tubes within the shell side of the evaporator. The Pure Steam distillate is collected and delivered to a deaerator to remove volatile components. A distillate pump then transfers the water from the deaerator to the condenser, which is cooled by feedwater to the unit. The distillate product water is then delivered to a downstream storage tank.

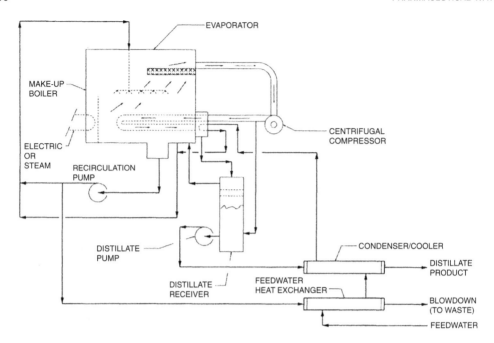

Figure 5.15 A vapor compression distillation unit with boiling outside tubes. *Source*: Courtesy of Aqua-Chem, Inc. (1996).

Standard design features for the unit should include the following:

- 316L stainless steel valves.
- 316L stainless steel, orbitally welded or sanitary ferrule tubing.
- All surfaces in contact with distillate product water should be 316L stainless steel, with a minimum mechanical polish of 15 to 25 Ra and electropolished.
- The feedwater heat exchangers, condenser, and blowdown heat recovery unit should be of a sanitary design, with a double tube sheet.
- The evaporator shell and tubes should be fabricated of 316L stainless steel.
- Surfaces in the deaerator; compressor ductwork; and all tubing in contact with distillate product, feedwater, and blowdown should be constructed of 316L stainless steel. The distillate pump and evaporator recirculation pump should be of a sanitary centrifugal type.
- Commercially available units employ a centrifugal compressor with enclosed titanium impellers. This particular design provides the physical strength required to compress the Pure Steam with relatively high efficiency. The Pure Steam and oil seals should be of noncontact type for successful, long-term operation. The compressor bearings should be journal type, lubricated by a pressurized oil system.

Distilled product water conductivity should be less than 1 μS/cm at 25°C when boiler feedwater is softened with a total hardness value of less than 1 ppm, expressed as calcium carbonate.

The unit does not require cooling water for operation. Oil, used as part of the vapor compressor, is cooled by a forced air cooling system.

The temperature of the Pure Steam in the evaporator section is 215 to 220°F. The suggested distillate water temperature is 85°C. While lower distillate water temperatures may be achieved, cooling below 85°C should only be considered for USP Purified Water applications.

The blowdown flow rate is 10% of the distilled water flow rate.

The centrifugal compressor should be provided with an enclosed vane impeller, labyrinth shaft seals, and journal bearings. Oil cooling is required. The oil cooling system should include a pump, an air cooler, a filter, a safety valve, and instrumentation.

The deaerator, positioned between the tube side outlet from the evaporator unit and the condenser, should be provided with a level control system.

The system must be provided with an acid "clean-in-place" system to facilitate scale removal from the evaporator and heat exchangers. Visual observation ports are provided to allow operating personnel to observe the evaporative process, not only to determine proper system operation but also to observe scale-forming precipitates.

The condensing unit for the system should have a hydrophobic vent filtration system.

The suggested instrumentation for the unit should include, but not be limited to, the following:

- A sanitary-type evaporator level sensing, transmitting, and control system to control the feedwater flow to the unit and the deaerator level control valve.
- The facility steam or electrically heated start-up unit in the evaporator should be provided with a level control system that will not introduce impurities or provide a location for bacteria proliferation.
- The feedwater piping/tubing to the unit should be equipped with an "emergency" valve to shut down the unit in the event of high evaporator level, which is associated with improper control by the feedwater modulating valve.
- The evaporator section should be provided with a drain valve positioned at the base of the unit.
- The top of the evaporator section should have a relief valve to eliminate potential overpressurization.

In addition to the above instrumentation, local temperature indicators or temperature sensors with indicating transmitters should be provided to monitor the following:

- Feedwater inlet temperature
- Feedwater temperature after the condenser heat exchanger
- Feedwater temperature after the blowdown heat exchanger
- Blowdown discharge temperature
- Distillate product outlet temperature
- Evaporator shell temperature
- Evaporator recirculating water temperature
- Lubricating oil temperature

Pressure indicators or pressure sensor with indicating transmitters should be provided for the following:

- Evaporator shell pressure
- Discharge of the distillate water transfer pump
- Discharge of the evaporator water recirculation pump
- Feedwater inlet pressure

Suggested alarm conditions should include, but not be limited to, the following:

- Evaporator shell vacuum condition
- Evaporator shell high-pressure condition
- Evaporator boiler subsection low water level
- Compressor lube oil low pressure
- Compressor lube oil high pressure
- Compressor "surge" (low compressor motor amperage draw)

The evaporator shell and heat exchangers should be insulated with high-density, chloride-free material covered with an aluminum jacket.

Pure Steam Generators
The discussion of design considerations associated with Pure Steam generators will, by definition, be structured around units producing Pure Steam that, when condensed, meets the chemical, bacterial, and bacterial endotoxin specifications for Water for Injection. Currently USP contains an *Official Monograph* for Pure Steam. The design essentially consists of the first effect of one of the three types of multiple-effect distillation units discussed earlier in this chapter. Obviously, the evaporator and vapor-liquid disengaging sections would be included. A condenser is not included.

Generally, heat exchangers are used to heat feedwater and obtain sensible heat from evaporator blowdown and/or facility steam condensate from the evaporator heating media. Heat exchangers should be of double-tube-sheet design.

It is suggested that all surfaces in contact with feedwater, water and facility steam in the evaporator section, and Pure Steam, including the vapor-liquid disengaging section, be fabricated of 316L stainless steel.

Generally, units are maintained in a "hot standby or ready" condition, capable of supplying Pure Steam on demand. Shutdown periods where the units are at ambient conditions should be minimized, but they may be employed for sporadic Pure Steam requirements when a number of days pass between Pure Steam demand. In any event, frequent thermal cycling (ambient temperature to hot operating temperature) of Pure Steam generators should be minimized.

The blowdown flow rate from the evaporator section of a Pure Steam generator will vary with system design, but it is generally 5% to 15% of the Pure Steam flow rate ("equivalent" condensate volume).

Feedwater quality requirements are similar to those for multiple-effect distillation units with the exception of additional monitoring required for specific regulatory agencies (AAMI/ ANSI, 2006 and British Standards Institute, 2006). The additional feedwater chemical specifications are provided in an attempt to insure that Pure Steam is free of noncondensable gases, water (liquid), and other impurities when used for sterilizing applications. Monitoring provisions should include in-line feedwater conductivity and the ability to sample the feedwater periodically to conduct appropriate laboratory analysis.

To reduce potential scaling of heat transfer surface areas, particularly surfaces within the vapor-liquid disengaging section, it is suggested that the total reactive silica concentration in the feedwater be less than 1 mg/L. To provide long-term successful operation, the feedwater should be of RO/CEDI quality. For units employed at facilities with vapor compression distillation units with softened feedwater, feedwater may be Water for Injection as discussed earlier in this chapter.

The chemical and microbial quality of the Pure Steam condensed water should be either directly monitored or periodically determined by collecting Pure Steam condensate at a point of use. For larger-capacity Pure Steam generators online, chemical monitoring for conductivity and TOC should be considered. For smaller capacity units, periodic collection of condensed point-of-use samples is appropriate. Continuous monitoring may require a dedicated, small-condensing unit or online instrument with integral condensing capability.

In addition to testing requirements set for in the USP *Official Monograph* for Pure Steam, additional testing should be performed periodically as outlined in the literature (AAMI/ANSI, 2006; British Standards Institute, 2006; and HTM 2031, 2007). Testing requirements have been discussed earlier and presented in Table 5.1. Again, the additional physical/thermodynamic test parameters and specification are based on sterilization considerations, summarized as follows:

- *Noncondensable gases*: If noncondensable gases, such as carbon dioxide, ammonia, etc., are present in Pure Steam (>3.5%), they will *not* condense on surface in a sterilization unit. They are incapable of transferring their latent heat to surfaces/components during the sterilization operation. Subsequently, the presence of noncondensable

gases, particularly with fluctuating concentration, could negate the effectiveness of a sterilization operation validated with Pure Steam without, or with lower concentration of noncondensable gases.
- *Pure Steam superheat*: Superheating of Pure Steam can be associated with heat introduction to "saturated" Pure Steam. After water (liquid) is completely converted to Pure Steam (gas), any additional method of heating the gas will result in an increase in gas (Pure Steam) temperature. Heating of Pure Steam above its saturation temperature at a set pressure greater than 25°C is not acceptable for sterilization operations. Before the Pure Steam can transfer its latent heat for sterilization it must cool from the superheated temperature to its saturated temperature. The time delay associated with this temperature drop, particularly for a gas (Pure Steam) could negate the effectiveness of a sterilization operation validated with Pure Steam at saturation temperature for a specific pressure.
- *Pure Steam thermodynamic quality ("dryness")*: As discussed on several occasions earlier in this chapter, Pure Steam (gas) must not contain water (liquid). Vapor-liquid disengaging sections for Pure Steam generators and distillation units are very critical components. Earlier in this chapter, enthalpy was discussed. Further Pure Steam thermodynamic quality was discussed. The Pure Steam thermodynamic quality should be ≥ 0.90 for sterilization of nonmetal containing "loads" and ≥ 0.95 for metal loads. Again, lower thermodynamic steam quality is associated with a lower "total enthalpy" value per volume of Pure Steam that will negate the effectiveness of a sterilization operation validated with Pure Steam without water (liquid).

All surfaces in contact with Pure Steam should be 316L stainless steel or Teflon. The stainless steel surfaces should be mechanically polished to 15 Ra and electropolished.

The Pure Steam pressure from the unit, and selection of a specific Pure Steam generator for a given application, is a function of the component(s) requiring Pure Steam (e.g., an autoclave). Pure Steam flow and pressure as well as the projected Pure Steam pressure loss in the delivery tubing to the component must be considered when a Pure Steam generator is selected. The pressure drop through the tubing from the Pure Steam generator to the "most remote" component requiring Pure Steam should be calculated at the maximum pressure requirement and flow rate for all components using Pure Steam. While Pure Steam generators are commercially available with discharge pressures as high as 100 to 125 psig, most Pure Steam generator requirements can be satisfied with a pressure of 30 to 75 psig.

All Pure Steam connections should be of a sanitary type. The use of threaded connections or compression-type fittings is *not* acceptable. However, "sanitary" ball valves may be employed in Pure Steam tubing. The use of conventional sanitary diaphragm valves in Pure Steam lines can present issues, particularly with regard to diaphragm "oscillation" in a throttling condition. This situation results in fairly rapid degradation of valve diaphragms and inability to provide Pure Steam flow/pressure regulation.

Pure Steam generator evaporator pressure vessel design, construction, and testing should be in accordance with the ASME Code for Unfired Pressure Vessels as discussed earlier.

A method for relieving overpressurization of the Pure Steam generator should be provided, preferably at the top of the evaporator section. The pressure relief device should not allow accumulation of water vapor. If installed in an area in contact with Pure Steam, particularly after the vapor-liquid disengaging section, the device should be of a sanitary type.

The evaporator column, heat exchangers, and interconnecting tubing, as well as the delivery tubing to points of use requiring Pure Steam, should be thermally insulated with chloride-free material and PVC jacket. Larger components, such as the evaporator column and heat exchangers, should contain aluminum sheeting around the chloride-free insulation. Obviously, insulation minimizes thermal loss from the unit to the atmosphere. In addition, particularly for surfaces in contact with Pure Steam, insulation will assist in maintaining the thermodynamic quality of the Pure Steam. While not specifically addressed in this chapter, it is extremely important that Pure Steam delivery tubing be equipped with provisions to remove condensate. Experience indicates that the thermodynamic quality of "delivered" Pure Steam is associated with improper delivery tubing routing, inadequate number of Pure Steam traps, improper location of Pure Steam traps, Pure Steam distribution system "dead legs," and inadequate Pure Steam trap maintenance/cleaning. The type and nature of the devices used to remove any Pure Steam condensation will be a function of the specific application, the length of the delivery tubing, the configuration of the delivery tubing, and other variables. Field inspection and observation of the "installed" Pure Steam distribution system is suggested. However, point-of-use Pure Steam thermodynamic quality testing during Performance Qualification testing may be required. While the condensate should exhibit properties of USP Water for Injection, its presence in Pure Steam may be detrimental to applications requiring Pure Steam as discussed earlier. Any device used to eliminate condensed Water for Injection in a Pure Steam line must be of a sanitary type, designed so that it will not compromise the chemical, bacterial, or bacterial endotoxin quality of the Pure Steam. The system design should also establish a proper feedwater flow rate and Pure Steam outlet pressure.

Monitoring provisions for the Pure Steam generators should energize an audible alarm and visual indicator for critical parameters. Suggested parameters triggering an alarm would include, but are not limited to, the following:

- High evaporator section water level
- High Pure Steam outlet pressure
- Low Pure Steam outlet pressure
- High feedwater conductivity
- High condensed Pure Steam conductivity (if installed)
- High condensed Pure Steam TOC (if installed)

If the unit is equipped with a Pure Steam condensing unit with a continuous conductivity measuring device, an alarm should also be provided for high Pure Steam conductivity and TOC as indicated.

Figure 5.16 graphically demonstrates Pure Steam pressure as a function of available facility steam capacity (lb/hr), for various Pure Steam delivery pressure requirements. Depending on the available pressure of feedwater to the unit and the required Pure Steam pressure, it is often necessary to provide a feedwater booster pump for the Pure Steam generator. Pretreated feedwater to a Pure Steam generator from a RO/CEDI system should include provisions for eliminating both back contamination and stagnant sections of tubing (dead legs). While a feedwater storage tank may be employed for this application, a "double block and bleed" valve/tubing arrangement, depicted in Figure 8.6, should be considered.

Condensing Units

A condensing unit may be "coupled" with a Pure Steam generator for low-volume daily demands of USP Water for Injection. This combination is appropriate for applications where small volumes of Water for Injection and noncontinuous Pure Steam are required.

It is suggested that the Water for Injection requirement (volume/day) be carefully considered and that a storage tank of adequate volume be provided to avoid nonsimultaneous operation of the Pure Steam generator for production of Pure Steam and Water for Injection. In other words, it is desirable to satisfy the Pure Steam requirements for Water for Injection to the condensing unit during "off-shift" hours or when Pure Steam will not be required for "other" applications, such as chamber and maintenance steam to an autoclave.

The condensing unit should exhibit all of the physical properties and characteristics discussed in this chapter for a "gravity discharge" distillation unit.

It is suggested that this technology may be applicable for Water for Injection use up to 100 gal/hr but not greater than 500 gal/day. The capacity of the condensing unit, in terms of gal/hr of Water for Injection, is a function of the Pure Steam outlet volumetric flow *and* condenser selection (heat transfer area, cooling water, etc.).

For smaller volume Water for Injection applications, where Pure Steam is also required, the capital cost of a Pure Steam generator and a condensing unit is significantly less than that of a separate distillation unit and Pure Steam generator.

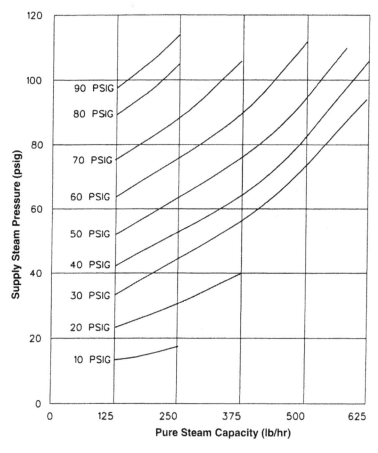

Figure 5.16 The relationship between supply steam pressure, Pure Steam pressure, and Pure Steam capacity. *Source*: From Mueller-Barnstead (1987b).

The condensing unit should be equipped with a double tube sheet of sanitary type. Interior surfaces of the condensing unit in contact with Pure Steam or Water for Injection should be 316L stainless steel or Teflon. Surfaces should be mechanically polished to a 15 Ra finish and electropolished. The condensing unit should be equipped with a heated and jacketed hydrophobic vent filtration system.

The parameters associated with gravity feed provisions for the delivery of Water for Injection to a downstream storage tank apply. The controls and instrumentation for the Pure Steam generator with condensing unit should include, but are not limited to, the following:

- Start and stop of the unit based on downstream tank level
- Lockout of the unit, where appropriate, when Pure Steam is required for other applications
- Preoperational (distilled product water flow) waste-to-drain cycle
- Pure Steam inlet pressure
- Cooling water inlet pressure
- Instrument air pressure
- Distillate water temperature with high- and low-temperature alarms
- Cooling water effluent temperature monitor with high-temperature alarm
- Flow or pressure switch in cooling water feed line with "low-flow" or "low-pressure" alarm
- Distilled water conductivity monitor with high-conductivity alarm and automatic divert-to-waste provisions

OPERATING AND MAINTENANCE CONSIDERATIONS

General operating and maintenance considerations are addressed for each item discussed in section "Design Considerations".

Single-Effect Distillation Units

The following items should be periodically verified or inspected:

- Distillate water flow rate at a pre-established supply steam pressure (after allowing adequate time for equilibration at the selected facility steam pressure).
- Proper function of the distillate water divert-to-waste system by adjusting the distillate product water conductivity set point.
- Monitoring of distillate product for Water for Injection specifications, including conductivity, TOC, total viable bacteria, bacterial endotoxins, and nitrates (EP).
- Inspection of the evaporator section of the unit. If scale formation is noted on heat transfer surface areas, clean the surfaces to remove the scale. If it is impossible to remove the scale, replace the steam heating coil. If electrical heating elements are used, periodically inspect the elements and replace, as necessary.
- Inspection of vapor-liquid disengaging section of the unit. This extremely important section of the distillation unit insures that water, potentially containing bacterial endotoxins, is not carried over with Pure Steam. A "roughening" of the surface within the vapor-liquid disengaging section, even without the appearance of scale-forming deposits, can have an adverse effect on proper system operation. The deposits should be removed or the vapor-liquid disengaging device replaced.
- Calibration of all meters, gauges, sensors, and so on every 6 to 12 months.
- Continuity of the tin-coated surface (if used in the evaporator section). If there is a loss in continuity of the tin-wiped surface, the surfaces may be appropriately treated with molten tin by the service organization of the distillation unit manufacturer.
- Remove and replace the condenser hydrophobic vent filtration membrane at least every six months. When replacing the membrane, inspect the "used" membrane for any deterioration due to elevated temperatures. If visible deformation and/or a "melting" appearance are noted, it is possible that impurities have entered the condenser. Inspection of the surfaces within the condenser should be considered if this condition is detected. Further, the vent filter heating provisions must be thoroughly reviewed.

Single-Effect Distillation Units with Condensate Feedback Purification

The operating and maintenance criteria outlined above for single-effect distillation units without condensate feedback purification should be performed as needed. In addition, the following items should be performed:

- Replace the rechargeable canisters when the conductivity increases above a preset value (suggested as 1 µs/cm) or once every three months. For smaller units, disposable cartridges may be used.
- If amines are present in the facility steam at the concentration discussed earlier in this chapter, periodically verify that amine compounds are not present in the distillation unit feedwater and, more importantly, the distilled product water.
- Include within the calibration program all gauges, monitors, and so forth in the condensate feedback purification portion of the system.
- Periodically (once each day or once each working shift) log the temperature in the condensate feedback purification system, specifically the feedwater temperature to the rechargeable (or disposable) canisters. Adjust the cooling water flow rate as required to insure that the feedwater temperature to the condensate feedback purification system does not exceed 120°F.

Multiple-Effect Distillation Units

The following items should be considered for multiple effect distillation units:

- If the unit is equipped with a feedwater system, periodically inspect and lubricate the feedwater pump. If the feedwater system includes a storage tank, replace the hydrophobic vent filter at least every six months.
- Periodically verify that the feedwater pump, if supplied, is operating properly by noting its discharge pressure. To determine proper feedwater pump motor operation, periodically verify the speed of the pump-motor combination, the external pump motor temperature, and the pump feedwater pressure.
- Log the temperature, pressure, and water level in each of the effects daily while the unit is operational.
- If the distilled product water conductivity is not recorded or displayed on a data logging system, log the value. Include the temperature of the distillate product water. This should be performed, as a minimum, at the beginning and end of each operating shift, or daily for single-shift-per-day operation.
- Periodically verify the distillate product water flow rate from the condensing section of the unit. Replace the condenser hydrophobic vent filter membrane at least once every three to six months. Inspect the used membrane for deformation and, if appropriate, pursue follow-up action if deformation is noted (discussed earlier in this chapter).
- Inspect the interior of the evaporator sections and clean as required. The cleaning operation should be performed to remove any scale deposits that will decrease the overall heat transfer within the unit and, subsequently, the distilled product water quantity and flow rate.
- Log the regulated facility steam pressure to the jacket of the hydrophobic vent filter installed on the condenser. This value should not exceed 2 to 5 psig. If the value increases above 5 psig, immediately attempt to readjust the supply steam regulator. If the pressure cannot be reduced below 5 psig, inhibit the steam flow and replace the supply steam regulator as soon as possible. Overheating of the hydrophobic vent filter will result in physical degradation of "support" material for the hydrophobic membrane material, principally polypropylene, resulting in potential introduction of "melted" plastic material into the condenser and loss of hydrophobic membrane filter integrity.
- Conduct a thorough periodic inspection of the vapor-liquid disengaging sections for each effect in the multiple-effect distillation unit once every 6 to 12 months or when a rapid or gradual increase in product water conductivity, above a preset value, is noted. The vapor-liquid disengaging sections may be cleaned in accordance with the distillation unit manufacturer's instructions. Generally, an acid solution is fairly effective at removing physically hard and highly adherent deposits. The "smoothness" of surfaces within the vapor-liquid disengaging sections of each effect is critical to the separation process.
- Calibrate all monitors, sensors, and other instrumentation every 6 to 12 months.
- Verify the proper function of the distillate product water divert-to-drain system by decreasing the conductivity set point on the conductivity meter. Product water should be diverted to drain and the makeup valve to the downstream storage tank should close.
- It is possible that trace concentrations of organic compounds, primarily associated with THMs such as chloroform, may be present in distilled product water. These compounds, if present in raw water, will not be entirely removed by feedwater pretreatment operations, such as ion exchange and/or reverse osmosis. It is strongly suggested that samples of distillate product water be collected quarterly, in special containers, for a "volatile organic analysis." The gas chromatography/mass spectrometry analysis should include a "library search" for THMs, such as chloroform, and, where applicable, amine-type compounds associated with the decomposition of chloramines, an alternative disinfecting agent to chlorine (see chap. 2).

Vapor Compression Distillation Units

After initial start-up of a vapor compression distillation unit and subsequent to any maintenance of the unit that requires disconnection of the electrical supply to the pump motors, verify the rotation of the pumps before allowing the pump motors to operate. During operation of the unit, check the compressor oil pressure and temperature at least once per day.

The following should be recorded on a daily basis or continuously monitored by a data logger (as appropriate), particularly when the unit is operational:

- Blowdown flow rate
- Distillate flow rate
- Feedwater flow rate
- Evaporator pressure
- Distillate product water pressure
- Feedwater pressure

Operation and maintenance items include, but are not limited to, the following:

- Verify operation of the feedwater flow rate controller by noting modulation of the feedwater valve during unit operation.
- Ensure that the distillate product water quality (conductivity) is continuously monitored and recorded or analog output transmitted to a data logger.
- If the conductivity of the distillate product water is not recorded or transmitted to a data logger, manually log the value, while the unit is operational, at least two to three times per day.
- Proper operation of the unit requires periodic scale removal from heat transfer surfaces within the evaporator section and heat exchangers. The acid cleaning procedure for scale removal should be performed when the distillate flow rate decreases below 10% of the design value. The distillate flow rate may gradually decrease, over time, during operation, which is associated with scale not removed during routine cleaning. It is suggested that the long-term reduction in system capacity (over a number of years) should not exceed 10% of the original distillate flow rate design value for the unit. If this occurs, it may be necessary to replace the heat exchangers and/or other components.
- The cleaning agent used to remove scale should be consistent with the vapor compression distillation unit manufacturer's recommendations. This material *may* be citric acid or sulfamic acid.
- The compressor and support accessories are an extremely important part of the vapor compression distillation unit. The "drive" to the compressor is generally by "V belts." The drive system and accessories should be inspected weekly and include a check of the tension on the belts. Proper tension is critical. Loose belts can cause "slippage," while tight belts can affect the drive mechanism on the compressor. The inspection should also include verification of belt alignment and a visual determination of cracks and/or fraying of the belts. Obviously, if this situation is noted, the belts should be replaced. During proper operation, with correct tension, there should be minimum vibration from the compressor. Excessive vibration *may* be associated with improper belt tension.
- The vapor compressor should be operated in a manner that ensures that the bearings are properly lubricated. Bearing inspection should be performed at least quarterly. The suggested replacement frequency for bearings is at least once every two years. In addition, the lubricating oil system must be properly maintained. It is suggested that synthetic-type oil be used. If synthetic-type oil is used, it should be replaced after approximately 8000 hours of unit operation (or once per year). When synthetic oil is used, the oil filter should be replaced at the same frequency. If mineral oil is used as the lubricant, oil and filter replacement should be performed after 2000 hours of operation (or quarterly). Oil seals should be inspected monthly. The entire compressor lubricating system should be thoroughly cleaned annually. Finally, the

compressor operating speed should be measured and recorded during unit operation once every one to two weeks.
- A maintenance schedule should be established for inspection and preventive maintenance of valves in the system. This would include the replacement of valve diaphragms. Gaskets and seals, including sanitary ferrules, should also be replaced as part of a preventive maintenance program. Gasket replacement, particularly on the evaporator section of the unit, should not be delayed until a noticeable leak from the unit occurs.
- While chemical cleaning will remove deposits from heat transfer surface areas, there are other components within the evaporator requiring inspection (e.g., the interior walls of the evaporator section). Rouging of surfaces may occur. This is a concern when the feedwater is from a softened water source where bicarbonate removal is not achieved. The carbon dioxide associated with bicarbonate equilibrium, not removed as part of the degasification system prior to the evaporator (where applicable), may require conversion to RO quality feedwater.
- During inspection of the evaporator section of the unit, the vapor-liquid disengaging section above the actual evaporator section should be inspected.
- The hydrophobic vent filter installed on the condensing unit should be replaced every three to six months.
- The various pump motors used in the system should be properly lubricated. The pump seals should be inspected and replaced as needed. The pump motor speed should also be checked periodically (once every 6 months) to verify proper operation.
- A periodic instrument calibration program should be established (once every 6–12 months).

Pure Steam Generator

The following operating and maintenance items should be performed:

- The Pure Steam pressure, during normal operation, should be verified periodically at a predetermined facility steam pressure. If the unit is not operating at the pre-established facility steam pressure, the pressure should be adjusted. Once the facility steam pressure is stabilized for a period of about 30 minutes, the Pure Steam pressure should be verified.
- The evaporator section of the Pure Steam generator should be inspected once every 6 to 12 months. If appropriate, scale on heat transfer surface areas should be removed.
- The vapor-liquid disengaging section of the Pure Steam generator should be inspected for siliceous deposits, as discussed earlier in this chapter.
- All heat exchanger surface areas should be inspected periodically, such as once each year or more frequently if heat exchanger fouling or scaling is suspected.
- A periodic verification of unit controls, operating, and monitoring functions should be conducted on a monthly basis. Instruments, controllers, and gauges should be calibrated once every 6 to 12 months. The blowdown rate from the Pure Steam generator should be verified once per month during unit operation.
- Gaskets in the Pure Steam distribution system should be replaced every one to two years. Gasket material should be stainless steel impregnated Teflon (Rubber Fab, 2010).

Condensing Units

The following maintenance items should be performed:

- Diaphragms in all valves within the system should be replaced on an annual basis. Valves should use Teflon or Teflon backed with ethylene propylenediene monomer (EPDM) diaphragms.
- All meters, gauges, and indicators should be calibrated once every 6 to 12 months.
- The hydrophobic vent filter should be replaced every six months.
- The distilled product water flow rate from the condensing unit should be measured once every six months. Obviously, this should be performed while the unit is operational at a pre-established Pure Steam supply pressure.

- Operation of the divert-to-waste system should be performed on a monthly basis by decreasing the product water conductivity set point to a value lower than the conductivity of the product water, thus creating a divert-to-waste situation.
- A thorough inspection of heat transfer surface areas should be conducted on an annual basis. This inspection should focus on scaling of the heat transfer surface area as well as an inspection for cracks or crevices associated with chloride stress corrosion.

REFERENCES

AAMI/ANSI. Association for the Advancement of Medical Instrumentation and American National Standards Institute. Comprehensive Guide to Steam Sterilization and Sterility Assurance in Health Care Facilities, ST79:2006, Arlington, VA, 2006.

Aqua-Chem, Inc. A Worldwide Network of Technology and Solutions—Pharmaceutical. Milwaukee: Aqua-Chem, Inc., Water Technologies Division, 1996.

British Standards Institute (BSI). Sterilization, Steam Sterilizers, Large Sterilizers, BS EN 285:2006, British-Adopted European Standard, ISBN: 0580486885, 86 pages, 2006.

Cal Research Inc. The Ellison Throttling Steam Calorimeter. Ransomville, NV: Cal Research, Inc., 1998.

Collentro A. Production Methods for USP Water for Injection: An Evaluation of the USP 27 Official Monograph Change, presented at Biopharma 2005, ISPE New England Chapter, Boston, MA, 2005.

Collentro W. Unpublished volatile organic scan information of USP/EP Water for Injection produced by both multiple effect and vapor compression distillation units, 2005–2009.

European Medicines Agency. Reflection Paper on Water for Injection Prepared by Reverse Osmosis, Document reference: EMEA/CHMP/CVMP/QWP/28271/2008, London, UK, 2008.

Finn-Aqua America, Inc. A Reliable Way of Producing High-Quality Distillate. Apex, NC: Finn-Aqua America, Inc., 1986:7.

Finn-Aqua America, Inc. Finn-Aqua High Technology for Pharmaceutical and Biotech Industries. Apex, NC: Finn-Aqua America, Inc., 1989a.

Finn-Aqua America, Inc. Multi-Effect Water Stills—U.S. Units. Apex, NC: Finn-Aqua America, Inc., 1989b.

FR. Human drugs—cGMPs for LVPs and SVPs. 21 CFR, part 221. Fed Regist 1976; 41(106): 22202–22219.

HTM 2031. Clean Steam for Sterilization, NHS Estates/Department of Health, England, 2007.

McCabe WL, Smith JC, Harriot P. Unit Operations of Chemical Engineering. 5th ed. New York: McGraw-Hill, Inc., 1993:389.

MECO. Piping Diagram—Hot Distillate. New Orleans, LA: Mechanical Equipment Company, 1991.

Meltzer TH, Livingston RC, Madsen RE, et al. Reverse osmosis as a means of water for injection production: a response to the position of the European Medicines Agency, PDA J Pharm Sci Technol 2009; 63(1):1–7.

Paul Mueller Company. Economically Expandable Distillation Systems—Barnstead Thermodrive™ Stills. Bulletin No. TDA02-038510M. Springfield, MO: Paul Mueller Company, 1985:7–10.

Mueller-Barnstead. Multiple Effect Stills. Bulletin No. MB-2103. Springfield, MO: Mueller-Barnstead, 1987a:10.

Mueller-Barnstead. Non-Pyrogenic Steam for Guaranteed Sterilization: Mueller®/Barnstead™ Pure Steam Generator. Bulletin No. MB-2101. Springfield, MO: Mueller-Barnstead, 1987b.

PDA. Technical Report No. 48 – Moist Heat Sterilizer Systems: Design, Commissioning, Operation, Qualification, and Maintenance", Parenteral Drug Association, ISBN: 978-0-939459-29-2, Bethesda, MD, 2010.

Rubber Fab. TUF-STEEL, A Full Line of World Champion Gaskets, Rubber Fab Technologies Group, Bulletin No. RF-160, 2010:1–12.

Tarry SW, Henricksen G, Prashad M, et al. Integrity testing ePTFE membrane filter vents. Ultrapure Water 1993; 10(8):23–30.

Vaponics. Thermevap™ Single Effect High Velocity Steam Stills, Technical Bulleting P/N 7352C. Rockland, MA: Vaponics, an Osmonics Company, 1984:1–2.

Vaponics. VSS Four Effects Still Flow Schematic. Rockland, MA: Vaponics, an Osmonics Company, 1989.

Vaponics. Multiple-Effect Thermevap™ Stills. Bulletin No. P/N 117350, Rev. A. Rockland, MA: Vaponics, an Osmonics Company, 1997a.

Vaponics. Thermevap™ Single Effect High Velocity Steam Stills. Bulletin No. P/N 73520. Rockland, MA: Vaponics, an Osmonics Company, 1997b.

Vaponics. Vaponics™ Pretreatment/Stills CFBP & Ion Exchange Product Information, 2010. Available at: http://www.atlanticmetalcraftco.com.

6 | Storage systems and accessories

THEORY AND APPLICATION
There are many technical reasons for including storage systems in pharmaceutical water systems. In less complex distribution systems, a storage tank may simply provide a convenient method of repressurizing compendial water prior to a distribution loop. For other applications, such as Water for Injection systems with distillation, storage systems are required to provide an atmospheric discharge point for distilled product water and to maintain and allow the recirculation of hot Water for Injection at an elevated temperature, generally greater than 80°C. While condensate receivers may also be used to collect distillation unit product water, the need for a storage tank is not eliminated due to the required recirculation of product water, generally at an elevated temperature to insure microbial control (FDA, 1986). For USP Purified Water systems, product water from single-pass or double-pass reverse osmosis (RO) unit, with or without polishing continuous electrodeionization (CEDI), may be slightly pressurized (10–30 psig). Without a storage tank for a USP Purified Water system, RO membranes will not tolerate rapid changes in the product water pressure associated with "point-of-use" draw-off, which, unfortunately, results in rapid loss of RO membrane integrity. As indicated earlier, a sound system design includes three loops: pretreatment, ion removal, and storage and distribution. There are several factors that demonstrate the technical superiority of system design, outlined in detail as follows:

- From a microbial standpoint, a dedicated storage system isolates the Purified Water or Water for Injection "generating" system components from the distribution loop. As discussed in previous chapters, it is extremely desirable to maintain total viable bacteria levels within the generating pretreatment section of the system at a value less than 500 cfu/mL (the suggested "Drinking Water" value in the *General Information* section of USP). Obviously, the 500 cfu/mL total viable bacteria level is significantly greater than typical Alert and Action Limits for Water for Injection. The 500 cfu/mL is also greater than "acceptable" total viable bacteria Alert and Action Limits for critical USP Purified Water applications, such as the production of topical solutions, antacids, inhalants, and ophthalmic solutions. Segregation of the water purification generating system from the storage and distribution system provides an air break with atmospheric pressure delivery of water from the water purification generating system to the storage and distribution system.
- The species of bacteria associated with water systems, including observed USP Purified Water systems are generally gram negative. Segregation of the Purified Water generating system from the storage and distribution system will minimize the introduction and proliferation of bacteria to the downstream distribution system. This can be accomplished by a number of techniques, such as the use of an in-line ultraviolet sanitization unit (254 nm), 0.1- or 0.2-µm membrane filtration and/or ultrafiltration within the distribution loop, immediately downstream of the distribution pump, for USP Purified Water systems. While the use of bacteria retentive filters "within" a USP Purified Water distribution loop is discouraged, it may provide desired water quality from a bacteria standpoint when coupled with a responsive periodic sanitization program. As discussed later, the storage and distribution system can be periodically sanitized as a "separate section" of the system, physically isolated from the water purification system generating components.
- USP Purified Water systems without storage tanks do not provide the air break necessary to separate the water purification system tubing from the distribution tubing. Bacteria will replicate, within a biofilm, against the direction of flow. Subsequently, "tankless" USP Purified Water systems that recirculate loop water

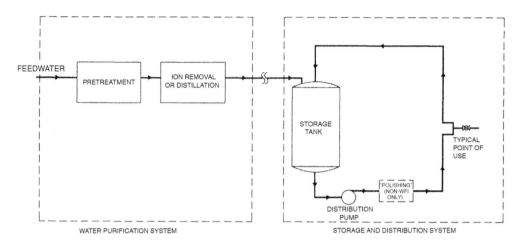

Figure 6.1 Segregation of the water purification system from the storage and distribution system. *Abbreviation*: WFI, Water for Injection.

back to the water purification generating system will require frequent sanitization and final filtration techniques, such as membrane filtration or ultrafiltration. Microbial control is difficult.
- For Purified Water systems, segregation of the generating system from the storage and distribution system minimizes or eliminates introduction of impurities to the stored and distributed water. If the Purified Water generating system is physically segregated from the storage and distribution system, any materials introduced into the water during the purification/generation process would be a "foreign substance or impurity," as defined in the *General Notices* section of the USP. While identification, control, and removal of foreign substances and impurities are required, components within the generating system frequently provide removal in a "controlled" manner. On the other hand, any material introduced after the final water purification unit operation associated with the production of USP Purified Water is considered an added substance, as indicated in the *Official Monograph* for Purified Water (Collentro and Zoccolante, 1994). The concept associated with segregating the Purified Water generating system from the storage and distribution system is demonstrated in Figure 6.1.
- Many pharmaceutical water purification systems, USP Purified Water, Drinking Water, and/or USP Water for Injection, require a large volume of water in a relatively short period of time. These "batching" applications provide a significant challenge to water purification system design, sizing, storage tank capacity, distribution pump selection, and diameter of the distribution tubing. As an example, water for a batching application may be required once per shift (1–3 shifts/day), or as infrequently as once each day, week, or longer. Unfortunately, this does not eliminate the need for delivering the large volumes of water required for the batching applications over a relatively short period of time. The frequency of the batching operation, the nature of the water purification unit operations (deionization, reverse osmosis, or distillation), with their associated makeup water flow rate, will dictate the size of the storage system. By its very nature, distillation generally provides a relatively slow makeup capability (flow rate). The sizing for membrane processes, such as reverse osmosis with CEDI generally require larger storage capacity than "older" systems employing ion exchange. Conversely, while the size of the storage tank may decrease for applications (e.g., USP Purified Water, Drinking Water) that employ deionization with a relatively rapid makeup capability, smaller volume storage systems are still required for batch applications.

- As discussed in chapter 7, many USP Purified Water systems and some Drinking Water systems use ozone for microbial control within storage and distribution systems. For systems using ozone for microbial control, the storage tank must provide a location for ozone to achieve bacteria destruction ("contact chamber") and also a vessel capable of continuous release of oxygen, the ozone decomposition compound. While the use of ozone is discussed in much greater detail in chapter 7, systems employing this technology are equipped with downstream dissolved ozone destruct in-line ultraviolet unit (prior to the distribution loop) to remove residual dissolved ozone during normal operation. Periodic sanitization of the distribution loop, performed as frequently as once per operating shift, can be achieved by simply inhibiting electrical power to the in-line ultraviolet units, allowing ozone to pass into the distribution loop for a period of 15 minutes to 2 hours depending on the residual dissolved ozone concentration in the recirculated water. Obviously, it would be extremely difficult to design a system using ozone for microbial control without a storage tank.
- The majority of Water for Injection systems maintain the stored and recirculated water at a temperature $\geq 80°C$ (a lower temperature may be used based on the application and/or the ability to control bacteria) (FDA, 1986). A method for providing temperature control in the storage and distribution system is required. Heating provisions (such as an external tank heating jacket) can be used to maintain the temperature of the water within the storage tank, and, with an appropriate temperature control system, the return water from the distribution loop at the coldest point within the storage and distribution system.
- Frequently, Water for Injection systems operate with hot storage and recirculating loops and "ambient" temperature subloops. The ambient temperature Water for Injection is frequently required to avoid thermal degradation of ingredients used during manufacturing. Most subloops simply feedwater to a pump that provides flow to points of use with return back to the suction of the pump. Periodic heating of the subloops (suggested at a frequency of at least once per day) should be performed. The heating can be provided by simply displacing subloop water with hot water from the main "hot" recirculating loop. While not generally suggested, some Water for Injection systems will use ambient storage and distribution systems. Occasionally, some of these systems will operate at ambient temperature for a significant portion of each working day (e.g., 20 hours) and undergo sanitization for the balance of the day by heating the storage and distribution system to a temperature greater than 80°C. A summary of various storage, distribution, and distribution subloop designs frequently used for Water for Injection systems is presented as follows:
  - Hot storage and distribution with no point-of-use cooling.
  - Hot storage and distribution with point-of-use heat exchangers to adjustable temperature value.
  - Hot storage, complete loop cooling to ambient temperature, distribution, and return loop heating.
  - Hot storage and distribution with ambient subloop(s).
  - Hot storage, "blending with ambient temperature loop return water," distribution, and "split of return water to the distribution loop pump suction and storage tank."
  - Cold (2–4°C) storage and distribution.
  - Hot storage, complete loop cooling to 2 to 4°C and return loop heating.
  - Hot storage and distribution with "cold" subloops.

    Obviously, there are several tank/loop arrangements that can be used. Each arrangement requires a storage tank. Unfortunately, depending on the demand for Water for Injection on a typical day, the limited storage and recirculation time period at elevated temperature may not be adequate to achieve desired microbial control. Each system requires an extended intense sampling and testing program, performance qualification (PQ), to demonstrate the ability to deliver, on a constant basis, point-of-use Water for Injection meeting the chemical, and bacterial endotoxin specifications as well as established total viable bacteria Alert and Action Limits.

Specifically, the frequency, duration, and temperature of the storage and distribution system to Water for Injection ≥80°C for bacteria control must be established.
- Occasionally Water for Injection storage and distribution systems will operate at ambient conditions on a batching basis. The basis of operation for these systems centers around the production, storage, and recirculation of a preestablished volume of USP Water for Injection; termination of the makeup process; cooldown to ambient temperature; point-of-use sampling of water; and the ultimate use of water for the batching operation. The ambient temperature Water for Injection remaining in the storage tank is continually recirculated and may be used for subsequent postbatch manufacturing requirements, such as initial or intermediate "clean-in-place" applications or the washing of components used during the manufacturing process. USP Water for Injection remaining in the storage and distribution system for more than 24 hours is drained (FDA, 1986). At the end-of-the-draining operation, the storage and distribution system is generally sanitized with Pure Steam prior to refilling with hot USP Water for Injection. The storage tank is an excellent location for introducing the Pure Steam for this application. Product manufactured with the ambient temperature Water for Injection is generally not "released" until the results of chemical, bacterial, and bacterial endotoxin analyses are obtained from samples collected at ambient temperature prior to using the water.
- Some Purified Water systems employ ambient storage and recirculation systems with periodic chemical (liquid) sanitization. The storage tank and/or distribution loop tubing/piping may *not* be of stainless steel construction. If the storage and distribution systems are of sanitary design, free of dead legs and equipped with zero dead leg valves, and if the USP Purified Water makeup contains less than 1 cfu/100 mL of bacteria, excellent microbial control is possible. Chemical sanitizing agents such as a 1% solution of peracetic acid and hydrogen peroxide not only destroy bacteria (in water and within a biofilm of tubing surfaces) but also remove biofilm. On the other hand, periodic hot water sanitization has limited effect on biofilm removal. The sanitization frequency, duration, and execution technique must be established to achieve required total viable bacteria Alert and Action Limits.
- Some Purified Water systems employ ambient storage and distribution with periodic hot water sanitization, as discussed previously. Bacteria control can be achieved with periodic hot water sanitization. However, while bacteria in an established biofilm are destroyed, biofilm removal is not achieved. Effective total viable bacteria destruction can be achieved, assuming that there are no dead legs (stagnant sections of tubing or other fittings greater than suggested three "pipe diameters" from the flowing hot water stream) when the recirculating Purified Water is heated to 90°C for at least 120 minutes. The "three pipe diameter length" should be determined using the smaller tubing diameter and measured from the center line of the tubing with flowing hot water, generally the larger diameter tubing. To achieve effective hot water sanitization at lower temperatures, the time period must be lengthened. The sanitization operation for the storage and distribution system, which routinely operates at ambient temperature, is achieved using a storage tank with heating jacket (or external heat exchanger). Occasionally, tank heating jackets may also be used for cooling applications, to remove Joule's heat from the relatively inefficient sanitary centrifugal recirculating pumps, thus ensuring that the stored and recirculated water temperature does not increase from ambient to incubation temperatures (e.g., 30–35°C). USP Purified Water system employing periodic hot water sanitization for bacteria control must be sanitized with a liquid sanitizing agent (peracetic acid and hydrogen peroxide) at least once every 12 months for biofilm removal.
- Purified Water systems equipped with provisions for periodic hot water sanitization may use either a heating jacket external to the storage tank or a dedicated heat exchanger within the distribution loop. Using an external heat exchanger does not eliminate the need for a storage tank that provides a location for water expansion and contraction during the sanitization operation. Tankless systems cannot be effectively sanitized because the Purified Water generating system components, including

piping, vessel linings, ion exchange resin, RO membranes (non-hot-water sanitizable), and so on, will not tolerate the suggested 90°C temperatures required for effective hot water sanitization.
- A limited number of USP Purified Water systems are designed to operate with cold storage and distribution systems ($\sim$2–4°C). While the proliferation of bacteria in the storage and distribution systems for cold systems is significantly less than that for ambient loops, operating data and published information indicate that periodic hot water sanitization is still required (Wong and Beria, 1993).
- In addition to benefits stated above, storage tanks provide a means of positive repressurization for USP Purified Water loops, which are required for delivery and recirculation with appropriate modulating-type back pressure devices (including instrumentation) installed in return tubing to the storage tank. Variable frequency drives for distribution pump motors may also be used or combined with a distribution loop return modulating valve.
- The reservoir of Purified Water within the storage tank also satisfies the "net positive suction head" (NPSH) requirements of the distribution pump, positioned downstream of the tank. Again, this presents a significant advantage when compared to tankless systems.
- The recirculating return tubing, equipped with a back pressure regulating system (modulating type), ensures that constant pressure is maintained at individual points of use within the distribution system, irrespective of the draw-off rate from the recirculating loop. Back pressure regulation for both Purified Water and Water for Injection distribution loops is extremely critical, since multiple operations, all requiring water, may be performed simultaneously at a facility. The manufacturing/production cycles can include a rapid demand for batching applications and lower flow rate, sporadic operations, which are associated with "support" operations. The flexibility of a storage and distribution system can be significantly expanded, where appropriate, by sequencing points of use in the distribution tubing based on operations requiring higher flow rates. This can be achieved through a central control panel and distribution pump motor variable frequency drive to provide significant variations in pump performance (flow rate and discharge pressure). This will be discussed further in section "Design Considerations".
- An additional benefit of storage systems is associated with the flexibility to provide multiple storage and distribution systems, considering water quality requirements and, more importantly, the potential for microbial and/or chemical back contamination. For example, a facility's manufacturing operations requiring Purified Water or Water for Injection will have defined microbial and chemical limits (bacterial endotoxin limits for Water for Injection systems). "Other" applications at the same facility, generally positioned in a defined physical area of the facility, may have less restrictive requirements, particularly from a total viable bacteria standpoint, than the manufacturing application. Housekeeping practices for the facility are generally well defined for manufacturing operations by standard operating procedures (SOPs). Conversely, laboratory applications, such as glassware washers or wash sinks, are generally not maintained with the same level of housekeeping diligence as manufacturing operations. Subsequently, while it is desirable to maintain a single validated USP Purified Water or Water for Injection system, dedicated storage and distribution systems for manufacturing and laboratory applications may be appropriate. Such systems would be extremely difficult to establish without storage tanks. Since the water demand for laboratory applications is generally small when compared to manufacturing operations, the capacity of a dedicated storage tank can be much smaller, the distribution tubing diameter can be significantly less, and sanitization frequency established for different total viable bacteria Alert and Action Limits can be less than that required for the manufacturing distribution loop, depending, of course, on the nature of the laboratory requirements. Another important consideration for multiple storage and distribution systems relates to the fact that an apparently minor problem, such as a small leak in a solenoid valve from a

recirculating line in the USP Purified Water or Water for Injection system feeding a glassware washer could, without multiple storage tanks, result in extensive microbial contamination of the entire facility system, including the manufacturing area. While it is fully acknowledged that total viable bacteria control is important for laboratory applications, the consequences of the bacterial contamination of a critical manufacturing operation, in monetary terms, can be considerable. Back contamination from a nondedicated storage and distribution system providing water to manufacturing and less-controlled support operations is further complicated by the fact that it may take three to four days before microbial contamination is identified due to the time delay associated with incubation of bacteria.

- It would be inappropriate to conclude a discussion associated with the application of storage tanks for pharmaceutical water purification systems without discussing non–stainless steel storage tanks, specifically tanks manufactured from plastic materials. There are many validated USP Purified Water systems that use plastic storage tanks, including biotechnology facilities, where certain point-of-use applications are highly sensitive to trace concentrations of metallic impurities such as iron, nickel, chromium, and molybdenum. Many smaller capacity systems, often for research or laboratory applications, use plastic storage tanks in lieu of stainless steel tanks. These systems generally operate in conjunction with distribution loop membrane filtration (or cartridge-type ultrafiltration) to control total viable bacteria levels within the storage and distribution system, extending the time period between sanitization cycles. Many of these systems, in addition to using plastic storage tanks, also use distribution piping constructed from plastic material, primarily unpigmented polypropylene or polyvinylidene fluoride (PVDF). If properly designed and maintained, plastic storage and distribution systems can exhibit point-of-use total viable bacteria levels similar to stainless steel storage and distribution systems.

DESIGN CONSIDERATIONS

While storage tank design would seem relatively simple, considering access to tank manufacturer's expertise, there are numerous critical items that must be considered for both the storage tank and related accessories.

Tank Size

The size of the storage tank is based on several factors. The literature contains information associated with compiling operating variables to develop a computerized projection for tank sizing. Storage tank size is generally determined at the beginning of the project, as part of the preparation of the System Basis of Design. Factors used in determining the size of the storage tank are the makeup water flow rate to the tank, the total daily volumetric demand at each point of use, the maximum instantaneous flow rate requirements at each point of use, the diversity of the demand (in terms of manufacturing cycle and other parameters), and available physical space for tank installation. For some applications, particularly where system enhancements are being performed, physical access to the area of installation, as well as related factors such as floor loading, must be considered when selecting a storage tank.

In preparing the Basis of Design, it may be necessary to increase the makeup water capacity rather than the size of the storage tank. However, particularly in the case of Purified Water systems where reverse osmosis (single or double pass) is used, or in the case of Water for Injection systems where single- or multiple-effect distillations are used, there are both technical and economic factors that must be considered. For example, RO systems used in a Purified Water system should not be oversized to the point where they operate for only a small fraction of a work day (<30%) if continuous recirculation provisions are not provided. While RO "recycle" and "flushing" cycles may be incorporated into RO system design, the efficiency of these operations (for microbial control) are poor. Recirculation within the pretreatment system to a RO unit should be considered but are not often used. If pretreatment component recirculation is employed, provisions should be included to remove Joule's heat, controlling temperature to maximize bacteria control. The Basis of Design is an extremely critical document because it justifies Purified Water and Water for Injection generating capability as a function of storage tank capacity for specific operating requirements.

For the majority of USP Purified Water systems, the technology employed for ion removal is RO coupled with CEDI. As discussed earlier, the use of deionization is limited to small capacity systems (rechargeable canisters), very high capacity systems with large instantaneous demand, or geographical locations where deionization remains the predominant ion removal technique.

Tank Dimensions

Once the volume of the storage tank has been established, the dimensions of the tank must be considered. Unfortunately, many sanitary stainless steel tank manufacturers have "standard products," especially for inside diameter and straight side height. Often, standard products "evolve" around calculations required to obtain a desired pressure (and vacuum) rating, consistent with the requirements of the ASME Code for Unfired Pressure Vessels, discussed later in this chapter. While custom fabrication may be costly, the greatest percentage of capital cost increase for a tank is associated with its diameter. The straight side height of the tank can generally be increased or decreased at minimal cost, provided that a standard diameter for the tank is selected, consistent with the diameter of an "ASME style" head.

For Water for Injection applications, feedwater generally enters the tank by gravity as discussed in chapter 5. The overall height of the tank and elevation of the inlet from the upstream distillation unit must be coordinated. Both elevations will be limited by the available height at the facility. The physical proximity of the distillation unit to the storage tank, for Water for Injection applications, is also an important consideration. The discharge tubing from the condenser on the distillation unit should be high enough to allow water to flow by gravity (through sanitary stainless steel tubing with a suggested slope of about 1 in./linear ft, depending on the capacity of the distillation unit) to the storage tank inlet. The "path" of the tubing from the distillation unit condenser outlet to the storage tank should generally be a straight section of tubing, containing a sample valve, distillation unit product water conductivity sensor, and "divert-to-drain" automatic valve. It is extremely important to note that the distillation unit product water quality, sampling, and divert provisions should not provide a location for the potential accumulation of stagnant water. Furthermore, the sampling valve, as well as the divert-to-drain valve, should not compromise Water for Injection quality by exposure to the atmosphere, defeating the function of hydrophobic vent filters installed on the storage tank and, in accordance with cGMPs, on the distillation unit condenser (chap. 5).

For Purified Water systems with "continuous recirculating" single- or double-pass RO systems and CEDI system, it is desirable to minimize "cycling" of the makeup to the tank. This may be associated with a "narrow" level control "band" for the tank. If possible, the tank length to diameter ratio should be increased to establish an adequate "level control band," which will minimize RO unit cycling required to support the distribution loop volumetric draw-off from the storage tank.

ASME Code for Unfired Pressure Vessels

It is strongly suggested that all sanitary stainless steel tanks be designed, constructed, and tested in accordance with the criteria established by the ASME Code for Unfired Pressure Vessels. Code requirements are most critical for tanks that will be heated (for normal operation or periodic sanitization) or Purified Water tanks that are ozonated. For Purified Water tanks that undergo periodic thermal cycling, it is suggested that a pressure rating of 30 psig and a full vacuum rating be specified. For Water for Injection storage tanks, where Pure Steam may be used for periodic sanitization, pressure ratings as high as 50 psig and full vacuum conditions should be specified.

ASME code–stamped vessels should have a Form U-1, "Manufacturer's Data Report for Pressure Vessels," which is provided by the *National Board of Boiler and Pressure Vessel Inspectors* of Columbus, Ohio. An example of an U-1 data sheet is presented in Figure 6.2. While the use of non-ASME code pressure vessels may be acceptable in some states, liability insurance coverage for a pharmaceutical firm will generally require ASME code–stamped vessels, irrespective of state requirements.

2. Manufactured for: ______ (Name and address of Purchaser)

3. Location of installation: ______ (Name and address)

4. Type: ______ (Horiz., vertical, or Sphere) ______ (Tank, separator, jkt. vessel, heat exch., etc.) ______ (Mfg's serial No.)

______ (CRN) ______ (Drawing No.) ______ (Nat'l Bd. No.) ______ (Year built)

5. ASME Code, Section VIII, Div 1 ______ Edition and Addenda (date) ______ Code Case No. ______ Special Service per UG-120(d)

Items 6-11 incl. to be completed for single wall vessels, jackets of jacketed vessels, shell of heat exchangers, or chamber of multi-chamber vessels.

6. Shell (a) No. of course(s): ______ (b) Overall length (ft & in): ______

Course(s)			Material	Thickness		Long. Joint (Cat. A)			Circum. Joint (Cat. A, B, & C)			Heat Treatment	
No.	Diameter, in.	Length (ft & in)	Spec./Grade or Type	Nom.	Corr.	Type	Full, Spot, None	Eff.	Type	Full, Spot, None	Eff.	Temp	Time

7. Heads: (a) ______ (Mat'l Spec. No., Grade or Type) H.T. - Time & Temp (b) ______ (Mat'l Spec. No., Grade or Type) H.T. - Time & Temp.

	Location (Top, Bottom, Ends)	Thickness		Radius		Elliptical	Conical	Hemispherical	Flat	Side to Pressure		Category A		
		Min.	Corr.	Crown	Knuckle	Ratio	Apex Angle	Radius	Diameter	Convex	Concave	Type	Full, Spot, None	Eff.
(a)														
(b)														

If removable, bolts used (describe other fastening) ______ (Mat'l Spec. No., Grade, size, No.)

8. Type of Jacket ______ Jacket closure ______ (Describe as ogee & weld, bar, etc.)

If bar, give dimensions ______ If bolted, describe or sketch.

9. MAWP ______ (internal) ______ (external) psi at max. temp. ______ (internal) ______ (external) °F Min. design metal temp. ______ °F at ______ psi

10. Impact test ______ (indicate yes or no and the component(s) impact tested)

11. Hydro., ~~pneu., or comb.~~ Test press. ______ Proof test ______

Items 12 and 13 to be completed for tube sections

12. Tubesheet: ______ Stationary (Mat'l Spec. No.) ______ Dia., in. (subject to press.) ______ Nom. thk., in. ______ Corr. Allow., in. ______ Attachment (welded or bolted)

______ Floating (Mat'l Spec. No.) ______ Dia., in. ______ Nom. Thk., in. ______ Corr. Allow., in. ______ Attachment

13. Tubes: ______ Mat'l Spec. No., Grade or Type ______ O.D., in. ______ Nom. Thk., in. or gauge ______ Number ______ Type (Straight or U)

Items 14-18 incl. To be complete for inner chambers of jacketed vessels or channels of heat exchangers

14. Shell (a) No. of course(s) ______ (b) Overall length (ft & in.) ______

Course(s)			Material	Thickness		Long. Joint (Cat. A)			Circum. Joint (Cat. A, B, & C)			Heat Treatment	
No.	Diameter, in.	Length (ft & in)	Spec./Grade or Type	Nom.	Corr.	Type	Full, Spot, None	Eff.	Type	Full, Spot, None	Eff.	Temp	Time

15. Heads: (a) ______ (Mat'l Spec. No., Grade or Type) H.T. - Time & Temp. (b) ______ (Mat'l Spec. No., Grade or Type) H.T. - Time & Temp.

	Location (Top, Bottom, Ends)	Thickness		Radius		Elliptical Ratio	Conical Apex Angle	Hemispherical Radius	Flat Diameter	Side to Pressure		Category A		
		Min.	Corr.	Crown	Knuckle					Convex	Concave	Type	Full, Spot, None	Eff.
(a)														
(b)														

If removable, bolts used (describe other fastening) ______ (Mat'l Spec. No., Grade, size, No.)

(A)

Figure 6.2 (**A**) Form U-1 manufacturer's data report for pressure vessels (front side). (**B**) Form U-1 manufacturer's data report for pressure vessels (back side) (Continued).

It is important to point out that the pressure and vacuum ratings mentioned above apply to the actual storage tank vessel. The ratings do not apply to heat transfer jacket(s) that may be positioned around the tank, discussed later in this chapter, which are generally designed for higher positive pressure ratings.

Full vacuum rating for thermally sanitized Purified Water storage applications is *not* an overdesign. Subsequent to periodic hot water sanitization, the storage and distribution system water temperature is decreased from a value of 80 to 90°C to ambient (e.g., 25°C). While numerous techniques may be employed for cooldown, such as an external heat exchanger and displacement of hot water with ambient temperature water, it is quite likely that the cooldown

FORM U-1 (Back)

16. MAWP _____ psi at max. temp. _____ °F. Min. design metal temp. _____ °F. at _____ psi.
    (Internal) (External)   (Internal) (External)

17. Impact Test _____
    (Indicate yes or no and the component(s) impact tested)

18. Hydro., pneu., or comb. test press. _____ Proof Test _____

19. Nozzles, inspection, and safety valve openings:

Purpose (Inlet, Outlet, Drain, etc.)	No	Diameter or Size	Flange Type	Material Nozzle	Material Flange	Nozzle Thickness Nom	Nozzle Thickness Corr	Reinforcement Material	How Attached Nozzle	How Attached Flange	Location (Insp. Open)

20. Supports: Skirt _____ Lugs _____ Legs _____ Others _____ Attached _____
    (Yes or No) (No.) (No.) (Describe) (Where and how)

21. Manufacturer's Partial Data Reports properly identified and signed by Commissioned Inspectors have been furnished for the following items of the report:
    (List the name of part, item number, mfg's. name and identifying number)

22. Remarks:

CERTIFICATE OF SHOP COMPLIANCE

We certify that the statements made in this report are correct and that all details of design, material, construction, and workmanship of this vessel conform to the ASME Code for Pressure Vessels, Section VIII, Division 1.
U Certificate of Authorization No. 24003   Expires: April 6   2001

Date _____ Name Precision Stainless, Inc. Signed _____
              (Manufacturer)   (Representative)

CERTIFICATE OF SHOP INSPECTION

I, the undersigned, holding a valid commission issued by the National Board of Boiler and Pressure Vessel Inspectors and/or the State or Province of MO
and employed by Commercial Union Insurance Company of Boston, MA have inspected the pressure vessel described in this Manufacturer's Data Report on _____ 19 _____ and state that, to the best of my knowledge and belief, the Manufacturer has constructed this pressure vessel in accordance with ASME Code, Section VIII, Division 1. By signing this certificate neither the Inspector nor his employer makes any warranty, expressed or implied, concerning the pressure vessel described in this Manufacturer's Data Report. Furthermore, neither the Inspector nor his employer shall be liable in any manner for any personal injury or property damage or a loss of any kind arising from or connected with this inspection.

Date: _____ Signed: _____ Commissions: NB 7376 "A"
                  (Authorized Inspector)   (Nat'l Board include Endorsement, State, Province and No.)

CERTIFICATE OF FIELD ASSEMBLY COMPLIANCE

We certify that the statements on this report are correct and that the field assembly construction of all parts of this vessel conforms with the requirements of ASME Code, Section VIII, Division 1.

U Certificate of Authorization No. _____ Expires: _____ 19 _____

Date: _____ Name _____ Signed: _____
                 (Assembler)   (Representative)

CERTIFICATE OF FIELD ASSEMBLY INSPECTION

I, undersigned, holding a valid commission issued by the National Board of Boiler and Pressure Vessel Inspectors and/or the State or Province of _____ and employed by _____ of _____ have compared the statements in this Manufacturer's Data Report with the described pressure vessel and state that parts referred to as data items not included in the certificate of shop inspection, have been inspected by me and to the best of my knowledge and belief, the Manufacturer has constructed and assembled this pressure vessel in accordance with ASME Code, Section VIII, Division 1. The described vessel was inspected and subjected to a hydrostatic test _____ psi. By signing this certificate of _____ neither the Inspector nor his employer makes any warranty, expressed or implied, concerning the pressure vessel described in this Manufacturer's Data Report. Furthermore, neither the Inspector nor his employer shall be liable in any manner for any personal injury or property damage or a loss of any kind arising from or connected with this inspection.

Date: _____ Signed: _____ Commissions _____
                  (Authorized Inspector)   (Nat'l Board incl. endorsement, State, Province and No.)

(B)

Figure 6.2 Continued

technique will result in a rapid "collapse" of water vapor above the water in the tank. This situation will create vacuum conditions. Implosion of a tank could occur if the full vacuum rating is not specified (Collentro, 1996).

Sanitary storage tanks should be equipped with an ASME approved design including dome-type top and an inverted dome bottom. The thickness of the tank sidewall and domed top and bottom are established by the ASME code, based on the specified tank dimensions as well as vacuum and pressure ratings.

Acceptable welding criteria are also established by the ASME specifications. The tank manufacturer should have an established quality assurance/quality control program with documented training and certification programs for procedures and equipment used during tank fabrication and assembly. Individuals performing welding must be properly trained. Documentation of the training program, including a "Certificate of Completion" should be available. Specifications prepared for storage tanks should include a definitive statement that the pharmaceutical company (ultimate owner of the tank) will have access to documentation as necessary.

Materials of Construction
- The use of 316L stainless steel as a tank material of construction is strongly suggested. The low carbon content of 316L stainless steel exhibits documented, improved welding characteristics (Mangan, 1991). As discussed earlier, plastic tanks may be used for a limited number of applications, which will be discussed later in this chapter.
- Material certifications for 316L stainless steel used in the fabrication of sanitary storage tanks should be provided. These certificates should include mill "heat numbers" and results from chemical analysis and physical testing. The validation process will require a documented "chain of custody" for the stainless steel material, from the time it is cast in the mill, rolled into sheets or plates, shipped to the tank manufacturer, and used for fabrication.

Tank Orientation
- Horizontal cylindrical storage tanks are discouraged except for applications where large volumes of Purified Water or Water for Injection must be stored in a physical area where available headspace is limited. The preferred configuration is a vertical cylindrical type. It is much easier to control potential tank corrosion (rouging) and minimize potential entrapment of stagnant water on upper surfaces within the tank, using spray ball(s), discussed later, in vertical cylindrical tanks than in horizontal tanks. A horizontal design also complicates issues such as level control based on pressure (head of water) and reduces velocity within the tank, potentially requiring a higher recirculating flow rate around the tank through the distribution loop. Horizontal cylindrical tanks are usually physically supported by external "saddles." To create a NPSH for the downstream distribution pumps, elevation of the tank, with associated height increase of the support saddles, may be required. Finally, for Purified Water systems using ozone for bacteria control, the surface area of the stored water in a horizontal tank is greater than a vertical cylindrical tank with the same capacity. This results in greater "outgassing" of dissolved ozone resulting in reduced dissolved ozone concentration in the stored water.

Heat Transfer
- For both Purified Water and Water for Injection applications, effective heating of the stored water (and recirculated water) can be achieved by positioning a heat transfer jacket around the exterior of the storage tank. The size (heat transfer area), type, and pressure rating of the heat transfer jacket should be established for specific applications.
- For Water for Injection applications, where hot (>90°C) distilled product water is available, the heat transfer area may be sized for "maintenance" heating only. However, as discussed earlier in this chapter, this is a function of the method of distribution loop operating criteria. As previously discussed, some pharmaceutical companies elect to employ thermal cycling of the Water for Injection storage and distribution system each day, operating the storage tank and distribution loop at ambient conditions for a portion of the day and then heating the stored and recirculated water during a preestablished time period (when there is no demand for ambient Water for Injection). This operating mode should be defined in detail in specific SOPs for the system. If thermal cycling of the tank and loop is required, the

heat transfer surface area of the jacket around the Water for Injection storage tank must be adequate to accomplish the required heating (for the thermal cycle operating mode) over a relatively short period of time. Generally, it is desirable to use relatively low pressure facility steam (30–50 psig) for this heating application.
- For Purified Water applications, feedwater to the tank from the Purified Water generating system will generally be at atmospheric temperature or lower when the feedwater source is from a surface water supply or contains cooling provisions for control of bacteria. If periodic hot water sanitization of the storage and distribution system is performed, a tank heating jacket may be used. This requires an adequate tank jacket heat transfer area to allow the sanitization process to be completed in a reasonable time period (e.g., 3–5 hours). The tank heating jacket may be employed for heating (latent heat from facility steam) and a sanitary shell and tube heat exchanger within the distribution loop for cooling (sensible heat from chilled or cooling water).
- Heating applications for storage tanks using facility steam are effective since latent heat, as opposed to sensible heat, is used. Generally, a "two-zone" heat transfer jacket is suggested, one positioned around the straight side portion of the tank and the second positioned around the base of the tank. This allows the upper heat transfer zone (of both zones) to be used during heat-up applications from ambient temperatures to 80 to 90°C (during initiation of the sanitization operation), while including the lower heat transfer zone, with the heat transfer area of about 20% to 25% of the upper heat transfer zone, for introducing "maintenance steam" once sanitizing temperature has been reached. This particular system design significantly increases the ability to control the tank temperature in a reasonable range by providing reduced heat transfer area for maintenance of steam requirements.
- There are inherent advantages associated with a two-zone heat transfer configuration. For applications where water is recirculated at ambient temperatures, the "lower" heat transfer zone may be used to remove Joule's heat input from the recirculating pump with facility cooling or chilled water. As discussed on several occasions in this text, small increases in water temperature (5–10°C above ambient of 20°C) will produce bacteria "incubation" temperature conditions. There are significant differences in the cooling requirements associated with hot water sanitization operations with facility steam and the removal of Joule's heat with facility cooling water or chilled water during normal operation. Because the lower heating zone of the tank has a reduced heat transfer surface area when compared to the sidewall zone, it is possible to use the side zone for heating (periodic sanitization) and the bottom zone for "trim" cooling.
- There are multiple types of external tank heating jacket systems available, including open channel, half-pipe arrangement, and dimpled heat transfer area. The selection of the heat transfer jacket design should consider the specific criteria, specifically required heat transfer area. Figure 6.3 demonstrates a half-pipe design, while Figure 6.4 demonstrates a dimpled jacket heat transfer arrangement. Table 6.1 presents operational characteristics associated with the half-pipe heat transfer design, while Table 6.2 presents the same information for a dimple-type heat transfer design.
- The half-pipe or dimpled heat transfer area can be used for two-zone (or multiple zone) heat transfer arrangements. A separate pressure rating (and temperature rating) developed in accordance with the ASME criteria must be specified for the heat transfer area. This information is supplied on tank manufacturer's drawings and on the U-1 data sheet discussed earlier.
- For certain applications, particularly where frequent and/or rapid thermal cycling conditions are anticipated, it is suggested that an external heat exchanger for cooling applications be considered. This suggestion is based on the fact that it may be physically impossible to provide adequate heat transfer area to accomplish cooldown in a reasonable time period using a heating zone around the storage tank. For example, to obtain adequate differential temperatures required for postsanitization cooldown of a Purified Water storage and distribution system, a significant flow rate of facility cooling or chill water may be required. The cooldown operation is based on

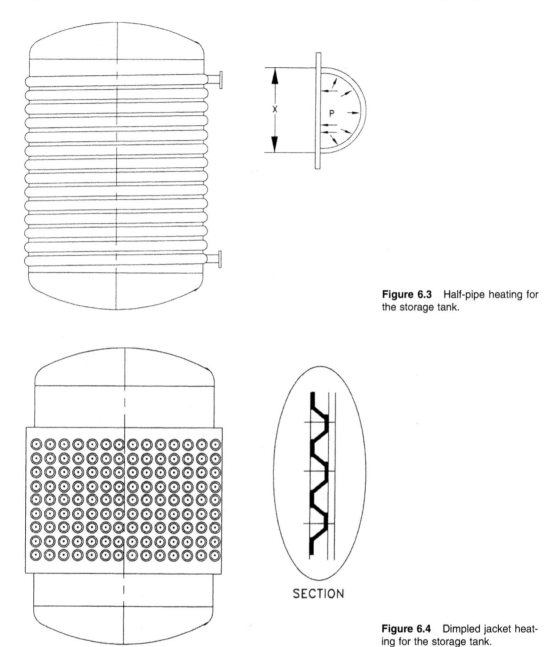

Figure 6.3 Half-pipe heating for the storage tank.

Figure 6.4 Dimpled jacket heating for the storage tank.

the removal of heat from the stored water by facility cooling or chill water. This involves transfer of sensible heat as opposed to latent heat (associated with heating for the sanitization application). To demonstrate the significant difference in required heat transfer area, 1 lb of water, heated 10°F (a good "limit" for the temperature increase of chill water for this application), removes only 10 BTU of heat. Conversely, 1 lb of steam at 30 psig will input about 925 BTU of heat. For many applications, it is inappropriate to "oversize" the heat transfer area around the tank for cooling application; it is more appropriate to install a separate heat exchanger within the distribution loop for postsanitization cooldown.

- In addition to the tank heating jacket limitations associated with potential thermal cycling, high chill water flow rate, and excessive heat transfer area requirements,

Table 6.1 Storage Tank Half-Pipe Heat Jacket—Advantages and Disadvantages

Advantages:
- Pipe thickness can be selected to allow operation with high pressure (e.g., 125 psig) facility heating steam or cooling media.
- Pipe diameter may be selected considering maximum acceptable pressure drop at the design flow rate.
- Heat transfer area adequate for chill water cooling requiring low differential temperature and high flow rate (sensible heat).
- Flexibility for "dual function" applications where heating and cooling is desired by using multiple "zones."
- Fair to good resistance to chloride stress corrosion and chloride pitting attack.

Disadvantages:
- Overall heat transfer coefficient relatively low requiring greater area, higher flow rate of cooling/heating media, and/or higher differential pressure.
- Cooling or heating media fittings or tank wall penetration can present installation, layout, or access issues.
- Costly, particularly for larger tanks (e.g., $\geq$10,000 gal)

Source: From Collentro (1995a).

Table 6.2 Storage Tank Dimpled Heat Jacket—Advantages and Disadvantages

Advantages:
- High overall heat transfer per unit area
- Cost about 50% of "half-pipe" design
- Easier to position and install around the straight sidewalls and inverted bottom dish of tank
- Can be thermally cycled without adverse effect

Disadvantages:
- Maximum operating pressure less than that of half-pipe design
- Flow path results in higher pressure drop for a given flow rate when compared with half-pipe design

Source: From Collentro (1995b).

there is another reason for considering an external heat exchanger for cooldown of the stored and recirculated water. Cooldown is performed subsequent to the elevated temperature sanitization operation to destroy bacteria. It is desirable to accomplish cooldown in a fairly rapid time period, particularly as the stored and recirculating water temperature approaches and enters the incubation temperature range for bacteria. Unfortunately, due to the thermodynamic properties involved (log mean temperature difference) (McCabe et al., 1993), the rate of cooldown decreases as the temperature approaches a suggested operating value of 20 to 22°C, which is demonstrated by the Table 6.3.

Finish

- An issue that must be considered in specifying a sanitary storage tank is the internal finish for the vessel. It is suggested that the finish on the interior of the vessel (mechanical polish and, where applicable, electropolish) be consistent with the interior finish of the distribution tubing. As discussed in chapter 9, there are several terms used to express the finish (smoothness) of stainless steel surfaces. The most common term for surface finish is Ra. Historically, mechanical finish was expressed by the term "grit." A comparison of these two designations is given in Table 6.4. Generally, Purified Water systems will use distribution tubing and an interior tank finish with Ra = 30 – 35 (180 grit), while Water for Injection applications will generally use Ra = 10 – 15 (240 grit) with electropolish.
- In selecting the interior finish of a storage tank, it is important to consider that the water velocity through the tank is extremely low. The use of spray ball(s), discussed later in this chapter, will assist in minimizing interior tank surfaces with stagnant water film (from water vapor of the stored water). As a general observation, it appears that accelerated rouging can occur in systems that use highly interior

Table 6.3 Effect of Changing Log Mean ΔT on Cooldown Rate

Cooldown rate $= UA\, (\Delta T)_{lm}$

Where
- $U =$ Overall heat transfer coefficient [BTU/(hr $\times$ ft^2 $\times$ °F)]
- $A =$ Heat transfer area in ft^2
- $(\Delta T_{lm}) =$ Log mean temperature difference

Tank temperature (°F)	(ΔT_{lm}) (°F)	"Cooldown factor"
185	145	1.00
165	125	0.86
145	105	0.72
125	85	0.59
105	65	0.45
85	45	0.31

Assumptions:
- U is constant over the entire temperature range
- Initial tank temperature $= 185°$F
- Chill water feed temperature $= 35°$F
- Chill water temperature increase $= 10°$F
- Tank volume $= 2500$ gal
- Tank recirculating rate $= 58$ gpm
- Chill water flow rate $= 100$ gpm
- Cooldown time period is approximately 5.5 hr
- Water temperature in the tank is uniform

Table 6.4 Classical "Grit" Finish Vs. Surface Roughness

Mechanical finish/polish (Grit)	Ra (Arithmetic average of surface roughness) (microinches)
120	4–50
150	30–35
180	20–30
240	15
320	10

polished distribution tubing and "rougher" polishing on the interior of the storage tank vessel. This is particularly true when the distribution tubing is also electropolished. While it is difficult to explain, it is suggested that a slight difference in the *electronic potential* between the interior surfaces of the distribution tubing and the interior of the tank walls, for the conditions stated above, will result in slow corrosion or rouging of the interior surfaces within the storage tank. Obviously, a similar situation could occur if the tank finish is superior to the finish on the interior of the stainless steel distribution loop tubing.
- If possible, it is desirable to inspect the interior of the tank during the mechanical finishing and polishing process at the tank manufacturer's facility. A careful inspection of the interior of the tank by a qualified inspector will indicate if a smooth finish has been obtained, or if mechanically removed material is simply "filling" adjacent "holes," thus creating the appearance of a smooth finish. An effective monitoring program can be achieved by preparing a specification that allows observation of the interior tank finish at the Ra $= 30 - 35$ and Ra $= 20 - 30$ level, particularly when the tank will be polished to a Ra $= 15$ finish and electropolished. The literature indicates that the following techniques may be used for inspection and classification of surfaces (ASME BPE-2009, 2009):
  - Borescope
  - Liquid penetrant

STORAGE SYSTEMS AND ACCESSORIES

- Profilometer
- Scanning electron microscope (SEM)

Generally, interior tank field finish measurements employ a profilometer, a device that determines a surface profile to determine "roughness."

- When the finishing process is completed, the interior of the tank may be passivated at the tank manufacturer's facility. The purchaser of the tank may elect to perform this operation as part of system passivation of the tank and distribution tubing. It is suggested that tank manufacturer's passivation be considered even if subsequent "installed" passivation will be performed. While the atmosphere in the tank may be protected during shipment (by the addition of water vapor removing desiccant material or plastic "barrier"), even slight corrosion can be accelerated once the tank is placed in operation. Passivation is described in the literature (Coleman and Evans, 1990; Grant et al., 1997; Balmer and Larter, 1993; Banes, 2010). This process increases the resistance to corrosion by using acid to enrich the metal surface closest to the water, with stainless steel component elements such as chromium that are less susceptible to corrosion.
- Electropolishing may be used in a similar fashion. The effect of this operation, including representative chemical equations, is presented in Figure 6.5. It is strongly suggested that electropolishing be limited to surfaces that have been mechanically polished to $\leq Ra = 10 - 15$ finish or finer.
- When storage tanks are provided with external heat transfer zones, the exterior of the heat transfer surfaces are generally treated with a protective coating. In addition, a 2-in. thick, chloride-free, nonfibrous insulating material is suggested. The entire system, including the heat transfer area and the insulation, is generally enclosed by an outer shroud that surrounds the straight side and lower inverted dish of the tank. Typical shrouds are constructed of 314L stainless steel. Since the shroud is visible, it is generally mechanically polished to $Ra \leq 30$ finish. The shroud and insulation not only provide an attractive tank external appearance but also provide personnel protection from heat transfer surfaces.
- The shroud must be constructed in such a way so as to accommodate sidewall and bottom wall penetrations for the tank. A tank sidewall "alcove" is often used for penetrations for a thermowell with temperature element and pressure element for

Electropolishing is an electrochemical process that removes surface atoms from the finished stainless steel surface. The surface acts as an anode. In this process, the concentration of iron on the stainless steel surface decreases, while the concentration of chromium increases. A smooth, chromium-rich, corrosion-resistant oxide layer is formed. This layer is generally 25–50 Å thick, controlled by the current and voltage used for the process. Properly prepared surfaces will exhibit a mirrorlike finish. Potential biofilm formation is minimized by the material's smooth finish and its resistance to oxidation and corrosion.

$$\text{Typical Reactions}$$
$$\text{Anode Reactions:}$$
$$\text{metal} - e \rightarrow \text{metal ions}$$
$$4\,OH + 4e \rightarrow 2H_2O + O_2$$
$$\text{Cathode Reactions:}$$
$$2H^+ + 2e \rightarrow H_2$$
$$\text{metal ions} + e \rightarrow \text{metal}$$

Figure 6.5 Electropolishing—chemical equations.

differential pressure/level determination). Usually, the domed top of the tank is not insulated, primarily due to the fact that there are multiple fittings positioned on the top of the tank that make fabrication of the shroud for this area extremely difficult. Also, the domed top of the tank does not contain external heat transfer provisions.

Tank Access
- Most tanks are provided with a sanitary-type, domed top–mounted access manway. For smaller tanks (<250 gal), this manway may be approximately 14 in. in diameter. The diameter of the manway for larger tanks is generally 18 to 20 in. The manway should consist of a collar and provisions for "swinging" the cover open and of sanitary design. The design should include a positive sealing mechanism that will not allow air introduction into the tank. This would include a tight-fitting gasket constructed of appropriate material and an adequate number of securing "lugs." While gasket material may vary for Purified Water applications, Teflon® "envelope" gaskets are preferred for Water for Injection applications. The number of sealing lugs and radial spacing should ensure that an airtight seal is maintained between the tank interior and the atmosphere.
- For some applications, it may be inappropriate to use a manway with provisions for "swinging" the cover open. This is primarily due to the fact that there may be inadequate overhead room to open the manway completely, thus limiting access to the interior of the tank. Obviously, it is possible to specify and obtain a sanitary manway without hinge provisions required for the swinging action. However, the weight of the access manway may present maintenance concerns, particularly during scheduled interior tank inspection.
- The domed top of the tank should contain, in addition to the access manway, several fittings required to support various operations. These fittings generally consist of a short extension section of tubing, terminating in a sanitary ferrule. The size of the ferrule should be determined for the specific application. The use of sanitary ferrules less than 1 to 1/2 in. is not suggested, since the section of tubing required for smaller ferrules may be inaccessible to water from the spray ball assembly. This situation is undesirable since it will decrease the ability of the spray ball to fully "wet" the surface of the smaller section of tubing associated with the smaller sanitary fitting.
- As a minimum, fittings at the top of the tank should include provisions for makeup water to the tank, a hydrophobic vent filtration system, a direct reading pressure gauge (compound type), a rupture disc, and a center position return line from the distribution loop adequately sized to allow the installation and removal of a spray ball assembly. For tanks that will be ozonated, thermally cycled, equipped with nitrogen blanketing provisions, or any other unit operations that will provide even a slight pressure or vacuum, a sanitary ferrule should be provided for a differential pressure sensor sensing probe (level control system probe). It is further suggested that two to three "spare" fittings of size(s) used for other top-positioned fittings be included.

Spray Ball System
- Reference has been made on several occasions to internal spray ball assemblies. Horizontal tanks, or large capacity vertical cylindrical tanks, may require multiple spray balls. The spray ball(s) provide multiple functions. One of the desired functions of an effective spray ball system is to insure that interior tank surfaces above the water level are exposed to a constant stream of water from the distribution loop return tubing. It is suggested that the tank manufacturer, familiar with the physical geometry of the tank, specifically the domed top and its multiple fittings, offers the most qualified source for providing the spray ball system. In theory, maintaining continuous exposure of the surfaces to water and avoiding alternate wetting and drying associated with exposure to water and water vapor, minimize the tendency for corrosion. However, experience indicates that both rouge and discoloration may

be noted on the interior tank wall in a "pattern" consistent with the discharge of the spray ball system. It is suggested that this observation is associated with long-term rouging of stainless steel surfaces in the storage and distribution system.
- A second function of the spray ball system is associated with microbial control. Water vapor will accumulate on the tank walls and domed top during normal operation. While the atmosphere in the tank should not contain bacteria, since displacement air enters the tank through a hydrophobic vent filter, bacteria may be present in the stored water and return water from the distribution loop. During hot water sanitization, it is critical for all water in the tank to reach sanitizing temperatures. By proper design, a spray ball system insures that all surfaces are continually exposed to water. During normal operation, "accumulated" water on the tank walls and domed top will be "rinsed" into the stored water. During sanitization, any accumulated water will be at elevated sanitization temperature.
- As discussed above, during periodic inspection of tank interior surfaces, discoloration and rouging may be noted on the tank top and sidewall surfaces. The location of the discoloration may coincide with the spray ball pattern, particularly for hot water. It may be suggested that the spray ball contributes to rouging rather than inhibiting corrosion associated with rouging. This misconception may be further substantiated by the presence of rouge on the spray ball and, occasionally, significant corrosion of the spray ball. Rouging may be occurring throughout the storage and distribution system, introducing material that is literally impinged on tank surfaces by the spray ball assembly. If excess spray ball corrosion is observed, alternative materials of construction, such as Teflon, should be evaluated.
- The spray ball should be constructed of 316L stainless steel. In selecting a spray ball, potential changes in the loop draw-off rate (sum of all point-of-use requirements) and, subsequently, the return line water flow rate should be considered. For operating flexibility, the spray ball "assembly" should be removable (through the top of the tank), allowing operating personnel to install an alternative spray ball(s) as necessary. Connection of the spray ball(s) to the section of return distribution loop tubing penetrating the domed top of the tank should employ a "pined-type" connection shown in Figure 6.6.
- As indicated earlier, it is suggested that the tank manufacturer is the most knowledgeable source for determining the size, type, number, and arrangement of spray ball(s) required for a specific application. Spray ball "coverage" should be specified in degrees (e.g., 180°, 270°, and 360°) demonstrated in Figure 6.7. The tank manufacturer should verify the specified coverage using riboflavin dye testing. The tank manufacturer should include documentation of riboflavin dye coverage in conformance to the indicated tank specification.

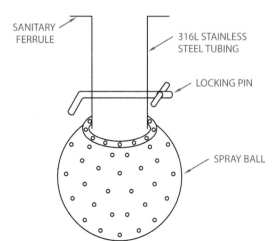

Figure 6.6 Pined spray ball assembly.

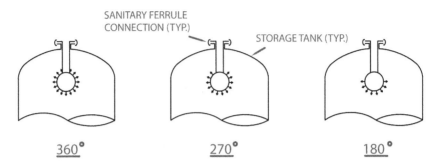

Figure 6.7 Spray ball "coverage".

- Purified Water systems employing ozone for microbial control should not be equipped with spray balls. This application, discussed in chapter 7, requires the use of "dip tubes" to avoid outgassing of ozone from return distribution loop water.

Sensing/Control Devices
- The tank lower sidewall and, for most applications, the domed top of the storage tank should contain provisions for a level-sensing device. A preferred sanitary level-sensing device uses pressure (or more appropriately differential pressure) sensors and transmitters. The pressure sensors are connected to the tank by sanitary ferrule connections. One sensor is physically positioned on the lower section of the straight sidewall of the tank suggested as 6 to 12 in. above the weld to the dish bottom. It is not suggested that the fitting be positioned in the inverted domed bottom at the base of the tank since the required sanitary ferrule on the tank could result in microbial accumulation and inability to completely drain the tank without removing the pressure element. While other level-sensing techniques are available, it is suggested that the reliability of a pressure-based system is extremely good. Other level-sensing techniques may be more "intrusive" to the stored water than pressure-based sensors. Some sensors may also be affected by turbulence within the tank. Obviously, the choice of level-sensing devices should be limited to components that are of sanitary design.
- A differential-type pressure sensing system for level control applications should be considered for any applications where, during any operating conditions, the pressure of the gaseous space is greater than, or less than, atmospheric pressure. Even small changes will significantly affect pressure-based level sensors. This condition will exist in any tank that is heated (normal operation or thermal sanitization) as well as any Purified Water tank that is ozonated.
- Tank pressure sensors should operate in conjunction with a level-indicating control systems, positioned in a central control. The analog signal from each pressure element is used to allow control of tank level and accurately indicate level. While the central control system processor provides valuable flexibility, such as changing the tank operating level set points using non-"hard-coded" provisions, a discrete controller, properly mounted on the face of a control panel and electrically connected to accessories such as an audible alarm, can provide functions similar to a central processor. However, a central system processor has inherent capability to "communicate" with facility data collection, recording, and alarm systems.
- Generally, a four-point level control system is employed as shown in Figure 6.8. The upper level set point represents a "high-high" alarm condition, which indicates a failure of the upstream Purified Water or Water for Injection generating system to operate within the control band or failure of the makeup valve to the tank. Immediate operator action is required to avoid overflow of the tank through the hydrophobic vent filtration system and/or the rupture disc system.

STORAGE SYSTEMS AND ACCESSORIES 239

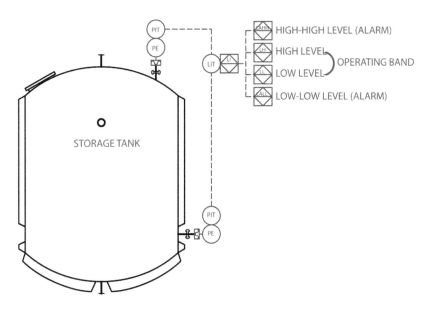

Figure 6.8 Four-point level control system.

- The two "middle"-level set points are used as an "operating band" for the storage tank. When water in the storage tank decreases to a "low"-level set point, the Purified Water or Water for Injection generating system will be activated, introducing makeup water to the tank. The flow of makeup water will continue until the level increases to a "high"-level set point. Generally, it is desirable to use approximately 60% to 75% of the tank volume for this operating band, decreasing the cyclic operation of Purified Water or Water for Injection system upstream components as well as operation of the makeup valve to the tank for Purified Water systems.
- The lowest level set point for the level control system is referred to as "low-low" level, which indicates excessive draw-off (beyond the Basis of Design) from the storage system. When the water level reaches this point, an audible alarm will sound. Simultaneously, electrical power to the downstream distribution pumping system will be inhibited. To avoid periodic cycling of the pump(s), power should not be restored until the water level in the storage tank increases to the "low"-level set point.
- The low-level set point should be high enough to ensure that the downstream pump does not cavitate during normal operation. This should be periodically verified by daily observation and data logging of the reading on a compound pressure gauge positioned in the feedwater tubing to the distribution pumping system.
- The lower straight side of the tank should contain a penetration with a thermowell connection. The thermowell, physically mounted on the tank and projecting far enough into the tank to ensure that its position is within the bulk recirculating water flow, should be used in conjunction with a "flush-fitting" temperature-sensing element, allowing water temperature monitoring and control within the tank during various thermal cycles.
- Unless the storage system and distribution loop are operated at ambient temperatures without periodic thermal cycling, a temperature-indicating control system must be provided. This system employs a temperature element installed in the lower straight side of the tank, discussed above. The temperature element is inserted in a thermowell such that it is flush with the interior surface of the well. The element is not in contact with stored Purified Water or Water for Injection. The temperature element and other temperature elements within the storage and distribution system operate in conjunction with a temperature-indicating control system that, similar to the level-indicating control system for the storage tank, is generally located in a

Purified Water or Water for Injection central control panel. The sensing and control system should be designed to operate in conjunction with a processor in the control panel. The design criteria for the temperature-indicating control system is directly related to the type of water (Purified Water or Water for Injection), the planned method of operation, and periodic thermal cycling requirements.

- The temperature-indicating control system generally includes four separate set points, similar to the level-indicating control system in philosophy. For normal operation of a hot Water for Injection system and for periodic hot water sanitization of a Purified Water system, the temperature-indicating control system would use similar monitoring and control techniques. Generally, two temperature elements are employed with a single controller. Temperature elements are positioned in the storage tank (or at the effluent of a heat exchanger installed in the associated distribution loop) and in the distribution loop return tubing (at the coldest point in the system). During a heat-up mode, or while maintaining elevated temperature for sanitization, the temperature-indicating control system regulates the flow of facility heating steam to the tank heat transfer jacket (or external heat exchanger) to a preset value for the sensor at the "coldest" point in the storage and distribution system. This objective requires "tuning" of the temperature control loop since the temperature at the outlet of the heating location cannot exceed a preset value. Incremental/repetitive cycling of the facility steam source may be required as storage and distribution loop temperatures equilibrate to the point where the desired temperature at the "coolest" point in the system is reached.
- Generally, an analog output is provided from the central control panel to a circular recorder, strip chart recorder, or data logger, clearly indicating the temperature value as a function of time. Postsanitization operations are controlled in a similar fashion using facility cooling or chill water and the temperature sensor at the hottest point in the system, as opposed to the coldest point in the system.
- The highest temperature set point for the temperature-indicating control system is designated as a high-high temperature alarm point, which is associated with a control malfunction, primarily excessive heat input. This would usually be caused by a loss of steam control to the heating jacket on the tank (or external heat exchanger). However, in certain situations, this alarm condition could be activated by mechanical heat input (Joule's heat) from the recirculating pump to a constantly recirculating system without any draw-off for an extended period of time. Sanitary centrifugal pumps are relatively inefficient. Depending on the extent of insulation on the distribution tubing, heat introduced by the pump could exceed the thermal requirements for "maintenance steam" associated with the normal operation of a hot Water for Injection system or recirculation during periodic sanitization of a Purified Water system. While activation of this alarm, associated with a steam regulating valve malfunction, can be corrected by closing a manual valve in the facility steam supply piping and eventually repairing the problem with the supply steam–modulating valve, the pump mechanical heat introduction problem is chronic in nature and is associated with improper system design.
- On the basis of field experience, another important item should be noted for storage and distribution system temperature control systems. Temperature control is based on the indicated tank or heat exchanger product water value. As indicated previously, "tuning" of the control system is critical. It should consider the fact that when a modulating facility steam supply valve closes the tank heating jacket or shell of the heat exchanger are full of saturated steam. The facility steam will continue to transfer its latent heat to water even though the facility steam valve is closed. This could result in high-high temperature alarms as well as cycling of system temperature values. In many cases it may be necessary to establish a tank or heat exchanger set point slightly lower than the desired system temperature value to avoid excursions.
- An operating temperature range is established by using high-temperature and low-temperature set points. Unlike the "on/off" function associated with the

level-indicating control system, the temperature-indicating control system, coupled with a signal from a central control panel, can provide a proportional signal, within the control band range, to a modulating steam valve feeding the heat transfer area surrounding the tank or external heat exchanger. This mechanism provides excellent control, assuming that mechanical heat introduction from the pump is not appreciable (>5°F/hr) and that the "residual" steam situation described above is considered.
- The low-low temperature set point indicates that the tank temperature (or the temperature at the coldest point in the storage and distribution system) has decreased below a preset value. This situation is generally associated with a malfunction of the facility steam–modulating valve to the heat transfer jacket surrounding the storage tank (or external heat exchanger).
- For applications where external heat exchangers are used for either heating or cooling, dedicated input from temperature elements installed downstream of the heat exchanger(s) and the coldest or hottest location in the system should be integrated to provide the desired temperature control. For example, assume that an external heat exchanger is provided for cooldown of a Purified Water storage and distribution system subsequent to periodic hot water sanitization. The temperature-indicating controller input from the temperature element monitoring the effluent of the external heat exchanger will rapidly indicate that the temperature has decreased to 25 to 30°C, the desired "normal operating" temperature value. However, cooling should continue until the temperature of the water at the "hottest" point in the system, the storage tank, decreases to the desired value.

Utility Parameters
- Utility parameters, such as facility supply steam conditions, and cooling or chill water temperature and flow restrictions (as well as the limit on chill water temperature increase) are extremely important to the successful operation of a heated or thermally cycled storage and distribution system. Facility steam supply and condensate return piping to the heating jackets surrounding the tank (or external heat exchanger) should include, but not be limited to, a "local" supply steam temperature indicator, a local supply steam pressure indicator, a supply steam relief valve (required by the ASME code), a steam-modulating valve, a manual supply steam isolating valve, a pneumatically actuated facility steam supply shutdown ball-type valve, a condensate steam trap, a local condensate pressure indicator, a local condensate temperature indicator, and a condensate isolation valve. Facility cooling water (or chill water) supply and return lines will also require comparable accessories. The facility supply steam–modulating valve and automatic shut-off valve should be selected to "fail" in a "safe" position—closed. On the other hand, cooling water or chill water supply valves should fail in an "open" position, a safe position.
- The discharge from a facility steam relief valve should be directed to a "safe area." Generally, relief is directed to the roof of the facility. It is suggested that the roof discharge contain a "double-elbow" piping arrangement, positioned at an elevation above roof level to avoid roof damage upon relief or restricted flow potential because of weather conditions. The discharge should also be protected against animals such as birds. Finally, to avoid rain water accumulation in the relief valve discharge piping to the roof with associated corrosion, a "drip pan" should be employed.
- The presence of water in facility steam is a concern. It is desirable to install a separate condensate trap in a bypass line from the main facility supply line to remove condensate. This will eliminate water hammer associated with two-phase flow in the coil surrounding the tank or heat exchanger, particularly during initial operation of the system. Further, this arrangement will reduce steam with a thermodynamic quality significantly less than 1.0 since it may originate from long steam supply lines to the water purification components.

Tank Support

- The upper domed top of the tank should be equipped with lifting lugs. The lifting lugs should support the weight of an empty tank and should be used to physically locate the tank into its installed position.
- Support legs should be provided for the storage tank. It is suggested that all but the smallest tanks be provided with at least four individual support legs. The legs should be of 304 or 304L stainless steel construction with base support plates. The support leg package should allow operating personnel to "level" the storage tank once it is installed at a facility. Generally, tanks are installed in an area that is sloped in order to provide for draining water to a local floor drain system. A tank provided with an ability to adjust the leg height (adapting to this slope) will eliminate "makeshift" provisions during installation.
- Tank support legs should be positioned such that the weight from the tank is transferred to the legs in a vertical direction. Using "saddles" that extend the diameter of the support legs to a dimension greater than the diameter of the tank is inappropriate, since this configuration imparts a horizontal force to the sidewall of the tank, which is undesirable. For a limited number of large-volume applications, tanks may be mounted from a "skirt" positioned around the straight side height of the tank. For applications using this tank support mechanism, appropriate reinforcement to the straight side of the tank must be provided.
- Support legs for storage tanks are designed to secure the vertical cylindrical vessels after installation. They are *not* intended to provide a means of support during the installation process—an operation that lifts the tank from a horizontal to a vertical position using the legs as a fulcrum point. If this feature is desired, the diameter of the legs must be increased. In addition, stainless steel supports must be added (generally in a horizontal configuration) to ensure that the legs will not snap off during the lifting process.

Fittings

- The orientation and location of both top-mounted and side-mounted fittings is extremely important. Fitting location should be established such that tubing interferences are minimal. In an example discussed earlier in this chapter, the tubing connection from the condensing section of a single- or multiple-effect distillation unit (Water for Injection application) and the storage tank should be free of elbows and fittings for physical connection to the tank, since distilled product water flow is by gravity.
- As discussed earlier, it may be desirable to install additional isolating valves and plant steam–modulating valves for tanks equipped with two or more heating zones. This is particularly applicable for periodic sanitization where rapid heat-up is desired, using all available heat transfer surface area. Maintenance steam is subsequently required, using only a portion of the heat transfer surface area (generally the lower heat transfer zone on the storage tank). For applications where heat-up and maintenance steam control are required for an existing tank with a single heating zone, dual steam supply lines may be considered. The parallel steam supply lines would contain smaller and larger piping lines and, more importantly, smaller and larger modulating steam control valves.
- The bottom inverted dish of the storage tank should contain a center-mounted sanitary ferrule connection for discharge to a downstream distribution pumping system. For versatility, it is suggested that the size of the sanitary ferrule be one size larger than the sanitary ferrule size on the suction side of the downstream distribution pump(s). This feature also assists in eliminating cavitation of the distribution pump.
- For Purified Water or Water for Injection applications where the storage tank is infrequently drained, it is suggested that a dedicated drain valve is not required on the inverted domed bottom of the tank. Alternatively, a sanitary ferrule connection

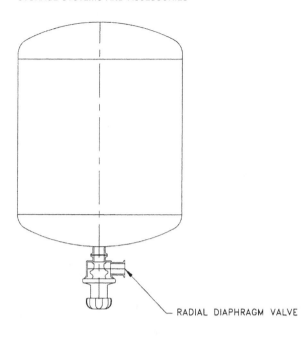

Figure 6.9 Zero dead leg valve at the base of a sanitary stainless steel tank.

and a distribution pump feedwater isolation valve with sanitary ferrule connections can be used (with appropriate hose) for infrequent tank draining operations.

- For some applications for Water for Injection systems, it may be necessary to drain the contents of the tank regularly, based on batching requirements, temperature considerations, and mode of operation. The requirement for the cyclical operation is outlined in an FDA Technical Inspection Guide (1986). For applications requiring frequent draining of the storage tank, a dedicated tank drain valve should be considered. On the basis of the nature of the storage system, and the fact that the valve is positioned at the base of the tank, "zero dead leg tank bottom/drain" valves should be employed (Fig. 6.9).
- The selection and location of valves on the makeup water line to a storage tank for both Purified Water and Water for Injection applications should be carefully considered. The malfunction of a tank level control system or a transient condition, such as a loss of electrical power, could produce a condition where tank overfilling is experienced. The makeup water line to the storage tank should contain an in-line conductivity cell (providing a signal to the conductivity-indicating control system). Provisions should be included to allow below quality water to be diverted to waste. It is suggested that a positive-acting valve be positioned directly in the makeup water line to the storage tank. This valve should be a sanitary diaphragm type, air-to-open, spring-to-close. It is further suggested that a second diaphragm valve be positioned in the divert-to-waste line. This valve can be spring-to-open, air-to-close. Should high-high tank level be experienced, even during a loss of electrical power, makeup water can be diverted to waste, eliminating potential overflow of the tank. This use of a "multiple valve" system is important. For Purified Water applications with tank feedwater from a single- or double-pass RO/CEDI "recirculating" system, a three-valve arrangement is suggested. This arrangement would include a tank makeup valve, divert-to-waste valve, and recirculation RO/CEDI loop valve. It is important that valve opening/closing include time delays to eliminate termination of RO/CEDI product water flow. For Water for Injection applications, the "double-valve" arrangement eliminates potential "flooding" of the distillation unit condensing section since distillate flow cannot be terminated instantaneously.

Tank Pressure
- The loop return line to a tank should contain a method of back pressure control. For applications with low recirculating flow requirements (<5–10 gpm), or for applications where there is a minimum number of points of use with predictable demand (draw-off rate), back pressure may be achieved by a spray ball or a spray ball operating in conjunction with a manual diaphragm valve installed in the distribution loop return line. A sanitary pressure gauge should be positioned upstream of the spray ball (or manual diaphragm valve), as appropriate, to indicate the return pressure. For most applications, it will be necessary to use a modulating-type back pressure regulating valve that is operating in conjunction with the signal from an upstream pressure element through a pressure-indicating controller and current-to-pneumatic (I/P) converter. The proportional signal from the I/P converter will establish the required modulation of the valve required to create a constant pressure at individual points of use during various demand conditions from the recirculating loop.
- The modulating back pressure regulating valve in the return tubing from the distribution loop should be air-to-close, spring-to-open. This is not only consistent with the desire to have the valve fail in a safe position but also to provide a second highly desirable function. Many Purified Water and Water for Injection distribution loops, due to improper initial design or facility expansions that have not considered the capacity of the recirculating loop, will experience classical "overdraw" conditions. This condition is encountered when point-of-use draw-off demands exceed the flow rate and pressure performance of the distribution loop pumping system. During this condition, a portion of the distribution tubing may be only partially filled with water. Bacteria containing atmospheric air can be introduced through a point-of-use valve during reduced or loss-of-flow conditions. To alert operating personnel of this situation, the pressure-indicating controller providing a signal to the I/P converter should also provide a signal to an audible alarm when the return line pressure decreases below a preset value.
- Conventional automatic diaphragm valves do not provide an effective method of controlling loop back pressure, since they tend to cavitate, particularly under low flow conditions without back pressure. This situation can be minimized by applying back pressure to the valve from an appropriately sized spray ball system. However, it is strongly suggested that a radial-type (or plug-type) diaphragm valve be used for this application. An example of the suggested type of valve is shown in Figure 6.10.
- The air supply line from the I/P converter to the back pressure regulating valve should be equipped with a relief valve. Many Purified Water storage and recirculating systems and Water for Injection systems are equipped with provisions for heating and cooling. With the exception of isolating valves for the distribution pump(s) and support components, manual or automatic valves should not be positioned in the distribution loop (not including the obvious point-of-use delivery valves that do not inhibit flow in the distribution loop) with the exception of the back pressure regulating valve. If overpressurization is experienced in the storage tank, the rupture disc will fail. However, if overpressurization is experienced within the distribution loop, excess pressure will be exerted on the back pressure regulating valve, increasing the pressure of the modulated air supply to the valve. An air relief valve provides a positive mechanical means of relieving pressure (to the Purified Water or Water for Injection storage tank), eliminating pressure buildup in the distribution loop tubing with associated catastrophic failure. This system provides a positive mechanical method of relieving pressure in a heated distribution system. It eliminates the requirement for the installation of a rupture disc (compound type) within a recirculating distribution loop that will be thermally cycled.
- A sanitary pressure gauge should be positioned on a sanitary ferrule provided on the domed top of the storage tank. This compound-type pressure gauge should be positioned such that it is visible from an operator-accessible location adjacent to the tank. If necessary, a large-diameter gauge can be specified for this application. Since

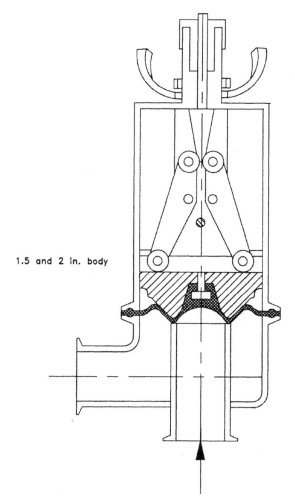

1.5 and 2 in. body

Figure 6.10 A plug-type diaphragm back pressure regulating valve.

the gauge is measuring, in general, very low positive pressure or slight vacuum, the range and increments of calibration should be appropriately specified.
- The selection of a rupture disc should be based on anticipated tank pressure (positive and vacuum) that will be encountered during normal operation. It is suggested that all sanitary stainless steel tanks that will be used in heated or thermally cycled applications be equipped with compound-type rupture discs that will fail on a preset positive pressure value or a preset vacuum condition.
- Since the relief action of a rupture disc will result in the intake or expulsion of gas and/or hot water, it is important to remember that, unlike hydraulic pressure, which requires a relatively small volume to relieve positive pressure (or vacuum), gaseous systems require a large opening to relieve pressure or vacuum. Subsequently, it is suggested that a rupture disc have a minimum diameter of approximately 3 in. The holder for a 3-in. rupture disc will use a 4-in. sanitary ferrule–type fitting for connection to a mating sanitary ferrule fitting on the domed top of the storage tank. All rupture discs used for storage tank applications should be provided with an electrical continuity "strip," capable of verifying the integrity of the disc. This should be accomplished by a remote indicator with alarm, readily visible and accessible to operating personnel.
- Rupture discs fail during "unacceptable transient or accident" conditions. The violent nature associated with these conditions may expose operating personnel to potential

Table 6.5 Vapor Pressure of Water at Various Temperatures

Temperature (°F)	Vapor pressure (psia)
150	3.7
160	4.7
170	6.0
180	7.5
190	9.3
200	11.5
212	14.7

Source: From Keenan and Keyes (1963).

hazards, unless the atmospheric side of the rupture disc is directed to a safe physical location. For most applications, this can be achieved by connecting a full size (4 in.) section of tubing, with appropriate fittings, to the atmospheric side of the rupture disc holder, directing the discharge (or suction) to a location at the base of the tank, about 6 to 10 in. above floor level. Obviously, the physical location of this tubing section should be remotely accessible to operating personnel. If this cannot be accomplished, it is suggested that a protective guide or railing be installed to limit routine operator access to the "discharge" location.
- As indicated earlier, the selection of the pressure and vacuum rating for a rupture disc is directly related to the application and sizing of the hydrophobic vent filter assembly. For Water for Injection applications, it is suggested that the pressure rating be considered at 50% of the tank design pressure or, if sanitization with Pure Steam will be performed, no greater than 75% of the design pressure. The vacuum rating for Water for Injection storage tank applications should be in the range of 10 to 20 in. of water.
- For Purified Water applications, where tanks will be thermally cycled, it is suggested that a rupture disc pressure rating of 50% of the maximum tank design pressure rating is appropriate. The vacuum rating should not exceed 15 to 20 in. of water. The rupture disc vacuum requirement for Purified Water storage tanks is based on the high vapor pressure of water at the suggested sanitizing temperature of 90°C and rapid condensation associated with a system excursion. Table 6.5 provides data demonstrating the vapor pressure of water at various sanitizing temperatures.

Hydrophobic Vent Filtration
- Pharmaceutical water storage tanks should be provided with a hydrophobic vent filtration system. A properly sized hydrophobic vent filtration system should be mounted on a sanitary ferrule connection positioned on the domed top of the tank. Physical location, in terms of elevation, is important. While it is not necessary to position the vent filter at the highest point on the domed top of the tank, it is *inappropriate* to provide tubing from a connection at the top of the tank to a lower elevation in order to facilitate access to the vent filter housing. As discussed below, condensation of water vapor in this system may result in bacteria generation in stagnant locations. Quite frequently, filters are positioned at a convenient access point for operating personnel but still above the domed top of the tank. Condensation of liquid in the tubing lines between the hydrophobic vent filter and the sanitary ferrule connection on the domed top of the tank is a concern. In certain cases, tubing is arranged in such a manner so as to create a "water trap," thus compromising the quality of the stored water as well as the function of the hydrophobic vent filtration system.
- For hydrophobic vent filtration systems used in Water for Injection or Purified Water applications where the storage tank will be thermally cycled, it is necessary to provide a method of eliminating condensation within the housing and the tubing between the housing and the storage tank. From a technical standpoint, the preferred

method of heating the hydrophobic vent filter, housing, and tubing is to employ a housing with a steam jacket. However, both hydrophobic vent filter manufacturers and regulatory investigators recognize the temperature limitations of common hydrophobic vent filter "support" material such as polypropylene. The suggested steam jacket pressure for proper operation is 2 to 5 psig. This may require multiple pressure-regulating "stages" from a facility steam line. Unfortunately, many operating systems do not maintain the indicated 2 to 5 psig pressure. Higher pressures result in a loss of membrane integrity as the filter support material melts (deforms). This is extremely undesirable, since atmospheric bacteria will enter the tank during draw-off. Furthermore, it is possible that plastic contaminants from the deformed or melted filter may be deposited directly in the storage tank, adversely affecting the chemical quality of the stored and recirculated water.

- On the basis of field experience, as an alternative to steam-jacketed housings, proper electric heat tracing may be considered. Unfortunately, electrically heated hydrophobic vent filtration housings exhibit nonuniform temperature characteristics. "Hot spots" are often observed with the same result (on hydrophobic vent filter membranes) as overheating from greater than 5 psig facility steam. Selection of heat tracing and/or blanket for a hydrophobic vent filtration system is critical. Improper selection and installation can result in hot spots, which can produce an electrical short circuit. Field observations indicate that this can result in stainless steel tubing "holes" in the area of the short circuit. Further, the short circuit inhibits desired heating of the system.
- Obviously, heat must be provided to the vent filter housing to vaporize any water entrapped from the tank associated with heating. Inadequate hydrophobic vent filter system heating will result in water condensation in the vent filter housing and tubing. This situation is often detected by repetitive failure of a compound-type rupture disc. It is another factor justifying the use of compound-type rupture discs for heated tanks or tanks that will be thermally cycled.
- On several occasions, it has been noted that hydrophobic vent filters are installed incorrectly in relation to the direction of flow. While this would appear to be a rather obvious item, the direction of filtration is from the bacteria containing atmospheric air, through the filter, into the storage tank. Displacement air, which does not require filtration, will flow from the storage tank to the atmosphere during an increase in the water level of the storage tank.
- It is important to remember that the hydrophobic vent filtration system provides a single function—the removal of bacteria from displacement air during tank drawdown. Occasionally, water purification systems, including storage systems, are installed in utility areas of a facility. Any mist, oil vapors, or other inorganic or organic vapors present in trace concentrations within the atmosphere will not be removed by the hydrophobic vent filter unless they exist as particulate matter. A strict interpretation of the USP *Official Monographs* for both Purified Water and Water for Injection considers these trace impurities as "added substances," since they are introduced after the production of Purified Water or Water for Injection. This situation can be addressed by using a "clean" nitrogen source (low pressure) to the feedwater side of the hydrophobic vent filtration system. Nitrogen blanketing systems will be discussed in greater detail later in this chapter.
- Considering the above comment as it relates to the installation of storage tanks in a "facility" area, it is important to remember that dust may be present in the atmosphere. Fine dust particles will settle on horizontal surfaces over time. It is suggested that the suction side of the hydrophobic vent filtration system be provided with a double-elbow arrangement that direct the gas flow in an upward vertical direction in or out of the filter, depending on whether the tank level is increased or decreased.
- It is suggested that installation of an isolation valve between the storage tank and the hydrophobic vent filtration system present a few challenges. The valve provides a large thermal mass that must be heated to eliminate the accumulation of water vapor.

Tank overpressurization and related rupture disc failure can occur if the valve is inadvertently left in a closed position after the hydrophobic vent filter is replaced. If new hydrophobic vent filter(s) are installed, sanitization of the tank and distribution loop for Purified Water applications should be considered. This would eliminate the need for an isolation valve.
- By far, the greatest design issue associated with hydrophobic vent filtration systems centers around proper sizing. For storage and distribution loops operating at ambient conditions, where thermal cycling will never be performed, the size of the hydrophobic vent filtration system can be based on the maximum draw-off rate from the storage tank. While the return and makeup water flow rate would appear to provide some compensation for the draw-off rate from the distribution pump, the actual design basis should reflect a draw-off associated with the maximum ability of the loop distribution pumping system to remove water from the storage tank, disregarding makeup and return.
- For hydrophobic vent filtration systems used in Water for Injection and Purified Water systems where thermal cycling will occur (or tank heating is part of normal operation), the instantaneous total collapse (condensation) of water vapor or steam above the minimum operating level of water in the storage tank must be considered. It is suggested that the situation should not produce a condition where a vacuum in excess of 10 in. of water will be created. In a Water for Injection system, a vacuum greater than 10 in. of water could literally suck water through the condensing section of the upstream distillation unit. It is highly improbable that the hydrophobic vent filter installed on the condensing unit of the distillation unit will be capable of compensating for the vacuum condition. For Purified Water applications, this will have an effect on water purification unit operations immediately upstream of the tank, specifically membrane processes such as reverse osmosis, CEDI, membrane filters, and ultrafiltration membranes. The potential consequences of undersizing hydrophobic vent filtration systems can be catastrophic. Figure 6.11 presents a representative calculation, demonstrating factors that should be considered in the design of hydrophobic vent filtration systems.

Conductivity
- For applications where distribution loops are operated at ambient conditions and significant tank draw-off is encountered as part of normal operation, it is possible that the "online" conductivity value of the stored and recirculated water may increase above the specified value for Purified Water and Water for Injection. Carbon dioxide will rapidly react with the stored ambient temperature water, producing the bicarbonate ion and the hydronium ion. Table 2.1 presents a chart of the equivalent conductance of ions, specifically ions associated with carbon dioxide adsorption, such as the bicarbonate and hydronium ions. The high equivalent conductance of the hydronium ion, when compared with other ions, results in a dramatic increase in conductivity, even at low concentrations. This situation can be avoided by maintaining a slight nitrogen pressure over the stored water, in the gas space of the storage tank. The situation may also be addressed by collecting periodic "grab" samples from points of use to verify compliance with the "laboratory" measurement for conductivity, a value greater than the online value.

Nitrogen Blanketing
- The use of nitrogen blanketing for a Purified Water or Water for Injection storage application, while not recommended, may be required for certain applications. Proper design of a nitrogen blanketing system is important. As indicated earlier, any trace materials introduced to a Purified Water or Water for Injection system would be considered as an added substance. Perhaps the "cleanest" source of nitrogen is from a dedicated liquid nitrogen supply or a dedicated, high-purity pressurized cylinder. The nitrogen should be supplied with a Certificate of Analysis and should be filtered through a hydrophobic vent filter, if applicable, to remove bacteria. Plant nitrogen is

Vent Filter Calculations
A. 175 gpm × 0.1337 ft² = 23.4 scfm
B.
$$A = 3.14\left[\frac{\left(10^1\right)^2}{2} + \left(10^1 \times 30^1\right)\right] = 1{,}099 \text{ ft}^2$$

C. temperature differential = 190°F − 85°F = 105°F
D. scfm = 1,099 ft² × .001 × 105°F = 115.4 scfm
E.
$$N = \frac{115.4 \times .019}{.019 \times 50 \times 0.5 \text{ psi}} = 4.6 \text{ elements needed}$$

F.
$$\text{element pressure drop} = \frac{115.4 \times 0.019}{0.019 \times 50 \times 6} = 0.35 \text{ psid}$$

G. housing pressure drop = < 0.1 psid
H. total assembly pressure drop = < 0.5 psid

Assumptions (tank volume = 10,000 gal)
  Tank dimensions are 10 ft deep by 30 ft long.
  The tank has dished ends.
  The maximum inflow is 150 gpm, the maximum outflow is 175 gpm.
  The Purified Water tank is to be periodically hot water sanitized at 190°F.
  Water is to be held at ambient temperature.
  The tank has a full vacuum rating.
  Rupture disks are to be used: They start to fail at 28 in. w.c.
  The size of the vent filter for the Purified Water tank is 14 in. w.c. (half of the rupture disk rating).

Figure 6.11 Calculations for sizing a hydrophobic vent filter.

generally inadequate for nitrogen blanketing applications of Purified Water or Water for Injection storage tanks, since it could contain trace quantities of undesirable impurities, such as oil, that would be considered an added substance.
- Nitrogen blanketing systems can be purchased with double pressure regulators to produce and regulate both the flow and volume required for effective over-pressurization of Purified Water and Water for Injection storage systems. An example of a typical nitrogen blanketing system for this application is presented in Figure 6.12.

Ozone for Microbial Control
The use of ozone for microbial control in Purified Water systems is discussed in detail in chapter 7.

Materials of Construction—Non–Stainless Steel Tanks
- The discussion associated with storage tanks to this point has addressed stainless steel units. As indicated earlier, plastic tanks or steel tanks lined with plastic may be

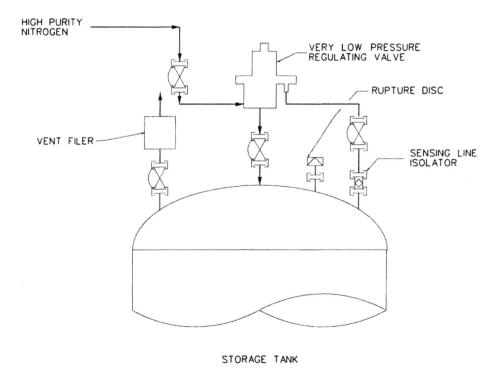

Figure 6.12 Nitrogen blanketing system.

used for selective USP Purified Water applications, particularly when trace concentrations of metallic impurities are undesirable. Generally, applications of this type require small volumes of water. It is suggested that storage tanks be vertical cylinders with conical bottoms. Because of the nature of the conical bottom, steel support stands should be considered. However, plastic tanks are commercially available with "integral plastic support structure."

- The suggested material of construction for Purified Water systems that will be chemically sanitized is unpigmented polypropylene. The interior surface of unpigmented polypropylene is extremely "smooth" unlike other plastic materials such as PVC (Gillis and Gillis, 1996). In fact, the surface finish is similar to polished stainless steel. Obviously, the nonporous, highly smooth finish of unpigmented polypropylene is an important consideration for microbial control within the storage system (Vess et al., 1993).
- Unlike stainless steel, unpigmented polypropylene storage tanks can be provided with custom fabrication features at a relatively low cost. Generally, commercially available conical bottom vertical cylindrical tanks are equipped with an undesirable loose fitting cover or "threaded access manway." However, the tanks can be purchased with a tight-fitting, full-diameter, gasketed cover. The bolts attaching the cover to the tank, as shown in Figure 6.13, are spaced in a manner to eliminate air leakage through the tank, in lieu of the hydrophobic vent filter provided on the tank, in an attempt to compensate for the fact that all plastics will exhibit some deformation with time.
- Alternatively, tanks may be constructed of PVDF lined steel or stainless steel. PVDF has the added advantage of withstanding sanitizing temperatures (unpigmented polypropylene cannot withstand sanitizing temperatures). However, the FDA's *Guide to Inspections of High Purity Water Systems* (1993) questions the release of fluoride ion from PVDF surfaces, particularly during initial operation and during application

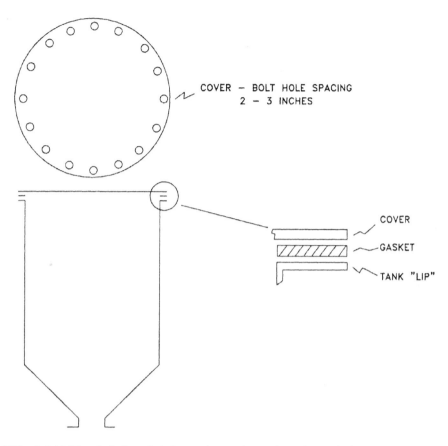

Figure 6.13 A tight-fitting, bolted, gasketed cover for a polypropylene storage tank.

with hot water. To take full advantage of the hot water sanitizing characteristics of PVDF, the tank may be combined with a PVDF distribution loop (chap. 9). However, in addition to concerns associated with the release of fluoride, there are additional issues that must be addressed regarding the use of PVDF. For example, the installation cost for PVDF distribution piping ("bead and crevice free") is comparable to that of stainless steel and requires frequent support (continuous for smaller diameter piping), due to the nonrigid nature of the material. The thermal coefficient of expansion associated with PVDF surfaces requires through design and installation considerations to avoid piping stress particularly in thermally cycled systems. For any system that will be thermally cycled, provisions must be included to insure that stress will not result in catastrophic failure of a distribution loop.
- Considering unpigmented polypropylene storage, the vertical cylindrical conical bottom tank should be equipped with a top-mounted hydrophobic vent filter. The vent filter housing can be of stainless steel or unpigmented polypropylene construction. If unpigmented polypropylene is used, the filter housing connection to the tank should be "sloped" and the filter housing "inverted," thus minimizing the potential for water accumulation. The filter housing-to-vent filter seal mechanism should employ a double O-ring seal.
- In general, a rupture disc is not used on an unpigmented polypropylene storage tank. Loss of feedwater control result in overfilling of the tank and leakage from the hydrophobic vent filter and other connections. Complete tank rupture, under high-level conditions, has never been observed directly by the author. While difficult to

achieve, an experience associated with vacuum conditions in a conical bottom, unpigmented polypropylene vertical cylindrical tank has been observed. A "sealed" tank was installed in a system, filled with water, and placed on-line without return water flow. The inlet to the hydrophobic vent filter inadvertently contained a threaded plug. The vacuum created within the tank was relieved by implosion of the flat cover into the tank. Actually, the flat cover provided a very large "rupture disc." On the basis of this experience, however, it is suggested that compound rupture disc assemblies may be considered for unpigmented polypropylene storage tanks. "Conservation vent" assemblies provide a method of relieving pressure but can compromise microbial control.

- Related to the above situation, several installations using plastic tanks have been observed that employ a unique pressure relief mechanism. These tanks are directly vented to the atmosphere through a hydrophobic vent filter. A fitting from the top of the tank directs pipe to a "loop seal" filled with water, water containing a chemical sanitizing agent, or alcohol. It is suggested that this loop-seal arrangement, while capable of reliving tank vacuum and pressure conditions, is inappropriate. Should a vacuum condition occur in the tank, liquid in the loop seal will be literally "sucked" into the tank. Furthermore, there is a vapor pressure associated with the material contained in the loop seal. This will lead, over time, to microbial introduction to the stored water. Again, it should be pointed out that any material introduced into the stored water in the tank would be considered an added substance.
- In general, an effective level control system for unpigmented polypropylene storage tanks is nonintrusive proximity switches. Figure 6.14 presents a representative drawing of a four-point level control system with details of the proximity switch

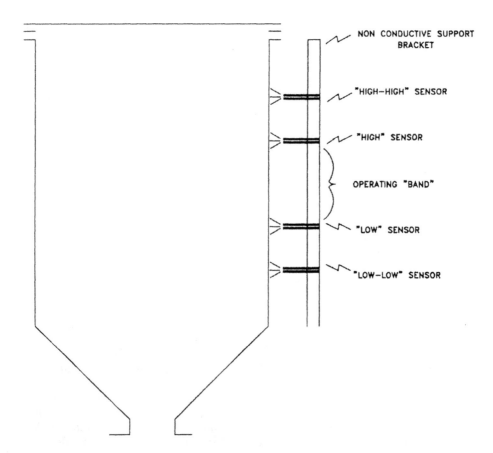

Figure 6.14 Proximity-type control system for an unpigmented polypropylene storage tank.

arrangement. Proximity switches provide an on/off function, on the basis of observation of the water level at a specific elevation within the tank.
- In comparing other required accessories for unpigmented polypropylene tanks, as compared to stainless steel tanks, a top-mounted pressure gauge is not required, a spray ball system is required, a back pressure regulating valve is required, and an additional fitting mounted on the top of the tank with cap/plug is required to perform periodic chemical sanitization. An effective sanitizing agent for unpigmented polypropylene is a 1% mixture of peracetic acid and hydrogen peroxide (Minntech Corp., 1997). Based on experience, this material appears to attack the biofilm that may form on the unpigmented polypropylene surfaces, allowing excellent microbial control if downstream membrane filtration or, preferably, ultrafiltration is used in a continuously recirculating mode, as discussed in chapter 9.

OPERATING AND MAINTENANCE CONSIDERATIONS

There are several operating and maintenance factors that should be considered for pharmaceutical water storage tanks. These factors are summarized below.

- An issue discussed previously centers around potential "overdraw" of the distribution loop recirculating "around" the storage tank. Occasionally, USP Purified Water systems will experience unexplained "spikes" in bacteria levels (and species of bacteria), conductivity, and total organic carbon (TOC). If a system displays these characteristics, it is suggested that a temporary portable recording device be connected to the control signal from the pressure sensor employed to control modulation of the return loop back pressure valve. If the distribution loop is not equipped with this valve, a portable temporary pressure sensor, with the ability to provide an analog signal (e.g., 4–20 mA), should be connected to an infrequently used point of use within the system. The results of continuous pressure monitoring for a brief time period should be recorded to verify system pressure as a function of time. If loop overdraw is experienced, it does not necessarily imply that a greater storage capacity and, more importantly, a larger-diameter distribution tubing system must be installed. It is quite possible that the impeller size for the distribution pump can be increased and a variable frequency drive added for the pump motor. Depending on the nature of the overdraw, it is possible that increased flow can be provided by the "enhanced" distribution pump with larger impeller size using the variable frequency drive coupled with either a flow rate indicator or pressure indicator installed in the distribution loop return tubing line.
- Integrity testing of the hydrophobic vent filtration system on the storage tank must be performed. Integrity testing is of particular concern when a tank is thermally cycled. A suggested frequency for integrity testing is once every one to three to six months, depending on the application, thermal cycles, and other operating variables. There are two methods of integrity testing for hydrophobic vent filters. One method verifies integrity using a procedure performed on the hydrophobic membrane in a laboratory or isolated condition (HIMA, 1982; Rowe et al., 1996). The second test method involves "in situ" testing (Tarry et al., 1993; Dosmar et al., 1992). From a technical standpoint, it would appear that in situ integrity testing of hydrophobic vent filtration system, using water, be considered. This particular test method insures that the installed hydrophobic vent filter-to-filter housing assembly is capable of removing bacteria from displacement air during tank drawdown. The in situ test method can be established as a nonintrusive test. From a practical standpoint, a disadvantage of in situ testing is that it must be performed on the filter housing, where physical access is often difficult. Further, the valves and accessories required for testing add significant thermal mass, requiring heating, to a system where accumulation of water vapor is unacceptable. For most applications, it is suggested that laboratory testing be performed in lieu of in situ testing.

- Periodic inspection of the tank interior and support accessories is required. As a minimum, the tank should be drained on an annual basis, and the interior of the tank inspected by access through the top-mounted manway. The inspection should include documentation of any "rouging," verification of structural integrity, and inspection for both chloride stress and chloride pitting corrosion.
- While the frequency for integrity testing of hydrophobic vent filters has been discussed, it is suggested that hydrophobic vent filters be replaced on a periodic basis. A conservative replacement frequency is once every six months. However, a shorter frequency may be required if the storage tank is installed in an area where high levels of particulate matter are present in the atmosphere.
- It is suggested that the tank rupture disc be inspected once every six months. The inspection should include careful review of disc integrity. Occasionally, rupture disc failure will occur without activation of the integrity alarm system. Although rare, this is the result of a small rupture on the disc that has not affected the electrical continuity of the integrity strip. For ozonated systems, it is suggested that the rupture disc be replaced annually and that the rupture disc holder be cleaned (passivated if necessary) annually. During inspection of the rupture disc, operation of the continuity integrity system should be verified by disconnecting the electrical connection to the disc and verifying alarm indication.
- Calibration of sensors and control instrumentation, such as temperature gauges, level control sensors, temperature control elements, and so on must be performed. The maximum time duration between calibration cycles should not exceed one year.
- Because of its importance to system operation, and considering that it can be removed if properly designed, the spray ball system for the tank should be inspected once every six months. This inspection should verify that excessive corrosion/erosion of the spray ball(s) has not occurred. It should be anticipated that a spray ball assembly will require replacement on a periodic basis (e.g., once every two to three years).
- A periodic replacement program for diaphragms in diaphragm valves and gaskets in sanitary ferrules should be considered. This replacement frequency should be structured around system design, operation, and materials of selection for the gaskets and diaphragms. Generally, systems that operate in a continuous, hot recirculating mode or ozonated system will require more frequent replacement of gaskets and diaphragms. Thermal cycling will have an effect on Teflon gaskets and diaphragms or Teflon gaskets and diaphragms with a "backing" material such as ethylene propylenediene monomer (EPDM). Viton or Teflon envelope gaskets should be considered for hot, thermal cycled, or ozone applications. Suggested diaphragm replacement frequency is every one to two years. Suggested sanitary ferrule gasket replacement frequency is once every two to three years. However, as indicated, the increased frequency may be required based on system operating conditions.
- Critical operating parameters should be logged or recorded on a daily basis. When available, measured parameters should be continuously "tracked" using a facility data logging-type system. Critical parameters include the storage tank pressure, the supply steam pressure and temperature to and from the jacket on the hydrophobic vent filter (or temperature of the external surface of the housing for electrically heated systems), the tank temperature, the return loop temperature and pressure, the tank level, the distribution pump feedwater and product water pressure, and the distribution pump "speed" when the pump motor is provided with a variable frequency drive (VFD).
- Depending on operating conditions (hot, ambient, or thermal cycling), as well as other parameters, such as the quality of the orbital welding for the distribution loop, stainless steel storage and distribution systems will require repassivation about once every two to three years. However, for certain systems, based on the nature of operation and the quality of welding, it is not uncommon for repassivation (with or without derouging) to be required more frequently.

REFERENCES

ASME-BPE-2009. Bioprocessing Equipment, An International Standard, The American Society of Mechanical Engineers, Section SF-5, New York, NY, 2009:120.

Balmer KB, Larter M. Evaluation of chelant, acid, and electropolishing for cleaning and passivating 316L stainless steel (SS) using ager spectroscopy. Pharm Eng 1993; 13(3):20–28.

Banes DP. Insights Into Rouge: Definition, Remediation, and Monitoring, Presented at Ultrapure Water Pharma – 2010, New Brunswick, New Jersey, 2010.

Coleman DC, Evans RW. Fundamentals of passivation and passivity in the pharmaceutical industry. Pharm Eng 1990; 10(2):43–49.

Collentro WV. USP purified water and water for injection storage systems and accessories, Part I. Pharm Technol 1995a; 19(3):78–94.

Collentro WV. USP purified water and water for injection storage system and accessories, Part II. Pharm Technol 1995b; 19(4):76–90.

Collentro WV. USP water for injection systems—case histories. Pharm Technol 1996; 20(3):95–125.

Collentro WV, Zoccolante G. Defining an added substance in pharmaceutical water. Ultrapure Water 1994; 11(2):34–39.

Dosmar M, Wolber P, Bracht K, et al. The water pressure integrity test—a new integrity test for hydrophobic membrane filters. J Parenter Sci Technol 1992; 46(4):102–106.

FDA. Inspection Technical Guide, No. 46, ORO/ETSB (HFC-133). Rockville, MD: Food and Drug Administration, Public Health Service, Department of Health and Human Services, 1986.

FDA. Guide to Inspections of High Purity Water Systems. Rockville, MD: Food and Drug Administration, Office of Regulatory Affairs, Office of Regional Operations, Division of Field Investigations, 1993.

Gillis RJ, Gillis JR. A comparative study of bacterial attachment to high-purity water systems surfaces. Ultrapure Water 1996; 13(6):27–36.

Grant A, Henon BK, Mansfield F. Effects of purge gas purity and chelant passivation on the corrosion resistance of orbitally welded 316L stainless steel tubing. Pharm Eng 1997; 17(2):94–109.

HIMA. Microbiological Evaluation of Filters for Sterilizing Liquids. Document No. 3, Vol 4. Washington, D.C.: Health Industry Manufacturer's Association, 1982.

Keenan JH, Keyes FG. Thermodynamic Properties of Steam. New York: John Wiley & Sons, 1963.

Mangan D. Metallurgical manufacturing and surface finish requirements for high purity stainless steel components. J Parenter Sci Technol 1991; 45(4):170–176.

McCabe WL, Smith JC, Harriot P. Unit Operations of Chemical Engineering. 5th ed. New York: McGraw-Hill, Inc., 1993:316–319.

Minntech Corp. Minncare—The Reverse Osmosis Membrane Disinfectant Designed to Protect Your System and You. Minneapolis, MN: Minntech Corporation, 1997.

Rowe P, Tingley S, Walker S. Hydrophobic membrane filters: an effective means of controlling biocontamination. Pharm Eng 1996; 16(1):44–52.

Tarry SW, Henricksen G, Prashad M, et al. Integrity testing ePTFE membrane filters. Ultrapure Water 1993; 10(8):23–30.

Vess RW, Anderson RL, Carr JH, et al. The colonization of solid PVC surfaces and the acquisition of resistance to germicides by water microorganisms. J Appl Bacteriol 1993; 74(2):215–221.

Wong PWK, Beria S. Cold U.S.P. Purified Water system in a pharmaceutical bulk manufacturing plant. Paper presented at: the 24th Annual ASQC-FDC/FDA Conference, 1 April in New Brunswick, New Jersey, 1993.

7 | Ozone systems and accessories

THEORY AND APPLICATION—GENERAL

Ozone, an unstable powerful oxidant, may be dissolved in water and used as a very effective disinfecting agent in Purified Water systems. Ozone is injected into the Purified Water storage tank as a gas or dissolved in water on the basis of the method of production. Ozone will destroy bacteria in the stored water. Subsequently, when properly designed, operated, and maintained, water from the Purified Water storage tank should be free of bacteria at a level of 1 cfu/100 mL or lower. Residual ozone from the tank, prior to the Purified Water distribution system, is removed by in-line ultraviolet radiation at a wavelength of 254 nm and intensity $\geq 100,000$ µW-sec/cm^2. Ultraviolet radiation converts ozone to oxygen, a substance present in Purified Water. Periodic loop sanitization is performed, with point-of-use demand inhibited, by de-energizing the in-line ultraviolet unit, increasing the dissolved ozone concentration, and allowing ozonated water to flow through the Purified Water distribution loop back to the Purified Water storage tank. For Purified Water applications where total viable bacteria control and absence of "objectionable organisms" are a concern, ozone treatment is the most applicable technology. However, as indicated, design, operation, and maintenance are extremely important as discussed within this chapter.

- As a gas, ozone is very hazardous. The Occupational Safety and Health Administration (OSHA) regulates parameters for gaseous ozone in industrial applications. Selective information from the OSHA requirements is presented in Table 7.1 (Compressed Gas Association, 2001).
- Ozone will destroy bacteria in Purified Water. A dissolved ozone concentration of 0.008 to 0.012 mg/L is required for complete destruction of bacteria with proper contact time.
- Ozone is an extremely powerful oxidant. The oxidative potential of compounds/radicals is expressed as its "electrochemical oxidation potential" (EOP). Table 7.2 presents the EOP for various disinfecting agents. The three disinfecting agents with greater EOP values than ozone are fluorine, a gas; the hydroxyl radical, which has an extremely short half-life; and atomic (or nascent) oxygen, which also has an extremely short half-life. Subsequently, ozone, with a half-life estimated at about 30 minutes and soluble in water (solubility constant about 10% that of oxygen), is the logical choice for bacteria destruction in Purified Water systems. Ozone exhibits about 50% greater EOP value than chlorine. The literature suggests that ozone is an at least 20 times more powerful oxidant than chlorine (Juras, 2005; Nebel, 1985; Nebel et al., 1973; and Rakness, 2005). Finally, ozone decomposes to oxygen, present in Purified Water.
- The literature indicates that ozone destroys bacteria by multiple mechanisms (Juras, 2005), presented as follows:
  - Direct oxidation and destruction of the cell wall with release of cellular constituents to the water.
  - Reaction with radical by-products of ozone decomposition in water, such as hydrogen peroxy (HO_2) and the hydroxyl radical (OH).
  - Damage to the constituents of the nucleic acids with the cells of the organism.
- The literature (Wallhauser, 1988) further suggests that the effectiveness of ozone for bacteria destruction is indicated by the fact that a sample containing 1000 cfu/mL of *Pseudomonas aeruginosa* will exhibit total destruction of the highly desirable organism in 90 minutes with water containing 0.2 mg/L of ozone or 180 minutes in water containing 0.025 to 0.050 mg/L of ozone. For most Purified Water systems requiring the absence of undesirable organism, detection of *P. aeruginosa* is generally a concern when a single colony is noted in a 100-mL sample or 1/100,000 the level indicated in

Table 7.1 Effects of Personnel Exposure to Ozone

Ozone concentration in air (mg O_3/L of air)	Observed effects
0.01–0.04	Threshold odor detection by humans
0.1	Maximum 8 hr average exposure limit
>0.1	Minor eye, nose, and throat irritation; headache; shortness of breath
0.5–1.0	Breathing disorders, reduction in oxygen consumption, lung irritation, severe fatigue, chest pains, dry cough
1–10	Headache, respiratory irritation, possible coma, possibility of pneumonia
15–20	Lethal to small animals within 2 hr
>1700	Lethal to humans in a few minutes

Source: From Compressed Gas Association, Inc. (2001).

Table 7.2 EOP of Various Oxidants

Oxidant	EOP (V)	EOP versus chlorine (ratio)
Fluorine	3.06	2.25
Hydroxyl radical	2.80	2.05
Nascent oxygen (O)	2.42	1.78
Ozone	2.08	1.52
Hydrogen dioxide	1.78	1.30
Hypochlorite ion	1.49	1.10
Chlorine	1.36	1.00
Chlorine dioxide	1.27	0.93
Oxygen (O_2)	1.23	0.90

Abbreviation: EOP, electrochemical oxidation potential.
Source: From Stanley (1999).

  the literature. Obviously, the time for destruction of the lower levels of P. aeruginosa in ozonated Purified Water Systems is much lower than the 90 to 180 minutes indicated.
- Destruction of bacteria by ozone in water is not only a function of concentration and time ($C \times T$) but several other factors, discussed later in this chapter. The literature suggests that the $C \times T$ factor for ozone is about 1.5 for 99.9% destruction of bacteria.
- As indicated, ozone is a very powerful oxidizing agent. It will destroy bacteria but also react with inorganic and organic impurities in the makeup water to the Purified Water storage tank. It will also react with 316L stainless steel tank materials of construction and accessories. This will result in the requirement for periodic Purified Water storage and distribution system derouging and repassivation. As a minimum, reverse osmosis (RO) should be used as the primary ion removal unit operation for feedwater to the Purified Water storage tank. Removal or control of the following from the makeup water to the Purified Water storage tank *must* be considered:
  - Trihalomethanes
  - Total organic carbon (TOC)
  - Total viable bacteria
  - Bacterial endotoxins
  - Metallic ions
  - Ammonia/ammonium.
- Ozone may be introduced to the Purified Water storage tank as a gas or dissolved in water. The theory and application of each technique are discussed later in this chapter.
- Finally, it is extremely important to discuss the effect of ozone on biofilm. Chapter 9 discusses biofilm. While biofilm and associated bacteria control can be achieved in Purified Water storage and distribution system with proper design, operation, and maintenance, initial inoculation of a distribution system with certain organisms, such as *Ralstonia pickettii*, may present problems as discussed in chapter 9. This is

supported by the literature (Adley and Saieb, 2005), which indicates a long-term *increase* in total viable bacteria (132%) after switching from chlorine for bacteria control to ozone (Escobar and Randall, 2001). It is strongly suggested that the increase is related to ozone penetration of the biofilm and associated "release" of bacteria from the biofilm to the "flowing" stream. Field experience indicates that the use of "service organizations" to perform ozonation of a nonozonated Purified Water storage and distribution system exhibiting bacteria excursions results in an increase in total viable bacteria subsequent to sanitization. Further, for existing systems "switching" to ozone for bacteria control from thermal sanitization and/or membrane filtration, biofilm removal, as discussed in chapter 9, is critical.

THEORY AND APPLICATION—GASOUS OZONE GENERATION

- Ozone, a gas, may be generated and dissolved in Purified Water by specific techniques. For Purified Water systems using gaseous ozone, dissolved oxygen passes between an electrode with ceramic "dielectric" and a grounded "earth" electrode, as shown in Figure 7.1. The space ("gap") between the ceramic dielectric and the grounded stainless steel plate is about 0.3 to 3.0 mL depending on the purity of oxygen (Rakness, 2005 and Vezzu et al., 2008). A high-current electronic field is established between the two electrodes. An electronic "discharge" (corona discharge) occurs through the oxygen gas. Oxygen is "diatomic" containing two atoms of oxygen for every molecule of oxygen. Oxygen is "split" into an unstable state by the electronic discharge per the following equation:

$$O_2(\text{oxygen molecule}) \rightarrow 2O(\text{oxygen atom})$$

The oxygen atoms quickly react with oxygen molecules producing ozone per the following equation:

$$O(\text{oxygen atom}) + O_2(\text{oxygen molecule}) \rightarrow O_3(\text{ozone})$$

The theoretical energy required for this reaction is 0.372 kW-hours/pound of ozone generated (Rakness, 2005). The actual energy required is much higher due to inefficiencies associated with ozone production. Ozone is not stable and will revert back to oxygen with a half-life of about 30 minutes per the following equation:

$$2O_3(\text{ozone}) \rightarrow 3O_2(\text{oxygen})$$

- The high-purity oxygen required for feed to the ozone generator may be supplied from a liquid oxygen source, high-pressure and high-purity oxygen cylinder, or

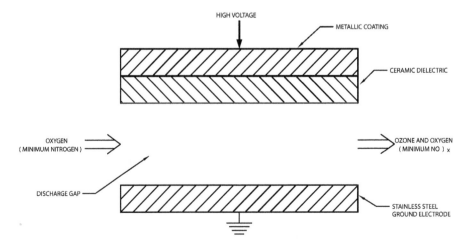

Figure 7.1 Gaseous ozone generation ("corona discharge").

OZONE SYSTEMS AND ACCESSORIES

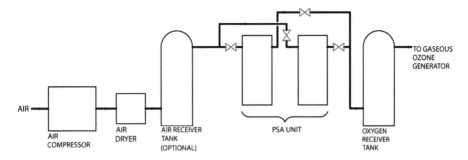

Figure 7.2 Oxygen generator—pressure swing adsorption (PSA) system with support equipment.

"pressure swing adsorption" (PSA) unit. The majority of Purified Water applications employ PSA technology. Figure 7.2 presents a drawing of a PSA system with accessories. The system generally consists of a dedicated air compressor generating oil-free air, dryer with desiccant and heaters, PSA unit consisting of dual chambers with molecular sieve to remove nitrogen, water, and organic material, and oxygen receiver. The oxygen concentration of the product gas should be $\geq 90\%$. The desiccant should be capable of reducing the dew point of air to at least $-60°C$ at atmospheric pressure.
- Operation of the oxygen generator is critical to the successful operation of the ozone generator. Nitrogen in the oxygen feed to the ozone generator can result in the production of nitrogen oxides. The nitrogen oxides can ultimately produce nitric acid in the Purified Water storage tank, lowering the pH and increasing water conductivity. Impurities in oxygen accelerating the production of nitrogen oxides in the ozone generator include the following:
  - Moisture (water vapor)
  - Organic compounds
  - Particles
- The concentration of ozone in the gas stream from the ozone generator is a function of several variables. The oxygen concentration in the feed gas to the unit is important. The literature states that an ozone concentration up to 10% may be generated when the feed gas oxygen concentration is 95% (Juras, 2005). Another source indicates that an ozone concentration of 15% may be possible with an oxygen feed gas oxygen concentration of 100% (Stanley, 1999). A third source indicates that an ozone concentration of 8% to 12% may be possible with an oxygen feed gas oxygen concentration of 90% (Rakness, 2005). Temperature, operating voltage, and impurities in the oxygen feed gas will also affect the ozone concentration.
- The techniques employed for gaseous ozone dissolution in water are discussed later in this chapter. However, it should be indicated that the ability to dissolve ozone in water is a function of the following:
  - Increase in the gaseous ozone concentration in the feed to Purified Water increases the dissolved ozone concentration.
  - Decrease in the physical size of the "bubbles" of ozone containing gas increases the dissolved ozone concentration.
  - Increased retention time of ozone gas in the Purified Water storage tank increases the dissolved ozone concentration.

THEORY AND APPLICATION—DISSOLVED OZONE GENERATION
- Dissolved ozone generation, also referred to as electrolytic ozone generation, uses Purified Water to generate ozone. The process is shown in Figure 7.3. Purified Water is introduced to a stainless steel chamber containing an anode, cathode, and semipermeable membrane. Water is introduced at the anode side of the chamber. The

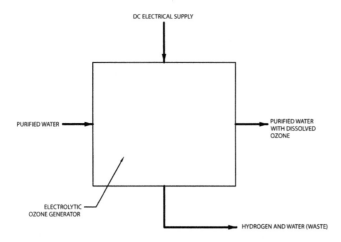

Figure 7.3 General electrolytic ozone generator.

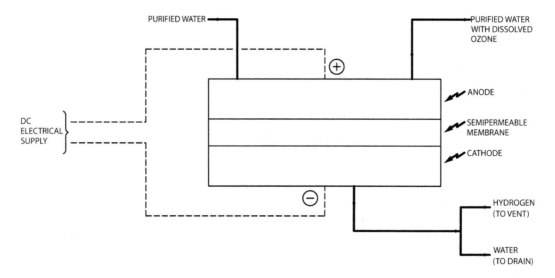

Figure 7.4 General electrolytic ozone generator flow streams.

applied high DC current to the anode and cathode "spits" water into "hydrogen" (H^+) and "oxygen" (O^-) ions. The hydrogen ions are attracted to the cathode and removed from the anode side of the chamber. They are unstable and quickly combine with another hydrogen ion to produce hydrogen gas (H_2). The remaining oxygen ions at the anode side of the cell quickly react with oxygen (O_2) to produce ozone (O_3). As indicated in Figure 7.3, the cell contains a single feed line, a product line, and a waste line. Figure 7.4 presents a representative drawing of the feed, product, and waste lines to and from the cell. The feedwater line is Purified Water. The product water line contains dissolved ozone gas, dissolved oxygen gas, and water. The waste line contains hydrogen gas and water. As shown, a U-type trap arrangement is employed to separate the hydrogen gas from water in the waste stream. While the water can be directed to waste, the hydrogen gas, while small volume, should be directed to "exhaust." It is lighter than air and could accumulate in the physical area above the electrolytic ozone generator to an explosive concentration.

- The anode, cathode, and semipermeable membrane are positioned in the stainless steel cell holder. While the operating life of the anode, cathode, and most importantly the semipermeable membrane will vary with cell manufacturer, they should be replaced periodically as part of a preventative maintenance program.

OZONE SYSTEMS AND ACCESSORIES

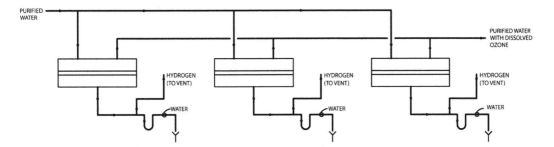

Figure 7.5 Multiple-cell dissolved ozone generator.

- The cells are of standard design with noncustom size and subsequent capacity. While the capacity of a single cell can be adjusted to a maximum value by increasing the applied amperage to a maximum value, the full amperage capacity cannot be exceeded. Subsequently, it is often necessary to purchase a unit with multiple cells. Figure 7.5 contains a representative flow diagram of a three-cell unit. Purified Water flow rate measurement and control to each cell is required. Cell temperature monitoring is critical since high temperature (>35–40°C) impacts cell operation and operating life of the semipermeable membrane. A "local" temperature indicator with "local" temperature switch should automatically inhibit operation of a cell if high-temperature conditions are detected. This can be performed by internal electrolytic ozone generator control or by control through a central panel with a processor.
- A single electrolytic ozone generator may be provided with three or four cells. If greater ozone capacity is required, multiple electrolytic ozone units can be used in parallel.
- The suggested source of Purified Water for electrolytic ozone unit operation is discussed later in this chapter. However, it should be indicated that a "dedicated" loop with pump "around" the storage tank for feed and ozonation of the stored Purified Water is *not* recommended.
- As discussed for gaseous ozone systems, concentration and contact time are the two major factors that must be considered for destruction of bacteria. It is important to note that the "target" dissolved ozone concentration for electrolytic ozone is about a factor of five to ten times lower than the concentration required for dissolved ozone produced from a gaseous ozone generator. While discussed further in section "Design Considerations" of this chapter, this must be considered when specifying the electrolytic ozone generator. It is suggested that several factors may contribute to this field observation. Perhaps the greatest factor is associated with the dissolved nature of ozone. As discussed in chapter 9, ozone is a gas dissolved in water. For gaseous ozone systems, a significant volume of ozone is present in the top of the Purified Water storage tank above the water level. "Out-gassing" of ozone is an important consideration. The measured dissolved ozone concentration in a system utilizing a gaseous ozone generator may actually reflect ozone that is not fully dissolved or ozone that is not uniformly distributed in water, lower at surface-to-water interface areas. While this situation cannot be explained, it has been observed on several occasions and must be considered during system design.
- Since oxides of nitrogen and production of nitric acid associated with gaseous ozone generation are not a concern, the pH and conductivity of Purified Water are not generally affected by the addition of dissolved ozone from an electrolytic ozone generator. However, increases in conductivity associated with the presence of trihalomethanes in the makeup water to the Purified Water storage tank have been noted as shown in Table 7.3.
- While gaseous ozone is present in the top of the Purified Water storage tank above the water level, the concentration appears lower than that for systems employing gaseous ozone generators.

Table 7.3 Purified Water Distribution Loop Conductivity from Total Trihalomethanes in Raw System Feedwater

Feed water total trihalomethane concentration (mg/L)	Conductivity (μS/cm at 25°C)
30	0.72
40	0.88
50	1.06
60	1.53
70	2.01
80	2.65

Notes: Values from field data graph at feedwater total trihalomethane values. No loop polishing. Average chloroform concentration about 74% of total trihalomethane value. Feedwater source disinfected with chlorine, not chloramines (distribution). Conductivity data collected without Purified Water distribution loop draw-off (equilibrium).
Source: From Collentro (2010).

- There are several technical considerations that should be considered when selecting electrolytic ozone generation as opposed to gaseous ozone generation, presented as follows:
    - From a safety perspective, there is no tubing/piping containing an elevated concentration of gaseous ozone.
    - There are no requirements for designing a method of effectively introducing ozone gas into Purified Water.
    - Water used for generation of dissolved ozone is Purified Water, eliminating potential introduction of impurities.
    - Operation of the electrolytic ozone generator cell under pressure conditions allows generation of dissolved ozone a high concentration.
    - Issues associated with generation of oxides of nitrogen and nitric acid are eliminated.
    - Maintenance requirements for the oxygen generator and accessories are eliminated.
- It should be noted that electrolytic ozone generator operation *often* requires that a continuous electronic current be supplied to the individual cells. Battery backup capability is included with some systems to provide a short-term solution to this requirement. However, it is suggested that a "supplemental" power source be included to avoid replacement of anode, cathode, and/or semipermeable membrane upon loss of electrical power.

DESIGN CONSIDERATIONS—GENERAL

- Selection of the method for ozone generation is important. It is suggested that electrolytic ozone generation requires less maintenance, is safer, and has numerous other advantages when compared with gaseous ozone generation for Purified Water application. For high-flow rate system, it is suggested that the use of extensive multiple electrolytic units is not appropriate. Gaseous ozone generation is more appropriate for these applications. Finally, it should be mentioned that maintenance of electrolytic ozone generation units requires a higher "skill set" than gaseous ozone generation. If preventative maintenance of the system is performed, under contract, by a service organization, personnel experience must be considered.
- The use of ozone for bacteria destruction in a Purified Water system requires complete removal of dissolved ozone prior to the first point of use in the distribution loop. This is achieved by use of an in-line ultraviolet unit at a wavelength of 254 nm and high ultraviolet radiation intensity (when compared with "normal" disinfecting intensity). As a minimum, the ultraviolet radiation intensity should be at least 100,000 μW-sec/cm^2 measured at a distance not greater than 1 cm from the source of the ultraviolet radiation throughout the ultraviolet sanitizing chamber.

- Microbial control within the distribution loop is maintained by periodic "shock" treatment of the loop (by shutting down the ultraviolet unit) and increasing the concentration of dissolved ozone maintained in the storage tank for a preestablished period of time. The duration of the sanitization operation, as well as the ozone concentration used during sanitization, is a function of the particular distribution loop configuration, ozone generation method, and other system characteristics.
- The Purified Water storage tank, distribution loop, and accessories must be constructed such that all surfaces in contact with ozone gas or dissolved ozone are 316L Stainless Steel. While a mechanical polish of 15 to 20 Ra and electropolish are preferred, 25 to 30 Ra mechanically polished systems without electropolish are acceptable. Derouging frequency for the nonelectropolished systems may be greater but generally are performed as preventative maintenance around infrequently scheduled shutdown of the system. Many systems with one or two shutdowns per year may not reduce the derouging frequency when electropolishing is specified. Valve diaphragm material should be Teflon® with ethylene propylenediene monomer (EPDM) backing. Gaskets and seals should be Teflon or Viton. While numerous elastomer catalogs indicate that EPDM is an acceptable material for use in ozonated systems, field experience indicates that EPDM pump seals and sanitary ferrule gaskets exhibit visual degradation with time. The portion of the EPDM sanitary ferrule gaskets in contact with ozonated Purified Water exhibits a transition from a black color to a dark brown color as well as small but visually apparent cracks. Field experience indicates that the cracks will eventually result in leaks. In addition to concerns associated with the compatibility of EPDM and ozonated Purified Water, there are concerns associated with potential leaching of elastomers oxidation products to the Purified Water. As indicated, Viton does not appear to exhibit the same characteristics. As a final note, both Viton and EPDM gasket material are black colored. However, there is a color "code" for elastomers. EPDM material will generally have a single green-colored dot, while Viton has a yellow-colored dot and a white-colored dot.
- The USP Purified Water storage tank should be of sanitary design, similar to that of a Water for Injection storage tank but without heating provisions. Insulation may be considered to avoid condensation of water on the outside wall of the tank as discussed in chapter 6. Fittings on the tank should include provisions for the following:
  - Differential pressure-type level monitoring and control
  - Top-mounted sanitary ferrule for Purified Water makeup
  - Top- or side-mounted tank access manway
  - Center positioned bottom outlet to a downstream distribution pump
  - Top-mounted rupture disk fitting
  - Top-mounted distribution loop return fitting(s)
  - Top-mounted vent fitting for feed to gaseous ozone destruct system and hydrophobic vent filter
  - Spare fittings
- It is critically important that the design of the Purified Water storage and distribution system considers the fact that the system is chemically sanitized and operates at ambient temperature. As indicated in chapter 9, *all* extensions from the loop are considered "dead legs." All tubing tees used in the orbitally welded 316L stainless steel loop should be "short outlet-type." All point-of-use valves and sample valves from the distribution loop should be zero dead leg type. Bacteria control within the loop is based on the following items:
  - Total destruction of all bacteria in the Purified Water storage tank prior to distribution.
  - Periodic distribution loop sanitization with dissolved ozone on a daily basis (minimum).
  - Elimination of dead legs in the Purified Water distribution loop.

- Elimination of *any* source of back contamination that could inoculate the Purified Water distribution loop with bacteria.
- It is important that total absence of dissolved ozone in the product water from the ozone destruct in-line ultraviolet unit be verified during normal operation. An antimicrobial agent cannot be present in Purified Water. Further, during distribution loop sanitization with ozone, end users must be aware of the fact that water from individual points of use cannot be used. The notification technique must be rigorous. Where possible, automatic actuators can be used and "locked out" during distribution loop sanitization, including a period of time after the ozone destruct in-line ultraviolet unit is reenergized. For manual point-of-use valves, a flashing beacon, low-pitched alarm, or other device can be employed to alert operating personnel of the loop sanitization condition. While the odor of dissolved ozone in Purified Water during loop sanitization will be obvious to operating personnel, point-of-use valves should not be opened.
- The effectiveness of periodic loop sanitization can be increased by decreasing the velocity (flow rate) through the loop during the sanitization operation. While the velocity should be adequate to maintain turbulent flow conditions, it is suggested that lower flow rates allow dissolved ozone to destroy any bacteria on tubing walls by ensuring uniformity of concentration throughout the cross-sectional area of the flowing Purified Water. To obtain the desired lower flow rate during distribution loop sanitization, the distribution pump motor can be provided with a variable frequency drive (VFD), controlled by the central panel processor.
- The solubility of ozone decreases with increasing water temperature. To maintain dissolved ozone levels adequate for destruction of bacteria in the Purified Water storage tank and during loop sanitization, the loop temperature should not exceed 35°C. If pump mechanical heat coupled with minimum water use from the loop results in an increase in Purified Water temperature >35°C, use of a trim cooling heat exchanger in the distribution loop should be considered.
- As indicated earlier in this chapter, gaseous ozone will be present in tubing/piping for system employing a gaseous ozone generator and in the "head space" above water in the Purified Water storage tank. Ozone has a molecular weight of about 48 Da, while air has a molecular weight of 29 Da. If a gaseous ozone leak occurs, the ozone will travel, by gravity, to floor level. Gaseous ozone sensors should be positioned at appropriate physical locations in the area of the system. Areas would include, but not be limited to, the following:
  - Base of the gaseous ozone generator, if employed, about 12 in. above floor level.
  - Physical area beneath tubing/piping used to transfer gaseous ozone, if applicable, to the Purified Water storage tank about 12 in. above floor level.
  - Area beneath the Purified Water storage tank about 12 in. above floor level.
  - Area around the distribution pump(s) and ozone destruct in-line ultraviolet unit, about 12 in. above floor level.

An alarm condition at any of the gaseous ozone monitors should terminate the flow of ozone (gaseous or dissolved) to the Purified Water storage tank until the situation prompting the alarm has been identified and corrected as appropriate. Most gaseous ozone sensors will detect other oxidizing substances such as sodium hypochlorite or peracetic acid/hydrogen peroxide vapors. If possible, the use of chemicals creating gaseous ozone monitor alarms should be limited to periods when system preventative maintenance is being performed. A calibration method should be established for installed gaseous ozone monitors. While many gaseous ozone monitors are provided with an "internal" response check feature, this does not provide a positive calibration response. It is strongly suggested that a battery-powered portable gaseous monitor be employed for routine ozone monitoring of the physical area of installation. Further, while intentional release of gaseous ozone is not suggested, a sample of Purified Water downstream of the distribution pump but prior to the dissolved ozone destruct in-line ultraviolet unit can be obtained in an

OZONE SYSTEMS AND ACCESSORIES

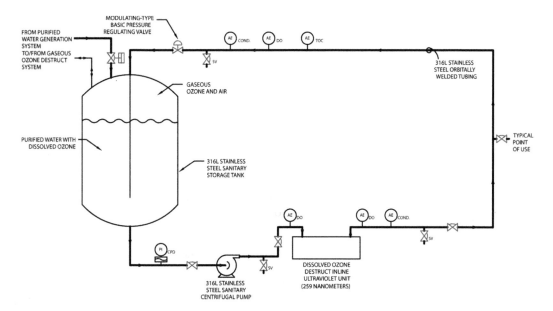

Figure 7.6 Generic ozonated Purified Water storage and distribution system with accessories.

open top beaker. The sample may be placed physically at the sensor of the installed gaseous ozone monitor. Simultaneously, a gaseous ozone response can be obtained with the indicated portable meter. The short half-life of ozone coupled with diffusion in the atmosphere presents calibration challenges. The indicated "gaseous ozone response" comparison provides verification of installed meter response.

- Since the Purified Water storage tank and the tubing from the tank to the ozone destruct in-line ultraviolet unit always contain ozonated water, it is suggested that water-sensing devices be positioned on the floor beneath the tank and the area from the tank to the pump and ozone destruct in-line ultraviolet unit. If water is detected, an audible and visual alarm should be energized. A water leak from the tank or a pump seal in an ozonated system must be identified as quickly as possible to avoid an ambient ozone alarm condition and associated leak of gaseous ozone.
- Figure 7.6 depicts an ozonated Purified Water storage and distribution system *without* ozone input capability (gaseous or electrolytic). This process flow diagram is provided to indicate the location of "grab" sample valves and in-line dissolved ozone sensors (if employed). As shown, dissolved ozone grab sample valves should be positioned in the distribution pump discharge tubing, ozone destruct in-line ultraviolet unit discharge tubing, and distribution loop return line tubing. All sample valves should be zero dead leg manual diaphragm type. Suggested valve size is 1/2 in. If employed, dissolved ozone sensors should monitor dissolved ozone at the three indicated locations. Field experience indicates that "sidestream"-dissolved ozone sensing and monitoring systems require frequent calibration and maintenance. It is strongly suggested that sidestream-dissolved ozone monitoring results be verified using grab sample results on a frequent basis. Portable grab sample test kits are available that provide rapid dissolved ozone data for comparison to sidestream monitor data (HACH, 2009a and HACH 2009b). Sidestream-dissolved ozone monitoring system and grab sample portable test kits must have a minimum sensitivity of 0.01 mg of O_3/L and maximum detection limit of 1.0 mg of O_3/L.
- If sidestream-dissolved ozone monitors are employed and properly calibrated, they can be used to automatically adjust ozone generator output during normal

operation and distribution loop sanitization. During normal operation, the monitor downstream of the distribution pump can be used to adjust the supply current to the gaseous or electrolytic ozone generators to a preestablished value adequate for complete destruction of bacteria in the Purified Water storage tank. During normal operation, the monitor downstream of the dissolved ozone destruct in-line ultraviolet unit and in the return tubing from the distribution loop should verify complete absence of dissolved ozone. During sanitization of the distribution loop, the monitor downstream of the distribution pump can "ramp up" the current to the ozone generator to obtain a preestablished dissolved ozone level for distribution loop sanitization. Loop sanitization time period can begin when the distribution loop return tubing dissolved ozone monitor value reaches a preestablished level. At the completion of the distribution loop sanitization cycle, the electrical current can be decreased to the ozone generator based on the dissolved ozone concentration downstream of the distribution pump. Further, formal loop sanitization can be terminated when the monitor after the dissolved ozone destruct in-line ultraviolet unit and in the return tubing from the distribution loop verify complete absence of dissolved ozone.
- For individual "cells" of an electrolytic ozone generator, sample valve(s) should be positioned in the product waterline. Samples from the valve may be used to verify the dissolved ozone concentration, checking the performance of each cell.

DESIGN CONSIDERATIONS—GASEOUS OZONE GENERATION
- Selection of a gaseous ozone generator for a Purified Water system is a function of several variables. Generally, the vast majority of units employed for total viable bacteria control in USP Purified Water systems will have an ozone output capacity expressed in grams per hour. Gaseous ozone from the ozone generator will be fed to the storage tank using an injector or sparging system, as discussed later in this section. The suggested operating dissolved ozone concentration for a Purified Water system using ozone from a gaseous generator is 0.1 to 0.3 mg/L during normal operation and 0.5 mg/L during distribution loop sanitization. The capacity of the gaseous ozone generator, assuming constant Purified Water storage tank level and constant dissolved ozone concentration, can be calculated as follows:

    Note: As indicated, the calculation presented below makes several assumptions. The primary assumptions are that the water level and dissolved ozone concentration in the Purified Water storage tank do not change during draw-off at distribution loop points of use. As indicated earlier in this chapter, bacteria destruction by dissolved ozone is a function of concentration and contact time. "Corrections" to the material presented below must be made to account for changes in tank level and dissolved ozone concentration. It is strongly suggested that the selection of the gaseous ozone generator be verified with a gaseous ozone generator manufacturer prior to specification or procurement.

$$\text{Theoretical ozone demand} = (0.5\,\text{mg/L} \times \text{maximum distribution loop flow rate}$$
$$\text{expressed in gallons per minute})$$
$$\times\,(3.785\,\text{L/gal} \times 60\,\text{min/hr}) \times \text{g}/1000\,\text{mg})$$

or

$$\text{Theoretical ozone demand (in g/hr)}$$
$$= (0.114) \times (\text{maximum flow rate})$$

The actual ozone generator capacity conservatively includes ozone outgassing in the tank, oxidation of inorganic and organic impurities, and other factors. The actual capacity can be calculated as follows (Schilling, 2010):

$$\text{Actual ozone demand (in g/hr)} = (0.23) \times (\text{maximum flow rate in gpm})$$

OZONE SYSTEMS AND ACCESSORIES

Table 7.4 Manufacturer's Data for Gaseous Ozone Generation Unit

Model no.	Maximum ozone output (g/hr)	Oxygen flow rate (L/min)	Variable output controller (%)	Power supply (V/Ø/Hz)	Power consumption (W)	Cooling water flow rate (L/min)
TOGC8X	8	5	20–100	115/1/60	600	N/A
TOGC13X	13	5	20–100	115/1/60	650	N/A
TOGC45X	45	5	20–100	115/1/60	1300	90

Notes: Feedwater gas: oxygen, compressor flow rate: 70 L/min at 45 psig.
Source: From Degremont Technologies (2009).

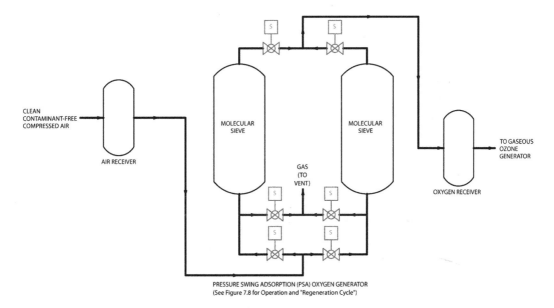

Figure 7.7 Gaseous ozone generator oxygen feed system.

For example, a Purified Water system with a maximum distribution flow rate of 100 gpm will require an ozone generator capable of providing 23 g/hr. Table 7.4 provides gaseous ozone manufacturer's data for a unit that may be considered for this application. Table 7.4 includes the oxygen flow rate (100%), electrical power requirements, and cooling water requirements for the ozone generator. Further, the unit contains variable output capacity, integrated oxygen concentrator, and compressor. Many gaseous ozone generators employ a separate air compressor, air dryer, PSA oxygen generation system, and oxygen receiver. The oxygen generator system can be selected, as outlined later, once ozone generator selection has been performed.

- The oxygen generation system feeding the gaseous ozone generator is important for long-term successful operation. The suggested oxygen feed system is depicted in Figure 7.7. The system consists of an air compressor, air dryer, and PSA oxygen generation system. The PSA unit employs two columns each containing a molecular sieve. One column is in operation, removing nitrogen from air at high pressure, while the other unit is releasing adsorbed nitrogen to waste at low pressure. Flow to the oxygen-enriched column is "switched" prior to breakthrough of nitrogen in the oxygen product gas. This process is demonstrated in Figure 7.8. The oxygen product stream is directed to an oxygen "receiver" that not only provides storage but also suppresses the pressure fluctuation in oxygen feed to the gaseous ozone generator. The oxygen generator capacity (product water flow rate) should be adequate for feeding the selected gaseous ozone generator. As indicated earlier, the minimum oxygen concentration of the product gas should be 90%. Gaseous ozone generator

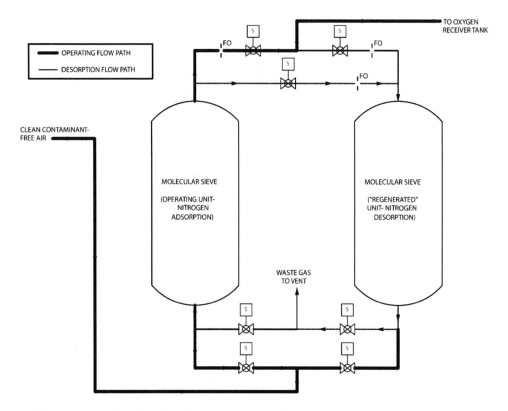

Figure 7.8 Pressure swing adsorption (PSA) oxygen generation cycle.

capacity and long-term successful operation increase with increasing oxygen-feed concentration above the 90% value.
- Figure 7.9 presents a flow diagram of a Purified Water system employing a sidestream loop with eductor for introduction of gaseous ozone. A dedicated pump recirculates Purified Water around the USP Purified Water storage tank. An eductor, positioned in the loop, draws the gaseous ozone into the recirculating Purified Water, delivering the ozone to the Purified Water storage tank. The ozone dissolves in the stored Purified Water, maintaining the desired concentration, compensating for dissolved ozone that is continuously removed by the dissolved ozone destruct in-line ultraviolet unit in the distribution loop. To enhance gaseous ozone transfer to dissolved ozone in Purified Water, the return tubing from the recirculating loop should be equipped with "extender tubes." The extenders tube terminal horizontal position should be below the Purified Water level. The level should not be too low since it may encourage "short circuiting" of ozone directly to the suction of the distribution pump. This situation not only decreases Purified Water storage tank ozone concentration but also reduces ozone contact time for bacteria destruction in the tank. As indicated earlier, some of the gaseous ozone may pass through the stored water to the top of the tank, above the Purified Water level. While this method of introducing ozone is acceptable, it is suggested that internal tank spargers provide an alternative method for gaseous ozone introduction to the Purified Water storage tank.
- Figure 7.10 presents a flow diagram of a Purified Water system employing spargers for introduction of gaseous ozone. The spargers are physically positioned inside the Purified Water storage tank. Suggested spargers consist of inert sintered material of flat disk-type configuration, shown in Figure 7.11. The disk(s) are positioned in a flat

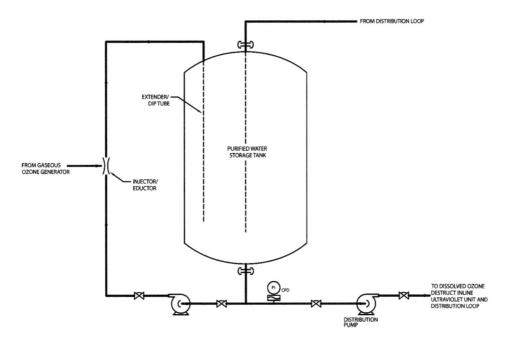

Figure 7.9 Gaseous ozone injection—recirculation loop and injector/eductor.

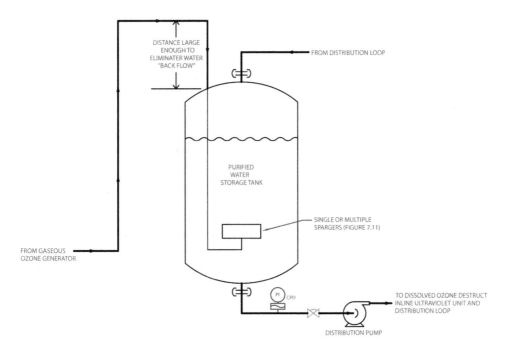

Figure 7.10 Gaseous ozone injection—direct tank feed with spargers.

316L stainless steel "plate" using a Teflon gasket, also shown in Figure 7.11. The sintered material produces extremely small "bubbles" of ozone that readily dissolve in water. If properly maintained with periodic sintered disk replacement, experience indicates that the system provides excellent gaseous ozone dissolution when compared to the eductor system discussed in the preceding text. However, the

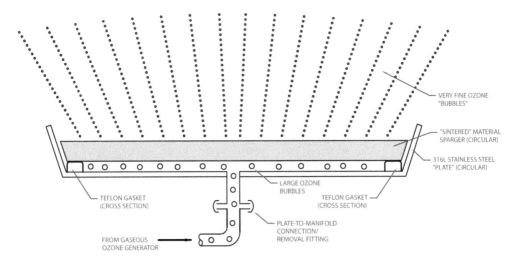

Figure 7.11 Typical gaseous ozone "plate-type" sparger.

spargers are not readily accessible, requiring shutdown of the Purified Water storage and distribution system for access. The horizontal level of the spargers in the tank should be carefully established. Contact time required for bacteria destruction and elimination of short circuiting to the distribution pump suction tubing, as discussed for the eductor system, is also a concern for the sparger system.

- As indicated earlier, the suggested normal operating dissolved ozone concentration from the Purified Water storage tank using gaseous ozone, measured prior to the dissolved ozone destruct unit, is 0.2 to 0.3 mg/L. The suggested dissolved ozone concentration, measured at the same location during distribution loop sanitization, is 0.5 mg/L. Daily distribution loop sanitization is suggested. The suggested minimum loop sanitization time period is 30 minutes. The dissolved ozone concentration measured from the sample valve in the distribution loop return tubing should increase during sanitization. However, it is quite likely that the dissolved ozone concentration will be less than the 0.5 mg/L value in the loop supply tubing. Subsequent to distribution loop sanitization, a "delay period," generally 5 to 10 minutes in length, should be established to ensure complete dissolved ozone removal from the distribution loop. The suggested delay period should be verified by the absence of dissolved ozone in distribution loop supply and return samples. If the loop distribution pump motor is equipped with a VFD, the flow rate should be decreased to produce a suggested velocity of 0.5 to 1.0 ft/sec.
- Gaseous ozone generator systems generate heat. Many units employed for Purified Water systems use air cooling for removal of heat. Larger gaseous ozone generator units may require cooling water for heat removal. Cooling water should be filtered and softened. Representative cooling water requirements are presented with the manufacturer's data in Table 7.4.
- Figure 7.12 contains a suggested vent system for an ozonated Purified Water storage tank. The vent system includes a rupture disk for pressure relief. During normal operation, the Purified Water level in the tank will increase and decrease with point-of-use demand and makeup water flow from the Purified Water generation system. The suggested vent system consists of a thermal gaseous ozone destruct system and hydrophobic vent filtration system. During tank "drawdown," displacement air is drawn into the tank. It is technically desirable to remove atmospheric bacteria from the displacement air. The hydrophobic vent filter achieves this objective. During tank makeup with "net" increase in Purified Water level, air and ozone will be displaced from the tank. Gaseous ozone *should* be removed. While materials such as manganese

OZONE SYSTEMS AND ACCESSORIES

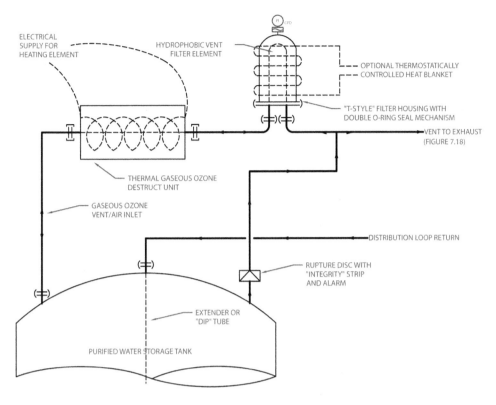

Figure 7.12 Purified Water storage tank gaseous ozone vent system.

Table 7.5 Gaseous Ozone Thermal Destruct Units—Manufacturer's Data

Model no.	Capacity (scfm/min)	Feed gas ozone concentration (weight %)	Product gas ozone concentration (mg/L)	Operating pressure (psig)	Electrical rating (kW)	Electrical power requirements (V/Ø/Hz)
ODT-003	2.0	<1.5	<0.1	<6.4	0.8	230/1/60
ODT-006	3.5	<1.5	<0.1	<6.4	1.8	230/1/60
ODT-012	7.0	<1.5	<0.1	<6.4	3.2	480/3/60
ODT-020	12.0	<1.5	<0.1	<6.4	5.3	480/3/60
ODT-030	18.0	<1.5	<0.1	<6.4	8.0	480/3/60
ODT-060	35.0	<1.5	<0.1	<6.4	16.0	480/3/60
ODT-120	70.5	<1.5	<0.1	<6.4	32.0	480/3/60
ODT-180	106.0	<1.5	<0.1	<6.4	44.0	480/3/60

Source: From Ozonia North America (2009).

dioxide and activated carbon may be used for removal, neither are suggested. Operating issues such as the requirement to remove water vapor from manganese dioxide and, more importantly, potential introduction of impurities from the media into the Purified Water storage tank are a concern. Table 7.5 contains information associated with thermal ozone destruct units. Essentially, the units employ a heated transfer tube to covert ozone gas to oxygen. The units operate at a temperature of about 750°F. A properly sized unit will thermally decompose ozone (to oxygen) to a concentration <0.10 mg/L. Field experience indicates that a conventional hydrophobic vent filter (Teflon membrane material with polypropylene "cage" and end caps) may be used in the discharge of the thermal gaseous ozone destruct unit without any deformation of polypropylene material or loss in filter integrity (6-month operating life). The vent system should include provisions for relieving overpressurization in the Purified Water storage tank. This is accomplished by use of a rupture disk as shown in Figure 7.12.

DESIGN CONSIDERATIONS—ELECTROLYTIC OZONE GENERATION

- Selection of an electrolytic ozone generator for a Purified Water system is a function of several variables. Generally, the vast majority of units employed for total viable bacteria control in Purified Water systems will have an ozone output capacity expressed in grams per hour. Dissolved ozone from the ozone generator will be fed to the storage tank using sidestream flow and mixing tee, as discussed later in this section. The suggested operating dissolved ozone concentration for a Purified Water system using ozone from an electrolytic generator is 0.03 to 0.05 mg/L during normal operation and 0.12 mg/L during distribution loop sanitization. The capacity of the electrolytic ozone generator, assuming constant Purified Water storage tank level and constant dissolved ozone concentration, can be calculated as follows:

Note: As indicated, the calculation presented below makes several assumptions. The primary assumptions are that the water level and dissolved ozone concentration in the Purified Water storage tank do not change during draw-off at distribution loop points of use. As indicated earlier in this chapter, bacteria destruction by dissolved ozone is a function of concentration and contact time. Corrections to the material presented below must be made to account for changes in tank level and dissolved ozone concentration. It is strongly suggested that the selection of the electrolytic ozone generator be verified with an electrolytic ozone generator manufacturer prior to specification or procurement.

$$\text{Theoretical ozone demand} = (0.12 \, \text{mg/L} \times \text{maximum distribution loop flow rate expressed in gallons per minute})$$
$$\times (3.785 \, \text{L/gal} \times 60 \, \text{min/hr} \times \text{g}/1000 \, \text{mg})$$

or

Theoretical ozone demand (in g/hr) = $0.0274 \times$ maximum flow rate

The actual ozone generator capacity conservatively includes some ozone outgassing in the tank, oxidation of inorganic and organic impurities, and other factors. The actual capacity can be calculated as follows (Schilling, 2010):

Actual ozone demand (in g/hr) = $0.036 \times$ maximum flow rate (in gpm)

For example, a Purified Water system with a maximum distribution flow rate of 100 gpm will require an electrolytic ozone generator capable of providing 3.6 g/hr. Figure 7.13 provides electrolytic ozone manufacturer's data for a unit that may be considered for this application. Figure 7.13 includes the electrical power requirements and number of cells required for the electrolytic ozone generator.

- As indicated earlier, the suggested normal operating dissolved ozone concentration from the Purified Water storage tank using dissolved ozone from an electrolytic generator, measured prior to the dissolved ozone destruct unit, is 0.03 to 0.05 mg/L. The suggested dissolved ozone concentration, measured at the same location during distribution loop sanitization, is 0.12 mg/L. Daily distribution loop sanitization is suggested. The suggested minimum loop sanitization time period is 30 minutes. The dissolved ozone concentration measured from the sample valve in the distribution loop return tubing should increase during sanitization. However, it is quite likely that the dissolved ozone concentration will be less than the 0.12 mg/L value in the

OZONE SYSTEMS AND ACCESSORIES

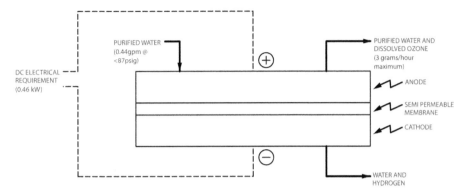

OZONE PRODUCTION (grams/hour)	NORMAL FEED WATER FLOW RATE (gpm)	MAXIMUM FEED WATER PRESSURE (psig)	MAXIMUM FEED WATER CONDUCTIVITY (μs/cm @ 25 C)*	ELECTRICAL RATING (kW)
3	0.44	87	20	0.46
6	0.88	87	20	0.86
9	1.32	87	20	1.27

Figure 7.13 Electrolytic ozone generation—manufacturer's data. *Source*: From Ozonia, 2009.

loop supply tubing. Subsequent to distribution loop sanitization a delay period, generally 5 to 10 minutes in length should be established to ensure complete dissolved ozone removal from the distribution loop. The suggested delay period should be verified by the absence of dissolved ozone in distribution loop supply and return samples. If the loop distribution pump motor is equipped with a VFD, the flow rate should be decreased to produce a suggested velocity of 0.5 to 1.0 ft/sec.

- Gaseous ozone destruct is required on the vent from the Purified Water storage tank. The gaseous ozone destruct system, including hydrophobic vent filtration system, should be similar to that for a system utilizing gaseous ozone generation, as discussed earlier.
- To provide stability to the electrolytic membrane process, it is suggested that the electrical supply to the electrolytic ozone generator be equipped with current/voltage stabilization provisions. This will extend the life of the anode, cathode, and membrane. Further, it provides protection during rapid electrical transients or "brown-outs," which could affect cell performance.
- The suggested technique for introduction of dissolved ozone from the electrolytic ozone generator into Purified Water is presented in Figure 7.14. Purified Water from the distribution loop return is directed to the electrolytic ozone generator prior to a distribution loop modulating-type back pressure regulating valve. The indicated modulating back pressure regulating valve maintains a minimum back pressure of 25 psig during maximum loop draw-off conditions. As an alternative, for Purified Water distribution loops with significant variations in return loop flow rate/pressure, employing a VFD on the distribution pump motor may consider feed from a zero dead leg valve positioned downstream of the dissolved ozone destruct in-line ultraviolet unit. Purified Water flows through one or more individual electrolytic cells in parallel. Cell product water, with dissolved ozone, is directed to a static mixer positioned downstream of the modulating-type valve to the return tubing of the Purified Water storage tank. The distribution loop return Purified Water, with dissolved ozone, is directed to a "dip tube," shown in Figure 7.15. The dip tube attempts to distribute the dissolved ozone uniformly over the entire cross-sectional

274 PHARMACEUTICAL WATER

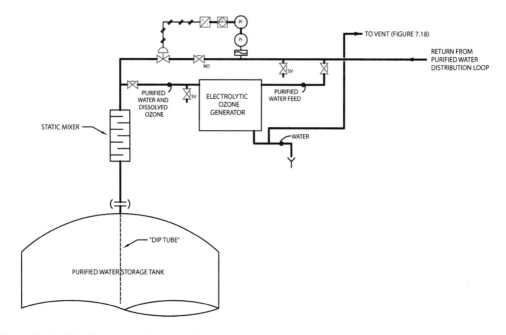

Figure 7.14 Dissolved ozone injection technique.

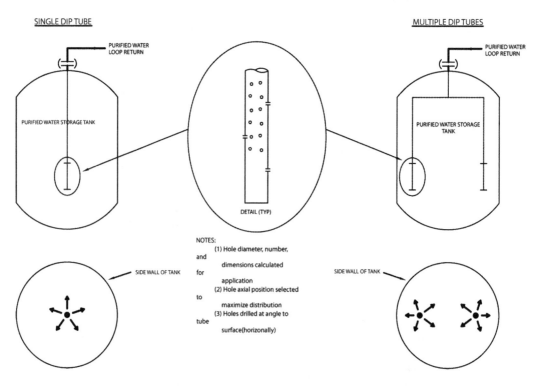

Figure 7.15 Suggested "dip tube" configuration design.

OZONE SYSTEMS AND ACCESSORIES

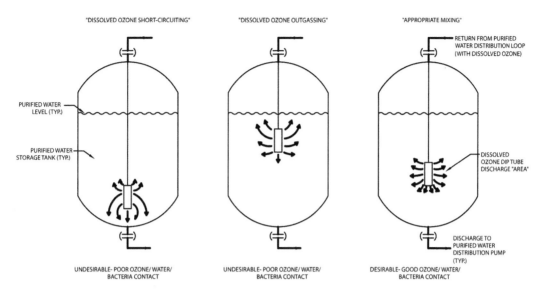

Figure 7.16 Dissolved ozone dip tube discharge location.

area and depth of the Purified Water in the tank. Multiple dip tubes may be employed for larger tanks. The location and horizontal position of the discharge should provide distribution *without* short circuiting, depicted in Figure 7.16. While it is desirable to discharge ozone, through the dip tube(s), at a low elevation in the tank (greater water "head" to minimize outgassing and water contact if outgassing occurs), this could result in dissolved ozone flow directly to the suction side of the Purified Water distribution loop pump, decreasing desired destruction of bacteria in the tank. A minimum Purified Water storage tank operating level must be maintained, through the central control panel, to provide adequate dissolved ozone contact time with Purified Water (bacteria) and to eliminate the possibility of dip tube discharge above the water level.

OPERATING AND MAINTENANCE CONSIDERATIONS

- The performance of dissolved ozone monitors, if employed, should be verified by measuring the dissolved ozone periodically. It is suggested that daily grab samples be obtained from each dissolved ozone sample valve in the Purified Water distribution loop for analysis. It is further suggested that dissolved ozone sensor/ transmitter/monitor periodic maintenance and calibration be performed every six months. The sensors should be installed from a zero dead leg sanitary valve in the distribution loop. Flow from the valve flows through the sensor and is directed to waste. The tubing from the zero dead leg valve to the sensor should be configured such that back contamination of the distribution loop from the sensor is eliminated even when loop flow/pressure is terminated.
- Inspection of the interior of the Purified Water storage tank, sparger(s) gaseous ozone extender tube(s), electrolytic ozone dip tube(s), interior of the first tubing elbow in the distribution loop downstream of the distribution pump, and zero dead leg point-of-use diaphragm valve weirs should be performed annually.
- Proper operation of the dissolved ozone destruct in-line ultraviolet unit is critical since *all* dissolved ozone must be removed prior to individual points of use.

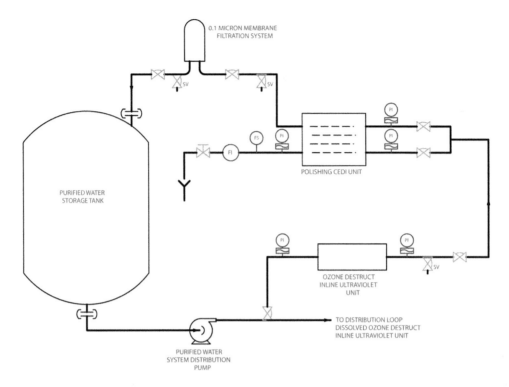

Figure 7.17 Optional ozone system polishing CEDI diagram. *Abbreviation*: CEDI, continuous electrodeionization.

Subsequently, it is suggested that ozone destruct in-line ultraviolet unit quartz sleeves, lamps, and O-rings be replaced every six months. While this would appear to be a highly conservative recommendation, the costs of the indicated consumables are extremely small when compared with the costs of lost "product" or "research" related to the presence of dissolved ozone in Purified Water at points of use.
- As indicated previously, the presence of trihalomethanes in the makeup to the Purified Water storage tank will result in an increase in storage and distribution loop conductivity as the trihalomethane compounds are oxidized to ionic material. A method of maintaining loop conductivity below the Stage 1 USP Purified Water Limit in *Physical Tests* Section <645> when techniques to remove trihalomethanes, present in the raw feedwater, are *not* employed in the system providing makeup to the Purified Water storage tank may include "loop polishing." Loop polishing is depicted in Figure 7.17. It is highly desirable to utilize polishing unit feedwater from a source that it constantly ozonated (during normal operation) with polished product water returned directly to the Purified Water storage tank. It is suggested that the loop polishing system be sanitized with a 1% solution of peracetic acid and hydrogen peroxide, as discussed in chapter 4, every 6 to 12 months. The continuous electrodeionization unit and support components, as shown in Figure 7.17, are not exposed to dissolved ozone.
- Purified Water distribution pump seals should be Viton elastomers. It is suggested that seal replacement be performed annually. While Viton exhibits excellent resistance to oxidation by dissolved ozone, it is constantly exposed to water containing dissolved ozone. Replacement of the seal during normal operation requires distribution loop sanitization (prolonged ozone exposure or 1% peracetic acid and hydrogen peroxide).
- Diaphragms in *all* valves exposed to dissolved ozone should be replaced annually. Diaphragm material is Teflon with an EPDM backing. While the cycle frequency for

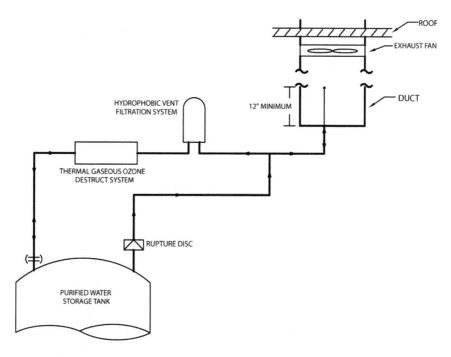

Figure 7.18 Desired gaseous ozone vent system.

each automatic and manual valve will vary, Teflon, even with a "soft" EPDM backing material, will exhibit a "history" of compression against the weir of a valve. Very small leaks in valve seating surfaces can result in creation of a vacuum that could draw contamination into the loop. Sanitary ferrule gaskets in constant contact with dissolved ozone should be replaced annually. Viton is the material of choice, although Teflon envelope gaskets may be used. Sanitary ferrule gaskets in cyclic contact with dissolved ozone should be replaced every one to two years.
- The gaseous ozone thermal destruct should be equipped with a temperature indicator/display. The reading should be included in daily log data for the system. Further, a visual inspection of the unit should be conducted every six months. It should be indicated that gaseous ozone thermal destruct units may emit an odor and even a small volume of "smoke" during initial start-up. This situation should not last longer than a few days. The hydrophobic vent filter in the vent system should be replaced every six months. If discoloration of the hydrophobic vent filter is noted, the atmospheric source of displacement air should be evaluated for impurities such as dust and oil. If appropriate, the location of the intake should be changed. Ideally, the exhaust/intake of the system should be from a vented "duct" as shown in Figure 7.18. This vent method allows release of gaseous ozone while providing air feed to the hydrophobic vent filter during tank drawdown. It avoids roof mounting of the exhaust/intake with resulting requirement for electrically heat tracing of tubing surfaces during freezing weather conditions. The hydrogen vent gas from an electrolytic ozone generator should be configured in the same manner. Finally, the rupture disk provides emergency relief of tank pressure. It avoids potential rapid relief of hydraulic pressure associated with obstruction in the vent system and/or loss of tank level control. Since the disk is constantly exposed to ozone gas, it should be replaced annually.
- Online conductivity monitoring of distribution loop supply and return water is recommended. Conductivity cells may exhibit deposition of rouge-like material over a period of time. This will affect the noted conductivity value since it changes the cell

constant (chap. 2). It is suggested that two conductivity probes be purchased for each online monitoring locations. The cells should be removed annually and replaced with a "cleaned and recertified cell." The removed cell should be returned to the instrument/cell manufacturer for cleaning, measurement of the post-cleaning cell constant, and returned with "Certificate of Calibration" with the new cell constant. Calibration of the conductivity system should be performed in accordance with criteria set forth in USP *Physical Tests* Section <645>.

- Online TOC measurement of distribution loop Purified Water is encouraged. To avoid long-term maintenance/repair issues, it is suggested that the zero dead leg valve feeding the TOC analyzer be of automatic diaphragm type. The valve should be closed during distribution loop sanitization. As an alternative, flow to the analyzer (from the "sidestream tubing") may be terminated, while sample water containing dissolved ozone flows to drain during loop sanitization. Monthly system suitability of the TOC analyzer is suggested. TOC analyzer calibration should be performed as set forth in USP *Physical Tests* Section <643>.
- Periodic preventative derouging and passivation of the Purified Water storage and distribution system should be scheduled. The frequency of these operations will vary with each system. However, it is strongly suggested that a proactive program be employed. This recommendation is based on the type and nature of chemicals employed based on the extent of rouging (Banes, 2010 and Roll and Petrillo, 2010). For example, derouging and repassivation conducted on a annual basis will generally be highly effective using phosphoric acid as the derouging chemical. However, storage and distribution system derouged in a "reactive" manner may require a citric/formic acid mix or sodium hydrosulfite. Use of "stronger" derouging agents may generate a highly obnoxious, pungent odor requiring evacuation of personnel without respiratory protection. While conservative, annual derouging and repassivation are recommended.
- The anode, cathode, and semipermeable membrane in the electrolytic cell will degrade with time. Degradation may not be associated with a loss in dissolved ozone production. It is suggested that the semipermeable membrane for each cell be replaced annually. For certain cells, this will require anode and cathode replacement.
- The dissolved ozone concentration from each cell in the electrolytic ozone generator should be measured periodically (1–2 weeks). Replacement of cell anode, cathode, and/or semipermeable membrane should be performed if a significant decrease (>25%) in dissolved ozone concentration is noted. Dissolved ozone measurement of grab samples may vary. Subsequently, at least three consecutive samples results should be used as the basis for replacement.
- As indicated, individual cells are provided with temperature indicators and switches. The temperature switches inhibit cell operating amperage if high temperature is noted. The temperature switches and temperature indicators for each cell should be calibrated at least annually. Further, the temperature indicator value for each cell should be noted and logged on a daily basis.
- The manway gasket on the Purified Water storage tank should be replaced annually. If frequent access to the interior of the tank is required, the replacement frequency should be greater, ensuring a tight seal to eliminate the possibility of leaking gaseous ozone to the environment.
- Most electrolytic ozone generators are provided with internal battery backup to maintain a current to the cells on unit shutdown. The batteries should be replaced annually.
- Subsequent to maintenance requiring loop shutdown and/or loop intrusive work (point-of-use valve diaphragm replacement, pump seal replacement, etc.), an extended dissolved ozone sanitization of the distribution loop should be performed. During the sanitization operation, all point-of-use values should be cycled upon (one or two at a time), and Purified Water with dissolved ozone directed to drain. The flow rate from the valve should be adjusted to provide a slow, steady stream of water, allowing contact time of dissolved ozone with valve surfaces.

REFERENCES

Adley C, Saieb F. Biofilm Formation in High-Purity Water: Rastonia Pickettii, A Special Case for Analysis, Ultrapure Water, January-February, Littleton, CO, 2005:14–17.

Banes DP. Insights Into Rouge: Definition, Remediation, and Monitoring, presented at Ultrapure Water Pharma, New Brunswick, NJ, May 22, 2010.

Collentro W. Unpublished data from New Jersey Pharmaceutical Facility, 2010.

Compressed Gas Association. Safe Handling of Ozone Containing Mixtures Including the Installation and Operation of Ozone Generating Equipment, product No. CGA P-34, Chantilly, VA, 2001.

Degremont Technologies. TRIOGEN TOGC8X, 13X, & 45X Ozone Packages, Triogen Limited, Degremont Technologies, Catalog Sheet No. DIST05101EN-V2-09/2009, February, 2009.

Escobar I, Randall A. Case study: ozonation and distribution system biostability. J Am Water Works Assoc 2001:77–89.

HACH. Pocket Colorimeter™, HACH Literature Manual 2596, Revision 2, HACH Company, Loveland, CO, 2009a.

HACH. Ozone Reagent LR, MR, and HR AccuVac® Ampules, Low Range, Medium Range, and High range, HACH Company, Loveland, CO, 2009b.

Juras D. Ozone Disinfection, Canadian Greenhouse Conference, Workshop, October 6, 2005, Leamington, ON, 2005:1–3.

Nebel C, Gottschling R, Hutchison R, et al. Ozone disinfection of industrial - municipal secondary effluents. J Water Pollut Control Fed 1973; 45(12):2493–2507.

Nebel C. Ozone: The Process Water Sterilant. Proceedings of the Pharm Tech Conference '85, September 10–12. Cherry Hill, NJ: Aster Publishing Corp., 1985:269–285.

Ozonia North America. Ozonia ODT™ Series Thermal Ozone Destruct Units, Degremont Technologies, Catalog No. DISO10202EN-V2-08/2009, August, 2009.

Rakness K. Ozone in Drinking Water Treatment: Process Design, Operation, and Optimization", American Water Works Association, ISBN 1-58321-379-1, Denver, CO, 2005.

Roll D, Petrillo P. Rouge: Monitoring, Measuring, & Maintenance in Water and Steam Systems, presented at Ultrapure Water Pharma, New Brunswick, NJ, May 22, 2010.

Schilling B. Personal Communication. August 6, 2010.

Stanley B. Electrolytic Ozone Generation and It's Application in Pure Water Systems. Duebendorf, Switzerland: Ozonia Limited, 1999.

Vezzu G, Merz R, Gisler R, et al. Evolution of Industrial Ozone Generation. Dubendorf, Switzerland: Degremont Technologies, 2008:1–16.

Wallhauser KW. Praxis der Sterilisation, Disinfektion – Konservierung. 4th ed. Stuttgart, Germany: Georg Thieme Verlag, 1988.

8 | Polishing components

INTRODUCTION
This chapter discusses various unit operations that could be positioned downstream of a storage system as "polishing" components. With the exception of distribution pumps, polishing components do not apply to USP Water for Injection systems.

DISTRIBUTION PUMPS
Chapter 3 summarized recirculation pump applications, primarily for the pretreatment section of a pharmaceutical water purification system. This chapter discusses polishing components and accessories that would generally be positioned downstream of a storage tank. Some of the parameters discussed for recirculation pumps will apply to items discussed in this chapter. However, it is important to emphasize that distribution pumps recirculate higher quality water, from both a chemical and microbial standpoint, than recirculation pumps in the pretreatment section of a system.

Theory and Application
Distribution/recirculation pumps, primarily positioned downstream of storage tanks for Potable Water, Purified Water, or Water for Injection should provide adequate flow and pressure to meet system requirements. The pump(s) should be capable of providing an adequate flow rate at an established discharge pressure to meet the criteria set forth in the Basis of Design for the system. For polishing applications, the flow and pressure characteristics are related and established by the maximum instantaneous draw-off from the distribution system, the velocity of water through the distribution loop, the associated pressure drop, the configuration of the distribution tubing establishing an "equivalent tubing length" and the tubing diameter.

The selection of a pump for providing distribution of Purified Water is generally limited to either centrifugal or multistage centrifugal type. Water for Injection applications employs a sanitary centrifugal-type pump. Most Purified Water applications will be provided with a sanitary centrifugal-type pump; however, certain systems will employ multistage centrifugal pumps where metallic surfaces in contact with water are of stainless steel construction. Potable Water applications may use either multistage centrifugal pumps or centrifugal pumps (sanitary or nonsanitary), depending on the application.

It is suggested that the pump selection should be consistent with the nature of the storage and distribution system. In other words, if a sanitary stainless steel tank is used for the system, a sanitary stainless steel distribution loop is appropriate. If periodic hot water sanitization is anticipated, pump selection should be limited to a sanitary centrifugal design. Generally, these applications are for Purified Water, while some Potable Water systems may also resemble the storage and distribution system described above. As previously indicated, sanitary centrifugal pumps should be used for all Water for Injection applications.

A typical sanitary centrifugal pump curve is shown in Figure 8.1. It should be noted that the pump curves provided by pump manufacturers represent a "family" of pumps, which are capable of supplying a broad range of flow rates at various pressures using different diameter impellers and drive motors at a fixed speed (rpm). It should also be noted that, in general, pump curves for sanitary centrifugal-type units are relatively "flat." In other words, a significant drop in discharge pressure for the pump is generally not observed for an increase in flow rate from the pump. The series of curves indicated on Figure 8.1 represents a specific pump with performance curves for various impeller sizes. As higher discharge flow rates and pressures are desired for a given pump and impeller size, the required pump motor horsepower size will increase. Pump selection should allow latitude to increase or decrease both the discharge flow rate and pressure for a given application. Subsequently, it is

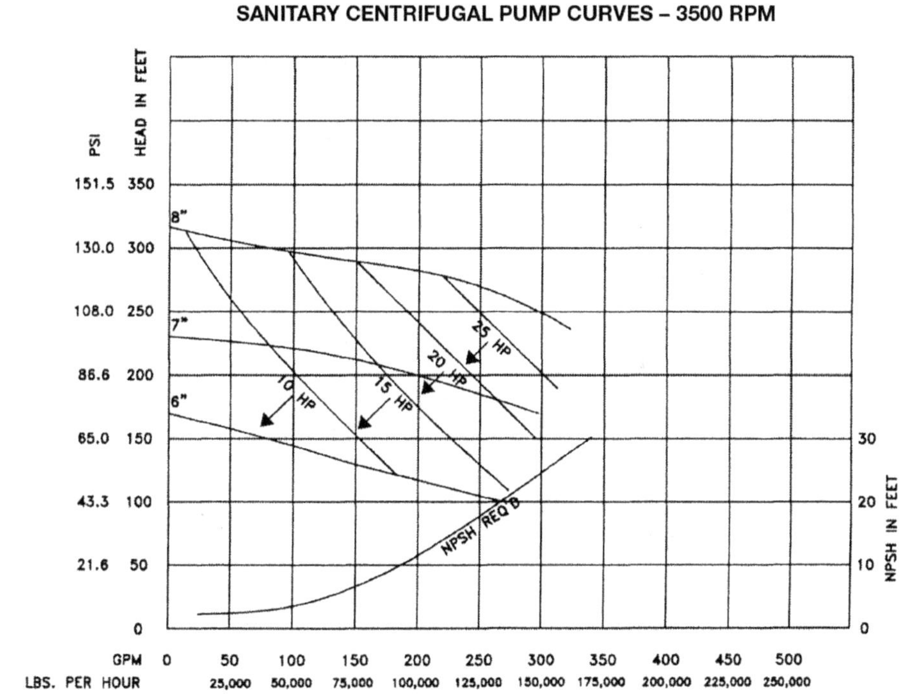

Figure 8.1 Typical "family" of sanitary pump curves. (The pump speed is 3500 rpm. Horsepower and net positive suction head requirements are not shown.) *Source*: From Tri-Clover, Inc. (1994).

undesirable to select a pump with the smallest or largest impeller diameter for a specific manufacturer's "size." As an example, if the smallest impeller size is selected for a specific discharge flow rate and/or discharge pressure and the requirements change (decrease) it will be impossible to decrease the size of the impeller. While the desired pressure and flow rate criteria may be obtained using the pump with "smallest" diameter impeller, a variable frequency drive (VFD) will be required for the pump motor. Conversely, if a pump is selected with the largest impeller size, and the discharge flow rate and/or pressure are higher than required, it is impossible to increase the parameters without installing a larger size pump.

The horsepower of the pump motor is also important. The horsepower should allow increases in the flow rate for a given impeller size. Again, this provides latitude regarding the changing demands for an installed pump. The key to selecting the proper pump (including impeller size and motor horsepower) is to provide "freedom of movement" on the pump curve (lower and higher pump discharge flow rate) as well as the ability, at a given flow rate, to increase or decrease the discharge pressure by increasing or decreasing the size of the impeller.

While centrifugal pump impeller size and drive motor selection are very important, it may be possible to provide the indicated freedom of movement, particularly for applications where pump discharge flow rates and pressures routinely change, by using a VFD. Figure 8.2 provides a family of curves for a sanitary centrifugal pump with fixed impeller size and motor horsepower as a function of pump speed (rpm). The use of a VFD for both multistage and single-stage centrifugal pumps is encouraged. For pretreatment applications, VFDs can provide flow variations required for backwash and regeneration of pretreatment components. For RO systems, VFDs can provide pressure changes required for RO membrane feed water as fouling occurs. Finally, for Purified Water and Water for Injection distribution systems, VFDs can respond to distribution loop flow and pressure "swings" associated with high flow rate draw-off at points of use.

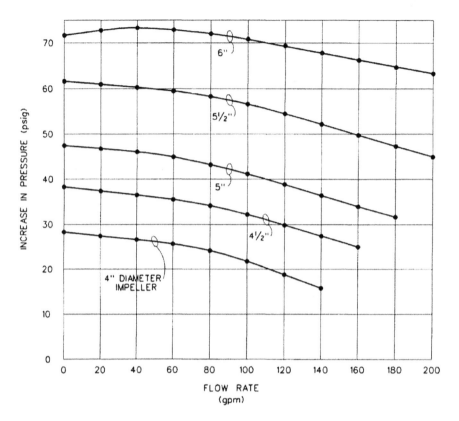

Figure 8.2 Pumps curves at different speeds.

For certain pump applications, the use of a pump motor "soft start" drive may be appropriate. The soft start initiates pump operation at a low speed, with resulting lower discharge pressure, and ramps up to full speed and pressure. This type of system may be applicable for applications feeding membrane processes such as reverse osmosis.

The pump curves shown in Figure 8.1 provide additional important parameters, such as the efficiency of the pump (a factor associated with potential Joule's heat input to the recirculating water), pump speed, and net positive suction head (NPSH) requirements to avoid cavitation. For most applications, pump selection should be entirely based on flow and pressure requirements. Unless the efficiency is extremely low and will ultimately affect operation, it should be a factor that, while important in determining heat input, can generally be "controlled" by proper system design (e.g., the use of "trim cooling"). For the same pump and impeller size, significantly different flow and pressure characteristics are observed at lower pump speeds as shown in Figure 8.2. Furthermore, a lower pump speed is generally associated with lower maintenance requirements and quieter operation. Table 8.1 is a representative calculation that demonstrates the importance of the pump NPSH. This value is expressed in absolute pressure, as opposed to gauge pressure. It should be considered when the pump elevation is above the water level in the tank, lengthily tubing section with multiple fittings (elbows) are in the pump suction line from the upstream tank, and undersized diaphragm valve(s) that decrease the pressure provided by the "head of water" in the upstream storage tank are used.

The Joule's heat input to the water system is a function of several variables, including the efficiency of the pump. This heat input is inversely proportional to the efficiency expressed as a fraction. As discussed in chapter 3, multistage centrifugal pumps operate at higher efficiencies than single-stage centrifugal pumps. Subsequently, a multistage centrifugal pump operating at

Table 8.1 Pump NPSH Calculation

Design calculation:

$$NPSH = hss - hfs - p$$

where
- NPSH is the calculated, in feet of water, which must be greater than the pump manufacturer's NPSH value for the selected pump,
- hss is the vertical distance, in feet, from the lowest anticipated upstream tank level to the centerline suction of the pump plus atmospheric pressure also expressed in feet of water,
- hfs is the pressure drop, expressed in feet of water, through pipe/tube, valve(s), and fittings from the tank discharge to the pump suction, and
- p is the vapor pressure of water, expressed in feet of water, at the temperature of the stored/repressurized water.

Existing installation:

$$NPSH = atm + hgs - p + hus$$

where
- atm is atmospheric pressure expressed in feet of water,
- hgs is the pressure gauge reading on the suction side of the pump expressed in feet of water, and
- hus is $\dfrac{(\text{velocity, expressed in ft/sec, at the pump suction})^2}{32.17 \text{ ft/sec}^2}$.

Abbreviation: NPSH, net positive suction head.
Source: From Perry et al. (1984).

an efficiency of 60% will introduce half of the energy (and increase the temperature of the recirculating water by 50% less) than a centrifugal pump operating at 30% efficiency.

Design Considerations

The feed water tubing to distribution/recirculation pumps installed downstream of a storage tank should be equipped with a compound-type pressure gauge. If the storage tank and distribution tubing are of sanitary type, the pressure gauge should be of sanitary type. For applications where the design is either nonsanitary, or a combination of sanitary and nonsanitary connections, a nonsanitary compound pressure gauge may be used. However, the gauge should be equipped with a diaphragm isolator to minimize the potential for bacterial growth within the pressure-measuring mechanism of the gauge (bourdon tube).

A manual diaphragm-type isolation valve should be positioned upstream of the pump. For distribution/recirculation pumps installed directly downstream of storage tanks, the isolation valve should be positioned downstream of the storage tank but upstream of the compound pressure gauge to facilitate gauge calibration without draining the tank. The selection of the valve for this application is critical, since it may restrict flow (and, more importantly, pressure) to the distribution pump, resulting in cavitation within the pump. The situation can be avoided by using a radial-type diaphragm valve or sanitary "plug-type" valve. As discussed in chapter 6, this valve, in combination with sanitary ferrule connections to the downstream pump and at the base of the storage tank, can be used with a section of hose to drain the storage tank, in lieu of providing a dedicated drain valve on the tank.

For certain Water for Injection distribution loop applications, a subloop may be established off the main distribution loop. This is common for Water for Injection systems where a hot *main* recirculation loop is maintained with an ambient temperature subloop (periodically heated to a temperature $\geq 80°C$). As discussed in chapter 3, an "inline" repressurization pump, such as the pump used for a subloop, should be equipped with feed water pressure sensing element, operating in series with the pump motor starter, to insure that adequate water flow (and associated pressure) are available to the inline repressurization/recirculation pump.

There are several items that should be considered during pump specification/selection. From a design standpoint, to satisfy system requirements, the following pump parameters should be specified:

- Feed water and discharge connection type and size
- Impeller size
- Operating speed and rpm (generally 1750 or 3500 rpm)
- Pump motor horsepower and type [totally enclosed, fan cooled (TEFC) or "open drip" (ODP)]
- Pump coupling mechanism to pump motor (close coupled vs. base mount)
- Casing drain (size and orientation)
- Air vent for multistage centrifugal pumps
- Seal selection (Purified Water vs. Water for Injection)
- Materials of construction
- Pump seal materials of construction
- Pump materials—material certification with "heat numbers" and analysis

As indicated above, the pump to motor seal connection is important. For sanitary centrifugal pumps for Purified Water applications, a "clamped in-seal/seat" mechanical-type seal with rotating seal components may be considered. The seal material may be silicone carbide, ceramic, or tungsten carbide. For Water for Injection applications, a double mechanical seal with positive Water for Injection continuous "flush" is strongly suggested. The double seal mechanism uses product water from the discharge of the pump to feed water to the area between the two seals continuously at a relatively low flow rate; it is subsequently discharged to a depressurized drain with an "air break" as shown in Figure 8.3. To extend the life of the mechanical seals used for this application and to minimize long-term pump maintenance requirements, it may be desirable to provide a small heat exchanger to cool the flush water. However, this heat exchanger should not be installed in a manner that would compromise the quality of the recirculating hot Water for Injection.

As discussed in chapters 6 and 9, certain applications may require a pump with VFD. This technique provides operation of pumps at preselected speeds (rpm), increasing or decreasing the flow and pressure characteristics for a given impeller size and motor horsepower. Figure 8.2, discussed earlier, is a family of curves for a specific pump and impeller size, clearly demonstrating the flexibility associated with using a VFD. This design option may provide flexibility for "batching" operations, particularly when "high flow rate end users" are positioned as the initial points of use in the distribution loop. Considering this flexibility, it may be possible to provide an initial section of distribution tubing at a larger diameter, with the majority of the distribution tubing at a smaller diameter associated with "normal" recirculating flow rate requirements.

Both centrifugal and multistage centrifugal pumps should be provided with feed water and product water isolation valves. The valves may be of a diaphragm type. The valves should be positioned in such a manner to allow isolation (or removal) of the pump for maintenance without fully draining the distribution loop or upstream storage tank (if applicable).

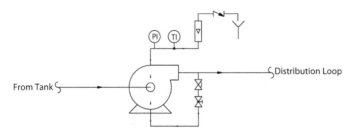

Figure 8.3 Sanitary pump: double mechanical seal with flush.

A pump casing drain may be provided. The pump casing drain may be used, with other low point drain valves in Purified Water or Water for Injection distribution loops, to insure complete draining of water from the pump feed water and product tubing as well as the casing. The casing drain should be equipped with a diaphragm valve with sanitary ferrule "cap."

The discharge tubing from the pump should contain a pressure gauge. In general, it is desirable to select a pressure gauge with a range such that the anticipated maximum pump discharge pressure will be approximately 50% to 70% of the "full-scale" reading.

The pump discharge tubing should be equipped with a sample valve. The sample valve should be of zero dead leg diaphragm type for sanitary distribution systems applications. As a minimum, diaphragm- or needle-type valves with bellow-type seal should be used for storage and distribution systems that are not full sanitary design (or a combination of sanitary and nonsanitary). Periodic sampling can be employed to determine the quality of water (chemical and microbial) within an upstream storage tank.

Many recirculation pumps installed downstream of storage tanks are equipped with product water check valves. The check valves eliminate the possibility of back flow, during pump shutdown, from the distribution loop to the storage tank, potentially flooding the area adjacent to the storage tank. While it is suggested that the concept of using check valves is well intended from a design standpoint, there are several technical flaws that should be considered. It is highly desirable to avoid the use system accessories that may fail (even after an extended period of time) and that may affect the overall quality of the water from a chemical and, more importantly, in the case of the check valve, microbial standpoint. It has been indicated that a manual isolation valve should be positioned in the product tubing from the distribution pump. In general, the volume of water in a distribution loop will be fraction of the volume of water in the storage tank. For most systems, under worst conditions, total draining of the loop (which will probably never occur due to the fact that the tubing is not directly vented) would introduce a limited volume of water. The high-level set point on the storage tank level control system could be programmed to provide adequate tank volume to collect water from the loop. It is suggested that distribution pump shutdown primarily occurs for or excursion such as "low-low" level in the upstream storage tank, or pump/motor failure. For periodic maintenance, electrical power to the pump motor is manually inhibited. Operating personnel can simply close the valve on the discharge side of the pump to avoid backflow of distribution loop water. Both pump/motor failure and a low-low tank level condition represent system excursions. Field experience indicates that the "sears" on "sanitary-type" check valves must be periodically replaced to insure the desired seal during the indicated system excursions. This does not appear to offset system microbial control considerations for Purified Water and Water for Injection distribution loops but may be useful for nonsanitary portions of a system such as pretreatment. Finally, the indicated casing drain on the pump is a low point location for the difficult to drain sections of the distribution pump, feed water tubing and discharge tubing. Using a check valve negates this convenient design feature.

Another design issue that should be addressed centers around "spare" pumps. Experience indicates that there are numerous methods for eliminating dead legs associated with installed spare pumps. It is suggested that installed spare pumps provide a location for bacteria to accumulate and proliferate. Figure 8.4 depicts a system that uses an "installed" but physically unconnected spare pump. In this system, the main feed water and discharge tubing is physically positioned in the centerline between two installed pumps. The primary operating pump is piped, using double elbows and sanitary ferrules, to the feed water tubing from the upstream tank and feed to the distribution loop. The second pump is equipped with sanitary caps (or flanges in the case of a multistage centrifugal pump) to avoid the contamination of interior surfaces. It is not physically "hard piped" to either the tank outlet or the distribution loop supply. The spare pump is, however, equipped with a dedicated motor starter with its "fuses" removed. Centrifugal pumps will operate for an extended period of time if they are *not* cycled. Long-term routine maintenance, such as the replacement of seals, will be required but can be scheduled. By using the arrangement shown in Figure 8.4, a single dedicated pump will be the primary operating unit. If failure occurs, the "installed spare pump" can be chemically or thermally sanitized. The double elbow system, with feed water and product water isolating

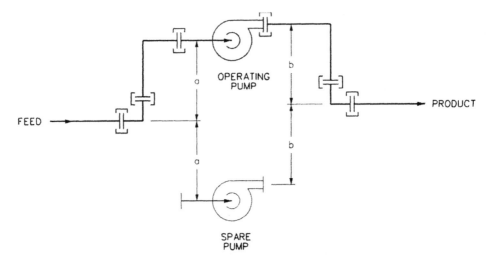

Figure 8.4 Process flow diagram—installed spare pump. (Gauges, sample valves, isolation valves, etc., are not shown. A dedicated motor starter is provided for each pump. Fuses are removed for the spare pump. Sanitary ferrules for the spare pump are provided with gaskets, caps, and clamps. The pump casing drain is not shown.)

valves, can be readily connected to the storage tank outlet and distribution loop inlet. Depending on the installed feed water and product water tubing arrangement, it may be possible to place the spare pump online without sanitization of the entire storage and distribution system. Experience indicates that a properly maintained pump may operate for a number of years without needing an installed spare pump.

Finally, related to the above, many Purified Water and Water for Injection distribution systems employ two pumps with isolation valves and check valves as shown in Figure 8.5. Water flow is through the "operating" pump in the normal flow direction. To avoid stagnant conditions in the fully installed spare pump, feed and outlet manual isolation valves are always open. "Holes" are drilled in the "disc" of the check valves to allow flow in the reverse direction, eliminating stagnant water conditions. However, reverse flow through the spare pump will result in both reverse rotation of the spare pump impeller and drive motor. It is suggested, based on field observations, that this situation results in both premature pump seal replacement and pump motor replacement (acting in a power "generating" mode with reverse flow). This situation is often further complicated by a "pump-cycling" program where the operating/spare pump function is changed periodically.

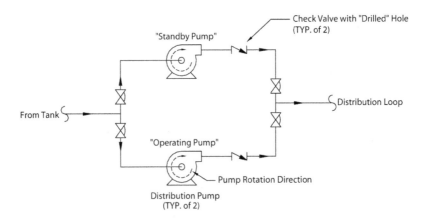

Figure 8.5 Dual distribution pumps with check valves.

Operating and Maintenance Considerations

Generally, a lubrication fitting is provided on the pump motor shaft. Periodic lubrication, as outlined in the operating and maintenance manual for the pump, should be performed.

The pressure reading on the compound gauge in the feed water line to the pump should be monitored periodically. When possible, monitoring should be performed when the tank level is in the lower range. The results of periodic pressure readings, when compared with the NPSH, *and* any unusual noise coming from the pump should be used to evaluate if pump cavitation is occurring. However, it should be emphasized that pump cavitation can occur without any detectable "unusual" noise.

It is suggested that system qualification include verification of the actual pump speed. It is further suggested that the pump speed be verified on a periodic basis (e.g., once every three to six months). The pump speed should be checked if erratic performance (specifically fluctuations in discharge pressure) or audible noise is noted.

Pump seals should be replaced as part of a preventative maintenance program. This is particularly true for single mechanical seals. Water leaking from pump seals can also be associated with air being introduced into recirculated water. The presence of water on the floor of any pharmaceutical water purification system indicates a problem that should be corrected immediately. If a dual pumping system is used (not recommended as discussed earlier), it may be necessary to establish a pump seal replacement program with greater frequency than that for continuously operating pumps without reverse flow. Planned cyclic operation of a pump (e.g., one pump operating for one week with another pump operating for the next week) may produce slight "drying" of the nonoperating pump seal, again resulting in increased pump seal replacement frequency.

It is strongly suggested that pumps be disassembled and inspected annually. The inspection should include direct observation or swabbing of internal surfaces to evaluate potential rouging and examination of the impeller. Field experience indicates that sanitary pump impellers may "shift" position in a horizontal direction over a period of time. Generally, this is associated with pump operating conditions. Unfortunately, this situation results in impingement of the pump impeller on the pump casing. This is a highly undesirable conditions since, if not detected, could result in stainless steel particle release to Purified Water or Water for Injection distribution loops. The particles are both difficult to detect and even more difficult to remove.

Frequently, newly installed pumps will exhibit extremely poor performance, characteristics, principally a low flow rate and, more importantly, a very low discharge pressure. This situation is usually associated with improper rotation of the pump motor. Pump motor rotation should be verified during installation of the pump and further verified during the IQ phase of system validation.

Pumps used in Water for Injection systems that are equipped with a flushing mechanism require periodic monitoring of the flow and temperature readings of the flush water. This insures proper performance of the pump.

INLINE ULTRAVIOLET UNITS

The use of inline ultraviolet units in pharmaceutical water purification pretreatment sections was discussed in chapter 3. Since ultraviolet units are also used as polishing components (microbial control) in some Potable Water and Purified Water systems, and since the application, design, and maintenance requirements have subtle differences from units used in pretreatment applications, it is appropriate to address polishing applications for ultraviolet units.

Theory and Application

Inline ultraviolet units are occasionally used as a polishing technique to control bacteria in Potable Water and Purified Water systems. The use of inline ultraviolet units, as a polishing technique in Water for Injection systems, is inappropriate and conflicts with acceptable pharmaceutical water system practices. The theory associated with inline ultraviolet units was discussed in chapter 3. For polishing applications, there are several limitations associated with their use for microbial control.

Most inline ultraviolet units are equipped with low-pressure mercury vapor lamps. If a lamp and its associated quartz sleeve break, mercury will be introduced into the recirculating water. At temperature conditions encountered for Purified Water or Potable Water recirculating loops, the mercury would exist as a liquid—insoluble in water. It is considered an "Added Substance" if introduced to Purified Water. The potential ramifications of mercury introduction into product, passing through a point-of-use valve, are significant. If catastrophic failure of a quartz sleeve and low-pressure mercury lamp were to occur, it is possible that the mercury could form an "amalgam" with stainless steel, literally coating the stainless steel surface with mercury. There would be little indication that this situation has occurred other than the noted failure of the quartz sleeve and mercury vapor lamp. However, the mercury from the amalgam with stainless steel would gradually be released to the recirculating water, over a period of time, as an Added Substance. Furthermore, it is possible that the mercury, after forming the amalgam with stainless steel, could weaken radially welds, potentially resulting in tubing rupture.

If an existing system uses ultraviolet units, it is strongly suggested that provisions be included to minimize, or eliminate, the potential for two-phase flow with associated water hammer, a condition that could result in quartz sleeve and lamp failure.

Inline ultraviolet units used for polishing disinfecting applications emit ultraviolet radiation at a wavelength of 254 nm and ultraviolet intensity of 30,000 to 35,000 $\mu W\ sec/cm^2$. As discussed in chapter 3, the ultraviolet radiation intensity, particularly after operation for a period of time, may not be adequate to provide total destruction of all pathogenic Gram-negative species of bacteria in Drinking Water and Purified Water systems. This is of significant concern when inline ultraviolet units are used as a polishing component, since there are no downstream components, with the possible exception of membrane filters, to provide effective removal of sublethally destroyed organisms. To complicate this situation, standard bacterial measurement techniques for point-of-use samples may not detect the sublethally destroyed organisms. Bacteria enumeration method including culture media, incubation time prior and incubation temperature may all be important to the detection of the sublethally destroyed organism or resulting "reactivated" bacteria. In fact, the sublethally destroyed organisms may not be detected using Membrane Filtration of a 100-mL sample, highly sensitive culture media such as R2A, incubation temperature of 22°C and 72- to 120-hour incubation time period. The literature suggests that sublethally destroyed organisms may only be detected with R2A culture media, membrane filtration of a 100-mL sample, with incubation for seven to eight days at 22°C (Collentro, 1995). While the presence of sublethally destroyed organisms may be a concern, it is highly unlikely that a typical USP Purified Water system PQ (or routine bacterial monitoring program) will be adequate to identify the sublethally destroyed species of bacteria. This matter has been addressed by various regulatory authorities (Munson, 1993b). Obviously, inline ultraviolet units should only be used for polishing applications, where appropriate, *and* where it can be clearly established by a responsive bacteria monitoring program, from samples obtained at points of use, that sublethally destroyed organisms are not present at a value greater than the established bacterial Alert and Action Limits for the system. The monitoring program should also verify that the level of pathogenic Gram-negative bacteria do not exceed, where appropriate, an established level.

To demonstrate the concerns presented above, the literature contains examples of the ability of sublethally destroyed organisms to survive for extended periods of time. Certain sublethally destroyed or reactivated bacteria rapidly replicate when exposed to nutrients or other conditions and exhibit accelerated growth when compared with the preultraviolet radiation exposed species of the particular organism. This is particularly true for *Berkholderia cepacia* (Corson and Petersen, 1975).

The FDA *Guide for Inspections of High Purity Water Systems* (1993) states that ultraviolet units should be considered as a technique capable of inactivating only 90% of bacteria in the feed water to a unit. For polishing applications, the 10% of bacteria passing through the inline ultraviolet unit will replicate and, subsequently, establish a biofilm on the interior walls of the ambient recirculated Purified Water distribution system. While it is possible that the required system sanitization frequency may be extended by the use of inline ultraviolet units for polishing applications, it is also suggested that the relative extension, in terms of the percentage increase in elapsed time between sanitization cycles, when compared with units without ultraviolet units, is small.

For certain Purified Water systems, inline ultraviolet units (254 nm) are used in conjunction with a distribution pump, polishing mixed bed deionization units, and final membrane filtration. Since mixed bed deionization units, particularly conventional rechargeable canisters, generally provide an excellent location for bacterial growth, it is suggested that it may be appropriate to position the inline ultraviolet unit prior to the mixed bed units, rather than after the units, in an attempt to destroy bacteria before they have an opportunity to enter the carbonaceous resin beds. The mixed bed units operate at a neutral pH and support rapid bacterial proliferation. Unfortunately, most inline ultraviolet units used in systems with mixed bed polishers are positioned after the mixed beds. Thus, the inline ultraviolet units are often "overwhelmed" with bacteria.

A limited number of Purified Water systems, employ inline ultraviolet units (185 nm and intensity >30,000–35,000 $\mu W\ sec/cm^2$) in conjunction with a distribution pump, polishing mixed bed deionization units, and final membrane filtration. At the increased ultraviolet intensity zone and/or the hydroxyl radical are produced. While both of these very powerful oxidants will inactivate bacteria, they must be removed prior to points of use. If not removed, the "antimicrobial" agent present in "delivered" conflicts with guidance provided in the *General Notices* section of USP. Unfortunately, the presence of an antimicrobial in point-of-use water samples results the *indication* of the total absence of bacteria.

Design Considerations

Inline ultraviolet units used in polishing applications, where the upstream storage tank, distribution pump, and distribution tubing are of sanitary design, should be equipped with sanitary ferrule inlet and outlet fittings. Flanges are not consistent with the sanitary nature of the system and should be limited to pretreatment applications or applications where sanitary storage and distribution loops are not employed.

Inline ultraviolet units should be provided with feed water and product water sample valves. The sample valves should be an integral part of the unit (if sanitary type) or installed within the feed water and product water tubing. Sample valves should be of diaphragm type. For nonsanitary applications, needle-type (with bellows-type seal) or diaphragm-type samples valves may be used.

Inline ultraviolet units are often provided with a threaded drain connection positioned at the base of the stainless steel sanitizing chamber. The system specifications should indicate that threaded fittings are unacceptable for this application. The drain connection presents a potential source of bacterial accumulation and proliferation, particularly when a nonsanitary threaded valve is connected to the fitting. The valve, threaded fittings, and water contained in the "holdup" volume of the components will not be exposed to adequate ultraviolet radiation to ensure bacterial destruction. While it is acknowledged that a drain connection is required for performing maintenance of the unit, a sanitary ferrule should be specified, positioned (physically) as close as possible to the sanitizing chamber. A sanitary diaphragm valve on the drain line, mating to the sanitary ferrule, may be provided for unit maintenance, which is performed about once every 6 to 12 months. The inline ultraviolet unit should be provided with feed water and product water isolation valves. When maintenance is required, the relative volume of water contained in the sanitizing chamber, feed water tubing, and product water tubing may be small. The drain connection should be accessible. For low flow rate systems (e.g., <20 gpm), it is suggested that a drain valve may not be necessary but can be replaced with a sanitary cap (and associated gasket and clamp). When maintenance is required, the clamp can be loosened to the point where water is allowed to flow to a collection pail positioned beneath the connection eliminating the potential dead leg associated with the drain valve.

As discussed in chapter 3, inline ultraviolet units are frequently provided with electric power cabinet required for operation of the low-pressure mercury lamps (including ballasts), physically positioned directly underneath the sanitizing chamber. When the units are drained, or if a small leak is experienced from the O ring(s) positioned at the end(s) of the quartz sleeves, water will drip onto, and potentially into, the electrical enclosure. For this reason and to facilitate periodic replacement of the quartz sleeves and lamps, it is suggested that units be equipped with a remote electrical power cabinet.

As discussed earlier, inline ultraviolet units are generally provided with a direct reading intensity meter. The meter contains either a direct reading analog or digital display. A review of manufacturer's operating instruction generally states that "calibration" is performed by using an adjustment screw on the meter to read "100%" after installation of new lamps and sleeves. It is suggested that periodic (e.g., every 6–12 months) calibration procedure should be performed with a source traceable to the National Institute of Standards and Technology (NIST). Experience indicates that there are numerous inline ultraviolet units operating with ultraviolet intensity meters providing indications that cannot be correlated to the actual dose rate. Observation of operating units indicates that the intensity meter reading is over the full scale, from values below zero to values above 100% "transmission." Subsequently, there does not appear to be a correlation between the observed values and the need for quartz sleeve and lamp replacement. As indicated earlier in this section, ultraviolet radiation intensity is critical, specifically as it relates to sublethal destruction of bacteria.

Related to the above, an analog signal is generally available from the ultraviolet unit intensity meters to a central monitoring and control panel. While the signal may be used to display the intensity for system ultraviolet units at a central location and provide an alarm indication, at a preset value, fluctuations in readings can result in "nuisance" alarms. It is suggested that the reading be displayed at the central control and monitoring panel but unit alarm condition be based on an available discrete output from the inline ultraviolet unit for "lamp-out" status. Finally, based on field experience, if multiple analog UV unit intensity signals are provided from two or more units in a system, control signal "isolators" are required in the analog signal wiring to the central monitoring panel.

Observation of inline ultraviolet units installed for both pretreatment and polishing applications indicates that a fair percentage of units are equipped with a permanently installed bypass. The bypass is generally equipped with a single manual valve, which is unacceptable. The dead leg created by this situation, in part, counteracts the effectiveness of bacterial destruction by the ultraviolet unit. The required maintenance for ultraviolet units is generally limited to semiannual or annual replacement of quartz sleeves, lamps, and O rings. Annual or semiannual calibration of the ultraviolet radiation intensity meter will also be required. However, maintenance, including calibration, should not take more than about one hour every 6 to 12 months. It is suggested that the bypass is unnecessary and cannot be technically justified. The sanitizing chamber of the inline ultraviolet unit is nothing more than an extension of the distribution tubing loop. If a quartz sleeve and/or lamp were to rupture, total loop shutdown would be required. The bypass does not provide a useful purpose. For operating units currently equipped with a bypass and single valve, it is suggested that the bypass be removed or replaced with a bypass containing a "double block and bleed" system. Obviously, the "blocking valves" should be positioned as close as possible to the recirculating loop feed water and product water connections, as demonstrated in Figure 8.6.

Ultraviolet lamps should not be energized when there is no flow through the units. This affects the ultimate performance of the unit. For inline ultraviolet units used in polishing applications, it is suggested that the power supply to the units be connected in series with the "control leg" of the power supply to the upstream distribution pump. As discussed in chapter 6, the low-low level set point on the upstream storage tank will automatically inhibit power to the distribution pump on low tank level. If the power to the distribution pump is manually inhibited, power will also be inhibited to the inline ultraviolet unit, avoiding the stagnant condition.

The effectiveness of inline ultraviolet units decreases significantly when an operational unit is exposed to hot water. For Purified Water applications employing hot water sanitization operation, it is important that power be inhibited to the inline ultraviolet unit when the temperature increases above a preset value and restored upon completion of the sanitization operation. For systems that are sanitized daily, the suggested frequency of replacement for quartz sleeves and lamps is six months. As indicated, a properly operating ultraviolet radiation intensity meter will allow operating personnel to determine the effect of hot water on unit performance. In an attempt to minimize the effects of hot water sanitization on ultraviolet

POLISHING COMPONENTS

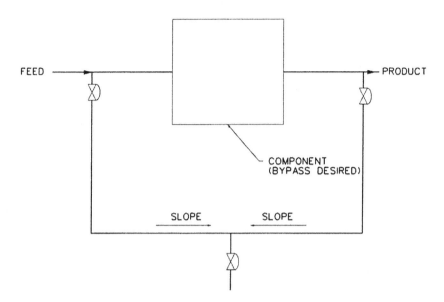

Figure 8.6 Double block and bleed tubing and valve arrangement. (All valves, particularly "blocking" valves are to be as close as possible to the "main flowing line." The "bleed" valve may be positioned at either end of the bypass line, allowing slope in one direction for draining.)

lamps and the electronics of the unit, it is suggested that the specification for units include an internal temperature sensing device with automatic power "cutoff" provisions. This is usually an optional item for UV units.

In chapters 6 and 9, the use of unpigmented polypropylene piping for storage and distribution systems is discussed. If inline ultraviolet units are used as polishing components for systems equipped with unpigmented polypropylene piping, the units should be equipped with feed water and product water stainless steel "transition" connectors ("light traps") that will inhibit exposure of the polypropylene surfaces to ultraviolet radiation (Fig. 8.7). Ultraviolet radiation will degrade unpigmented polypropylene and other plastic piping materials such as PVC and CPVC. Exposure of plastic piping to ultraviolet radiation may lead to "elution" of organic decomposition compounds and long-term failure of piping material.

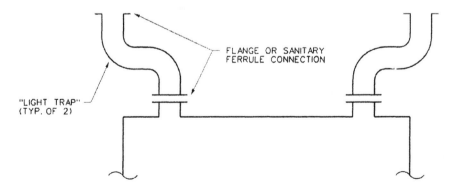

Figure 8.7 Typical ultraviolet sanitation unit "light trap." (The materials of construction are 316L stainless steel. The diameter, width, and height of the curvature are critical for a specific unit.) *Source*: From Aquafine Corporation (1993).

Operating and Maintenance Considerations

Generally, ultraviolet unit manufacturers suggest that the lamps be replaced every 8000 to 9000 working hours (approximately one year). However, the author strongly suggests that lamp replacement be considered after six months of operation. Since the quartz sleeves may ultimately absorb some of the ultraviolet radiation from the lamps and accumulate impurities in raw water such as rouge and other fine particles, it is also suggested that quartz sleeves should be replaced each time lamps are replaced. The time required to perform suggested "manual" cleaning of quartz sleeves is probably greater than the cost of new sleeves and much more effective. O rings, providing the seal between the sanitizing chamber and the quartz sleeves, should be replaced every time the quartz sleeves are replaced. A minimum of one spare set of quartz sleeves, lamps, and O rings should be considered for replacement upon catastrophic failure associated with a system excursion such as severe water hammer.

Ballasts are contained in the electric circuitry providing power to the ultraviolet lamps and affect the overall performance of the units. A routine maintenance program for replacing the ballasts should be established. A suggested replacement frequency is once every two to three years. For many units, two lamps are powered by single ballast. Again, spare ballast should be maintained to avoid loss of ultraviolet radiation intensity upon failure of ballast.

Calibration of the ultraviolet radiation intensity meter should be performed periodically as discussed in the previous section. This includes adjustment of the "100% transmission" value each time lamps and sleeves are replaced.

During routine maintenance and calibration of the unit, the indicated optional high water temperature cutoff provision should be verified. When the water temperature reaches the high cutoff temperature set point (e.g., 110–120°F) UV unit power should automatically be inhibit.

During operation, two-phase flow should be entirely eliminated. Water hammer associated with two-phase flow will have potentially catastrophic effects on the unit, specifically failure of the quartz sleeves and, ultimately, the lamps inside the quartz sleeves. It is extremely important that operating personnel vent any air present in the system prior to ultraviolet unit operation. For systems with a storage tank and distribution pump prior to the unit, this can be accomplished by closing the feed water valve to the unit and venting air from system tubing using sample valves or other appropriate vent provisions.

Since most ultraviolet units are equipped with electronic power supplies below the sanitizing chamber, and since leaks at O-ring seals may introduce air, containing bacteria, into the recirculating loop, any observed leaks should be repaired immediately.

Unit sanitizing chambers are generally constructed of 316 or 316L stainless steel. For most applications, the chamber is mechanically polished and electropolished. If physically possible, the sanitizing chamber should be inspected using a boroscope or similar device annually to ensure that rouging or other corrosive mechanisms (chloride stress corrosion or chloride pitting attack) are not occurring within the chamber. There are other advantages of inspecting the sanitizing chamber. Foreign objects have been encountered in the sanitizing chamber because it provides a trap for material. Occasionally, when a quartz sleeve is being removed from a unit, a section of the sleeve may break and fall into the chamber. Operating personnel may not note this situation. In addition to pieces of quartz sleeves, other foreign material, such as resin beads and sections of valve diaphragms and sanitary gaskets have been observed in sanitizing chambers. If observed, the contaminants should be removed before placing the unit back into operation.

POLISHING ION REMOVAL

Rechargeable mixed bed "canisters" and are occasionally used in polishing systems primarily for Purified Water applications. Ion exchange would *not* be used as part of the storage and distribution system for Water for Injection. The chemical specifications for Purified Water include conductivity. There are several techniques that can be used to provide feed water to the upstream storage tank with a conductivity meeting the Purified Water specification. It is strongly suggested that ion removal, including polishing be performed prior to the Purified Water storage tank. RO coupled with CEDI can produce water with conductivity approaching that of the "ion-free" value of 0.055 microsiemens/cm at 25°C. Subsequently, the use of ion exchange polishing within the Purified Water distribution system should not be required and

Table 8.2 Summary of Commercially Available Standard Rechargeable Canisters

Diameter (in.)	Height[a] (in.)	Resin capacity (ft^3)	Flow Rate[b] (gpm)
6	20	0.25	0.75
8	20.5	0.45	1.0
8	34	0.79	3.0
8	48	1.2	5.0
14	50	3.6	10.0

[a]Height dimensions are overall vertical value.
[b]Flow rate based on mixed ion exchange resin.
Source: From Siemens Water Technologies Corporation (2010).

is technically unattractive, as indicated, considering microbial control and system sanitization requirements.

Theory and Application

Occasionally, laboratory systems, in addition to requiring Purified Water at points of use, must meet additional specifications for water such as those established by the American Chemical Society (ACS), the National Committee of Clinical Laboratory Standards (NCCLS), and, most importantly, the American Society for Testing and Materials (ASTM, 2006). The required conductivity/resistivity *may* be more restrictive than the Purified Water requirements. The use of top inlet and bottom outlet rechargeable canisters with controlled mixed bed resin, as discussed in chapter 4, *may* be considered for these applications.

Design Considerations

As discussed, while not recommended, standard rechargeable mixed bed canisters may be used in polishing applications. Table 8.2 summarizes commercially available "standard" canisters with diameter, approximate height, and mixed ion exchange resin volume. Figure 8.8 demonstrates the flow path and distribution arrangement for standard rechargeable canisters. The flow characteristics through the canisters, as discussed in chapter 4, are limited by the bottom distributor connected to a "riser" tube from the top-mounted inlet/outlet fitting. It is impossible to provide an appropriate distribution system, ensuring a complete "plug flow" situation through the rechargeable canister, with a top inlet and outlet arrangement. It is highly unlikely that uniform velocity through the ion exchange resin in the canister can be achieved with a single distributor. This fact is verified by the observed ion exchange capacity (actual vs. theoretical) of the resin. This indicates that channeling is occurring and that the entire volume of resin within the canister is not effectively exposed to water. From an operating and maintenance standpoint, it is obvious that the design of rechargeable canisters is dictated by the requirement for "mobility" for the physical movement, shipment, and transfer of the canisters, as opposed to the technical attributes necessary to promote good ion exchange characteristics *and* bacterial control.

The flow properties associated with the standard canisters, presented in Table 9.2, will generally require the use of multiple canisters operating in parallel. Furthermore, most "service deionization organizations" will attempt to compensate for limitations associated with flow characteristics within the units by placing two units in series. As a result, if a specific polishing application requires five units operating in parallel to support the feed water flow (e.g., 50–70 gpm), the rechargeable canister system will consist of 10 canisters, arranged in five "banks" of 2 canisters each in series, to support the flow rate. For this specific application, with a flow rate of 50 gpm, the 10 standard canisters would contain about 35 ft^3 of mixed bed ion exchange resin. This same application could be supported by a single, custom-fabricated ion exchange vessel containing about 4 to 5 ft^3 of mixed ion exchange resin with a top inlet and bottom outlet connection. In other words, about one-tenth of the resin volume for "standard design canisters" is required to achieve the same projected product water quality using a single, custom-fabricated polishing unit. In fairness, it should be indicated that most service deionization organizations are simply not equipped to "handle" custom canisters. Due to the

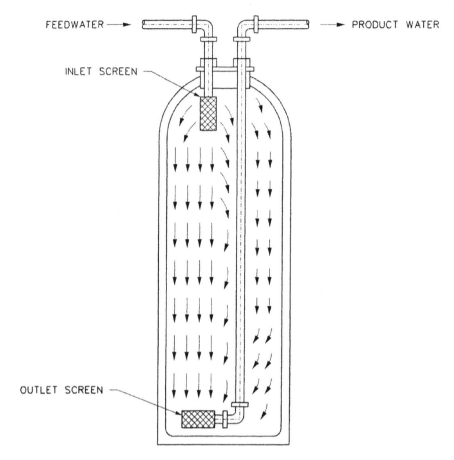

Figure 8.8 Flow diagram of a "standard" rechargeable canister. [The length of the arrows indicates the relative velocity (flow rate) on the basis of distributor arrangement.]

nature of the regeneration equipment required, regenerant waste volume treatment, and required discharge waste permits, a significant number of service deionization organizations have limited competition and, subsequently, limited incentive to consider the use of custom-fabricated, harder-to-handle, rechargeable canisters.

Another factor that limits the flow rate through standard rechargeable canisters is the physical size of the connections on the top of the unit. In fact, when the feed water opening is inspected, it is about half the diameter of the feed water connection (tubing/pipe size). Perhaps this size reduction creates an orifice to limit the flow rate in accordance with the flow requirements presented in Table 9.2.

An issue that must be considered for any polishing application is bacterial control. The nature of the flow characteristics through standard polishing mixed resin canisters encourages the proliferation of bacteria. Obviously, it is undesirable to have "stagnant" areas in the polishing portion of a system (i.e., within the storage and recirculation loop). The flow pattern through a standard unit, as exhibited in Figure 8.8, clearly demonstrates anticipated stagnant areas within the resin bed during routine operation.

Another issue associated with rechargeable canisters relates to the method of providing feed water and collecting product water from the units. To facilitate the installation and removal of the canisters, connections are generally of the "quick disconnect" type. Many of the quick disconnect fittings will use "hose barb"–type fittings with "worm-type" clamps, which provides an excellent area for microbial proliferation. This condition has prompted written citations from regulatory officials. Other quick disconnect–type hose connections employ flat gasket seal mechanisms that also provide a location for bacterial growth. The hoses should be periodically sanitized (chemically). Hose materials of construction should not introduce any materials into the water, since, at this point in the system, the material introduced would be considered as an Added Substance.

Flow balancing through multiple polishing canisters operating in parallel must also be addressed. The pressure drop through each canister will not be exactly the same. To maximize the efficiency of ion exchange using this standard canister arrangement, flow rate indicating meters with regulating valves (e.g., diaphragm or needle valves) must be positioned in the feed water to canisters operating in parallel. In addition, while a common feed water pressure gauge and sample valve may be provided to demonstrate proper polishing system performance, in accordance with good pharmaceutical system design, pressure gauges and sample valves should be positioned in the product water line from each of the "lead" and "trailing" rechargeable mixed bed canisters. The ion exchange capacity of resin within a specific canister, considering the nature of the flow characteristics within the canisters, cannot be accurately predicted. Conductivity monitoring at the effluent of both the lead and trailing canisters should be considered. All of the indicated accessories should be of sanitary type to minimize/eliminate a location for bacteria to accumulate and replicate. As indicated, as an alternative, the 50-gpm polishing flow rate could be supported using a single "custom" canister with about 4 to 5 ft^3 of ion exchange resin. This alternative system would require a single feed water pressure gauge, feed water sample valve, product water pressure gauge, product water sample valve, and product water conductivity cell.

If "standard" rechargeable canisters are used, as discussed above, specified. If a standard type of hose is supplied by the rechargeable service deionization organization, the materials of construction must be evaluated to determine potential organic elution characteristics.

Figure 4.26 depicts a custom-fabricated, rechargeable, mixed bed deionization unit specifically designed for polishing applications. The diameter of the unit is selected to maintain a face velocity in the range of 20 to 30 gpm/ft^2 of cross-sectional bed area. The suggested bed depth (straight side height of the unit) is 36 in. The canister contains top-mounted feed water and bottom-mounted product water connections, with domed top and bottom. Flat plates containing appropriately positioned stainless steel distributors are included to create a fixed bed (chap. 4). Units may be fabricated of 316L stainless steel. The design of the unit can be such that resin is removed through a top port (and top plate fixing the resin bed) or on the lower straight side of the unit. Positioning of the resin removal port should be carefully considered to minimize potential dead legs with related bacterial growth. Historically, similar units have been commercially available as "standard products" (Vaponics). Highly unique operating characteristics have been observed. While the units are designed for operation at the 20 to 30 gpm/ft^2 of cross-sectional bed area face velocity for polishing applications, they demonstrate an ability to provide excellent removal of ionic material. As a result, it is *not* necessary to use two polishing canisters in series. Since the ion removal characteristics for the custom units are significantly better than for "standard" canisters, it may be appropriate to consider the use of "virgin" resin for recharging the units as discussed in chapter 4. The unit can be equipped with sanitary ferrule feed water and product water connections. Further, the system can be designed to eliminate the requirement for hoses, assuming that the feed water and product water tubing connections are also provided with sanitary triclamp connections. As previously indicated, bacterial proliferation within the units is minimal. For polishing applications, units may be recharged based on bacterial control considerations rather than exhaustion of the mixed bed resin. Experience indicates that units, similar to that shown in Figure 8.8, may operate for one to three months without exhibiting *any* increase in bacterial levels through the units. In fact, during initial operation, a decrease in total viable bacteria levels may be observed. The reduction of bacteria by ion exchange resin, for similar applications, has been documented (Collentro, 1997). Obviously, there are significant advantages associated with using this arrangement when compared with classical fiber-glass-reinforced polyester or vinyl ester top inlet and top outlet "standard" polishing rechargeable canisters.

If rechargeable mixed bed polishers are used in a Purified Water storage and distribution application, provisions should be included for periodic sanitization. This is difficult to execute. Mixed bed resin cannot be exposed to sanitizing temperature water (>80°C).

As a result of the inability to predict the capacity of "standard" polishing mixed resin canisters accurately, the service deionization organization, or the pharmaceutical operation using the canisters, will frequently store "spare canisters" at a facility to avoid "shutdown." Stagnant canisters may be located in a relatively warm facility area, in a stagnant condition, for a period of days or weeks. When placed online, bacteria levels in the product water will initially be "too numerous to count" and will remain at a high value throughout the operating cycle.

As suggested above, a properly validated Purified Water system requires appropriate monitoring of feed water and product water for the individual operations. Monitored parameters are specific to the nature of the unit operation, but they should always include total viable bacteria. Data generated from this monitoring program are plotted as a function of time, used to establish required maintenance for a specific unit operation. This ensures that chemical and bacterial levels at individual points of use from the Purified Water distribution system are not adversely affected. This is of particular concern for polishing rechargeable mixed bed units since they are physically "close" to the individual points of use, separated only by downstream membrane filtration or ultrafiltration.

All polishing canisters installed in distribution loops should be provided with adequate venting provisions to displace atmospheric air after installation. The inability to vent atmospheric air subsequent to canister installation defeats the purpose of the hydrophobic vent filtration system (installed on the upstream storage tank).

A final design consideration that should be considered is the inclusion of a preoperational "rinse-to-drain" system. Complete conversion of anion resin to the hydroxide form and cation resin to the hydronium form during the offsite regeneration process cannot be assumed. Further, product water TOC may also be a concern as discussed in chapter 4. Subsequently, a preoperational rinse-to-drain operation must be conducted for the polishing system before it is placed in "service." This can be accomplished "external" to the polishing system, but requires a source of water with a quality (chemical and microbial) similar to that of the feed water in the Purified Water system. Obviously, using a point of use to provide feed water to the canisters is highly undesirable, since microbial back contamination from the carbonaceous ion exchange bed to the recirculating Purified Water loop will occur.

Operating and Maintenance Considerations
In general, polishing mixed resin rechargeable canisters should be replaced when the total viable bacteria increase through the units exceeds a preestablished value. It is highly likely that total resin bed exhaustion will *not* occur when replacement is required to control system bacteria levels.

The selection of the distributors for rechargeable canister should consider the potential production of resin "fines." The resin used in the polishing mixed bed units should contain a minimal amount of resin fines. One of the advantages of using "virgin" resin, as discussed previously, relates to the fact that the new resin will, depending upon selection, contain less resin fines than resin that has been regenerated a number of times.

It is important to establish a method for maintaining "control" of resin within rechargeable canisters. It is extremely important to remember that the nature of a "regulated" pharmaceutical system dictates that the "owner" maintain control of *all* unit operations within the system. There is no guaranteed method of maintaining control of extremely small resin beads contained within rechargeable canisters that are physically transferred, regenerated, and returned to a facility. Rechargeable canisters may be regenerated by one of two techniques.

- Remove the resin, blending it with resin used for "other" applications. Separate the anion resin from the cation resin, regenerate the separated resins, remix the resins, and provide regenerated resin in randomly selected canisters for return to the pharmaceutical facility.
- Provide regeneration within dedicated canisters, clearly labeled with the pharmaceutical company's name, thus ensuring that the pharmaceutical client's resin and canisters are not mixed with resin and canisters from *any* other company.

Obviously, the first technique indicated above is extremely undesirable since it represents "loss of control." While the second technique indicated above provides the required control, there are some technical disadvantages. In-canister regeneration is associated with minimum ability to provide bed expansion. This may result in poorer contact between bacteria-destroying regenerant acid and caustic, potentially resulting in a regenerated canister with higher bacterial levels than canisters where resin has been separated, regenerated, and remixed. In summary, however, the ability to control in-canister regeneration, by ensuring that a pharmaceutical facility will have dedicated canisters and resin, far outweighs the technical issues associated with bacterial destruction during regeneration.

When in-canister regeneration is employed, two "sets" of rechargeable canisters are required. The dedicated canisters should clearly contain the name of the pharmaceutical company, as well as a unique model and serial number.

The resin regeneration facility should be inspected by pharmaceutical facility personnel at least once a year, or more frequently if problems are observed. This inspection should be performed by a qualified individual or a consultant to the pharmaceutical facility. The inspection should document the procedures maintained by the regeneration organization. It is possible that regulatory authorities will request specific information associated with the regeneration service organization's ability to meet critical parameters. Subsequently, SOPs, quality assurance or quality control program, and service "log" for regeneration should be prepared directly by the service organization and transferred to the pharmaceutical organization. It may be necessary to supplement the procedures from the service organization with specific documents to demonstrate a controlled condition.

Hoses used for rechargeable canisters should be periodically sanitized. The hoses generally used for conventional, polishing, rechargeable, mixed bed canisters will not withstand hot water sanitization temperatures ($\geq 80°C$). Subsequently, chemical sanitization is required. It is important to remember that any sanitization agent used for the hoses must be completely removed prior to use. Tests must be established to indicate that residual disinfectant is not present since the sanitizing agent would be an Added Substance as defined in the *Official Monograph* for USP Purified Water.

Periodic calibration of conductivity meters and cells is required. The calibration procedure should be performed in accordance set forth in the USP *Physical Tests* section for conductivity determination.

FINAL MEMBRANE FILTRATION

Membrane filtration is employed in many pharmaceutical water systems. Its use is not restricted to just polishing water filtration applications. For example, many systems Purified Water systems, Pure Steam generator feed water systems, and multiple-effect distillation unit feed water systems consist of single-pass RO with polishing CEDI. In this design, membrane filtration is often installed downstream of the CEDI system. While not suggested, the use of final membrane filtration in Purified Water distribution loops is still employed even for systems with stainless steel storage and distribution systems that can be periodically hot water sanitized. Systems using chemical sanitization rely on membrane filtration to "extend" the time period between sanitization cycles. Purified Water systems employing ozone, discussed in chapter 7, do not require membrane filtration for microbial control in distribution loops. As a unit operation, the selection of membrane filtration must be carefully evaluated. The "sterile" nature of the product water, implied by the 0.2-µm rating "assigned" to a particular membrane filter, can result in misapplication of the technology. As a technique for removing bacteria, ultrafiltration, discussed later in this chapter, has unique advantages for specific application when compared with membrane filtration.

The use of membrane filtration, with the exception of terminal "sterile fill process applications" (non–water purification system related) is generally limited to Purified Water systems. Hydrophilic membrane filtration *cannot* be used in Water for Injection *storage and distribution systems*. Point-of-use membrane filtration, while acceptable for limited applications where bacteria are of little concern, is not acceptable for use in Purified Water systems. When these units are employed in Purified Water distribution systems, while providing desired water quality (chemical and bacterial levels) at a specific point of use, they back contaminate the remainder of the distribution system. Depending on the application, systems using point-of-use membrane filters prompt regulatory action, specifically where bacterial control is desired at other points of use in the system. Obviously, this is associated with the fact that any point-of-use membrane filter will introduce bacteria into a recirculating loop, creating a situation where bacteria levels at other points of use are adversely affected, resulting in the need for frequent sanitization. Since it is impossible to predict the nature and rate of bacterial contamination, point-of-use membrane filtration (or membrane filtration coupled with other water purification polishing applications) presents an "out-of-control" scenario, which is inconsistent with the nature of a qualified or validated system. Attempting to achieve microbial control at point of use provides a single unit operation. Subsequently, membrane failure results in a complete loss of microbial control.

The scope of membrane filtration technology is significant when process filtration is considered. The material presented in this chapter is limited to basic applications, primarily for "bulk" Purified Water.

Theory and Application

Membrane filtration is employed as a method of eliminating and controlling bacteria in many USP Purified Water systems. Membrane filtration employed for microbial control in Purified Water distribution system can result in significant system bacteria excursions unless properly designed, installed, monitored, and maintained. Membrane filtration should not be used within a storage and distribution system (or at any other point downstream of a distillation unit or RO unit) in a Water for Injection application.

It is strongly suggested that membrane filters for microbial control in Purified Water systems be selected with a 0.1-μm rating, in lieu of 0.2 or 0.45 μm. It is also strongly recommended that the characteristics of the particular membrane filter employed be evaluated. The "micron rating" for a membrane filter will vary with manufacturer and even with membrane filter "lots" from a filter supplier. While the majority of Purified Water system that use membrane filtration have a 0.2-μm rating, experience indicates that a properly selected 0.1-μm rating will exhibit superior bacteria removal. It is extremely appropriate to discuss the "0.2-μm" designation of these "sterilizing" filters, since it is relates to the limitations of membrane filtration as a final unit operation (within a polishing system) for pharmaceutical applications.

The criteria used to establish the 0.2-μm rating evolved from the "Health Industry Manufacturers Association" (HIMA) bacteria challenge test procedure in 1982. The HIMA requirements were adapted to by ASTM to the current ASTM Standard F 838-05 (ASTM, 2005). The ASTM "Standard Test Method for Determining Bacterial Retention of Membrane Filters Used for Liquid Filtration" is to "challenge" the membrane with a suspension of *Brevundimonas diminuta*, American Type Culture Collection (ATCC)-19146 at a concentration of 10^7 organisms/ cm^2 of effective filtration area at a maximum differential pressure across the test filter of 30 pounds per square inch gauge (psig) and a flow rate of 0.5 to 1.0 gpm/ft^2 over the effective area. If the filtrate from the membrane filter, tested in accordance with the above criteria, indicates the absence of bacteria, the membrane filter is *assigned* a 0.2-μm rating. Obviously, individuals using 0.2-μm membrane filters must be aware of this test method *and* its relationship to a specific application. Since the actual "membrane" used for testing contains bacteria, membrane material produced in the same manner as the tested material is also assigned the 0.2-μm rating.

The mechanism for the removal of bacteria through a 0.2-μm membrane filter has been projected as either direct interception or inertial impaction (Osumi et al., 1991). There are numerous "pores" in the membrane filter material. When a pore, a convoluted path through the membrane filter, exhibits a diameter, within the changing diameter of the "tunnel," that is small enough to retain bacteria, direct interception is applicable. Inertial impaction is associated with the removal of bacteria on "tunnel walls," which is associated with changes in flow direction of water through a pore. From a practical standpoint, it is important to remember that the 0.2-μm rating does *not* imply that the pore size is <0.2 μm; it merely indicates that the membrane filter has passed the stated ASTM bacterial challenge. The literature contains numerous articles discussing the "grow-through" of bacteria (Wallhauser, 1983; Howard and Duberstein, 1980; Christian and Meltzer, 1986).

It appears inappropriate to engage in a detailed discussion expressing the postulated mechanisms for bacteria to pass through a 0.2-µm filter. However, it is important to emphasize the fact that commonly encountered species of bacteria in a pharmaceutical water purification system, such as *Pseudomonas aeruginosa*, *Ralstonia pickettii*, and *Pseudomonas cepacia*, particularly when exposed to low nutrient environments (conductivity <0.1 microsiemens/cm and TOC values <50 mg/L), will be capable of "passing" through a 0.2-µm filter. The ability of bacteria to adapt to a low nutrient environment, by maximizing surface-to-volume ratio (Gould, 1993), not only effects the ability to measure the number of colony forming units per unit volume (Colwell and Hug, 1995) but is also a very important factor to consider when selecting a final filtration membrane. To demonstrate this, one team of researchers exposed *P. aeruginosa* to a sublethal dose of penicillin, producing "L-forms" of bacteria, physically describing the shape of bacteria during microbial counting. It is suggested that these L-forms of bacteria represent a transition state of the rod-shaped *P. aeruginosa* to a coxcidal (spherical) form (Thomas et al., 1991). Thomas et al. indicate that 0.2-µm filters are incapable of removing the L-forms of *P. aeruginosa*, while 0.1-µm filters are more effective in removing the bacteria itself. This supports the results of earlier studies conducted in the mid-1980s that employed epifluorescent microscopy for the detection bacteria (direct counting; Collentro, 1989). It would further support the highly effective ability of ultrafiltration to remove bacteria, since the stated pore size for ultrafiltration is a "real" value rather than an "assigned" value.

As indicated earlier, membrane material selection and membrane manufacturer are critical to the assigned micron rating as it relates to the actual ability to remove bacteria present in Purified Water systems. It is suggested that there are several critical membrane filter performance characteristics that can be used to project the ability to remove L-forms of bacteria and organisms such as *R. pickettii* that may be found in 0.2-µm filter product water. Smaller pore size should be associated with lower flow rate through a membrane filter with the same surface area as one with larger pore size. Some 0.1-µm membrane filter manufacturers will employ a conventional "bubble point" test to determine the pore size relative to a 0.2-µm filter. If the bubble point pressure for the 0.1-µm filter is twice that of a 0.2-µm filter this may be used as one method of projecting the 0.1-µm rating. However, a second method should also be employed. This would, as an example, consist of successful ASTM bacteria challenge testing with both *B. diminuta* and *R. pickettii*.

While there are numerous opinions regarding the extent of bacterial penetration through 0.2-µm membrane filters, it appears that the initial penetration, within one to three days after the installation of a "new" filter, is about 40% of the membrane depth (Brock, 1983). While it is quite possible that additional penetration may take much longer to occur, it should be emphasized that the extent of penetration is a direct function of the species of bacteria present, the pore characteristics (and "manufacturing" techniques) of the membrane filter, and, perhaps most importantly, the purity of the water—effecting the size of the bacteria (extent of "shrinking" as they maximize the surface-to-volume ratio).

It would be inappropriate to present a discussion associated with membrane filtration and its relationship to both the species and size of bacteria present without referring to specific comments made by highly qualified individuals. Selected comments are presented as follows:

- "In your initial testing of your system, your start-ups, your initial phases of validation, you may want to try a R2A media; you may want to count your plates at 48 hours and then at 5 days too. Does it make a significant difference? At 5 days, do I get different organisms coming up then I did at 48 hours? ... You need to know the impact of those findings on your products" (Munson, 1993a).
- "For Purified Water Systems, in addition to the total microbial count, the specific organisms present must be known and their effect on the product assessed. You must use microbial test methods that are designed to detect microorganisms in chemically purified water and are capable of recovering sub lethally injured organisms if UV disinfection is used" (Munson, 1985).

- "Bacteria invoke a number of survival mechanisms in the transition from feast to famine. They minimize cell size in order to maximize surface-to-volume ratio. By becoming small round cells or very thin long cells, they minimize the amount of cellular material they need to synthesize, while maximizing their contact with scarce nutrient. Many bacteria adapted to poor nutrient conditions pass easily through 0.22 micron sterilizing filters" (Gould, 1993).
- "It is, indeed, reasonable to expect that the tighter filters will more likely restrain the passage of smaller organisms, both by sieve retention and by adsorption. Thus, the use of tighter filters may well be appropriate where smaller size organisms are the concern as indicated by experimental investigations" (Meltzer and Jornitz, 2006).
- "Because bacteria in the viable but non-cultural state are reduced in size and frequently in the coccoid form, they will pass through 0.45 micron and, in some cases, 0.2 micron filters. Furthermore, under very low nutrient conditions, the viable but non-culturable state may be predominant. Preparation of water for pharmaceutical and biotechnology applications requires consideration of the presence of bacteria in a non-culturable, dormant stage, especially for preparation of solutions of injectables or topical reagents" (Colwell and Hug, 1995).

The intent of citing these quotations is to emphasize the importance of determining if membrane filtration (with a 0.1- or 0.2-μm "designation") is appropriate for a given application, or if a "tighter" membrane, such as ultrafiltration, should be considered.

When evaluating the use of membrane filtration for pharmaceutical applications, the microbial quality of the water required for the specific application is extremely critical. The nature of the product, as well as the established bacteria Alert and Action Limits for a given application as well as the requirements for the absence of certain pathogenic species of bacteria, must be considered. While it is suggested that total bacteria levels (vs. total viable bacteria levels) are more important for semiconductor applications, it should be noted that the ASTM "type E-1" semiconductor water specifications include a 10 cfu/1000 mL limit for "small line-width" applications (ASTM, 2007). The same specifications, demonstrated in Table 8.3, specify an extremely low TOC level and ion-free water, which is an obvious low

Table 8.3 Electronics and Semiconductor Industries—Type E-1 Water Requirements

Parameter	Specification
Resistivity at 25°C (online)	18.1 MΩ-cm
TOC	5 μg/L
Dissolved oxygen (online)	25 μg/L
Residue after evaporation (online)	1 μg/L
Particles/L (0.1–0.2 μm range)	1000
Particles/L (0.2–0.5 μm range)	500
Particles/L (0.5–1.0 μm range)	200
Particles/L (1.0 μm range)	<100
SEM particles/L (0.1–0.2 μm range)	1000
SEM particles/L (0.2–0.5 μm range)	500
SEM particles/L (0.5–1.0 μm range)	100
SEM particles/L (1.0 μm range)	<50
Bacteria	5 cfu/100 mL
Total silica	5 μg/L
Dissolved silica	3 μg/L
Anions and ammonium by IC: ammonium, bromide, chloride, fluoride, nitrate, nitrite, phosphate, sulfate	0.1 μg/L
Metals by ICP/MS: aluminum, barium, calcium, chromium, copper, iron, lead, lithium, magnesium, manganese, nickel, potassium, sodium, strontium, zinc	0.05 μg/L

Notes:
- For applications where "line width" = 1.0–5.0 μm.
- Specifications are guidelines.
- Types E-1.1 and E-1.2 specifications are *more* restrictive.
- Boron may be monitored as an "operational parameter" for ion exchange beds.

Source: From ASTM (2007).

nutrient environment. For effective microbial control, which recognizes the potential reduction in the size of bacteria, state-of-the-art semiconductor facilities exclusively use final ultrafiltration in lieu of membrane filtration. If the product water from a pharmaceutical water purification system, specifically USP Purified Water systems, exhibits the chemical characteristics of Semiconductor Grade Water *and* if bacterial control is desired, it is suggested that ultrafiltration or ozonation be considered in lieu of membrane filtration.

For applications where bacterial Alert and Action Limits are low due to the nature and use of the final product (e.g., topicals, ophthalmics, antacids, inhalants, etc.) *and* the chemical characteristics of the USP Purified Water provides adequate nutrients (organic and inorganic) for bacteria to exist in a "normal" state, as opposed to an "adapted" state in a low nutrient environment, it is suggested that 0.1-µm membrane filters be considered in lieu of 0.2-µm membrane filters. Experience indicates that a properly selected 0.1-µm filter has a 10 to 100 times greater chance of removing bacteria than a "standard" 0.2-µm membrane filter (Collentro, personal data). This is particularly true after the membrane filters have been exposed to bacteria present in the feed water and have been "in service" for a period of time long enough to establish typical "grow-through" characteristics of membrane filtration.

Membrane filters used for pharmaceutical applications should be provided with a "Validation Guide." This Guide should contain certified information that clearly indicates the ability of the membrane to pass the ASTM microbial challenge discussed earlier. Additional documentation clearly demonstrating that materials used in constructing the membrane filter have passed "cytotoxicity testing" (NAMSA™ 1997) must also be provided. An example of a typical document associated with cytotoxicity testing is shown in Figure 8.9. Additional material in the Validation Guide should address requirements set forth in USP *Biological Tests* section <88> for "Biological Reactivity Tests, In Vivo" for class VI plastics, Fiber Release per 21CFR 210.3(b), Indirect Food Additive per 21CFR177-182 and ISO 10993-17 "Biological Evaluation of Medical Devices—Part 17;" "Methods for the Establishment of Allowable Limits for Leachable Substances" with reference to ISO 10993-17, "Identification and Quantification of Degradation Products from Polymer Medical Devices."

Design Considerations
Selecting the final filtration system includes not only selecting the appropriate membrane filter but also the membrane filter housing. For Purified Water systems where microbial control is desired, it is suggested that the final filter housing be of sanitary type, constructed of 316L stainless steel, and mechanically polished. Electropolishing is generally desired.

It is suggested that a "T-style" housing has technical advantages when compared with other housing arrangements. This particular arrangement, with base-mounted inlet and outlet connections, introduces the water in a flow direction parallel to the membrane filter surface. This flow path minimizes any shear forces on the membrane surface and, over the operating life of the membrane filter, potentially results in slower penetration of bacteria through the pores of the membrane filters when compared with a housing that provides a "dispersed" flow in a direction perpendicular to the membrane filter surface. An additional advantage of a T-style housing is that the inlet and outlet fittings (containing a 90° elbow) are physically in the same horizontal plane, which facilitates installation.

A positive sealing mechanism is desired between the membrane filters and the filter housing. A suggested filter configuration is single open ended, with "226" double O rings "bomb-finned" top connection, and "Code 7" bottom connection. The Code 7 configuration provides locking tabs on the membrane filter. The locking tabs fit in grooves within the filter housing. When positioned into the housing within these grooves, the membrane filters are rotated and "locked" into position, thus ensuring a positive seal. The bomb-finned top of the membrane filter is generally positioned in a "guide plate." This plate does *not* contribute to the sealing mechanism associated with the Code 7 arrangement and the double O rings at the base of the membrane filter housing, but ensure that there is minimal horizontal movement of the membrane filter element(s) during operation. This is particularly important when 20-, 30-, or 40-in. membrane filters are used. An example of the sealing mechanism and membrane filter arrangement is shown in Figure 8.10.

Test Article: Clariflow™ Cartridge
Test Article Size Used: 4 g
Experimental Procedure:

    A monolayer of L-929 cells was grown to confluency and exposed to an extract of the test article prepared by placing the test article in 20 mL of 5 percent Minimum Essential Medium and extracting at 37°C for 24 h. Duplicate MEM aliquots were used as negative controls. The positive control was extracted at 37°C for 24 h and tested using an end-point titration procedure. After exposure to the extracts, the cells were examine microscopically for cytotoxic effect (CTE). The presence (+) or absence (−) of a confluent monolayer, vacuolization, cellular swelling and crenation, and the percentage of cellular lysis were recorded.

CTE Score	Microscopic Appearance of Cells
Nontoxic (N)	A uniform, confluent monolayer, with primarily elongated cells, and discrete intracytoplasmic granules present at 24 h. At 48 h and 72 h, there should be an increasing number of rounded cells as the cell population increases and crowding begins. Slight or no vacuolization, crenation, or swelling should be present.
Intermediate (I)	Cells may show marked vacuolization, crenation, or swelling. Cytolysis (0–50 percent) of cells that results in floating cells and debris in the medium may be present. The remaining cells are still attached to the flask surface.
Toxic (T)	Greater than 50 percent of all cells have been lysed. Extensive vacuolization, swelling, or crenation is usually present in the cells remaining on the flask surface.

Results

	Confluent Monolayer	Vacuolization	Swelling	Crenation	% Lysis	CTE Score
24 h	+	−	−	−	0	N
48 h	+	−	−	−	0	N
72 h	+	−	−	−	0	N

Extract Conditions: Test: Clear

                                     Controls: Clear

The positive control, SCG #2, was toxic at a dilution of 1:16 at 24 h.

The negative controls were acceptable (nontoxic) under these extraction conditions.

The results and conclusions apply only to the test article tested. No further evaluation of these results is made by NAMSA™. Any extrapolation of these data to other samples is the responsibility of the sponsor.

Conclusion:	Nontoxic
Date Prepared:	9-16-94
Date Terminated:	9-20-94
Record Storage:	All raw data pertaining to this study and a copy of the final report are to be retained in **designated NAMSA archive files.**
Test Facility:	North American Science Associates, Inc., California Division

Figure 8.9 Cytotoxicity—MEM elution—MG023.

    The membrane filter housing should be provided with a top-mounted sanitary ferrule fitting (suggested at 1.5 in.). This fitting can be used for nonintrusive integrity testing of the membrane filter elements, discussed below.

    The final filtration system should be provided with appropriate accessories, such as sanitary diaphragm-type isolation valves, sanitary sample valves, sanitary pressure gauges, a

POLISHING COMPONENTS 303

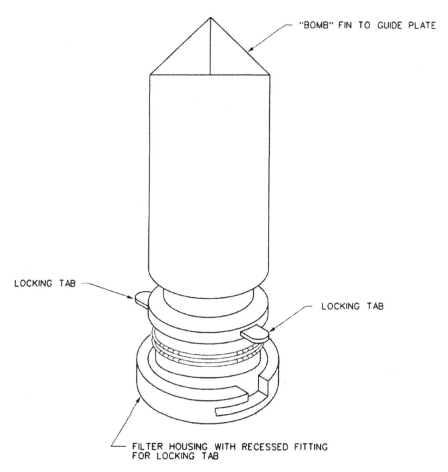

Figure 8.10 Code 7 filter housing to membrane filter sealing arrangement. (The bomb fin is a "guide" only. Double O rings form the seal from the membrane filter to the housing. The sealing lock is achieved by rotating the membrane filter after positioning the locking tabs.)

sanitary vent valve at the top of the filter housing, and an integrity test fitting positioned on top of the filter housing.

All filter housings used for validated Purified Water systems should be provided with a Validation Guide. Operating conditions should be in accordance with data provided in the Guide.

The flow rating for membrane filters (per 10-in. equivalent) will vary with the manufacturer and the nature of the filter membrane material matrix. However, for membrane filters used for pharmaceutical applications, it is suggested that 0.2-µm filters be rated at a flow rate of 3 to 5 gpm/10 in. equivalent, and 0.1-µm membrane filters be rated at a flow rate of 2 to 3 gpm/10 in. equivalent. Most membrane filter manufacturers and, more importantly, local representatives "stock" 10-, 20-, and 30-in. membrane filters. While filter housings using 40-in. long membrane filters are available, 40-in. membrane filters are less accessible.

In selecting a membrane filter housing it is desirable to minimize the diameter and increase the height. In other words, a membrane filter housing using six 10-in. membrane filters will not only be more expensive (from a capital cost standpoint) than a filter housing containing three 20-in. filter elements, it also doubles the number of membrane filter-to-filter housing seal mechanisms required.

As suggested earlier, a Code 7 arrangement with single open-ended membrane filters will be provided with a guide plate. The guide plate should *not* be secured/tightened in a manner that exerts a force on the individual membrane filters in either a vertical or horizontal direction since this force *may* interfere with the ability to provide a proper seal with the double O rings positioned at the base of the membrane to the filter housing.

General experience with different types of membrane filters provided by different membrane manufacturers indicates significant variation in performance. It is highly suggested that membrane selection for a particular application consider all potential sources of membrane filters. It is quite possible that a membrane filter with a cost double that of another membrane filter with a similar capability may operate for a period of three to four times that of the less expensive filter, based on product water bacteria levels. Furthermore, the materials of construction of the membrane filter should be consistent with the application. Certain membrane materials, while providing excellent bacterial removal capability, may be incapable of withstanding hot water sanitizing temperatures or exhibit limited capability for thermal cycling. Certain final filter membrane materials can withstand 20 to 40 hot water sanitization cycles without adversely affecting the performance (integrity) of the membrane filter.

Experience indicates that a well designed Purified Water system would employ a storage system and periodically sanitized distribution system of 316L stainless steel sanitary design, with reverse osmosis within the upstream water purification system of the storage tank. If 0.1-μm membrane filters are installed downstream of the Purified Water distribution pump after the tank, the product water should exhibit total viable bacteria levels in the range of <1 to 5 cfu/100 mL when cultured with R2A media using Membrane Filtration of a 100-mL sample, incubated for 120 hours at 30°C to 35°C. Furthermore, the product water should be free of pathogenic species of bacteria in the 100-mL sample.

As indicated earlier, membrane filtration for Purified Water systems, has extremely limited "point-of-use" application. It is *strongly* suggested that every attempt be made to eliminate the direct installation of membrane filtration at individual points of use, since the "dirty" side of the membrane (from a microbial standpoint) will release bacteria into the recirculating Purified Water loop, resulting in the formation of a biofilm, contamination of other points of use, and frequent storage and distribution loop sanitization. If a specific point of use requires membrane filtration (e.g., generally laboratory applications), or if membrane filtration in combination with other polishing techniques is required for a specific point of use, it is *strongly* suggested that a small tank and pump be considered for the application, providing an air break and eliminating microbial back contamination of the recirculating Purified Water distribution loop.

Operating and Maintenance Considerations

Sanitary pressure gauges should be installed in the feed water and product water tubing (to and from) the membrane filter housing. It would be inappropriate to suggest that membrane filter replacement will be determined only by a noticeable increase in the differential pressure through the membrane filtration system. In fact, the noted pressure drop during normal operation is associated with change in water flow path through the filter housing, not the membrane filters. Unless *significant* membrane filter bacteria loading has occurred, it is doubtful that a noticeable increase in the pressure drop will be observed over the life cycle of the membrane filters. It is equally important to point out that the differential pressure drop values should *not* be used as a criterion for replacing the membrane filters. Replacement frequency should be established by periodic sampling of the product water from the membrane filtration system *or* periodic nonintrusive diffusive flow integrity testing of the filtration system.

At a minimum, membrane filters should be replaced every six months. More frequent replacement may be determined by trending the operating data, principally effluent total viable bacterial levels as a function of operating time.

Periodic nonintrusive integrity testing should be considered for membrane filtration system. The technique suggested for integrity testing is diffusive flow, executed at a pressure below the bubble point for the membrane filter elements. The configuration of components required to perform nonintrusive integrity testing is shown in Figure 8.11. To perform this test accurately, an automated test unit should be employed. The test requires isolation of the filter housing by closing the feed water valve, draining the "upstream" side of the filter housing

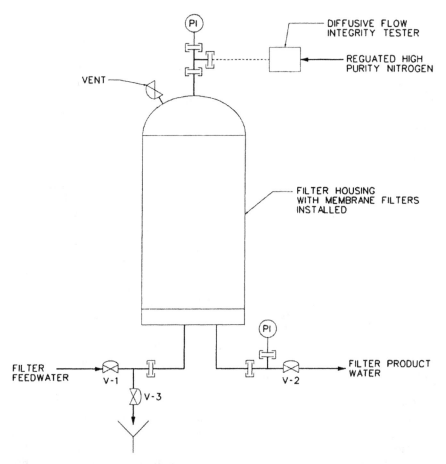

Figure 8.11 Membrane filter diffusive flow integrity test configuration. [The filter housing must be drained (feed water side) prior to the test. During the test, valves V-l and V-3 are closed. Valve V-2 is open. The filter housing "void" volume and the tubing volume downstream of valve V-l must be known. The "diffusive" nitrogen pressure should be below the "bubble point" valve, which is provided by the membrane manufacturer for the specific membrane employed.]

assembly, determining the upstream volume, establishing a high-purity nitrogen pressure on the upstream side of the membrane filters at a value below the bubble point, and determining (using the computerized tester) the amount of nitrogen passing through the membranes at a specific manufacturer recommended pressure. Generally, this is accomplished by measuring the nitrogen pressure drop over a relatively short period of time (5–15 minutes). The test period should not be too long, since the membrane filters may "dry out," thus allowing a greater amount of nitrogen to pass through the pores at the diffusive flow test pressure, yielding incorrect data. While this particular method may sound very complicated, it is very easy to execute if the proper accessories have been included in the design of the system. This technique is also extremely relevant for pharmaceutical applications because it is nonintrusive and does not require the opening of any valve positioned downstream of the filter assembly to the atmosphere. It is strongly recommended for Purified water system using membrane filtration in the distribution loop.

The frequency of integrity testing for membrane filters should be based on the desired product water bacteria levels, the feed water bacteria levels, the system bacteria Alert and Action Levels, and the frequency of system sanitization. It is suggested that the membrane filters be integrity tested at least once every two to four weeks, or after each hot water sanitization operation.

For Purified Water system distribution loop applications periodic membrane filter replacement should be performed. The maximum time between membrane filter replacements should be once every three months. Filter replacement is associated with a "breach" in the system, since the product water from the filtration system will be exposed to the atmosphere and the housing is physically open with personnel contact. It is strongly suggested that personnel replacing membrane filter elements be equipped with lab coats, hair covers, and sterile disposable gloves. It is important that the serial number of each membrane filter installed in the system be recorded on an appropriate data sheet for the system. It is also important to remember that proper venting of the filter housing is required to avoid "two-phase" flow, which may result in the potential loss of integrity for the installed membrane filters.

Field experience indicates another item of concern during membrane filter replacement. It is important to consider the fact that the outside of the membrane filter(s) *and* water from the outside of the membrane filter(s) will contain bacteria. Further, any feed water retained on the base plate of the filter housing will also contain bacteria. Adequate time should be allotted to allow complete draining of feed water from the membrane filters and housing prior to removal of the "old" membrane filters. While undesirable, it may be necessary to spray isopropyl alcohol (USP) on the base plate of the filter housing to avoid contamination of the product/clean end of a membrane filter inserted into the base of the housing.

It is suggested that the bottom O-ring gasket for the membrane filter housing be replaced annually. A spare gasket should be maintained at the facility to avoid interruption in the operation of the system due to gasket failure.

While the use of a double O-ring seal mechanism without locking tabs may be employed, membrane filter elements with flat gasket or "knife-edge" seal mechanism are inappropriate for any application where bacterial control is desired.

ULTRAFILTRATION

As a unit operation installed in a Purified Water distribution system, ultrafiltration is a highly effective method of controlling system bacteria levels. Continuously ozonation of the storage tank with periodic brief sanitization of the distribution loop with ozonated water, discussed in chapter 7, also provides an effective method of controlling storage and distribution system bacteria levels. Ultrafiltration provides an effective alternative method of providing excellent microbial control.

A properly maintained hollow fiber ultrafiltration system installed in the Purified Water distribution system should be capable, on a routine basis, of maintaining total viable bacteria levels in the range of <1 to 2 cfu/100 mL, free of pathogenic species of bacteria, when incubated with R2A media at 30°C to 35°C using Membrane Filtration of a 100-mL sample and 120-hour incubation time period. Both proper system design and membrane selection are critical. The "pore size" for an ultrafiltration membrane is a "true" value, not an "assigned" value (e.g., membrane filters). However, ultrafiltration membranes will generally exhibit nonuniform pore size distribution. Subsequently, the pore size is generally stated as a nominal value. One would anticipate that a typical Gaussian distribution would be associated with this nominal value, with a certain number of pores of greater size than the nominal value and a certain number less than the nominal value. This is consistent with observed pilot plant study results. Pilot studies conducted with hollow fiber ultrafiltration membranes with a stated nominal pore size of 0.005 μm in ultrahigh purity, low TOC water indicate that certain species of bacteria that have entered into a mechanism to maximize the surface-to-volume ratio (with associated decrease in size) may pass through the membrane (Collentro, 1993). On the other hand, similar results from operating systems using hollow fiber ultrafiltration membranes with a smaller nominal pore size rating of 0.001 μm did *not* indicate the presence of bacteria in a defense mechanism (Collentro, 1993) in product water. With proper ultrafiltration membrane selection, comparable microbial results may be achieved (with proper design of the storage and distribution system) to those observed for storage and distribution systems that undergo thermal cycling daily or for storage and distribution systems that rely on ozone for microbial control.

Theory and Application

The results of numerous pilot studies and extensive operating experience using ultrafiltration in a distribution loop clearly demonstrate that a hollow fiber configuration exhibits superior performance to a spiral wound configuration. This is based on observed product water bacterial levels, bacterial endotoxin reduction, colloidal removal, and removal of organic material (Collentro, 1993). A review of ultrafiltration units installed at pharmaceutical facilities indicates that the majority of systems use hollow fiber membranes.

One of the primary advantages of hollow fiber ultrafiltration, when compared with spiral wound (or other) membrane configurations, is the ability to provide a highly effective internal cleaning operation, referred to as recycle or "fast flushing." This technique, provided by the arrangement of the fibers, allows back flushing of the membranes with ultrafiltered water—a highly effective cleaning operation. In general, this operation can be performed in <5 minutes, including a post–fast flush final rinse operation.

Historically, ultrafiltration membranes have been manufactured from either acrylic or polysulfone material. Polysulfone membranes, in particular, have exhibited excellent performance characteristics. Alternative hollow fiber ultrafiltration membranes are manufactured from ceramic material. While the ceramic material is capable of withstanding both sanitizing and even steam temperatures, the membrane exhibits a higher pressure drop than polysulfone. Polysulfone hollow fiber membranes are commercially available with hot water sanitization capability at temperatures of 80°C to 90°C without long-term adverse effects (Robinson, 1998).

Within this section, the terms *differential pressure* and a more technically accurate designation, *transmembrane pressure*, will be used interchangeably. Transmembrane pressure is the difference between the average of the feed water pressure and wastewater pressure to and from an ultrafiltration membrane *and* the product water pressure from the ultrafiltration membrane. In other words, the transmembrane pressure actually refers to the average pressure on the "dirty" side of the membrane minus the product water pressure associated with the "clean" side of the membrane.

A hollow fiber ultrafiltration membrane is generally assigned two critical performance parameters, which should carefully be considered in selecting a membrane. These parameters are molecular weight cutoff, expressed in daltons (d) and the nominal pore size rating expressed in micrometers. For most Purified Water applications using polishing ultrafiltration, it is suggested that the hollow fiber ultrafiltration membrane have an assigned molecular weight cutoff of 10,000 d and a nominal pore size of 0.001 µm. It appears that the assigned values are extremely conservative *and*, coupled with the characteristics of hollow fiber ultrafiltration ("ripening" process), actually provide a much "tighter" molecular weight cutoff and pore size rating than the stated values.

When one considers the pore size distribution of an ultrafiltration membrane, it should be obvious that the statistical probability of bacteria, even in a low nutrient environment, passing through a pore in the membrane is extremely remote. Subsequently, it is difficult to detect total viable bacteria immediately downstream of a polishing ultrafiltration unit at the 1 cfu/100 mL level.

The ability of ultrafiltration to remove bacterial endotoxins, as demonstrated by numerous pilot studies (Collentro unpublished data; Collentro, 1993) and extensive operating data, is actually greater than that for a single- or multiple-effect distillation unit. Typically, a hollow fiber ultrafiltration membrane with a molecular weight cutoff of 10,000 d (0.001 µm pore size) will exhibit >5 log reduction in bacterial endotoxins. As discussed earlier in this text, bacterial endotoxins will aggregate in relatively high-purity water (e.g., <1 microsiemens/cm conductivity). This aggregation process produces bacterial endotoxins with a molecular weight much greater than 10,000 d. While ultrafiltration is not an approved unit operation for the production of USP Water for Injection, when used in a polishing mode (within the Water for Injection distribution loop), it is capable of producing water meeting the chemical, bacterial endotoxin, and total viable bacteria limits for Water for Injection. Subsequently, it provides an excellent method of producing water for active pharmaceutical ingredients ultimately used to manufacture "injectable" solutions.

For certain applications where colloidal removal is important, ultrafiltration is a valuable polishing technique. Ultrafiltration provides excellent reduction of colloids, particularly those

Table 8.4 Decrease in Conductivity Associated with Post Mixed Bed Deionization Unit Ultrafiltration—Operating Data

Location	Colloidal silica (μg/L)	TOC (μg/L)	Conductivity (microsiemens/cm at 25°C)
Ultrafiltration unit feed water	18	110	0.130
Ultrafiltration unit product water	2	40	0.057
Ultrafiltration unit feed water	16	160	0.118
Ultrafiltration unit product water	3	60	0.059
Ultrafiltration unit feed water	20	165	0.142
Ultrafiltration unit product water	5	50	0.061

Notes:
- "Sets" of values are obtained for individual days.
- Data represent average values for a day.
- Raw feed water to the system is from a river water source. Pretreatment includes clarification.
- Study performed during period of heavy rain with significant "runoff" into the river.

Source: From Collentro (1993).

of silica, aluminum, and iron (Kunin, 1980). Colloids of silica may be important since the colloid, in a similar manner to reactive silica, may be "carried over" with steam, affecting the overall performance of a Pure Steam generator or multiple-effect distillation unit. The use of ultrafiltration for colloidal removal obviously applies to systems that use deionization for ion removal. (note that reverse osmosis is a colloidal specific process).

The ultrafiltration process is not capable of *directly* removing inorganic material. However, research has demonstrated the ability of ultrafiltration to decrease the conductivity of water, especially high-purity water, and to remove "complex" inorganic compounds (Dvorin and Zahn, 1987). Table 8.4 demonstrates the decrease in conductivity associated with the use of ultrafiltration in an application after deionization. Colloidal material will exhibit a slight negative charge. For many high-purity water systems (conductivity <0.1 microsiemens/cm), with ion removal achieved by ion exchange and feed water from a surface source, it is possible to note a dramatic decrease in conductivity with ultrafiltration. It is speculated that the negative charge associated with a colloid is balanced by a positive charge attributed to the hydronium ion. As ultrafiltration removes the colloid with the negative charge, it is suggested that the pH of the solution may change as hydroxide ion reacts with "excess" hydronium ion to produce high-purity, electrically neutral, product water.

Another valuable function of ultrafiltration is associated with the *indirect* removal of undesirable inorganic compounds, specifically the chloride ion complexed with organic material. Table 8.5 illustrates data from a unique pilot study. During the pilot, duplicate ultrafiltration unit feed water and single product water samples were obtained. The chloride and sulfate concentrations of one of the feed water samples were analyzed by ion

Table 8.5 Removal of Complexed Organic Inorganic Ions by Ultrafiltration

Parameter and treatment technique	Ultrafiltration unit feed water	Ultrafiltration unit product water
TOC	173 μg/L	29.1 μg/L
Conductivity	0.077 microsiemens/cm at 25°C	0.058 microsiemens/cm at 25°C
Chloride ion (preautoclave)	0.83 μg/L	<0.01 μg/L
Chloride ion (postautoclave)	128.5 μg/L	3.5 μg/L
Sulfate (preautoclave)	0.94 μg/L	<0.1 μg/L
Sulfate (postautoclave)	398.4 μg/L	7.2 μg/L

Notes:
- Autoclave temperature is 250°F.
- Autoclave time period is about six hours.
- Feed water to ultrafiltration unit from deionization system (no membrane process such as RO).
- Raw feed water to the system is from a municipal water supply from surface source.

Source: From Dvorin and Zahn (1987).

chromatography, which is capable of indicating trace concentrations of the ions. The results (Table 8.5) indicate that there is no chloride or sulfate ion present. However, when the other feed water sample was heated for a period of time, chloride ion and sulfate ion were detected at significant concentrations, compared to the undetectable concentrations prior to heating. Samples of ultrafilter product water, after heating, did not indicate the presence of chloride or sulfate ion. This test procedure was duplicated on several occasions with similar results. The results clearly indicate that NOM present in the deionized ultrafiltration unit feed water thermally decomposes, releasing chloride and sulfate ions. These experimental results were verified by other researchers (D'Auria et al., 1987). The importance of this particular observation to pharmaceutical applications relates to systems using ion exchange for ion removal, that provide feed water to Pure Steam generators and multiple-effect distillation units where boiling occurs in pressurized vessels at elevated temperatures. A feed water analysis or conductivity measurement may indicate that little, if any, chloride ion, capable of producing both chloride stress corrosion and chloride pitting attack of austenitic (300) stainless steel, is present. However, the data in Table 8.5 clearly indicate that chloride ion will be generated, by thermal decomposition of NOM for feed water systems without RO, resulting in the ultimate failure of stainless steel surfaces. This phenomenon has been observed at several facilities using both multiple-effect distillation units, *and* Pure Steam generators manufactured of 316L stainless steel. It implies that ultrafiltration is a required unit operation for feed water to multiple-effect distillation units or Pure Steam generator units from a raw surface water supply using ion exchange for pretreatment of feed water.

Ultrafiltration units installed as polishing components in Purified Water systems where the raw water feed water is from a surface water supply and ion exchange is the ion removal technique will provide a decrease in the TOC concentration. Depending on the design of the system, this process may exhibit a dramatic decrease in TOC levels. Table 8.6 demonstrates that the TOC reduction appears to be a function of the transmembrane pressure drop through the ultrafiltration membranes. It is highly suggested that the removal of organic material, as well as other material such as bacteria, bacterial endotoxins, colloids, and so on, will be enhanced by an increase in the transmembrane pressure drop and associated accumulation of "filtered" material on the membrane surface. As implied by its name, ultrafiltration will exhibit typical properties of any filtration system, such as a depth filtration unit. As discussed in chapter 3, there is a noticeable ripening effect associated with the filtration process. This results in an increased ability to remove particulate matter (in this case, organic material, bacteria, and other impurities), as previously filtered material accumulates on the dirty side of the membrane. The ripening process would decrease the effective pore size from a nominal value of 0.001 μm (for the ultrafiltration membrane discussed in this section). It is further suggested that the actual molecular weight cutoff will also be reduced during this ripening process.

There are two techniques for cleaning hollow fiber ultrafiltration membranes. One method is performed frequently (one to three times per day), depending on the "loading" of the ultrafiltration membranes. This process is referred to as fast flush or recycle. It is most appropriate to design an ultrafiltration system such that the individual hollow fiber

Table 8.6 TOC Reduction as a Function of Ultrafiltration System Transmembrane Pressure Drop

Feed water TOC (μg/L)	Product water TOC (μg/L)	Transmembrane ΔP (psid)	Product water flow rate (gpm)
195	90.0	5.0	2.6
195	39.1	10.0	3.7
195	18.3	30.0	5.0
195	8.0	50.0	6.8

Notes:
- The ultrafiltration membrane material was polysulfone, hollow fiber, with 20-mm inside diameter.
- All data were collected in an eight-hour time period, allowing at least 30 minutes for stabilization after changing transmembrane ΔP.
- The membrane manufacturer's maximum recommended transmembrane ΔP was 30 psid.

Source: From Collentro (1995).

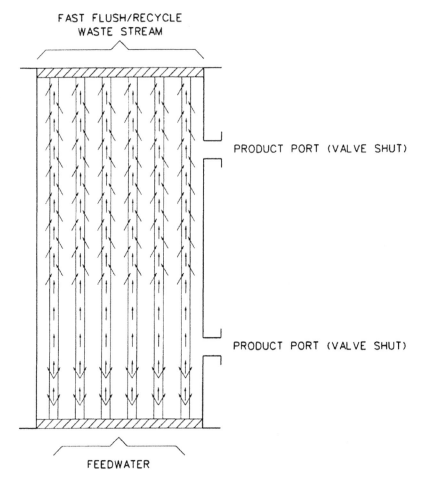

Figure 8.12 Fast flush flow path—hollow fiber membrane. (Excellent cleaning is achieved when the feed water flow rate per element is two to three times the "normal" operating rate. A dedicated waste flow path is required for high fast-flush flow rates.)

membranes are arranged in multiple banks of equal numbers of modules. During the fast flush operation, feed water flow is inhibited to one bank of membranes. This doubles the feed water flow rate to the other set of membranes. The product water valve is closed. The waste line, rather than flowing through a restricted valve, passes without regulation, directly to drain. This valve arrangement and increase in feed water flow rate results in the removal of entrapped material on the dirty side of the membrane, as demonstrated in Figure 8.12.

During this process, water will flow through the hollow fibers in the lower portion of the individual membranes. However, as water passes upward through the hollow fibers, assuming normal product water flow radially outward from the membranes, it enters the "shell" of the membrane until the pressure in the shell is greater than the pressure inside the fiber. This will occur at approximately 10% to 20% of the individual membrane length above the base (feed water port) of the membrane. As the ultrafiltered water passes back through the membrane, in the reverse direction of normal flow, accumulated material is removed by the "backwash" effect and is flushed to waste. This process is unique to hollow fiber ultrafiltration (as compared to spiral wound ultrafiltration). It is a highly effective cleaning method that can generally be performed in 15 to 30 seconds. By dividing the ultrafiltration membranes into two banks of membranes (also associated with effective chemical cleaning), it is possible to double the flow rate to the inside of the hollow fibers, increasing the pressure drop through the membranes and decreasing the amount of

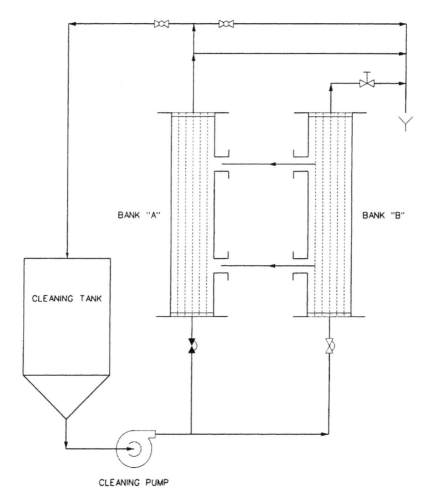

Figure 8.13 Ultrafiltration unit bank-to-bank cleaning schematic. (The waste stream from the "bank" being cleaned can be recycled to the cleaning tank or fed to waste. The procedure from bank "A" to bank "B" cleaning is similar. For clarity, only limited valves and pipings are shown in this schematic.)

ultrafiltration membrane area that will not be effectively cleaned (at the base of the membrane) during this operation.

A second cleaning mechanism for hollow fiber ultrafiltration is chemical cleaning of the membranes in a back flush mode, which is accomplished by using bank-to-bank cleaning (Fig. 8.13). In this process, a cleaning (or disinfecting) agent is introduced to one bank of membranes. Valves are arranged such that the product water from the initial bank of membranes flows directly to an adjacent bank of membranes in a back flush direction. This will displace material on the dirty side of the second set of membranes, flushing the displaced material to waste. It is extremely important to recognize the benefits associated with this particular operation. There are numerous Purified Water systems for both pharmaceutical and biotechnology applications that require low bacterial endotoxin levels. This cleaning method, when performed for the removal of bacteria, will not introduce dead Gram-negative bacteria to the clean side of the bank of ultrafiltration membranes. Since freshly ultrafiltered water with a disinfecting agent is used to remove bacteria from the membranes, the "time delay" required to ensure that the residual disinfectant agent has destroyed all bacteria in the chemical feed tank used for this application is not highly critical. Chemical cleaning to remove other contaminants, such as organic material, may require a residual disinfecting agent (e.g., sodium hypochlorite) and caustic solution (e.g., sodium hydroxide). For polishing applications where

Table 8.7 Heat-up Limitations—Polysulfone Hollow Fiber Ultrafiltration Membranes

Temperature (°C)	Maximum feed water pressure (psig)	Maximum transmembrane pressure (psid)
<10	75	60
10.25	65	50
25–50	60	45
50–80	50	35

Note: The suggested heat-up rate ≤2°C/min.
Source: From A/G Technology (1998) and Robinson (1998).

the primary ion removal technique is reverse osmosis, or if the raw feed water source is from a groundwater supply, minimal organic material would be anticipated. Subsequently, chemical cleaning would be limited to the use of a disinfecting agent. For storage and distribution systems constructed of stainless steel, effective sanitization may be achieved by heating the stored and recirculated Purified Water (and water passing through the ultrafiltration unit) to >80°C. If this method of sanitization is employed, the heating rate (°C/min for polysulfone constructed modules) should be consistent with manufacturer data (Table 8.7).

Unlike reverse osmosis used as a primary ion removal technique, ultrafiltration used as a polishing technique requires a limited continuous waste flow. For applications where the Purified Water generating system contains deionization and the feed water source is from a surface water supply, the waste flow rate may be as low as 5%. For applications where the Purified Water generating system contains reverse osmosis, or double-pass reverse osmosis, the polishing ultrafiltration unit waste flow rate may be as low as 1% to 2%. When polishing ultrafiltration units are employed solely for the removal of bacteria, it is possible to reclaim the waste stream, feeding it to a break tank at the beginning of the system. Residual disinfectant agent is introduced to the break tank for bacterial destruction. As discussed in chapters 2 and 3, adequate contact time in the break tank is required to allow the residual disinfecting agent to destroy all anticipated species of bacteria present in the ultrafiltration unit waste stream.

Regardless of the materials of construction of the storage and distribution system, the ultrafiltration unit will provide excellent removal of bacteria, significantly extending the elapsed time between required sanitization cycles.

Design Considerations

The selection of the specific hollow fiber ultrafiltration membranes required for each application is critical to successful operation. Specifically, the module diameter, length, materials of construction (membrane and support accessories), transmembrane pressure drop, and ability to undergo hot water sanitization should be considered. System design should insure that the "flux" rate through the membranes, during normal operation, is consistent with membrane manufacturer recommendations. Further, the parallel bank arrangement of membranes, discussed earlier in this chapter, should not contain an excessive number of individual modules. If required, multiple ultrafiltration units should be provided to ensure proper operation. If multiple units are provided, it should be clearly established that adequate water will be available for the "fast flushing" operation. If constant flow through the distribution loop is required, multiple units with "excess" capacity may be required to allow fast flushing without affecting the recirculation flow rate.

As discussed earlier, membranes should be arranged in two banks for each individual ultrafiltration unit. In other words, each ultrafiltration unit should contain an even number of membranes to provide adequate fast flushing water flow and to allow "bank-to-bank" chemical cleaning/sanitization.

The fast flushing operation should be performed at a flow rate twice the normal operating flow rate. Further, system design should ensure that there is no restriction in the waste line from the bank of modules during the fast flushing operation. Specifically, system design should include larger diameter waste piping than the normal operating waste piping from a bank of membranes. The normal operating waste piping should be equipped with a valve to regulate the waste flow rate to approximately 1% to 5% of the normal feed water flow rate.

Chemical cleaning of the hollow fiber ultrafiltration unit will require a chemical cleaning tank and pump. As demonstrated in Figure 8.13, cleaning should be performed in a bank-to-bank flow path so that bacteria and/or bacterial endotoxins are not introduced to the clean side of a membrane. This operation can be executed in a recirculating mode, with the waste being fed back to the cleaning tank, or in a mode where cleaning solution is continuously directed to drain. The latter method of operation is preferred. It is suggested that a fast flush operation be performed prior to and subsequent to chemical cleaning of a bank of membranes.

Where applicable, polishing ultrafiltration units should be designed for hot water sanitization. This operation may be performed by slowly heating the feed water to the membranes in a normal operating flow path configuration. The heat-up rate for water passing through the membranes must be consistent with manufacturer guidelines. In addition, many polysulfone hollow fiber ultrafiltration membranes and other "plastic" support accessories may have a reduced pressure rating at higher sanitizing temperatures (when compared with ambient temperature operation). This also means that the design pressure at these higher temperatures must be considered.

A critical factor associated with the proper design of an ultrafiltration unit is associated with the "rinse-to-drain" connector from the product line of the ultrafiltration unit. A major design flaw of many ultrafiltration units centers around the fact that the waste line from the product water side of the unit is physically connected to the normal waste line from the unit. Obviously, bacteria will be present, at a very high level, in the waste line from the ultrafiltration unit and will grow against the direction of flow to the connector from the product side of the unit. When the valve is physically opened, connecting this bacteria-laden waste line directly to the product line, bacteria will be introduced into the product line. The replication of bacteria will occur, significantly decreasing the observed effectiveness of ultrafiltration operation, particularly in light of the fact that the membrane surface area is extensive, which provides a large growth area for bacteria. This major design flaw can be totally eliminated by providing a dedicated drain line from the product divert-to-waste line of the ultrafiltration unit to a depressurized drain with air break.

The physical size of piping/tubing and valves for the feed water and wastewater lines of the ultrafiltration unit should consider the fast flushing requirements in addition to the normal operating flow requirements. Further, a "dual" waste line will generally be required (two waste lines operating in parallel) from each bank of membranes. One waste line would contain a smaller section of tubing with a diaphragm or needle valve to control the normal operating waste flow. The second waste line would contain a larger valve and piping to allow the flow of fast flush water.

Some ultrafiltration units are designed to include dedicated "reverse fast flushing" provisions. This design compensates for the fact that a fraction of the ultrafiltration membrane, at the lower part of a module, will not be cleaned by the fast flush operation. While this design feature has technical merit, it is suggested that the piping arrangement to perform this operation introduces either extensive dead legs or piping lines with a "double block and bleed" arrangement to eliminate bacterial growth. As an alternative to reverse fast flushing provisions, particularly when the primary function of the ultrafiltration unit is to control Purified Water distribution loop bacteria levels, the membranes can be periodically rotated 180°. While this "breaches" the integrity of the storage and distribution system, this operation can be performed prior to periodic system sanitization.

The ultrafiltration membranes are sensitive to "water hammer" and associated "two-phase flow." Subsequently, every attempt should be made to ensure that the membranes are properly vented and that the low-low level sensor on the upstream storage tank will not allow the pump to introduce air *and* water to the ultrafiltration unit. If deionization units are installed upstream of the ultrafiltration unit, it is critical that provisions are included to ensure that air is vented from the units prior to providing feed water flow to the ultrafiltration unit.

The transmembrane pressure drop through the ultrafiltration membranes is critical, since it is the primary method of ensuring membrane integrity. Manufacturer's literature for the specific membranes will clearly define a maximum allowable transmembrane pressure. Exceeding this pressure drop may result in a loss of membrane integrity. While there may be thousands of hollow fibers within a particular membrane, the failure of a single fiber will introduce bacteria to the clean side of *all* other fibers. Subsequently, effective bacterial control cannot be achieved if a single fiber has been damaged. To ensure that the transmembrane pressure is not exceeded, it is suggested that the system design include individual feed water and product water pressure sensors with transmitters to provide signals to a differential pressure controller with alarm provisions. Because of the critical nature of this unit parameter, it is further suggested that high transmembrane pressure not only activate an audible alarm but also inhibit flow through the unit, eliminating potential damage to the hollow fibers.

The ultrafiltration unit should be equipped with appropriate accessories including, but not limited to, feed water, product water, and wastewater pressure sensors; feed water and wastewater flow meters; a feed water temperature indicator; and feed water, product water, and wastewater sample valves. It is desirable to minimize the components positioned in the product tubing from the ultrafiltration unit. Subsequently, it is suggested that the feed water and wastewater flow meters can be used to provide (by difference) the product water flow, as opposed to installing a direct reading flow meter in the product water line. Obviously, a sanitary flow meter, at high cost, could be installed without affecting the product water quality from a microbial standpoint.

The feed water and product water lines to and from the unit should be equipped with conductivity cells, which should operate in conjunction with a single, non-temperature-compensated conductivity monitor. Conductivity monitoring is important for successful operation and determines the effectiveness of the rinsing operation after chemical cleaning. In addition, product water conductivity monitoring, with appropriate set point and a product water "divert-to-waste" valve, can be used to ensure that below-quality water is not fed to the distribution loop. Sanitary-type conductivity cells are required, particularly for the ultrafiltration unit product tubing.

The ultrafiltration unit should be equipped with provisions for in situ testing of the membranes. This can be accomplished for most hollow fiber membranes by installing transparent sections of piping or tubing in the wastewater connections from each membrane. For units that will be hot water sanitized, materials capable of withstanding the elevated temperatures should be used, such as Pyrex®. For units that will be chemically sanitized, transparent sections of plastic piping are adequate.

Operating and Maintenance Considerations

Product water monitoring is critical in establishing the proper operation of any ultrafiltration unit. While chemical testing, continuous conductivity monitoring, and bacterial endotoxin monitoring are all important, the primary indication for determining the operating status of the unit is total viable bacteria. Feed water and product water total viable bacteria levels should be determined and plotted as a function of time. The sanitization frequency for both the ultrafiltration unit and storage and distribution system will be dictated by the product water bacteria levels from the ultrafiltration unit.

Periodic sanitization should be performed with a frequency determined by ultrafiltration unit product water bacteria levels, using either a chemical sanitizing agent or hot water. The chemical sanitization process should use bank-to-bank cleaning, discussed earlier. Hot water sanitization should be performed in accordance with the membrane manufacturer's instructions.

Feed water and product water conductivity monitoring is important. While product water from the ultrafiltration unit may meet the Purified Water conductivity specification, an increase in the conductivity from the feed water to the product water is generally associated with a problem in the ultrafiltration unit. By monitoring the feed water and product water conductivity to, and from, the unit and recording the values, higher product water conductivity values associated with a simultaneous increase in feed water conductivity values can be identified. This eliminates premature response to a high product water conductivity value associated with an increase in the feed water conductivity value.

It is strongly suggested that the design of the system should, as a minimum, include *automatic* provisions for fast flushing the unit. It is further suggested that this operation be performed at least once every 24 hours. Generally, it is desirable to execute this operation during "off shift" or "nonpeak" hours, avoiding "loss of flow" or "reduced" flow to individual points of use. As discussed earlier, the fast flush operation, including the "rinse-to-drain," cycle, takes about five minutes to execute. For most operations, a schedule can be arranged to perform this operation without affecting manufacturing operations. If possible, the fast flushing operation should be performed two or three times a day, assuming that it does not interfere with production operations. Any automatic valves used in the assembly of the ultrafiltration unit should be designed to be "slow acting." This can be accomplished by installing regulating valves in the air supply lines to each of the automatic valves or selecting a diaphragm valve with appropriate slow opening and closing characteristics. It is suggested that using inline regulating valves in the supply air line to each automatic valve is a more reliable way of ensuring that the valves slowly open and close. Related to this item, manual valves installed on the unit for normal operation, fast flushing, and/or chemical cleaning *must* be slowly opened or closed. Water hammer, associated with the rapid operation of either a manual or an automatic valve, is highly undesirable. Ball valves should not be used for this application because of microbial control considerations. Butterfly valves can be opened very quickly when compared to diaphragm valves and are not recommended. This is particularly true for manual butterfly valves. Operating personnel must be aware of the catastrophic consequences of closing or opening valves rapidly.

Reverse fast flushing of the individual ultrafiltration modules has been discussed earlier in this chapter. It is suggested that reverse fast flushing provisions may be inappropriate for polishing ultrafiltration systems used for pharmaceutical applications. It is further suggested that, in lieu of reverse fast flushing, individual modules be disconnected (clamp-type fitting) and rotated 180° periodically. This should ensure that the entire membrane area is effectively cleaned. The frequency of module "rotation" is a direct function of microbial levels in the feed water to the unit (assuming that organic and colloidal matter is not present). This operation should be conducted prior to sanitization of the storage and distribution system, since it is an intrusive operation to the distribution tubing. It is suggested that the individual modules be inverted at least once every year. More frequent inversion of the modules may be required if the transmembrane pressure drop increases with time, demonstrating an inability to clean the lower section of the modules effectively by periodic fast flushing.

It is suggested that integrity testing of the hollow fibers in each membrane be performed annually. Test procedures for performing the integrity test should be provided by the equipment supplier and incorporated into the maintenance manual for the system. During the integrity testing procedure, it is not uncommon to observe the presence of *small* "bubbles" of testing gas (generally high-purity nitrogen). Membrane failure would be demonstrated by a rapid (and highly obvious) flow of nitrogen from a broken fiber.

In general, ultrafiltration units installed for polishing applications, particularly downstream of storage tanks with feed water from a Purified Water generating system containing an RO unit, operate extremely well. There is a tendency to become complacent regarding the operation and maintenance of the units. Periodic maintenance, fast flushing, chemical cleaning (where appropriate), and hot water sanitization (where appropriate) should be performed. Experience indicates that ultrafiltration membranes may have a life ≥ 5 years when positioned in a distribution loop RO quality feed water. However, it is suggested that membranes be replaced after a five year period when installed in systems using reverse osmosis or double-pass reverse osmosis as the primary ion removal technique. A membrane life of three years would be appropriate for units operating with a Purified Water generating system including deionization, particularly where the feed water source is from a surface water supply. However, these recommendations are only "general." The actual membrane replacement frequency should be based on the conditions experienced specifically feed water and product water microbial levels. For example, if the Purified Water generating system contains a single-pass RO unit with "standard" polishing mixed beds canisters high bacteria levels in the feed water to the unit would be anticipated.

REFERENCES

A/G Technology. UF/MF Operating guide—operating parameters. Needham: A/G Technology, 1998.
Aquafine Corporation. Aquafine Ultraviolet equipment, models SL-1, MP-2-SL, CSL-4R, CSL-6R, CSL-8R, CSL-12R, CSL-8R/60, CSL-10R/60, CSL-12R-60, and CSL-24R. Valencia: Aquafine Corporation, 1993.
ASTM. Standard test methods for determining bacterial retention of membrane filters utilized for liquid filtration, ASTM Designation F838-05, West Conshohocken, PA: American Society for Testing and Materials, 2005.

ASTM. Standard specification for reagent water, ASTM Designation D1193-06, Federal Test Standard 7916, West Conshohocken, PA: American Society for Testing and Materials, 2006.

ASTM. Standard guide for Ultra-Pure Water used in the electronics and semiconductor industries, ASTM Designation D5127-07, West Conshohocken, PA: American Society for Testing and Materials, 2007.

Brock TD. Membrane Filtration: A user's guide and reference manual. Madison: Science Tech, Inc., 1983:16–17.

Christian DA, Meltzer TH. The penetration of membranes by organism grow-through and its related problems. Ultrapure Water 1986; 3(3):39–44.

Collentro WV. Nonculturable viable organisms. Tutorial at the Pharmaceutical Manufacturers Association Water Quality Committee Conference, 6–8 February in Orlando, 1989.

Collentro WV. An overview of present and future of technologies for semiconductor, pharmaceutical, and power applications. Ultrapure Water 1993; 10(6):20–31.

Collentro WV. Microbial control in purified water systems—case histories. Ultrapure Water 1995; 12(3):30–38.

Collentro WV. Reduction of total viable bacteria using high velocity fixed mixed resin canisters. Pilot study results—unpublished, 1997.

Collentro WV. Unpublished data from 09/90–06/98.

Collentro WV. Personal data – experience using 0.1 μm filters in lieu of 0.2 μm filters – unpublished data, September 1997–October 2010.

Colwell RR, Hug A. Viable but nonculturable bacteria and their implications for water purification. Ultrapure Water 1995; 12(3):67–75.

Corson LA, Petersen NJ. Photoreactivation of *Pseudomonas cepacia* after ultraviolet treated waters. J Clin Microbiol 1975; 1(5):462–464.

D'Auria GD, Itteilag T, Pastrick R. The impact of reverse osmosis on makeup water chemistry at millstone unit two nuclear power station. Proceedings of the First Annual High Purity Water Conference and Exposition, 12–15 April in Philadelphia, 1987:1–6.

Dvorin R, Zahn J. Organic and inorganic removal by ultrafiltration. Ultrapure Water 1987; 4(9):44–46.

FDA. Guide to inspections of high purity water systems. Rockville: Food and Drug Administration, Office of Regulatory Affairs, Office of Regional Operations, Division of Field Investigations, 1993.

Gould MJ. Evaluation of microbial/endotoxin contamination using the LAL test. Ultrapure Water 1993; 10 (6):43–47.

Howard G Jr., Duberstein R. A case of penetration of 0.2 mm rated membrane filters by bacteria. J Parenter Drug Assoc 1980; 34(2):95–102.

Kunin R. The role of silica in water treatment—part 2. In Amber-Hi-Lites, No. 165. Philadelphia: Rohm and Haas Company, 1980.

Meltzer TH, Jornitz MW. Pharmaceutical Filtration: The Management of Organism Removal, ISBN: 1-930114-77-X. Bethesda, Maryland: Parenteral Drug Association and River Grove, Illinois: Davis Healthcare International Publications, LLC, 2006.

Munson TE. FDA views on water system validation. Proceedings of the Pharm. Tech Conference '85, 10–12 September. Cherry Hill: Aster Publishing Corporation, 1985:287–289.

Munson TE. Water systems for the 21st century. Presentation at the Pharm Tech Conference '93, 20–22 September in Atlantic City, 1993a.

Munson TE. Excerpts from the Pharm Tech Conference '93, panel discussion, 20–22 September in Atlantic City. *The Gold Sheet* 26 1993b; (12):2 (published by F-D-C Reports, Inc., Chevy Chase).

NAMSA™. Cytotoxicity—MEM elution—MG 023. Lab No. 94C 20614 00, December 18, 1997. Newbury Park: PTI Technologies, Inc.

Osumi M, Yamada N, Toya M. Bacterial retention mechanisms of membrane filters. Translated from Pharmaceutical Technology—Japan 7 (11):1–7. Reprinted by permission of Pall Filtration Company, East Hills, 1991.

Perry RH, Green DW, Maloney JO. Perry's Chemical Engineer's Handbook. 6th ed. New York: McGraw-Hill, Inc., 1984:6-4-6-6.

Robinson G. Personal communication, June 17, 1998.

Siemens Water Technologies Corporation. Water purification through service deionization, Technical Brochure No. HPSV-SDIdr-BR-0510, 2010.

Thomas AJ, Dürrheim HH, Alport M, et al. Detection of L-forms of *Pseudomonas aeruginosa* during microbiological validation of filters. Pharm Technol 1991; 15(10):74–80.

Tri-Clover Inc., Pump Data Sheets—Triflo pump, model 218, 6 3/4" impeller. Kenosha: Tri-Clover, Inc, 1994.

Vaponics. Personal experience with VLT-150, VLT-300, and VLT-400 stainless steel rechargeable canisters (general sales information). Rockland: Vaponics, a Division of Osmonics.

Wallhauser KH. Grow-through and blow-through effects in long-term sterilization processes. Die Pharmzeutische Industrie 1983; 45(5):527–531.

9 Distribution systems—design, installation, and material selection

INTRODUCTION

This chapter discusses distribution systems—an important part of any pharmaceutical water system. The materials of loop construction are one of the important considerations. Certainly, 316L stainless steel should be used for all Water for Injection distribution systems. It is suggested that 316L stainless steel should be considered for any system where microbial control is a consideration, including Purified Water systems and certain active pharmaceutical ingredient water systems. As indicated in an earlier chapter, the most effective method of eliminating bacteria within a storage and distribution system is to provide hot distilled water to a storage tank, maintain the stored and recirculating water at a temperature $\geq 80°C$ continuously, and eliminate the potential for back contamination. However, the indicated conditions are not plausible for most Purified Water systems and many Water for Injection systems. Periodic thermal sanitization of Purified Water and Water for Injection systems can provide bacteria "control." Specifically, it has been indicated that total viable bacteria destruction can be achieved, within a properly designed system, when the storage and recirculated water is heated to 90°C for a time period of two hours.

It is important to understand that periodic hot water sanitization at a temperature $\geq 90°C$ of a distribution system where bacteria are present may destroy bacteria (control), while the resulting dead gram-negative bacteria will adhere to the interior of tubing walls resulting in formation of a biofilm. While the biofilm primarily consists of lipopolysaccharide (LPS), bacteria from the flowing stream will adhere to the biofilm where they will replicate. Effective removal of biofilm is achieved by chemical sanitizing agents. While biofilm formation and control is discussed later in this chapter, it is important to note that all Purified Water systems (with the exception of properly operated ozonated systems—see chap. 7) and, to a lesser extent, thermally cycled Water for Injection systems will require periodic chemical sanitization with a material such as a 1% solution of peracetic acid and hydrogen peroxide.

Obviously, Purified Water distribution systems for critical applications, such as the production of ophthalmic products, topicals, inhalants, and antacids, will require bacterial control and, subsequently, periodic sanitization (ozone or hot water supplemented by chemical sanitization agent). Chapter 8 discusses technically inferior alternative methods of achieving bacterial control in "polishing components" positioned downstream of the distribution pump but prior to the distribution loop. Chapter 7 discusses the benefits and limitations of ozone as a sanitizing agent for Purified Water systems.

In this chapter, alternate distribution loop piping materials will also be discussed. The advantages and disadvantages of each type of material will be presented.

During a discussion of distribution loop design considerations, an emphasis will be placed on the requirement to provide full "serpentine-type" distribution systems. The use of "dead-ended" systems, or systems with long dead legs, is unacceptable and will have a significant effect on both chemical quality and points-of-use bacteria levels. In designing a pharmaceutical water purification system, several factors must be considered. This chapter discusses unit operations, from pretreatment through storage and polishing systems. One of the factors influencing system design is overall cost. Chapter 13 discusses system validation, emphasizing that a properly designed pharmaceutical water purification system must be capable of providing "in-specification" water to points of use, 100% of the time. This is associated with proper system design, operation, and proactive preventative maintenance. Occasionally, due to budget restraints, certain system design "compromises" are required. It is imperative to remember that the distribution system is the most important part of the system since it is the terminal operation that delivers water to the individual points of use. With a proper sanitization regime and frequency, the distribution systems can, to a certain extent, compensate for higher than desired total viable bacteria levels in the feedwater to the upstream storage system.

THEORY AND APPLICATION

The materials of construction and method of loop assembly, in order of preference, are as follows:

- 316L stainless steel, orbitally welded
- 304L or 316 stainless steel, orbitally welded
- Polyvinylidene fluoride (PVDF) with "seamless/bead and crevice-free" joints
- Unpigmented polypropylene with seamless/bead and crevice-free joints
- Unpigmented polypropylene or PVDF with socket-welded joints
- Unpigmented polypropylene or PVDF with sanitary ferrule connections
- Polyvinyl chloride (PVC) or copolymer of PVC (CPVC) ("Drinking Water system" only)

Obviously, there are many other potential combinations of materials and joining techniques available. For example, a Drinking Water system with welded connections (nonorbitally welded) would allow hot water sanitization and, for this application, would be superior to any heat-welded or solvent-welded plastic materials. Plastic distribution systems are discussed later in this chapter. PVDF has been selected because of its ability to withstand higher temperatures. Unpigmented polypropylene has been selected based on its very smooth interior surface and good chemical stability with very low levels of "extractables," similar to PVDF but without the ability to withstand hot water sanitizing temperatures. If other plastic materials are considered for a system, it is strongly suggested that a thorough review of potential extractables, particularly for the "installed" piping and fittings (after heat fusion or solvent welding), be conducted.

Distribution loops for Purified Water and Water for Injection should be designed in a full serpentine-type arrangement. With a single loop, this should not present any problems. For long distribution loops, with associated high-pressure drops, it may be necessary to repressurize at some point in the serpentine loop to provide adequate pressure for recirculation. While in-line repressurization is possible, repressurization with a storage tank and distribution pump may have advantages, particularly as it relates to ensuring adequate flow and pressure of delivered water to individual points of use. For certain distribution systems, principally facilities with multiple "floors," it may be appropriate to consider individual serpentine loops that provide water to individual floors or a "group" of floors. These loops will be provided water from a single "header" and return water to a depressurized common return line ("downcomer"). This method will be discussed later under section "Design Considerations." The use of "subloops" from a single serpentine-type main recirculating loop may also be considered, particularly for ambient Water for Injection applications. Subloops may be employed when isolation is required. These loops may employ a satellite tank with an associated satellite subloop recirculation pump. Other subloops may not use a satellite tank, withdrawing water from the main loop and recirculating it back to the suction side of the subloop distribution pump. Subloops may be used to operate ambient recirculating loops from a main hot recirculating loop. Subloops may also be employed to isolate certain areas or buildings of a facility. This may be appropriate for minimizing the possibility of chemical and/or microbial back contamination. It may provide a means of isolating operations producing different products or laboratory points of use from hard-piped manufacturing points of use.

The use of dead-ended loops or "ladder loops" is unacceptable for Purified Water or Water for Injection applications. Figure 9.1 depicts both a dead-ended loop and a ladder loop. Drinking Water applications may use either dead-ended or ladder loops, provided that each distribution systems demonstrate the ability to deliver water meeting the National Primary Drinking Water Regulations (NPDWR) as defined by the U.S. EPA at points of use. In addition, any dead-ended or ladder loops used for a Drinking Water system must also be able to maintain microbial control to all points of use. As previously discussed, the suggested maximum total viable bacteria level for Drinking Water systems, at point of use, is 500 cfu/mL, which is consistent with the recommended Action Limit in the *General Information* section of USP.

DEAD-ENDED DISTRIBUTION SYSTEM

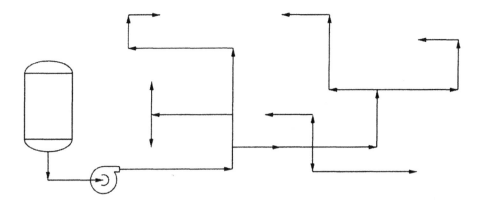

"LADDER TYPE" DISTRIBUTION "LOOP"

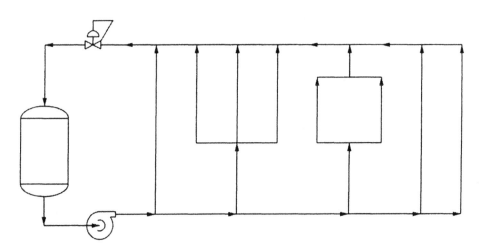

Figure 9.1 Examples of dead-ended and ladder-type distribution systems. (Flow arrows are for design purposes only. The direction of flow may reverse or system may be stagnant.)

The proper use of a serpentine-type distribution loop will have a dramatic effect on the pressure drop and associated velocity through the system. For a number of years, it has been suggested that a velocity in the range of 5 to 7 ft/sec was desired to minimize the formation of biofilm on the interior of distribution tubing walls. In fact, 6 ft/sec, commonly used as a design basis for Purified Water and Water for Injection distribution system, is based on non-pharmaceutical, "classical" applications. For industrial applications, not related to specific pharmaceutical water system distribution systems, 6 ft/sec is considered to minimize pipe size, based on economic considerations, while providing a moderate pressure drop. It is strongly suggested that velocities as low as 2.0 to 3.0 ft/sec are adequate for pharmaceutical water distribution systems. In fact, the velocity should be such that turbulent flow is maintained through the distribution loop during *all* operating conditions, including the "return" section of the loop. Table 9.1 presents data for velocity, flow rate, tube diameter, and fluid flow parameter, Reynolds number used to determine turbulent flow rate. The literature supports this suggestion (Gray, 1997; Mittelman, 1991, 2010).

To select the proper tubing (or piping) size, it is necessary to establish the maximum anticipated draw-off rate from the distribution loop. A "return factor" must also be

Table 9.1 Reynolds Number—Turbulent Flow Parameters

The Reynolds number is a dimensionless term/equation that can be used for calculating the flow profile through tubing. The Reynolds number can be calculated from the following equation:

$$N_{Re} = \frac{Dv\rho}{\mu}$$

where
- N_{Re} is the Reynolds number (dimensionless)
- D is the inside diameter of the tubing/piping (feet)
- v is the velocity of the water inside the tubing (feet/second)
- ρ is the density of water at the water temperature (pounds/cubic foot)
- μ is the absolute viscosity of water (pounds/foot × seconds).

Water flow inside tubing can be described as laminar, transition region, or turbulent depending on the Reynolds number. It is highly desirable to maintain flow in the turbulent region where the velocity over the cross-sectional area of the tubing is uniform, to minimize biofilm formation. While the exact value of the Reynolds number for various flow states depends on factors such as the smoothness of the interior tubing/piping wall, accepted Reynolds number for each flow state are presented as follows:

$$\text{Laminar flow: } N_{Re}(2100 - 2300)$$
$$\text{Transition region flow: } N_{Re} = (2100 - 4000)$$
$$\text{Turbulent flow: } > (4000)$$

Representative calculation:
- Flow rate = 30 gpm
- Tubing size = 1½ in.
- Water temperature = 25°C (77°F)

Determining terms in stated units:
- D = 1.402 in. = 0.1168 ft (Table 9.2)
- v = 6.2 ft/sec (Table 9.2)
- ρ = 62.19 lb/ft³ at 25°C
- μ = 0.898 centipoises at 25°C = 6.048 × 10⁻⁴ lb/ft × sec

Calculating:

$$N_{Re} = \frac{0.1168 \times 6.2 \times 62.19}{6.048 \times 10^{-4}}$$

N_{Re} = 74,560 (turbulent flow)

---

established. To determine the required flow rate to the distribution loop, the maximum instantaneous draw-off rate should conservatively be multiplied by a factor of 1.3 to 1.5. The resulting flow rate serves as the basis for selecting the tubing/piping size as indicated in Figure 9.1. The "equivalent length" of the distribution loop and maximum feedwater pressure may dictate a larger piping diameter. This does not present a problem unless velocity within the distribution loop decreases below the Reynolds number for turbulent flow. Table 9.2 summarizes the velocity, flow rate, and pressure drop through stainless steel tubing as a function of tubing diameter. For certain situations, the actual flow rate may require an "adjustment" within the suggested 1.3 to 1.5 return factor, based on a desire to select readily available material. For example, a system, using the technique suggested above, may require 3.5-in. stainless steel tubing. This material may not be readily available. Subsequently, larger (4-in.) or smaller (3-in.) tubing should be selected based on an evaluation of pressure drop and turbulent flow.

For plastic piping systems, this situation may be compounded by the maximum pressure rating of larger diameter tubing/piping. In other words, it may be necessary to increase the tube/pipe diameter for plastic distribution systems in order to maintain adequate pressure in the return section of the loop at maximum draw-off rate, since lower distribution loop feedwater pressures are required (considering the pressure drop) to comply with the maximum pressure rating of the plastic piping material. Unfortunately, many plastic tubing/piping systems require periodic chemical sanitization, a process more effective than hot water sanitization for microbial

Table 9.2 Velocity, Flow Rate, and Pressure Drop Through Stainless Steel Tubing as a Function of Tube Diameter

Tubing size (in.)	Tubing inside diameter (in.)	Flow rate (gpm)	Velocity (ft/sec)	Pressure drop (feet of water per linear ft of tubing)
1	0.902	10	5.0	0.12
1	0.902	15	7.5	0.25
1	0.902	20	10.0	0.43
1.5	1.402	15	3.1	0.04
1.5	1.402	20	4.2	0.06
1.5	1.402	25	5.2	0.08
1.5	1.402	30	6.2	0.105
1.5	1.402	35	7.3	0.135
1.5	1.402	40	8.3	0.170
2	1.87	25	2.9	0.025
2	1.87	30	3.5	0.035
2	1.87	35	4.1	0.04
2	1.87	40	4.7	0.05
2	1.87	45	5.3	0.063
2	1.87	50	5.8	0.073
2	1.87	60	7.0	0.10
2	1.87	80	9.3	0.16
2.5	2.37	50	3.6	0.022
2.5	2.37	60	4.4	0.03
2.5	2.37	80	5.8	0.05
2.5	2.37	100	7.3	0.075
2.5	2.37	120	8.7	0.105
3	2.870	80	4.0	0.02
3	2.870	100	5.0	0.03
3	2.870	120	6.0	0.04
3	2.870	140	6.9	0.05
3	2.870	160	7.9	0.07
3	2.870	180	8.9	0.08
3	2.870	200	9.9	0.10
4	3.834	140	3.9	0.013
4	3.834	160	4.4	0.015
4	3.834	180	5.0	0.02
4	3.834	200	5.6	0.025
4	3.834	220	6.1	0.028
4	3.834	240	6.7	0.035
4	3.834	260	7.2	0.040
4	3.834	280	7.8	0.045

Source: Courtesy of Tri-Clover.

control but more labor intensive. If the plastic tubing/piping diameter is increased to the point where the velocity does not result in turbulent flow, accelerated biofilm formation, with related bacterial growth, will be encountered. As indicated earlier, destruction of bacteria within an established biofilm as well as biofilm removal can be achieved with proper chemical sanitization. To extend the time between chemical sanitization cycles for plastic tubing/piping systems, it may be appropriate to consider polishing membrane filtration or ultrafiltration whenever possible.

Table 9.3 presents information associated with the maximum pressure rating of various types of plastic tubing/piping. The pressure rating for plastic tubing/piping is significantly affected (i.e., reduced) by an increase in size. Compounding this situation is the additional weight of water within the tubing/piping, which places added stress on joints, particularly for heat-welded or solvent-welded systems.

Table 9.4 contains information associated with temperature limitations for plastic tubing/piping. Using published information, the data in the table attempt to present the effect

Table 9.3 Maximum Pressure Rating for Various Plastic Piping as a Function of Diameter and Temperature

Socket-welded unpigmented polypropylene at 73°F

Schedule 80 pipe size (in.)	Pressure rating (psig)
0.5	150
1	150
1.5	150
2	150
3	115
4	115

Source: From Enpure™ (2007a).

SYGEF™ PN 16 PVDF Standard Pipe (0.630 in., 0.984 in., 1.575 in., 1.969 in., 2.953 in., and 4.331 in. outside diameter

Temperature (°F)	Pressure rating (psig)
68	232
86	224
104	202
122	176
140	156
158	137
176	119

Abbreviation: PVDF, polyvinylidene fluoride.
Source: From George Fischer (2009a).

Table 9.4 Selected Temperature Data for Plastic Pipe

Temperature (°F)	Unpigmented polypropylene "pressure derating factor" (multiplier)	PVDF "pressure derating factor" (multiplier)
70	1.00	1.00
80	0.97	0.95
90	0.91	0.87
100	0.85	0.80
120	0.75	0.68
140	0.65	0.58
160	0.50	0.49
180	Notes	0.42

Notes:
- Socket-welded piping
- Unpigmented polypropylene is only rated for low pressure (<20 psig) drain application at 180°F and is not recommended for use above 180°F.

Abbreviation: PVDF, polyvinylidene fluoride.
Source: Courtesy of Chemtrol (2001a).

of both pipe diameter and internal pressure on the temperature limitations of specific plastic tubing/piping materials. The data presented in Table 9.4 clearly indicate that it is impossible to use unpigmented polypropylene for a system that will be hot water sanitized at a temperature $\geq 80°C$. While PVDF may be acceptable, the tubing/piping layout and associated installation are critical due to the high thermal coefficient of expansion, discussed below. The literature also contains references to an increase in extractables from PVDF during operation at elevated temperatures (Harfst, 1994; FDA, 1993).

Table 9.5 Thermal Coefficient of Expansion—Calculated Effect on a Section of PVDF Piping

"Installed" length of pipe (in.)	Increase in length of pipe (in.)	New length of pipe (in.)	Temperature (°F/°C)
1200	0	1200	70/21.1
1200	0.94	1201	80/26.7
1200	1.86	1202	90/32.2
1200	2.81	1203	100/37.8
1200	3.73	1204	110/43.3
1200	4.67	1205	120/48.9
1200	5.59	1206	130/54.4
1200	6.54	1207	140/60.0
1200	8.40	1207	160/71.1
1200	9.34	1209	170/76.7
1200	10.26	1210	180/82.2
1200	11.21	1211	190/87.8

Formula:

$$\Delta L = L \times \Delta T \times \delta$$

where
- ΔL is the differences in length (expressed in inches) between the installed pipe at 70°F and the length of the pipe at the stated temperature.
- L is the original length of pipe, 100 ft or 1200 in. for this example.
- ΔT is the difference in temperature between the installed pipe and the indicated temperature expressed in °C.
- Δ is the coefficient of linear expansion for the pipe expressed in inches per linear inch of pipe $\times$ °C.
- Piping mater is "SYGEF" (PVDF).
- New pipe length rounded off to nearest inch.

Abbreviation: PVDF, polyvinylidene fluoride.

Source: From George Fischer (2009b).

Most plastic tubing/piping exhibits a high thermal coefficient of expansion. In general, this is not a concern for unpigmented polypropylene and PVC since they are generally used in ambient systems that are periodically chemically sanitized. This is a concern for materials such as PVDF that could be installed in a "hot" recirculating system, or in a thermally cycled system that is periodically hot water sanitized. PVDF has been found to be an excellent tubing/piping material for many semiconductor applications. Table 9.5 contains data and calculations indicating the result of PVDF's high thermal coefficient of expansion on a linear section of distribution tubing/piping. For systems that will be thermally cycled, the high thermal coefficient of expansion must be addressed, either by using relatively short tubing/piping runs or installing expansion "loops." It is suggested that either approach results in an increased number of fittings, which will undergo significant stress, particularly during repetitive thermal cycles. The integrity of any distribution loop is important. "Pinhole" leaks in any plastic piping system can result in the reduction of atmospheric air, which contains bacteria, into the distribution system. While the effect of the high thermal coefficient of expansion of PVDF may not appear as important for a hot system, disregarding effects associated with potential leaching of undesirable substances, it is suggested that a hot loop is never *continuously* hot. In other words, while thermal cycling of the loop may occur infrequently, cooldown must be considered. Further, the loop cannot be installed hot, presenting another complicating factor associated with the use of hot PVDF distribution systems.

While it is very inexpensive (when compared with all other distribution piping material), PVC should not be considered as a distribution loop piping material for applications where there are *any* concerns for bacterial control. When observed under a microscope, the interior surface of PVC piping is extremely porous (Vess et al., 1993) and virtually impossible to completely sanitize. The literature discusses exposure of numerous PVC loops to bacteria. Subsequent to exposure, the loops are drained and thoroughly sanitized with several different, highly effective sanitizing agents. After the individual loops are refilled with bacteria-free water, excessive bacteria levels are encountered in a relatively short period of time. The data presented within this study clearly indicate that bacterial growth on PVC surfaces, even when polishing membrane filtration or ultrafiltration is employed, will present problems.

Tubing selection for USP Water for Injection systems should be limited to orbitally welded 316L stainless steel. Existing systems using austenitic stainless steel, even systems without orbital connections, should be periodically inspected. The inspection should include nondestructive testing of welds and an evaluation of the accumulation of metallic oxides (rouging) on the interior of pipe walls. Further, an intense microbial monitoring program must be maintained.

Both USP Purified Water and Drinking Water systems may use distribution loop of plastic material, ranging from PVC to PVDF. There is an increasing trend to use 316L stainless steel USP Purified Water distribution loops, which resemble Water for Injection distribution loops. This is consistent with the increased use of ozone for microbial control as well as periodic hot water sanitization.

Extractables from plastic tubing/piping, particularly PVDF, which may be exposed to high temperatures and/or thermally cycled, are a concern. Validation documents, specifically IQ documentation, should include information clearly demonstrating the manufacturer's testing data for extractables. Furthermore, it is strongly suggested that recirculating water in distribution loops using PVDF operating at elevated temperatures or being thermally cycled be tested periodically. The sampling and test program should include gas chromatography/mass spectrometry analysis for undesirable organic compounds.

Dead legs within a distribution loop, including point-of-use valves, will contribute to bacteria levels. Historically, a dead leg was incorrectly defined as any area greater than six-"pipe diameters" from the centerline of the larger diameter tubing/piping to the end of the smaller diameter "branch." In fact, the six-pipe diameter "rule" only applied to hot systems, *not* systems operating at ambient temperatures. For an ambient system, dead legs should be eliminated to the greatest extent possible. The literature discusses dead legs for point-of-use valves (ASME BPE, 2009a). Further, the literature contains guidelines for high-purity water and clean steam systems (ASME BPE, 2009b). The literature defines a dead leg based on a "L/D" ratio greater than 2:1, where L is the length from the interior of the tubing in the distribution system and D is the interior diameter of the "extension leg." Figure 9.2 provides a sketch of a tubing tee with associated example of L/D calculation. The use of zero dead leg–type diaphragm valves at points-of-use, sample location and any branch from the distribution system is recommended for both USP Purified Water and Water for Injection distribution loops. A typical zero dead leg valve is depicted in Figure 9.3.

Finally, another item should also be discussed as it relates to *effective* chemical sanitization of distribution loops, reverse osmosis/continuous electrodeionization (RO/CEDI) loops, and

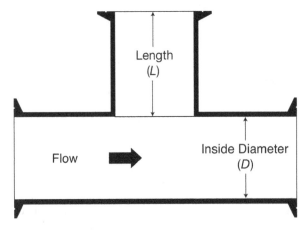

Tubing "branch" dead leg in a hot system defined as $L/D > 2:1$ where
L = Distance from inside tubing wall to end of "branch"
D = Inside diameter of tubing

Figure 9.2 Calculation of "L/D" for dead legs in hot recirculating distribution systems.

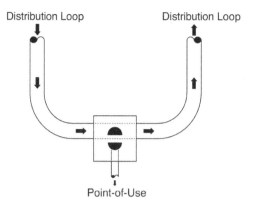

Figure 9.3 Typical cross-section of zero dead leg valve.

pretreatment recirculating loops. Chemical sanitization is performed by introducing an oxidizing agent to the stored and recirculated water. The solution is generally recirculated for a preestablished period of time to obtain the desired bacteria destruction, such that the storage and distribution system (or recirculating loop) can operate for a reasonable time period before resanitization is required. Experience indicates that a 1% mixture of peracetic acid and hydrogen peroxide is a very effective chemical sanitizing agent when the sanitization operation is executed in a responsive manner. Chemical sanitization should be executed as follows:

- It is highly desirable to introduce the chemical sanitizing agent into the storage tank feeding the loop. The volume of material introduced should be adequate to obtain the indicated 1% solution of peracetic acid and hydrogen peroxide. While other chemical sanitizing agents, including dissolved ozone, may be employed, experience and the literature clearly indicate the superior properties of the indicated solution for both destruction of bacteria and removal of an established biofilm (Mazzola, 2002; Mazzola, 2006; and Collentro, 2010).
- After introduction of the sanitizing agent, the recirculation (or repressurization) pump is employed to equilibrate the concentration of sanitizing agent throughout the distribution loop (including components if applicable). All valves that deliver water from the loop are cycled upon. Tests strips are employed to verify the concentration of sanitizing agent from each valve. Further, sanitary ferrule "caps" or similar fittings are loosened to allow sanitizing agent to enter any potential "low" flow areas. When complete, sanitizing agent, at the desired concentration, should be in direct contact with all wetted surfaces in the system.
- The sanitizing agent should be recirculated for a time period of 15 to 30 minutes.
- Upon completion of the initial circulation period of the sanitizing agent, power to the recirculation/repressurization pump should be terminated. The loop is in a stagnant mode. During this critical time period, sanitizing agent will "diffuse" into the biofilm, destroying bacteria. The biofilm material is "loosened." A portion of the biofilm may be fully or partially oxidized. The driving force for transfer of sanitizing agent from stagnant water to the interior walls of tubing, valves, fittings, etc. is concentration difference. This process will not occur without stagnant conditions. The duration of this stagnant time period is a function of system total viable bacteria levels and the "established" nature of the biofilm. It is suggested that a minimum 1- to 2-hour stagnant time period be considered, extended as appropriate.
- Power to the recirculation/repressurization pump is restored. Recirculation for a time period of 15 to 30 minutes should be performed.
- The sanitization solution is removed from the recirculation system by directing water to drain. Figure 9.4 presents a suggested method of displacing the disinfecting agent from the storage and distribution system to reduce the time required for removal of disinfecting agent.

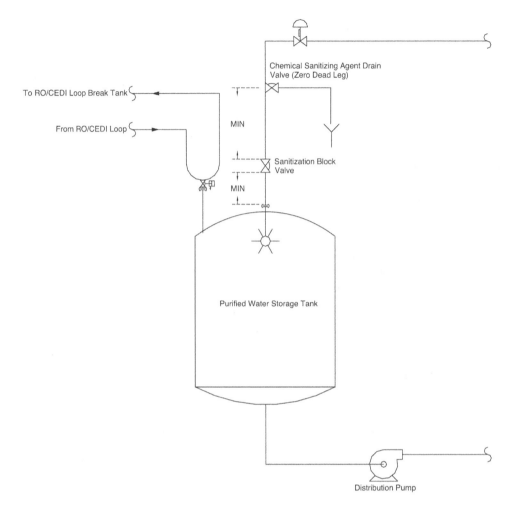

Figure 9.4 Valve configuration for displacement of chemical disinfection agent for the storage and distribution system.

- Tests strips are employed to verify reduction and ultimate complete removal residual disinfecting agent at all points in the system.
- The storage and distribution system (or loop) is returned to normal operation.

THEORY AND APPLICATION—BIOFILM

If "sterile" quality water is fed to a Purified Water or Water for Injection distribution system, bacteria would not be present assuming that it is not introduced into the distribution loop through a source such as back contamination. For reference purposes only, the word sterile implies a source of water from the Purified Water or Water for Injection storage tank that is continuously free of bacteria in a 100-mL sample. This may be a valid assumption for continuously hot recirculating Water for Injection systems or Purified Water system employing ozone for bacteria destruction with rigorous operating and maintenance requirements. Unfortunately, many USP Water for Injection systems employ reduced temperature loops (or subloops) that operated at reduced temperature for a fraction of the day. Further, Purified Water systems, with the exception of ozonated systems discussed in chapter 7, will exhibit the presence of bacteria in water from the storage tank. If bacteria are present in the water from a storage tank or in a subloop, a biofilm will be present.

The literature indicates that biofilm experimental results are somewhat contradictory because of the complex nature of the subject (Meltzer and Jornitz, 2006). It is beyond the scope

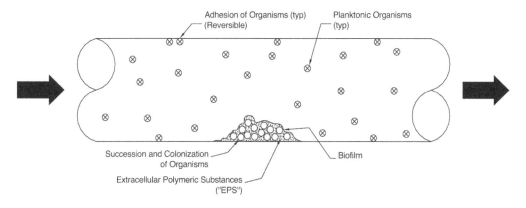

Figure 9.5 Biofilm formation. *Source*: From Mittelman (2010).

of the chapter to provide theory associated with the method of biofilm formation, its ability to "protect" bacteria in a thin layer on solid surfaces (tubing walls), nutrient production, ability to replicate, periodic bacteria "shedding," and the significant ability to resist destruction by sanitizing agents. The material presented below provides a brief history with references for biofilm, important practical factors to consider, and two actual examples of biofilm influence on total viable bacteria with corrective action. It is suggested that sanitization procedures for partial or seemingly total removal of a biofilm provide important information related to bacteria control in compendial water systems by removal of the "source," biofilm.

Biofilm is a thin layer (1 μm) of material on a surface such as the interior walls of distribution tubing. A simplified example of an accepted mechanism for biofilm formation is presented in Figure 9.5. The literature suggests that the interior walls of distribution tubing become "conditioned" with organic material from the recirculating water in the distribution loop. Bacteria in the water stream (planktonic) are attracted to the conditioned interior walls of the distribution loop tubing. The bacteria undergo "primary" adhesion on material in the biofilm, extracellular polymeric substance (EPS). At this stage, the literature suggests that the organism can adhere to the biofilm surface or return to the flowing water stream (Mittelman, 2010). If the organism adheres, it replicates in the biofilm. Additional EPS is generated from the organism resulting in "irreversible" attachment of the organism. Bacteria will periodically "shed" from the biofilm, enter the flowing stream, and be detected in samples from the distribution loop.

The following items present general characteristics of biofilm:

- It is suggested that greater than 99% of all bacteria exist in biofilm (Costerton et al., 1987).
- The "Center for Disease Control" estimates that the source of 65% of bacterial infection is associated with biofilm (Potera, 1999).
- Biofilm is not uniformly present on the interior walls of tubing but spread in diverse "patches."
- The material found in biofilm at different locations on tubing walls will vary.
- EPS may contribute up to 80% to 90% of the total organic carbon (TOC) present in biofilm (Fleming et al., 2001).
- Biofilm exists on potable water distribution piping (municipal water) and may present a health risk without proper treatment (EPA, 2002).
- The rate of biofilm growth, proliferation of bacteria within the biofilm, and release of bacteria to the flowing stream increase with bacteria levels in the free stream (makeup water bacteria level to the storage tank and postdistribution pump).
- If turbulent flow is maintained in the distribution loop, biofilm formation and growth are not significantly affected by velocity. It is suggested, however, that higher velocity (6–7 ft/sec) may produce a slight to moderate increase in the rate of biofilm

growth when compared with distribution loop velocities of 2 to 3 ft/sec. It is suggested that this observation may be associated with increased "exposure" of the biofilm to organic material present in the recirculated water at the higher velocity.
- Higher distribution loop water TOC values appear to accelerate biofilm growth.
- The mechanical polish for stainless steel tubing and surface roughness for plastic tubing appear to have a minor effect on biofilm growth.

Biofilm provides a protective "shield" that protects bacteria from destruction by sanitization agents. The literature suggests that the EPS secreted by biofilm bacteria provides a physical barrier to sanitizing agents (Thein, 2001; Lewis, 2001; Costerton et al., 1995; Meltzer and Jornitz, 2006). An interesting theory is that sanitizing agent is "deactivated" in the outer layers of the biofilm faster than it flows into the biofilm (Prakash, 2003). It is strongly suggested that this observation provides a significant *key* to effective methods for removing biofilm and controlling bacteria. Chemical sanitization with a 1% mixture of peracetic acid and hydrogen peroxide, discussed in this chapter, has provided excellent results. The sanitization technique employs a three-step sequence with both dynamic and stagnant flow conditions as well as extended time period (when compared with sanitization manufacturer's recommendation). This is supported by the literature which indicates that organism destruction by peracetic acid is superior to both hydrogen peroxide (without peracetic acid) and formaldehyde (Meltzer and Jornitz, 2006).

The literature indicates that bacteria tend to attach quickly to hydrophobic surfaces such as PVDF and Teflon® compared to stainless steel or glass (Pringle et al., 1983). Other researchers indicate that biofilm formation on plastic tubing/piping surfaces is slower than that for stainless steel. The information would support the fact that the use of plastic tubing/piping material does not offer an advantage for biofilm control when compared with stainless steel.

On the basis of field experience and supported by the literature (Adley et al., 2005), a particular organism appears to offer the greatest challenge for both biofilm control and associated total viable bacteria levels. This organism is *Ralstonia pickettii*. In fact, as discussed further below, biofilm containing *R. pickettii* is resistant to ozone. *R. pickettii* has been identified in biofilm on plastic piping of the U.S. Space Shuttle Drinking Water System. The water, generated by fuel cells during orbit, has a total viable bacteria limit of 1 cfu/100 mL. On two separate occasions, both *R. pickettii* and *Burkholderia cepacia* were noted at levels of 55 cfu/100 mL and less than 200 cfu/100 mL. *R. pickettii* has been identified as the reason for recall of saline products (MMWR, 1998), contamination in medical devices (Donlan, 2001), and in bloodstream infections (Roberts et al., 1999). It appears that *R. pickettii* produces unique qualities for biofilm formation in a nutrient-starved environment. Table 9.6 provides data from the literature associated with the source and attributes of *R. pickettii*.

As indicated earlier, a practical method of demonstrating the effects of biofilm on recirculating water total viable bacteria levels is to provide operating data. The first example represents microbial contamination of a well-controlled (total viable bacteria) system subsequent to storage and distribution system repassivation. The storage and distribution systems are shown in Figure 9.6. Feedwater to the Purified Water storage tank is from a conservatively designed recirculating pretreatment system with feed to a recirculating RO/CEDI system. Makeup water total viable bacteria levels to the Purified Water storage tank continuously indicate less than 1 cfu/mL [membrane filtration of a 1-mL sample in 99 mL of sterile water, 0.45 μm filter disc, plate count agar (PCA) culture media, 30 to 35°C incubation temperature, and 48-hour incubation time period]. Further, makeup water samples indicate the absence of *Pseudomonas aeruginosa* and total coliform in 100-mL samples. Hot water sanitization of the 316L stainless steel distribution loop (orbitally welded with zero dead leg point-of-use valves) is performed weekly (2 hours at 90°C). Chemical sanitization with a 1% solution of peracetic acid and hydrogen peroxide is performed once every six months. During three years of operation prior to the excursion, samples from about 20 laboratory points of use (collected in a "rotating arrangement" of four samples per workday, 5 days/wk) indicated total viable bacteria less than 1 cfu/mL, absence of *P. aeruginosa* in 100-mL samples, and absence of positive coliform in 100-mL samples. The online distribution loop conductivity

DISTRIBUTION SYSTEMS—DESIGN, INSTALLATION, AND MATERIAL SELECTION

Table 9.6 Data from the Literature Associated with the Sources and Attributes of *Ralstonia Pickettii*

Bergey's systematic/determinative:
- Tentatively considered by Pickett and Greenwood to be etiologically significant in humans
- Opportunistic animal pathogen

Bailey & Scott's:
- Environmental organism
- Not considered part of the normal human flora
- Mode of transmission not known, but probably involves human contact with heavily contaminated medical devices or substances encountered in the hospital setting
- Uncommon cause of infection in humans
- Very little known about virulence factors, if any, they exhibit

MCM:
- Recovered from a variety of clinical specimens
- Infrequent cause of bacteremia, meningitis, endocarditis, and osteomyelitis
- Identified in several nosocomial outbreaks due to contamination of intravenous products, "sterile" water, saline, chlor hexidine solutions, respiratory therapy solutions, and intravenous catheters
- Associated with pseudobacteremias and asymptomatic colonization
- May be recovered from respiratory tracts of cystic fibrosis patients

Abbreviation: MCM, Manual of Clinical Microbiology.
Sources:
Bergey's Determinative: Holt JG, Krieg NR, Sneath PHA, et al. (ed). Bergey's Manual of Determinative Bacteriology. 9th ed. Baltimore, MD: Lippincott Williams & Wilkins, 1994.
Bergey's Systematic: Krieg NR (ed.). Bergey's Manual of Systematic Bacteriology. Vol. 1. Baltimore, MD: Lippincott Williams & Wilkins, 1984.
MCM: Murray PR, Baron EJ, Pfaller MA, et al. (ed.). Manual of Clinical Microbiology. 7th ed. Washington, D.C.: ASM Press, 1999.
Bailey & Scotts: Forbes BA, Sahm DF, Weissfeld AS (ed.). Bailey & Scott's Diagnostic Microbiology. 11th ed. St. Louis, MO: Misby, 2002.

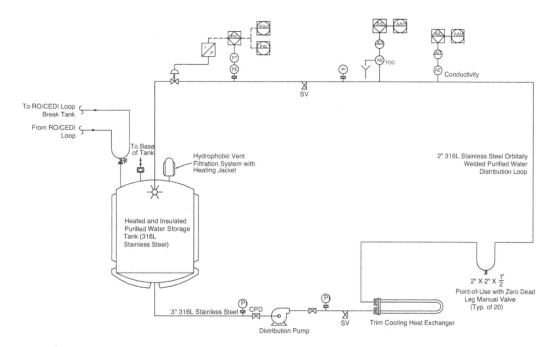

Figure 9.6 Bacteria excursion after passivation—storage and distribution system.

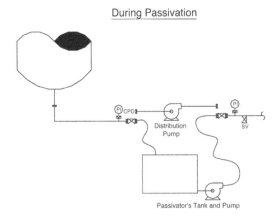

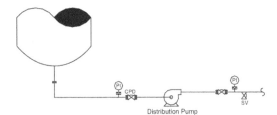

Figure 9.7 Bacteria excursion after passivation—stagnant section of distribution system.

averaged about 0.60 µS/cm and TOC value about 25 µg/L. To assist in distribution system microbial control, design included a cooling heat exchanger that maintained the recirculating water temperature at a value of about 20°C.

Inspection of the internal surfaces of the storage tank indicated that repassivation was required. Unfortunately, the passivation operation was conducted with an "external" tank provided by the passivation organization. A portion of the loop, indicated in Figure 9.7, was not passivated but allowed to not only sit in a stagnant condition for two days but also exposed to bacteria from both the atmosphere and personnel during connection/disconnection of temporary hoses. Subsequent to passivation and rinse, hot water sanitization of the storage and distribution system was conducted for four hours at 90°C. Since intrusive operations had been performed, point-of-use samples were collected daily for a period of two weeks. These data are presented in Table 9.7. The data clearly indicate that the system had been contaminated with bacteria. It is extremely interesting to note that it took about six days for the bacteria to be observed in point-of-use samples. Once observed, the bacteria dramatically increased. This would support a theory that point-of-use bacteria are from bacteria shed from biofilm and not from bacteria "replicating" in the flowing water stream.

A chemical sanitization of the loop (per the three-step procedure outlined in this chapter) was performed with a 1% solution of peracetic acid and hydrogen peroxide. Postchemical sanitization point-of-use bacteria data are presented in Table 9.8. It should be indicated that point-of-use bacteria results subsequent to the intense monitoring exhibited in Table 9.8 were consistent with data obtained for the three-year period prior to the repassivation operation.

The second example represents microbial contamination of a well-controlled (total viable bacteria) ozonated system subsequent to distribution system expansion. The storage and distribution systems are shown in Figure 9.8. Makeup water from a RO/CEDI System flows to the ozonated Purified Water storage tank. Feedwater to the tank exhibits total viable bacteria less than 1 cfu/100 mL (membrane filtration of 100 mL through a 0.45 µm filter disc, R2A culture media, 120-hour incubation time period, and 30 to 35°C incubation temperature).

Table 9.7 Operating Data—Microbial Excursion Subsequent to Passivation

Sample point number	Days after passivation	Total viable bacteria[a] (cfu/mL)	Pseudomonas aeruginosa per 100 mL (present/absent)	Total coliform[b] bacteria per 100 mL (present/absent)
1 through 20	1	<1	Absent	Absent
1 through 20	2	<1	Absent	Absent
1	3/4/5	36/~520/>5700	Absent[c]	Absent[c]
2	3/4/5	49/~560/>5700	Absent[c]	Absent[c]
3	3/4/5	36/~640/>5700	Absent[c]	Absent[c]
4	3/4/5	36/~690/>5700	Absent[c]	Absent[c]
5	3/4/5	36/~360/>5700	Absent[c]	Absent[c]
6	3/4/5	36/~420/>5700	Absent[c]	Absent[c]
7	3/4/5	49/~450/>5700	Absent[c]	Absent[c]
8	3/4/5	36/~780/>5700	Absent[c]	Absent[c]
9	3/4/5	36/~850/>5700	Absent[c]	Absent[c]
10	3/4/5	36/~580/>5700	Absent[c]	Absent[c]
11	3/4/5	36/~430/>5700	Absent[c]	Absent[c]
12	3/4/5	49/~490/>5700	Absent[c]	Absent[c]
13	3/4/5	36/~580/>5700	Absent[c]	Absent[c]
14	3/4/5	36/~510/>5700	Absent[c]	Absent[c]
15	3/4/5	36/~470/>5700	Absent[c]	Absent[c]
16	3/4/5	36/~410/>5700	Absent[c]	Absent[c]
17	3/4/5	49/~560/>5700	Absent[c]	Absent[c]
18	3/4/5	36/~460/>5700	Absent[c]	Absent[c]
19	3/4/5	36/~470/>5700	Absent[c]	Absent[c]
20	3/4/5	36/~560/>5700	Absent[c]	Absent[c]

Notes:
- Sampling terminated upon receipt of the third day results.
- System sanitized with a 1% mixture of peracetic acid and hydrogen peroxide on days 6 and 7 after passivation.

[a]Total viable bacteria determined by heterotrophic plate count of a 1.0-mL sample, PCA agar, 30 to 35°C incubation temperature, and 48-hour incubation time period.
[b]Total coliform indicates "confirmed" value—absence of fecal coliform and *Escherichia coli*.
[c]Data indicate "absence" on all three sample dates.

Dissolved ozone from an electrolytic ozone generation system is introduced to the tank, maintaining a concentration of 0.05 to 0.06 mg/L during operation and 0.10 to 0.20 mg/L during daily sanitization (45 minutes). The distribution loop online TOC unit indicates 2 to 5 µg/L. The distribution loop supply and return online conductivity measurements indicate an average value of 0.06 µS/cm. Water is delivered to about 35 points of use. Historically, point-of-use total viable bacteria levels were not detectable in the 100-mL samples. Observation of any bacteria colonies, per standard operating procedures (SOPs), required Gram stain and identification by RiboPrinter® (Tables 9.9 and 9.10).

Four additional points of use were added to the distribution system. During installation of the new 316L stainless steel tubing and zero dead leg valves, clean rooms at the facility were "open" for maintenance. Heavy construction with floor removal and ground excavation was performed as part of the expansion of the facility requiring distribution loop expansion. Subsequent to completion of work, the new facility area was "environmentally" sanitized. On the basis of the magnitude of the change in the distribution loop, a 20-day "Change Control Performance Qualification" (every point-of-use every day) was developed. Prior to execution, the expanded loop was sanitized with dissolved ozone at a concentration of 0.200 to 0.300 mg/L for a time period of 30 hours. During the sanitization operation, all point-of-use valves were slowly opened and flushed for a time period of about two minutes every hour.

Table 9.10 provides data for the first 10 days of Performance Qualification (PQ) execution. Organism identification by RiboPrinter indicated *R. pickettii*. The storage and distribution loops were resanitized with ozone at a concentration of 0.40 mg/L for a time

Table 9.8 Operating Data—Microbial Excursion Subsequent to Chemical Sanitization with 1% Mixture of Peracetic Acid and Hydrogen Peroxide

Sample point number	Days after passivation	Total viable bacteria[a] (cfu/mL)	Pseudomonas aeruginosa per 100 mL (present/absent)	Total coliform[b] bacteria per 100 mL (present/absent)
1 through 20	8	<1	Absent	Absent
1 through 20	9	<1	Absent	Absent
1 through 20	10	<1	Absent	Absent
1 through 20	11	<1	Absent	Absent
1 through 20	12	<1	Absent	Absent
1 through 20	13	<1	Absent	Absent
1 through 20	14	<1	Absent	Absent
1 through 20	15	<1	Absent	Absent
1 through 20	16	<1	Absent	Absent
1 through 20	17	<1	Absent	Absent
1 through 20	18	<1	Absent	Absent
1 through 20	19	<1	Absent	Absent
1 through 20	20	<1	Absent	Absent
1 through 20	21	<1	Absent	Absent
1 through 20	22	<1	Absent	Absent
1 through 20	23	<1	Absent	Absent
1 through 20	24	<1	Absent	Absent
1 through 20	25	<1	Absent	Absent
1 through 20	26	<1	Absent	Absent
1 through 20	27	<1	Absent	Absent

Note:
Sampling frequency returned to every point of use once per week after "day 27."
[a]Total viable bacteria determined by heterotrophic plate count of a 1.0-mL sample, PCA agar, 30 to 35°C incubation temperature, and 48-hour incubation time period.
[b]Total coliform indicates "confirmed" value—absence of fecal coliform and *Escherichia coli*.

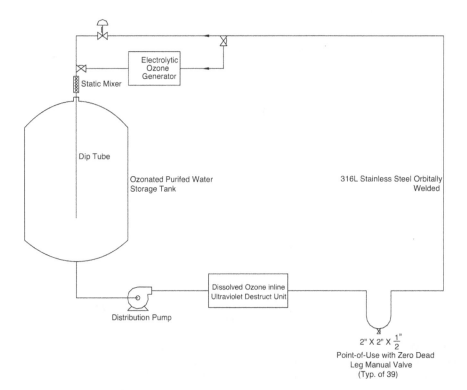

Figure 9.8 Ozonated storage and distribution system with bacteria excursion.

DISTRIBUTION SYSTEMS—DESIGN, INSTALLATION, AND MATERIAL SELECTION

Table 9.9 Purified Water System Total Viable Bacteria Excursion After Addition to Distribution Loop—Initial PQ Data

Sample point no.	TVB results day 1 (cfu/100 mL)	TVB results day 2 (cfu/100 mL)	TVB results day 3 (cfu/100 mL)	TVB results day 4 (cfu/100 mL)	TVB results day 5 (cfu/100 mL)	TVB results day 6 (cfu/100 mL)	TVB results day 7 (cfu/100 mL)
1	<1	<1	3	<1	1	<1	<1
2	<1	<1	<1	<1	<1	<1	<1
3	<1	<1	1	<1	<1	1	<1
4	<1	<1	<1	<1	<1	<1	<1
5	<1	<1	4	4	3	14	53
6	<1	1	1	2	2	4	9
7	<1	2	2	<1	20	1	7
8	17	9	67	10	22	9	55
9	3	9	1	1	54	33	27
10	<1	1	2	1	1	1	<1
11	<1	<1	2	1	<1	1	<1
12	114	440	53	98	71	161	105
13	19	62	31	66	133	15	6
14	100	470	139	51	25	29	26
15	<1	<1	4	6	3	2	23
16	<1	1	1	2	2	7	2
17	1	1	2	3	<1	5	4
18	137	55	5	18	33	52	152
19	<1	3	12	34	6	2	9
20	<1	1	<1	5	<1	1	6
21	<1	3	1	<1	<1	<1	2
22	<1	7	3	<1	1	<1	1
23	<1	<1	<1	<1	<1	<1	<1
24	<1	<1	<1	<1	<1	<1	1
25	<1	<1	1	<1	<1	<1	<1
26	<1	<1	<1	<1	<1	<1	<1
27	<1	<1	<1	<1	1	<1	<1
28	<1	<1	<1	<1	<1	<1	20
29	<1	<1	<1	<1	<1	<1	7
30	<1	<1	<1	2	<1	<1	3
31	<1	<1	<1	<1	<1	1	8
32	1	<1	<1	<1	<1	<1	8
33	<1	1	1	<1	<1	<1	<1
34	<1	1	<1	<1	<1	<1	1
35	<1	<1	<1	<1	<1	<1	<1
36	<1	1	<1	5	<1	<1	9
37	<1	<1	<1	<1	<1	<1	<1
38	<1	<1	1	2	<1	<1	11
39	<1	<1	<1	1	<1	<1	<1

Notes:
- Total viable bacteria (TVB) by membrane filtration of a 100-mL sample, R2A culture media, 30 to 35°C incubation temperature, and 120-hour incubation time period.
- Sampling terminated after day no. 7.
- Organism identification by RiboPrinter: *Ralstonia pickettii* (Table 9.10).

Abbreviation: PQ, Performance Qualification.

period of about 70 hours. Table 9.11 provides data for the second attempt of PQ execution. Organism identification by RiboPrinter continued to indicate *R. pickettii*. Ozone introduction was terminated. Storage tank– and distribution loop–dissolved ozone were removed by the in-line ultraviolet unit downstream of the storage tank. The storage and distribution loops were resanitized with a 1% solution of peracetic acid and hydrogen peroxide (three-step process) for a time period of about 62 hours. Table 9.12 provides data for the successful execution of the PQ.

Table 9.10 Purified Water System Total Viable Bacteria Excursion After Addition to Distribution Loop—Organism Identification

---

Identification by MIS
- Library number: "TSBA50"
- *Ralstonia pickettii* with "Sim Index = 0.845"
- *Erwinia chrysanthemi* biotype II with "Sim Index = 0.626"
- Oxidase result: Positive (rules out *E. chrysanthemi*-biotype II)

Identification by RiboPrinter
- Gram stain result: gram-negative rods
- Microorganism identification: *R. pickettii*
- RiboGroup: "227-323-S-3"
- Enzyme: EcoRI

Notes:
- Microbial Identification System (MIS) does not provide definitive organism identification.
- RiboPrinter provides positive identification, if possible, to by comparison to a "library" of data from DuPont.
- RiboPrinter can also provide a "DNA-type" pattern for the organism(s) to allow additional sampling at suspected points of contamination and comparison of RiboGroup.

RiboPrinter may also be used to verify the consistency or change in flora of bacteria in samples from various points in the system.

This case history provides very important information, summarized as follows:

- While the solubility of ozone in water is greater than 10× that of oxygen in water, obvious "outgassing" of ozone occurs as discussed in chapter 7. While outgassing in the storage tank is well documented, the amount of nondissolved ozone associated with outgassing in distribution tubing *may* be a factor related to contact of dissolved ozone with material in biofilm.
- Another factor related to ozone "effective" destruction of biofilm relates to the time that it takes for ozone to "diffuse" from the mainstream to material in the biofilm. Conservatively, the half-life of ozone is about 25 to 30 minutes depending on pH. During the diffusive flow from the flowing stream to the material in the biofilm, the ozone concentration is continuously decreasing by not only half-lives but also reaction with "shielding" biofilm material such as EPS.
- Sanitization with ozone is performed in a dynamic mode (full recirculation). Sanitization with a 1% mixture of peracetic acid and hydrogen peroxide is a three-step process (per the procedure in this chapter) providing a long stagnant time period, where diffusive flow by concentration difference occurs. As the 1% solution of peracetic acid and hydrogen peroxide is consumed by reaction with biofilm material, the diffusive flow of sanitizing agent to the biofilm increases since the concentration gradient increases.
- The absence of bacteria subsequent to extended time period effective chemical sanitization with a 1% solution of peracetic acid and hydrogen peroxide *and* the absence of bacteria in the feedwater to the distribution loop indicate that biofilm removal *may* be possible.
- The resistance of *R. pickettii* in biofilm to ozone sanitization is consistent with material in the literature. It would confirm the fact that it is the "marker" organism for determination of biofilm removal and bacteria control.

DESIGN CONSIDERATIONS

Since there is a significant difference in the design considerations for stainless steel distribution loops compared to plastic distribution loops, this section will address each material of construction separately.

Table 9.11 Purified Water System Total Viable Bacteria Excursion After Addition to Distribution Loop—Restart of PQ After Extended Loop Ozonation

Sample point no.	TVB results day 1A (cfu/100 mL)	TVB results day 2A (cfu/100 mL)	TVB results day 3A (cfu/100 mL)	TVB results day 4A (cfu/100 mL)	TVB results day 5A (cfu/100 mL)	TVB results day 6A (cfu/100 mL)	TVB results day 7A (cfu/100 mL)
1	<1	<1	<1	<1	1	<1	2
2	<1	<1	<1	<1	<1	1	<1
3	<1	<1	1	1	<1	<1	<1
4	<1	<1	<1	<1	<1	<1	1
5	<1	<1	<1	2	1	4	12
6	<1	<1	<1	2	1	3	6
7	<1	<1	<1	4	3	2	4
8	<1	<1	2	1	4	4	8
9	<1	<1	<1	3	8	14	10
10	<1	<1	<1	<1	1	1	4
11	<1	<1	<1	<1	<1	<1	<1
12	<1	84	128	68	90	48	86
13	<1	2	3	12	28	44	36
14	<1	3	7	8	25	18	22
15	<1	<1	<1	<1	2	3	2
16	<1	<1	6	10	12	8	12
17	<1	<1	<1	1	3	2	8
18	<1	8	15	12	20	110	64
19	<1	3	12	10	16	8	32
20	<1	<1	<1	1	<1	1	2
21	<1	<1	<1	<1	<1	<1	1
22	<1	<1	<1	<1	1	<1	<1
23	<1	<1	<1	<1	<1	<1	<1
24	<1	<1	<1	<1	<1	<1	<1
25	<1	<1	<1	<1	<1	<1	<1
26	<1	<1	<1	<1	<1	<1	<1
27	<1	<1	<1	<1	1	<1	<1
28	<1	<1	<1	1	<1	2	2
29	<1	<1	<1	<1	<1	<1	<1
30	<1	<1	<1	2	<1	<1	<1
31	<1	<1	<1	<1	<1	1	1
32	<1	<1	<1	<1	<1	<1	<1
33	<1	<1	<1	<1	<1	<1	<1
34	<1	<1	<1	<1	<1	<1	2
35	<1	<1	<1	<1	<1	<1	1
36	<1	<1	<1	2	<1	<1	2
37	<1	<1	<1	<1	<1	<1	1
38	<1	<1	<1	1	<1	<1	4
39	<1	<1	<1	<1	<1	<1	<1

Notes:
- Total viable bacteria (TVB) by membrane filtration of a 100-mL sample, R2A culture media, 30 to 35°C incubation temperature, and 120-hour incubation time period.
- Sampling terminated after day no. 7A.
- Organism identification by RiboPrinter: *Ralstonia pickettii*

Abbreviation: PQ, Performance Qualification.

Stainless Steel Distribution Loops

The accepted type of stainless steel tubing is 316L stainless steel. It is suggested that orbital welding of fittings, tubing sections, valves, etc. be performed for a distribution loop to the greatest practical extent. When sanitary ferrules are required, they should be orbitally welded to tubing ends and secured with a gasket (of appropriate material of construction and clamp. The chemical composition of 316L stainless steel should consider that orbital welding will be

Table 9.12 Purified Water System Total Viable Bacteria Excursion After Addition to Distribution Loop—Restart of PQ After Extended Chemical Sanitization

Sample point no.	TVB results day 1B (cfu/100 mL)	TVB results day 2B (cfu/100 mL)	TVB results day 3B (cfu/100 mL)	TVB results day 4B (cfu/100 mL)	TVB results day 5B (cfu/100 mL)	TVB results day 6B (cfu/100 mL)	TVB results day 7B (cfu/100 mL)
1	<1	<1	<1	<1	<1	<1	<1
2	<1	<1	<1	<1	<1	<1	<1
3	<1	<1	<1	<1	<1	<1	<1
4	<1	<1	<1	<1	<1	<1	<1
5	<1	<1	<1	<1	<1	<1	<1
6	<1	<1	<1	<1	<1	<1	<1
7	<1	<1	<1	<1	<1	<1	<1
8	<1	<1	<1	<1	<1	<1	<1
9	<1	<1	<1	<1	<1	<1	<1
10	<1	<1	<1	<1	<1	<1	<1
11	<1	<1	<1	<1	<1	<1	<1
12	<1	<1	<1	<1	<1	<1	<1
13	<1	<1	<1	<1	<1	<1	<1
14	<1	<1	<1	<1	<1	<1	<1
15	<1	<1	<1	<1	<1	<1	<1
16	<1	<1	<1	<1	<1	<1	<1
17	<1	<1	<1	<1	<1	<1	<1
18	<1	<1	<1	<1	<1	<1	<1
19	<1	<1	<1	<1	<1	<1	<1
20	<1	<1	<1	<1	<1	<1	<1
21	<1	<1	<1	<1	<1	<1	<1
22	<1	<1	<1	<1	<1	<1	<1
23	<1	<1	<1	<1	<1	<1	<1
24	<1	<1	<1	<1	<1	<1	<1
25	<1	<1	<1	<1	<1	<1	<1
26	<1	<1	<1	<1	<1	<1	<1
27	<1	<1	<1	<1	<1	<1	<1
28	<1	<1	<1	<1	<1	<1	<1
29	<1	<1	<1	<1	<1	<1	<1
30	<1	<1	<1	<1	<1	<1	<1
31	<1	<1	<1	<1	<1	<1	<1
32	<1	<1	<1	<1	<1	<1	<1
33	<1	<1	<1	<1	<1	<1	<1
34	<1	<1	<1	<1	<1	<1	<1
35	<1	<1	<1	<1	<1	<1	<1
36	<1	<1	<1	<1	<1	<1	<1
37	<1	<1	<1	<1	<1	<1	<1
38	<1	<1	<1	<1	<1	<1	<1
39	<1	<1	<1	<1	<1	<1	<1

Notes:
- Total viable bacteria (TVB) by membrane filtration of a 100-mL sample, R2A culture media, 30 to 35°C incubation temperature, and 120-hour incubation time period.
- Intense sampling continued for an additional 20 days with no bacteria present

Abbreviation: PQ, Performance Qualification.

performed. Table 9.13 provides data for the chemical composition of 316L stainless steel that will be orbitally welded. These data provide an excellent reference for specification of the quality of 316L stainless steel.

Sections of 316L stainless steel tubing and tube-to-fitting connections are generally orbitally welded (unless sanitary ferrules are used). Dimensions with acceptable tolerances for weld and clamp fittings are provided in the literature (ASME BPE, 2009c). Orbital welding is performed by an automated process, without the addition of "filler" material, in an inert gas atmosphere. The criteria for acceptance or rejection of orbital welds are presented in chapter 11.

Table 9.13 Chemical Composition of 316L Stainless Steel for Automatic Welding

Composition (%):
- Carbon (maximum): 0.035
- Chromium: 16.00–18.00
- Manganese (maximum): 2.00
- Molybdenum: 2.00–3.00
- Nickel: 10.00–15.00
- Phosphorous (maximum): 0.045
- Silicon (maximum): 1.00
- Sulfur: 0.005–0.017

Source: Courtesy of Bioprocessing Equipment, 2009.

All 316L stainless steel tubing employed for an application should not be "accepted" for installation into a distribution system unless the material is accompanied by certificates clearly indicating the quality of the material. Specifically, the mill heat numbers, with chemical and physical analysis, must be provided *with* delivery/receipt of all 316L stainless steel material. A compendial water distribution system should *not* contain any 316L stainless steel tubing, fittings, etc. unless definitive information associated with the chemical and physical properties of the material is available. Accelerated corrosion/rouging of stainless steel may occur when 316L stainless steel is "mixed" with other austenitic stainless steels in a storage and distribution system as a result of differences in "corrosion potential."

All 316L stainless steel tubing and fittings received for installation into the distribution loop should be physically separated from other stainless steel or metallic material at a facility. Material handling and control are extremely important. A dedicated section in chapter 11 discusses material handling and control for stainless steel.

Orbital welding machines, used to weld stainless steel sections of tubing and fittings, should clearly demonstrate the ability to produce properly welded joints with a smooth interior surface. A document "standard operating procedure" in the form of an "orbital welding machine procedure" must be provided prior to initiation of orbital welding. Further, the orbital welding machine ability to generate acceptable welds (excluding individual welder qualification) must be verified and documented. Welding procedures should be consistent with criteria specified in ASME BPE (2009d).

The individual(s) operating the orbital welding machines must be qualified to preestablished "standards." Numerous "professional society" qualification programs, such as ASME, are available. After successfully completing a qualification program, the individual welder(s) receives a Certificate of Course Completion. The certificate, along with documentation for the orbital welding machine, should be available prior to initiating welding for the project and should be clearly displayed at the area of installation.

To ensure that the orbital welding machine and the qualified welder are capable of producing "acceptable" welds, it is strongly suggested that test welds be performed at the beginning and end of each workday (or shift, if multiple shifts are used). The test welds should be performed using all tubing sizes that will be welded on a particular day. Each test weld should be clearly labeled and maintained in a secured area. The test welds should not be discarded after the distribution tubing installation is completed *or* after the system is validated. They should be retained for inspection by regulatory personnel at any time after system installation. Test weld acceptance criteria are presented in chapter 11.

While it is strongly suggested that a visual inspection each orbital weld be performed in accordance with ASME B31.3, the literature suggests that a minimum of 20% of all orbital welds shall be inspected (ASME BPE, 2009e). Obviously, due to system "closure," some "blind" welds will not be accessible for inspection. The inspection process, as a minimum, should be a thorough boroscopic internal type, examining the quality of the interior finish of the weld. Videotape recording, which provides a permanent record of this inspection, may be used, particularly if a "third-party" visual inspector is not used. Visual inspection of welds by an individual employed directly by the tubing installer (or under subcontract to the tubing installer) may be challenged. The third party inspector should be qualified to one or more acceptable standards (chap. 11).

If the inspector determines that a specific weld is unacceptable, both the orbital welding machine and individual performing the orbital welding should immediately terminate execution of any additional welding. The individual welder, using the welding machine performing the unacceptable weld, should immediately perform additional test welds. The inspector should review the test welds. If the test welds are acceptable, the welding machine and individual welder may continue to perform orbital welding. However, if two sequential acceptable test welds cannot be generated, the inspector should remove both the orbital welding machine and individual welder from the "roster" of equipment and qualified welders at the facility until an investigation is performed and the cause of the unacceptable welds identified.

There are several reasons for rejecting the quality of an orbital weld. Criteria may include, but are not limited to, one or more of the following:

- Observation of *any* pinholes
- Irregular or erratic weld bead appearance
- Observation of thermal cracking
- Observation of dark brown, black, or rainbow color around the weld bead
- Weld bead thickness at narrowest point less than 50% of the weld bead thickness at widest point
- Lack of fusion
- Misalignment of welded sections of tubing
- Welding "overrun" due to improper operation of orbital welder
- "Restarts" of orbital welder or welding "skips" around perimeter
- Noticeable convex weld bead (>10% of tubing wall thickness) above base stainless steel material
- Noticeable concave weld bead (>10% of tubing wall thickness) below base stainless steel material

The mechanical finish of a weld (along with the interior finish of tubing and fittings) should be selected based on the application. In general, a highly mechanically polished finish with electropolish is used for Water for Injection distribution systems, while the degree of polish for Purified Water distribution systems *may* be lower. The interior of stainless steel surfaces is mechanically polished to a specified finish. Historically, the mesh size of the abrasive material used in the final polishing step determined the mechanical polish. The mesh size was classified in units of "grit" related to the number of "scratches" per unit area of the abrasive material. This process resulted in nonuniform mechanical polish rating. Subsequently, surface finish achieved by mechanical polish is generally determined by instrumentation and expressed in Ra, the arithmetic average of the surface roughness expressed in microinches. RMS, the root mean square, of the average surface roughness in microinches is infrequently used. Mechanical polish inspection methods using instrumentation such as a profilometer may be used to accurately determine the Ra value. The relationship between grit, RMS, and Ra is presented in Table 9.14. As indicated earlier, the interior surface finish of tubing, fittings, and valves will affect the rate of biofilm growth but will not inhibit biofilm formation if bacteria are present in the water. Subsequently, lowly mechanically polished and poorly welded distribution loops will require more frequent sanitization than loops with a higher degree of polish and properly executed and inspected orbital welds.

Table 9.14 Relationship Between Grit, RMS, and Ra for Stainless Steel Tubing

Grit size	RMS	Ra
120	58	52
180	34	30
240	17	15
320	14	12

Abbreviation: RMS, root mean square.
Source: From Bigelow and DiVasto (1986).

Upon completion of distribution loop installation, hydrostatic or pneumatic pressure testing should be performed. The method of testing, test pressure, and acceptance criteria should be specified. The literature states that orbitally welded tubing system should be pressure tested (ASME BPE, 2009f).

Installed storage and distribution systems must be "passivated" prior to start-up. This process increases the resistance to corrosion by using acid to enrich the metal surface closest to the water with a stainless steel constituent, chromium, which is less susceptible to corrosion. The 316L stainless steel tubing surfaces contain 67% to 69% iron and 16% to 18% chromium, a chromium:iron ratio of about 0.24 (Kilkeary and Sowell, 1999). Further, the literature indicates that surfaces of orbital weld "beads" could exhibit chromium:iron ratios less than 0.15 (Roll, 1997). The passivation process removes iron in a thin layer (≤ 50 Å) at the surface of tubing, welds, etc. As iron is removed, the layer is enriched with significantly less corrosive chromium. Effective passivation will result in surface chromium:iron ratios of 2.0 or greater. The passivation process includes a precleaning operation with a warm alkaline solution, rinse, treatment with an acid, and final high-purity water rinse. Historically, nitric acid had been exclusively used for passivation (Coleman and Evans, 1990). However, citric acid coupled with a chelating agent such as ethylenediaminetetraacetic acid (EDTA) has demonstrated the ability to provide a well-passivated storage and distribution system (Balmer and Larter, 1993; Coleman and Evans, 1991). A field test, such as a "Ferroxyl Test for Free Iron," should be used to verify the effectiveness of the passivation operation (ASTM, 2006a).

Passivating 316L stainless steel storage and distribution systems will significantly inhibit but not eliminate rouge formation. It is suggested that repassivation of storage and distribution systems be performed every —one to three years based on periodic inspection of system surfaces. If excess rouging is noted, derouging, a more aggressive chemical treatment technique, may be required (Baines, 2010; Roll, 2010).

The interior of the distribution and storage systems may also be electropolished. Electropolishing is generally used for the interior of tubing surfaces with a mechanical polish of Ra less than 15 to 20. Electropolishing is an electromechanical process that removes surface atoms from the finished stainless steel surface. The surface acts as an anode. In this process, the concentration of iron on the stainless steel surface decreases, while the concentration of chromium increases. A smooth, chromium-rich, corrosion-resistant oxide layer is formed, generally 25 to 50 Å thick, controlled by the current and voltage used for the process. Properly prepared surfaces will exhibit a "mirror-like" finish. As indicated earlier, biofilm growth, not formation, is decreased on electropolished surfaces. A typical reaction associated with the electropolishing process is presented in Figure 6.5.

A properly installed stainless steel distribution loop should be sloped to allow complete draining. For certain systems, particularly USP Water for Injection distribution systems, complete loop drainage may be augmented by the ability to sanitize the loop with Pure Steam. Loops may be sloped in one direction to drain or may be configured such that sections drain in multiple different directions from high points. The sloped loops should be equipped with appropriate high point connection and a low point drains. As indicated, multiple drains may be used in "crowned" loops. A generally accepted criterion for sloping stainless steel tubing is 1/8 in./linear ft. However, it is suggested that the degree of sloping may be a function of the application and the diameter of the tubing. For smaller diameter tubing, it may be appropriate to slope lines at greater than 1/8 in./linear ft to compensate for "surface tension" that will inhibit gravity draining. For connections that rely on a total gravity feed, such as the connection from a single or multiple effect distillation system condensing unit to a storage tank, sloping with values as high as 1 to 2 in./linear ft may be appropriate. This ensures that there is no "holdup" of Water for Injection in the line from the condensing unit to the storage tank.

When specifying the layout of a distribution loop, it is generally more practical to provide a "plan view" with the intended "general" tubing arrangement/routing. A detailed dimensional drawing, particularly one attempting to show three-dimensional views, is often inappropriate. The tubing installer will generally "work around" other utility tubing/piping at a facility and, using previous experience, install the loop in the most efficient fashion. While distribution tubing installation should be reviewed by the pharmaceutical company or the company preparing the specification for the project, responsibility for providing detailed information (e.g., dimensioned isometric drawing, weld numbers, and heat numbers) associated with loop installation belongs to the tubing installer (chap. 11). As a minimum, the information should include isometric drawings that show the following three elements:

- The elevation of all points within the distribution system above a reference point. Elevations at the beginning and end of horizontal, vertical, or other sections of tubing, along with the tubing dimensions, are required.
- The mill heat numbers for each section of tubing and all fittings used in the installation must be provided.
- Identification of all welds in the system with reference to "weld numbers." The weld numbers should correspond to documentation associated with welder qualification, welder machine conditions (including printout where available), and the tubing inspector's observations. This material will be required for a properly prepared IQ of the distribution system.

Seamless Plastic Distribution Systems

High-purity water requirements for semiconductor applications employ plastic piping systems with a smooth inner surface at heat-welded joints. As previously indicated, suggested plastic tubing/piping materials of construction for pharmaceutical water systems are limited to PVDF and unpigmented polypropylene material. Refinements in the method of executing welding, to a certain extent mimicking the orbital welding of stainless steel, minimize raised and depressed surfaces or crevices at all welds, thus providing a system where desired microbial control can be achieved. The smoothness and hydrophobic nature of the plastic piping surfaces, without any interior discontinuities associated with fusing the tube/pipe and/or fittings, is technically attractive. There are some important items associated with plastic tubing/piping systems that should be addressed. The pressure, temperature, and thermal coefficient of expansion characteristics associated with plastic tubing/piping have been discussed earlier in the chapter. For smooth, interior-welded, plastic tubing/piping distribution systems, the material and installation cost is the same as for a 316L stainless steel distribution system. Thus, the technical merit of using a plastic tubing/piping system cannot be justified by cost savings.

There are certain critical applications, primarily USP Purified Water for biotechnology applications, where extremely low concentrations of trace metallic impurities are detrimental. For these applications, it may be appropriate to use smooth interior plastic tubing/piping systems.

Smooth interior plastic tubing/piping systems require verification of the reproducibility of weld quality. Test welds, performed at the beginning and end of each workday (or shift), similar to those discussed for 316L stainless steel tubing, should be specified. Individuals installing this type of plastic tubing/piping should meet, as a minimum, the welder qualification guidelines outlined in a prepared specification. The distribution loop installer should maintain a quality control/quality assurance program with specific training and testing requirements for individual welding machines and individual welders.

Physical support is required for plastic tubing/piping systems. The suggested spacing between support "hangers" for both unpigmented polypropylene and PVDF, as a function of pipe size, is presented in Table 9.15. For most systems, however, it is suggested that continuous support of plastic tubing/piping be considered. This is particularly true for PVDF systems operating at elevated temperatures or periodically thermally cycled. Proper continuous support of plastic tubing/piping should ensure that horizontal sections of piping are entirely in a straight plane. Quite often, installers will include fittings, with larger diameters than the tubing/piping, within the continuous support "channel." This places unnecessary stress on the welds associated with the fitting(s) since the connection is not in the same plane, resulting in an upward slope of the tubing/piping as it enters and/or exits from the fitting. A properly installed loop should include "cutouts" in the support channel for fittings that have a larger diameter than the tubing/piping.

Table 9.15 Suggested Spacing of Support for Unpigmented Polypropylene and PVDF Piping as a Function of Pipe Size

Schedule 80 unpigmented polypropylene Pipe size (in.)	Support spacing at 73°F (linear in.)
0.5	28.8
0.75	32.4
1	37.2
1.5	45.6
2	51.6
3	64.8
4	74.4

Source: Courtesy of Enpure™ (2007b).

Schedule 80 SYGEF (PVDF) Pipe size (in.)	Support spacing at 176°F (linear in.)
0.5	29.53
0.75	33.46
1	35.43
1.5	45.28
2	47.24
3	57.09
4	62.99

Abbreviation: PVDF, polyvinylidene fluoride.
Source: From George Fischer (2009c).

As discussed previously for 316L stainless steel distribution loops, it is highly desirable to ensure full drainability of the distribution loop by providing appropriate slope, suggested as 1/8 in./linear ft. This is a significant challenge for plastic material loops because the tubing/piping is not "rigid" in nature (like stainless steel) and is supported within the channel (for horizontal sections of tubing/piping). Some tubing/piping installers will suggest that sloping is not critical because a "clean" pressurized gas can be used to displace entrapped water from the distribution tubing/piping. Most plastic tubing/piping manufacturers provide highly specific warnings indicating that gas should not be used in the tubing/piping (Enfield, 1993). Sanitizing with steam is not an alternative because it is a gas and exceeds the operating temperatures suggested for unpigmented polypropylene and the pressure/temperature requirements for PVDF.

The maximum operating pressure as a function of temperature for various sizes of PVDF and unpigmented polypropylene piping has been presented in Tables 9.3 and 9.4. System design should consider the maximum allowable operating pressure for a given application.

While orbitally welded 316L stainless steel distribution loops are routinely "leak checked" at the end of installation, many plastic tubing/piping loops are not. To verify the complete integrity of an installed plastic tubing/piping loop appropriately, it is suggested that both a hydrostatic test and water pressure hold test (for a minimum time period of 4 hours at the design pressure of the tubing/piping) should be performed. Unfortunately, the pressure hold test, the most accurate indicator of a leak in a plastic tubing/piping loop, should be conducted at ambient conditions. Small fluctuations in the ambient temperature may contribute to the observed pressure since water is noncompressible. During the pressure hold test, the temperature of the water should be monitored to ensure that it does not influence the test results. For PVDF loops that will be operated at high temperatures or thermally cycled, operating personnel should perform testing at the elevated temperature, ensuring that leaks are not noted. It is inappropriate to conduct the pressure hold test at an elevated temperature without highly accurate temperature control and a mechanical means of pressure/vacuum "relief."

As discussed earlier, primarily for PVDF, the tubing/piping "layout" should be considered for distribution loops operating at elevated temperatures or, more importantly, thermally cycled. The relatively high thermal coefficient of expansion for PVDF must be considered in the design/layout of the distribution loop.

Socket-Welded (Heat-Welded) Plastic Distribution

While not recommended for Purified Water or Water for Injection distribution systems, particularly where microbial control is desired, socket-welded plastic tubing/piping is occasionally employed. Heat-fused/welded plastic joints present unique problems. Figure 9.9 shows four sketches that attempt to demonstrate the nature of the problems. This process requires skilled personnel who have performed the heat welding operation several times and are subsequently uniquely qualified to obtain required results. Figure 9.9A demonstrates a situation where the heat welding is performed correctly. A smooth surface exists without a "ridge" or crevice. Figure 9.9B depicts a situation where "overengagement" has occurred, resulting in generation of a crown. This may be referred to as a bead by individuals performing the heat welding. Experience indicates that beads may not present serious problems with regard to microbial control. Assuming that the axis of the fitting is perpendicular to the centerline of the tubing/piping, the ridge or bead should have relatively uniform height and thickness. While this may provide an area for bacterial growth, it is a less serious situation than that associated with "underengagement" (Fig. 9.9C). This situation is associated with tubing/pipe that has not been installed far enough into the fitting, resulting in a defined crevice—a highly undesirable condition. The crevice provides a location for bacteria to accumulate and replicate. Proper installation of heat-welded plastic tubing/piping should ensure that crevices are not present. Finally, Figure 9.9D depicts a situation where "overengagement" is so great that the ridge or bead contains so much material (mass) that it "rolls over," resulting in formation of a crevice. This crevice will present even greater problems than the crevice in Figure 9.9C since it may have areas that are highly inaccessible to chemical sanitizing agents (or, in the case of PVDF, hot water). This may result in the inability to totally remove biofilm, a highly undesirable condition. While manufacturers of plastic tubing, piping, and fittings continue to develop automated

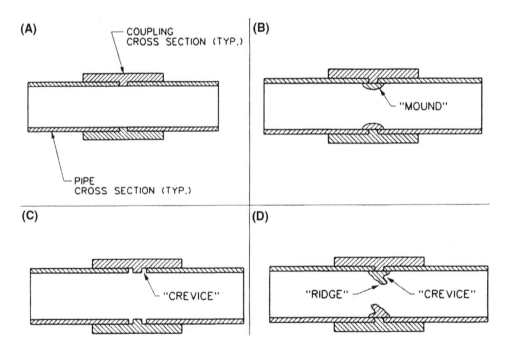

Figure 9.9 Examples of socket-welded pipe connections. (**A**) Proper weld; (**B**) overengagement; (**C**) underengagement; (**D**) overengagement with ridge.

Pipe size (in.)	(Copper coil resistance (Ω)	Fusion time (sec)	Single-weld fusion power (W)
0.5	0.17–0.18	113	44
0.75	0.22–0.23	113	54
1	0.28–0.31	113	66
1.5	0.39–0.43	113	83
2	0.47–0.51	113	103

Source: From George Fischer (1995).

Figure 9.10 Parameters for "Copper Coil" heat socket welding of unpigmented polypropylene.

machines to ensure that overengagement or underengagement does not occur, these conditions are common in systems using socket-welded plastic material.

Some manufacturers of heat-welded plastic fittings provide an internal heating coil, which is intended to allow operating personnel to ensure that proper engagement has occurred. Once engagement is verified, an electric current is connected to the leads of the internal heating coil, as shown in the chart in Figure 9.10, at a preset current, voltage, and time, to achieve a "proper" heat weld. This particular method of engaging pipe to fittings has certain advantages over a conventional "socket" heat-welded system. However, experience indicates that crevices may also occur in this type of system. Furthermore, there is concern associated with overheating or underheating the weld. Insufficient heating by the electric coil will result in improper fusion required for system integrity. Overheating results in penetration of the heating coil, generally copper, into the inner surface of the pipe wall, which is subsequently in direct contact with the recirculating water—a highly undesirable situation. As previously indicated, plastic piping systems may be considered for biotechnology applications where extremely low concentrations of trace metals in water are a concern. The introduction of trace amounts of copper for these applications is particularly undesirable.

In general, concerns associated with the preparation of test welds, welding machine operation, installer qualification, physical support of the tubing/piping, pressure rating of the tube/pipe as a function of diameter and temperature, effects of a high thermal coefficient of expansion, tube/pipe slope, and postinstallation hydrostatic testing, as well as pressure hold testing, apply to heat-welded socket plastic tubing/piping systems.

While not suggested for Purified Water distribution systems, PVC or CPVC may be used in "Softened Water" or Drinking Water systems. PVC may also be considered as a socket-welded plastic material. PVC fittings are generally solvent welded. Proper solvent welding requires cleaning prior to the application of an adhesive. Both the piping surface and the connection surface (valve, fitting, etc.) must be properly cleaned with glue applied before joining the surfaces. The amount of glue applied should not create deposits on the interior of the piping. Furthermore, proper engagement is also important for PVC piping to avoid crevices. As indicated earlier, PVC in distribution loops for pharmaceutical systems is seldom employed due to the high porosity of PVC material, with the resulting inability to control bacterial growth as a function of time after sanitization (Vess et al., 1993). PVC is used extensively for piping in pretreatment components and, in certain cases, deionization systems, "Softened Water systems," and RO units. It is strongly suggested that product water piping from any membrane process capable of removing bacteria be of stainless steel, unpigmented polypropylene, or PVDF material, not PVC. If PVC is used in pretreatment equipment, threaded fittings should be either eliminated or minimized. Flanges should be used in lieu of threaded fittings and PVC couplings. The flange connections, while providing a location for bacterial growth, are generally superior to both threaded fittings and unions. Any flanges used in a PVC system should have a gasket that extends to the inner diameter of the flange, eliminating a crevice at the "flange-to-flange" fitting.

Any connections from stainless steel to plastic fittings, particularly fittings exposed to pressures greater than 30 to 40 psig, should be by flange-to-flange. Threaded plastic fittings

Table 9.16 Sanitary Plastic Tubing—Pressure Rating as a Function of Diameter

Pressure ratings at 72°F Tubing size (in.)	Unpigmented polypropylene (psig)	PVDF (psig)
0.75	150	230
1	125	150
1.5	125	150
2	125	150
3	50	75

Temperature correction factor for pressure ratings		
Temperature (°F)	Unpigmented polypropylene	PVDF/KYNAR
72	1.00	1.00
100	0.85	0.90
125	0.65	0.80
175	Not recommended	0.60

Abbreviation: PVDF, polyvinylidene fluoride.
Source: Courtesy of Sani-Tech (1996).

connected to stainless steel by threads, particularly for relatively soft material such as unpigmented polypropylene, will "slip" over time, resulting in catastrophic failure of the piping system and a rapid release of a significant volume of water.

Plastic Distribution Systems with Sanitary Ferrule Connections

Plastic tubing/piping systems with sanitary ferrules appear to offer an excellent alternative to heat welding or the sophistication and cost associated with smooth inner tube welding of either unpigmented polypropylene or PVDF. In these systems, the ends of lengths of tubing are "formed" into a sanitary ferrule. Sanitary ferrules for two sections of tubing, or tubing and fittings, are joined using a gasket and plastic clamp. Ideally, this should provide fittings similar in nature to a stainless steel sanitary ferrule with a gasket and clamp. Unfortunately, there are limitations associated with this type of system. A limitation relates to the pressure reading for the sanitary tubing, clearly demonstrated as a function of tubing diameter in Table 9.16. Another disadvantage of this system is related to improper ferrule preparation and required tubing support. Improper plastic sanitary ferrule preparation is demonstrated in Figure 9.11. The face of all sanitary ferrules must be exactly perpendicular to the centerline of the tubing. If any sanitary ferrule is formed such that it is not perpendicular to the centerline of the tubing, a crevice will occur when the fitting is joined (Fig. 9.12). It is possible to purchase tubing with "preformed" sanitary ferrules. These prepurchased sections of tubing with sanitary ferrules, manufactured in the tubing supplier's facility, will be of excellent quality since they are prepared by highly experienced personnel in controlled conditions. However, sanitary ferrules for spool pieces with a length smaller than a standard length must be formed in the field (i.e., on-site). Unless this operation is performed by a highly qualified, experienced installer, with an appropriate heat-forming tool, the probability of obtaining the desired results is uncertain, particularly as the diameter of the tubing increases. Complicating the situation is the tubing support, which has greater effects on *tubing* systems using sanitary ferrules than plastic *piping* systems discussed earlier in this chapter. Unless the tubing is continuously supported, and the supports contain "slots" for the sanitary ferrule joints and clamps, allowing the tubing to lie "flat" in the support material, external physical pressure placed on the connection will result in a crevice (Fig. 9.13). Obviously, the conditions presented above must be carefully considered in the design and specification of a distribution system using sanitary plastic ferrules.

Another issue associated with the use of sanitary plastic ferrules relates to connection to stainless steel ferrules, particularly when stainless steel clamps are used and the tubing size is greater than 1.5 to 2 in. The connection between the "hard" stainless steel sanitary ferrule and the "softer" plastic fitting, with pressure exerted by a stainless steel clamp, may result in

DISTRIBUTION SYSTEMS—DESIGN, INSTALLATION, AND MATERIAL SELECTION   **345**

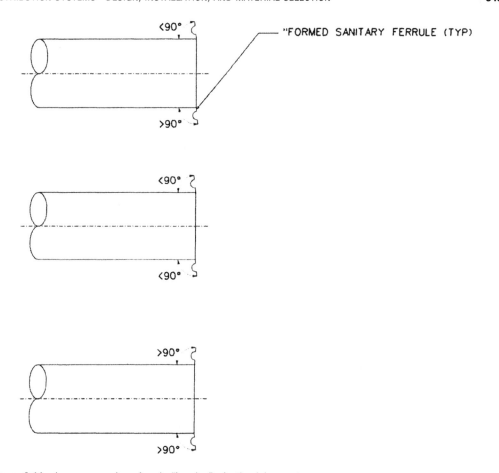

Figure 9.11 Improper sanitary ferrule "forming"–plastic piping systems.

deformation of the plastic tubing, particularly at higher pressures. This deformation could cause catastrophic failure of the plastic sanitary ferrule. If this type of fitting is required, and the pressure ratings for the tubing diameter required are satisfactory, it is suggested that a plastic clamp be considered for joining the stainless steel sanitary ferrule to the plastic sanitary ferrule.

To reiterate, a properly specified and installed sanitary ferrule tubing system will demonstrate excellent microbial control characteristics for a distribution loop. However, this can be achieved only when a conscientious effort is maintained by the specifying engineer, tubing installer, and tubing inspector. Related to comments presented above, continuous support should be used for all sanitary ferrule plastic tubing systems. The continuous support should include slots to allow the sanitary clamps to protrude in a manner that allows the tubing to lie totally "flat" in the channel.

Additional Design Considerations for All Distribution Tubing/Piping Materials

As discussed earlier, Purified Water and Water for Injection distribution loops should be of full serpentine design. Drinking Water distribution systems may be "dead ended"; however, this design should ensure that there is adequate residual disinfectant present in the water to control bacteria below the 500 cfu/mL suggested Action Limit for Drinking Water stated in the *General Information* section of USP. Full serpentine loops present the greatest challenge in achieving a predetermined flow rate, based on the pressure drop associated with friction loss through tubing/piping and fittings.

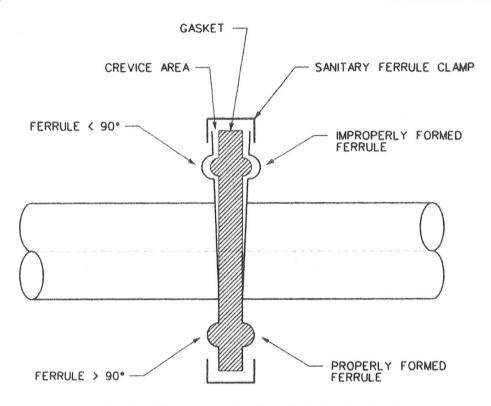

Figure 9.12 Crevice produced by joining improperly heat "formed" plastic sanitary ferrules.

The use of in-line repressurization and "satellite" storage tanks with repressurization pump(s) is discussed in chapter 6. It may also be appropriate to configure two or more serpentine-type loops, operating in parallel, around one or more storage tanks. Again, this will eliminate the pressure drop associated with a lengthy single distribution loop. The maximum operating pressure values for plastic tubing/piping, as a function of diameter, were presented in Table 9.3. While stainless steel tubing generally exhibits a much higher pressure rating than plastic tubing, particularly for larger diameters, pressure limitations may be associated with diaphragm valves in stainless steel tubing systems. As a result, it is suggested that it is inappropriate to design a distribution system to operate at a pressure greater than 100 to 110 psig.

Previously, it was suggested that the maximum flow rate required through a distribution loop should be determined by multiplying the maximum anticipated instantaneous draw-off rate by a factor of 1.5. Using this loop flow rate, the pressure drop through the serpentine distribution loop can be determined by calculating the "equivalent length" of the distribution tubing for a given tubing size. The equivalent length is associated with pressure drop through the linear feet of tubing plus the pressure drop, expressed as equivalent lengths of straight tubing, through fittings, such as elbows, tees, and reducers, as well as any in-line valves. The calculated equivalent length for a specific flow rate and tubing size can be used to determine the pressure drop through the loop in psig (Table 9.17). The tubing size may be increased or decreased to obtain an appropriate pressure drop, considering the flow rate through the tubing.

While it has been suggested that a maximum recirculation pump discharge pressure of 100 to 110 psig should be considered (lower values for plastic tubing/piping systems where appropriate), loop back pressure and the pressure drop through a spray ball in the return tubing/piping line to the storage tank should also be considered. Conservatively, the total return line pressure, including the pressure to the spray ball in the storage tank, should be ≥ 25 psig.

PROPER SUPPORT

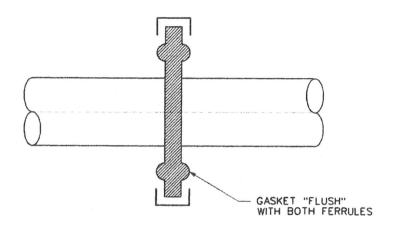

IMPROPER SUPPORT

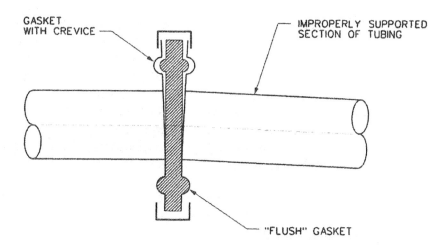

Figure 9.13 Crevice produced by improper support of sanitary ferrule tubing.

This results in a maximum available pressure drop through the distribution loop of 75 to 85 psig. If stainless steel tubing diameter or plastic tubing/piping diameter cannot be increased to allow the desired flow rate within the indicated pressure limitations, at a velocity greater than 2 to 3 ft/sec, the use of multiple parallel loops, in-line repressurization, or satellite tanks with repressurization pump(s) should be considered. If a facility consists of multiple stories (floors), the encountered pressure drop is complicated by the head of water required to reach the upper floors, particularly when the water purification systems, including the storage tank, are located on the lower floor. This situation is discussed below.

A slight modification to distribution loop design can compensate for "head loss" associated with the delivery of water from the basement area or lower floor of a building to the upper stories of a building (Fig. 9.14). A single "full-diameter" section of tubing can be used to deliver water to individual floors, or a combination of floors, extending to the top floor of the facility. After physically transversing an appropriate distance on the top floor of the facility, a back pressure regulator is positioned in the oversized main delivery loop. Tubing downstream of this valve will return to the spray ball in the storage tank. Subloops may be established on individual floors or

Table 9.17 Calculation of Equivalent Length and Pressure Drop for Specific Plastic Piping, Flow Rate, Size, and Configuration

Assumptions:
- Water flow rate: 30 gpm
- Piping material: Schedule 80 unpigmented polypropylene
- Type of connection: Socket weld
- Length of piping in loop: 1200 linear ft
- Number of 90° elbows in loop: 38
- Number of piping tees in loop: 12
- Piping size: 1.5 in.

Calculation:
- Piping pressure drop: 3.48 psig/100 linear ft
- 90° elbow equivalent length: 4.0 ft
- Piping tee with flow through "run" side equivalent length: 2.7 ft
- Piping: 1200 ft × 3.48 psig/100 ft = 41.76 psig
- Elbows: 38 elbows × 4.0 ft = 152 × 3.48 psig/100 ft = 5.29 psig
- Piping tees: 12 × 2.7 = 32.4 × 3.48 psig/100 ft = 1.13 psig

Total pressure drop: 48.18 (~48 psig)

Equivalent lengths:
- Tubing: 1200 ft
- Elbows: 152 ft
- Piping tees: 32.4 ft
- Total: 1384.4 ft

Source: Courtesy of Enpure™ (2007c).

on multiple floors, based on the pressure drop for full serpentine loops. A booster pump is positioned from the main "riser" loop, feeding the individual serpentine loops providing water to serpentine loops serving the individual floors, again based on the calculation of equivalent length. A back pressure regulating valve will be positioned in the section of tubing at the end of each individual subloop, which will flow to the return main tubing to the storage tank (Fig. 9.14). The size of the individual subloops and the selection of the individual subloop recirculation pumps would be calculated in a manner similar to that presented in the discussion associated with equivalent lengths above. While this method may seem cumbersome, it is an extremely effective technique of ensuring full serpentine-type distribution loops in multiple story buildings while maintaining "acceptable" operating pressures.

The previous information clearly defines acceptable methods of obtaining distribution in a loop while minimizing bacterial proliferation. Figure 9.15 illustrates a ladder-type loop that is often employed for distribution. This loop is technically flawed because it is not serpentine. It allows flow reversals, potential flow stagnation, and preferential "overdraw" at certain points, with associated inability to simultaneously deliver water to other points of use. This type of loop should *not* be employed for the distribution of USP Purified Water or Water for Injection.

Using distribution pump motors with variable frequency drives (VFDs) can provide a significant degree of flexibility for applications with multiple points of use, particularly when one or more points of use are "batch demands" with rapid fill requirements. The distribution loop layout may be arranged to configure, where possible, the point(s) of use where the high flow rate requirement(s) is (are) the initial "draw-off" point(s). The tubing diameter can be increased through these points of use and reduced in size for lower flow rates, as required for the other point-of-use applications. The points of use with lower flow rate requirements are unaffected by the increase in flow rate (and pressure) during upstream higher flow rate requirements. A back pressure regulating valve, positioned upstream of the storage tank, assists in "tuning-type" distribution loop pressure control during this operation. There are many advantages associated with this system, such as maintaining higher velocities through the majority of the distribution loop during normal operation *and* minimizing the length of tubing (and number of fittings) required for the higher flow rate applications.

DISTRIBUTION SYSTEMS—DESIGN, INSTALLATION, AND MATERIAL SELECTION 349

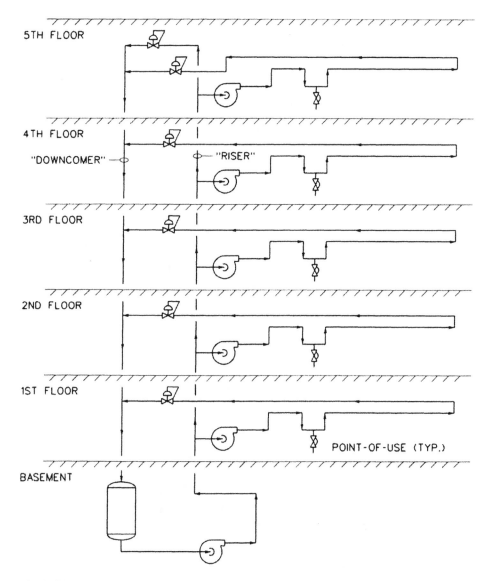

Figure 9.14 Distribution loop design—multistory application. (Individual "loops" may provide water to multiple floors if the pressure drop is not excessive. The top floor can operate without a pump if adequate pressure from the basement pump is available.)

Support equipment for the distribution loop should include a properly selected back pressure regulation system and storage tank return line spray ball. These accessories are discussed in chapter 6.

Loop "overdraw," another critical consideration, is also discussed in chapter 6. It is extremely critical to eliminate loop overdraw since it results in bacterial contamination of the stored and recirculated water through a points of use with insufficient flow rate requirements during maximum draw-off conditions.

It has been suggested, on several occasions, that the distribution loop flow rate should ideally be about 1.5 × the maximum design draw-off flow rate. Tables 9.18 to 9.20 contain information associated with flow rate, velocity, and pressure drop for 316L stainless steel tubing and selected plastic tubing/piping.

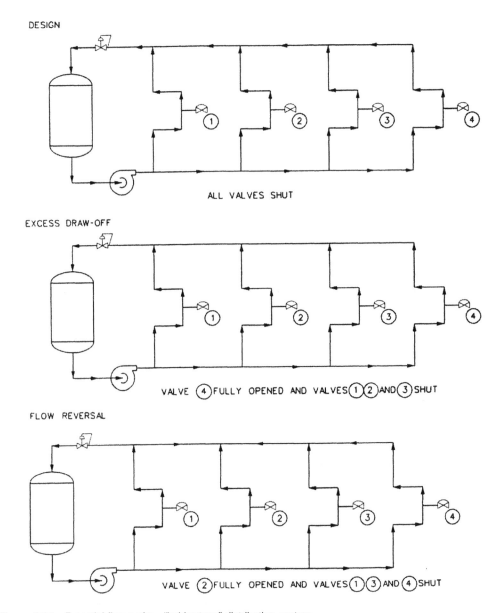

Figure 9.15 Potential flow paths—"ladder-type" distribution system.

The selection of point-of-use valves and accessories that may be connected to the valves, such as volumetric or mass totalizing meters and delivery hoses, must be carefully considered to eliminate potential back contamination of the distribution loop. The use of membrane filters, at points of use, is not appropriate or acceptable for bacteria control in USP Purified Water or Water for Injection systems. The use of laboratory "polishing loops" (which can significantly back contaminate the main USP Purified Water distribution loop) is also strongly discouraged. Some researchers require water with extremely high resistivity values and extremely low TOC values (<10 µg/L). Research applications can require ASTM Type 1 water, as summarized in Table 9.21. For such applications, it is suggested that a dedicated pretreatment system capable of processing raw water, providing pretreated water to the point-of-use polishers, be installed, rather than connecting the point-of-use polishing unit to the main recirculating loop.

Table 9.18 Flow Rate, Velocity, and Pressure Drop Vs. Pipe Size—Schedule 80 Socket-Welded Unpigmented Polypropylene

Pipe size (in.)	Flow rate (gpm)	Velocity (ft/sec)	Pressure drop (psig/100 ft of pipe)
0.5	1	1.48	0.97
0.5	3	4.43	7.43
0.5	5	7.38	19.12
0.75	5	3.92	4.10
0.75	7	5.48	7.64
1	7	3.26	2.16
1	10	4.86	4.18
1	15	6.99	8.86
1.5	20	3.75	1.65
1.5	25	4.69	2.49
1.5	30	5.62	3.48
1.5	35	6.56	4.64
1.5	40	7.50	5.94
2	30	3.35	0.99
2	35	3.91	1.31
2	40	4.46	1.68
2	45	5.02	3.09
2	50	5.58	2.54
2	60	6.07	3.56
2	70	7.81	4.74
3	70	3.49	0.67
3	75	3.73	0.75
3	80	3.98	0.85
3	90	4.48	1.06
3	100	4.98	1.29
3	125	6.22	1.95
3	150	7.47	2.73
3	175	8.71	3.63

Source: Courtesy of Chemtrol (2001b).

Conventional gooseneck faucets for the delivery of water from either a Purified Water or Water for Injection distribution loop should not be employed. They provide an unacceptable dead leg of stagnant water immediately after the point-of-use valve. When the valve is opened to deliver water, the "main" distribution tubing will be exposed to the biofilm associated with the stagnant water. This applies to plastic "recirculating" gooseneck faucets, particularly when the method of connecting the gooseneck faucet to a recirculating loop is considered (e.g., compression fitting). While customized plastic construction recirculating gooseneck faucets are commercially available, it is strongly suggested that *full* recirculation be provided. The distribution tubing may be physically positioned either directly above or directly below the laboratory workstation. A diaphragm valve of stainless steel or plastic construction (depending on the distribution loop material of construction) can be installed, minimizing the potential for bacterial growth. In addition to providing a much cleaner (from a microbial standpoint) valve at the point of use, the discharge side of the diaphragm valve does not contain the "hose barb" type of fitting normally included with a gooseneck faucet (stainless steel, nickel-plated brass, or plastic construction).

Unfortunately, many laboratory end users will install a section of nonsanitary hose to the hose barb connection on a gooseneck faucet. In addition to bacterial concerns associated with loop back contamination from the hose barb connection, the physical length of the hose is such that it often extends into a sink. In many cases, the hoses become a *permanent* fixture, are never or infrequently sanitized, and, on certain occasions, extend into a sink containing chemicals used to wash glassware. Obviously, the water in the sink may have significant chemical and microbial impurities that could result in serious back contamination problems to the main distribution loop. If the valve on the gooseneck faucet is slightly open or, when closed, exhibits

Table 9.19 Flow Rate, Velocity, and Pressure Drop Vs. Pipe Size—Schedule 80 Socket-Welded SYGEF (PVDF)

Pipe size (in.)	Flow rate (gpm)	Velocity (ft/sec)	Pressure drop (psig/100 ft of pipe)
0.5	1	1.00	0.38
0.5	2	3.18	1.37
0.5	5	5.01	7.50
0.75	5	2.93	2.03
0.75	7	4.10	3.78
0.75	10	5.85	7.31
1	10	3.56	2.18
1	15	5.34	4.62
1	20	7.11	7.87
1.5	25	3.40	1.15
1.5	30	4.08	1.61
1.5	35	4.76	2.14
1.5	40	5.44	2.74
1.5	45	6.12	3.41
2	40	3.24	0.78
2	45	3.65	0.97
2	50	4.05	1.17
2	60	4.86	1.64
2	70	5.67	2.19
2	80	6.48	2.80
2	90	7.29	3.49
3	70	2.78	0.39
3	80	3.18	0.49
3	90	3.57	0.62
3	100	3.97	0.75
3	125	4.96	1.13
3	150	5.96	1.59
3	175	6.95	2.11

Abbreviation: PVDF, polyvinylidene fluoride.
Source: From George Fischer (2009d).

a small "leak," water in the sink may be aspirated into the main loop. If the hoses are used for research applications at points of use, a sanitary-type fitting should be added to the discharge side of a diaphragm valve. The sanitary hose length should be adequate enough to provide water for routine operations, but *not* long enough to extend into the sink basin where glassware washing occurs. Further, dedicated hoses should be provided in "pairs," with clearly defined operating procedures for periodic sanitization and cleaning.

For manufacturing points of use from either Water for Injection or Purified Water distribution system, it is strongly suggested that dedicated "hard-piped" connections be considered. The suggested configuration to a batching vessel is shown in Figure 9.16. This arrangement provides a direct connection between the point-of-use "vessel" and the distribution loop. When Purified Water or Water for Injection is required for a manufacturing operation, the point-of-use valve (which may be either a manual or automatic valve) is opened, delivering the desired volume (or weight) of water to the processing tank. The same connection may be used for postmanufacturing cleaning of the processing tank. It is suggested that hard-piped systems, similar to those shown in Figure 9.16, be provided with a sampling valve, at an appropriate location, such that the sample is representative of the chemical and microbial quality delivered to the processing vessel. The sample valve may be positioned immediately upstream or downstream of the point of use, or through a sterile access port on the delivery valve, as shown in Figure 9.16. If sterile access ports with associated sample valves are used, the ability to sanitize the sample valves (during periodic hot water sanitization) or maintain hot conditions consistent with acceptable L/D ratio, discussed earlier, must be considered.

Table 9.20 Flow Rate, Velocity, and Pressure Drop Vs. Tube Size—Sanitary Unpigmented Polypropylene Tubing

Tubing size (in.)	Flow rate (gpm)	Velocity (ft/sec)	Pressure drop (psig/100 ft of tubing)
0.75	3.0	3.91	4.87
0.75	3.5	4.56	6.47
0.75	4.0	5.21	8.29
0.75	4.5	5.86	10.31
1.0	6.0	3.34	2.23
1.0	7.0	3.90	2.96
1.0	8.0	4.46	3.79
1.0	9.0	5.02	4.71
1.0	10.0	5.57	5.72
1.5	16.0	3.55	1.46
1.5	20.0	4.44	2.20
1.5	24.0	5.33	3.08
1.5	28.0	6.22	4.10
1.5	32.0	7.11	5.25
1.5	36.0	8.00	6.53
1.5	40.0	8.89	7.94
2	30.0	3.56	1.01
2	35.0	4.15	1.35
2	40.0	4.74	1.72
2	45.0	5.34	2.14
2	50.0	5.93	2.60
2	60.0	7.12	3.65
3	80.0	4.01	0.76
3	90.0	4.51	0.95
3	100	5.01	1.15
3	120	6.01	1.62
3	140	7.01	2.15
3	160	8.01	2.75

Source: Courtesy of Sani-Tech (1996).

Table 9.21 ASTM Reagent Grade Water Specifications

Parameter	Specification
Conductivity at 25°C	0.06 µS/cm
Total silica	3 µg/L
TOC	50 µg/L
Chloride	1 µg/L
Sodium	1 µg/L
Total viable bacteria	≤10 cfu/1,000 mL
Bacterial endotoxins	<0.03 EU/mL

Notes:
- Type I reagent grade water.
- Bacteria and bacterial endotoxin limits for "Type A" water.

Abbreviation: TOC, total organic carbon.
Source: Courtesy of ASTM (1993).

It would be inappropriate to discuss hard-piped points of use and sample provisions without discussing the "physical" sample location. The sampled water quality *must* be representative of the delivered water quality (chemical and microbial). An increasing number of facilities with batching tanks in "clean room" areas elect to extend tubing from the indicated sterile access port sample valves to a "gray space" (nonclean room) area. Sampling personnel

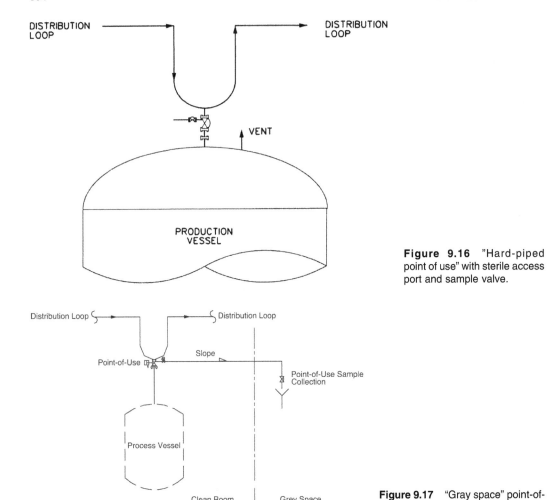

Figure 9.16 "Hard-piped point of use" with sterile access port and sample valve.

Figure 9.17 "Gray space" point-of-use sampling.

access to the sterile area is often "restricted." Figure 9.17 depicts a typical gray space sample arrangement. The length of tubing from the sterile access port valve to the actual sample discharge point may be 10 to 20 ft long and of narrow diameter. As discussed earlier in this chapter, biofilm may accumulate on tubing surfaces exposed to bacteria and will *not* be removed by periodic hot water flushing (prior to sampling). Fortunately, the sampling arrangement depicted in Figure 9.13 will affect the attributes of the actually delivered water in a "conservative" manner. In other words, it would be expected that TOC, conductivity, total viable bacteria, and bacterial endotoxin results associated with gray space samples would all be higher than the values for delivered water at the sterile access port valve.

One of the complicating factors associated with a hard-piped, point-of-use delivery system is the requirement to accurately determine the volume (or mass) of Water for Injection or Purified Water that is introduced, as an ingredient, to the manufacturing vessel. There are numerous ways to address this requirement. One method is to employ a "load cell" on the tank, a method of determining the weight of material in the tank. Unfortunately, this nonintrusive measuring technique is not applicable to many manufacturing applications for several reasons (physical orientation of tank, other ingredients, area of installation, etc.). Alternative methods include intrusive detection of the level within the processing vessel, visual inspection of the level in the vessel, and installation of a sanitary volume or mass totalizer within the tubing/piping between the point-of-use valve and the processing vessel. If a mass flowmeter is used, it may be appropriate, particularly for Water for Injection systems, to use a double block and bleed system, discussed earlier in this chapter. The double block and bleed system can be used with a hot recirculating main Water for Injection (or Purified Water) distribution loop to presanitize the meter and section of tubing between the processing tank and the point-of-use valve. For ambient temperature distribution loops with point-of-use mass flowmeters using a double block and bleed system, one of the "blocking valves," preferably the valve between the mass flowmeter and the main recirculating Water for Injection (or Purified Water) loop, can be equipped with a sterile access port that can be used with Pure Steam to provide sanitization.

This discussion of point-of-use volume (or mass) measuring techniques provides a *limited* presentation of options for some applications. The selected method of achieving the desired result is a function of the nature of the product being manufactured as well as the nature of the manufacturing operation. While several references have been made to "sanitary" conditions, many Purified Water systems may use tubing/piping and mass flow measuring techniques that are not "sterile" but adequate for the application. Unfortunately, the degree of latitude for Water for Injection systems is extremely limited.

Many applications requiring Purified Water for high-volume manufacturing operations configure processing vessels in a particular physical location of the facility. To facilitate withdrawal of product from the base of the processing vessels, Purified Water is generally supplied through a point-of-use valve to a "platform area." A single platform area could include one or two tanks or a dozen or more tanks. There may be multiple reagents added to a particular processing vessel as part of the manufacturing process for a particular product. Access manways may be included on the top of the processing vessels to facilitate the addition of reagents. Agitators and blenders may also be included as part of the accessories for the processing vessels. The use of Purified Water for these applications presents a challenge, particularly with regard to the quality of water *delivered* to each processing vessel *and* elimination of both chemical and microbial back contamination of the main recirculating loop.

A single platform area may be equipped with one or two points of use from a Purified Water distribution loop, which is servicing multiple processing tanks. Hoses, generally equipped with sanitary ferrule end connections, are used to deliver water from a point-of-use valve to the processing vessels.

Again, the volume (or mass) of Purified Water added as part of multiple manufacturing operations to produce multiple products is critical. Subsequently, it is not uncommon to use a single volumetric or mass totalizer downstream of the point-of-use valve. The material of construction for the delivery hose is also important. The material should be compatible with cleaning and sanitization requirements used to ensure repeated delivery of water without adversely affecting either the "main" Purified Water distribution loop (through back contamination) or the efficacy of the product manufactured in the vessels. Hot water sanitization, periodic autoclaving, and/or chemical sanitization of the hoses, in accordance with established SOPs, should be considered. If either of these techniques is used for microbial control, it is recommended that delivery hoses be manufactured of platinum-cured, medical grade–reinforced silicone with 316L stainless steel sanitary ferrule end connections. As a minimum, at least a single pair of hoses should be provided for a particular manufacturing platform. If multiple processing vessels are positioned in a defined physical area (with a single point-of-use valve), where isolation of any of the ingredients used for one product from another product is critical, multiple "pairs" of hoses should be considered.

SOPs should include requirements for the proper handling, cleaning, and sanitization of hoses and volumetric or mass totalizer within the delivery tubing, as well as all other surfaces in contact with Purified Water from the point-of-use valve between the Purified Water distribution loop and the processing vessel. The procedure should include, as applicable, the chemical cleaning of hoses, particularly if hoses are exposed to potential chemical back contamination in the manufacturing area. This is especially true when hoses, or sections of hoses, are exposed to contaminating surfaces, such as the floor of a manufacturing area. In addition, hoses should be properly labeled to clearly identify their intended application. It is also suggested that a hose-mounting bracket for each pair of hoses be included at the physical area of use. Finally, it should be emphasized that hoses present one of the greatest challenges, in terms of controlling chemical and bacterial purity, not only within the main Purified Water distribution loop but also at points of use. When hoses are used, the chemical and bacterial quality of the water should be periodically verified at the *end* of the hose feeding the processing vessel ("delivered" water). This would supplement data, such as online conductivity and TOC values, in the return line from the Purified Water distribution loop. Point-of-use quality is of particular concern to regulatory personnel when hoses are used. The distribution loop return conductivity and TOC values may *not* represent the water quality at the point of use, depending on "housekeeping" practices and strict adherence to applicable SOPs.

Valves used in the Purified Water distribution loop, including point-of-use valves, should be of diaphragm type. Zero dead leg valves should be employed for point-of-use valves as well as any valve where water is withdrawn from the distribution loop. Flow from the Purified Water distribution loop for the "zero dead leg" valves passes directly to the "recirculating loop" side of the diaphragm, as shown in Figure 9.18. This type of valve presents significant advantages for systems that are routinely operated at ambient temperature (and periodically sanitized).

There are two important items that should be addressed for diaphragm valves used for pharmaceutical applications. The first item is associated with conditions that promote bacteria back growth across the diaphragm and weir of the valve. Throughout this chapter, the importance of maintaining bacterial control on the discharge side of point-of-use valves has been emphasized. In chapter 4, the ability of bacteria to grow "against the direction of flow" is discussed. If a point-of-use diaphragm valve is opened, discharging water to tubing *with* bacterial contamination, it should be fully anticipated that bacteria will grow over the weir and diaphragm of the point-of-use valve and back into the recirculating distribution loop. If, for a

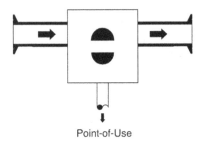

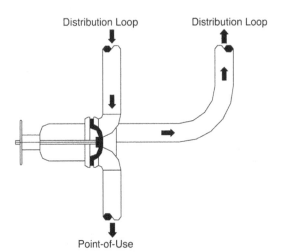

Figure 9.18 Examples of "zero dead leg" valves.

particular application, it is impossible to segregate a section of tubing, which *may* contain bacteria, from the point-of-use valve, a double block and bleed system should be considered. This system should be equipped with provisions for hot water or chemical sanitization of the section of tubing between the two blocking valves. If chemical sanitization is used, adequate postsanitization rinsing is critical.

The second item that should be addressed relates to the placement of diaphragm valves in the main recirculating distribution loop. In general, the loop distribution pump from the storage system will be provided with a product water isolating valve of diaphragm type. Point-of-use valves would be configured from the distribution loop, *not* positioned in the serpentine flow path. The back pressure regulating valve for the distribution loop should be provided with an actuator of air-to-close, spring-to-open type, failing in an open position. If a manual valve is positioned within the "run" section of the distribution loop, with the exception of the valve located downstream of the distribution pump, particularly for heated or thermally cycled loops, catastrophic failure of the distribution loop tubing could occur if valves are manually closed, isolating a section of tubing containing hot water. Automatic diaphragm valves will generally "open" at a pressure (90–125 psig) below that for catastrophic failure. Significant pressure or vacuum conditions can occur during heating or cooling of any isolated section of the distribution loop. This is particularly true for plastic construction distribution loops, but it is also true for stainless steel construction distribution loops. If this situation cannot be avoided, due to the configuration and requirements for the main distribution loop, a "positive" means of mechanical relief of pressure and/or vacuum must be provided. For sanitary applications, including Purified Water and Water for Injection applications, the method of relief must be a compound-type rupture disk. Use of a relief valve/vacuum breaker should be limited to nonsanitary applications where bacterial control is less critical.

Historically, a limited number of point-of-use *manual* diaphragm valves have been used with thermal probes. The valves are installed at points of use in thermally cycled distribution systems. The internal temperature sensor, with electrical feed, inhibits valve opening when the water in the main distribution loop is at an elevated temperature (e.g., 110–120°C). The greatest percent of "industrial" accidents occurring in the United States each year are associated with personnel exposure to hot water. Many Purified Water systems, and some Water for Injection systems designed for thermal cycle operations, are equipped with an automated system to ensure that points of use cannot be used during hot water sanitization. The ability to provide a system that will not allow operator exposure to hot water during the hot water sanitization cycle should be considered for personnel protection.

As discussed in chapter 6, distribution systems may be equipped with automatic point-of-use valves. Operation of the pneumatic actuators can be controlled by locally mounted, individual solenoid valves. All of the solenoid valves can be electronically connected to a central control panel that can limit the number of point-of-use valves that open at a particular time. This is an excellent concept for distribution loops that historically have experienced "system overdraw" conditions, resulting in bacterial contamination of the distribution loop and associated frequent sanitization requirements. However, the design considerations should include a 50% "excess" return factor at the maximum draw-off rate from the distribution system. While an automated valve arrangement is appropriate for existing systems, the Basis of Design for a new system, considering the suggested 50% excess return factor, may eliminate the requirement for automated point-of-use valves. It is suggested that one of the primary disadvantages of the automated system is related to the nature of multiple manufacturing processes at a particular facility. In certain cases, it may be extremely inappropriate to limit water flow to a particular operation in the "middle" of a critical manufacturing step. A properly designed new storage and distribution system should provide water to *all* end users, as needed, not affecting the nature of manufacturing activities or any other activities requiring water.

Previously, certain applications that required sanitary ferrules were discussed. Connections to storage systems and sanitary distribution pumps generally require installation of tubing (and valves) with sanitary ferrules. It is suggested that it may be appropriate to use point-of-use valves with sanitary ferrule connections for many applications. As a minimum, when hoses are employed in a manufacturing area, the discharge side of the point-of-use diaphragm valve should be provided with a sanitary ferrule. However, sanitary ferrules, and associated gaskets and clamps, require maintenance. Gaskets should be replaced in accordance with established preventative maintenance procedures. It is not uncommon for a sanitary clamp to require periodic tightening, particularly if a loop is thermally cycled. For distribution systems with plastic tubing/piping, sanitary ferrules will be necessary. Sanitary ferrules for any plastic tubing/piping system present a challenge, particularly with regard to an established operator inspection program to ensure that the clamps joining the plastic sanitary ferrules are tight. For both stainless steel and plastic systems, every sanitary ferrule in a flowing section of tubing/piping represents a potential source of bacteria input from atmospheric air. Permanently welded connections should be considered, in lieu of sanitary ferrule connections, wherever possible.

Many distribution systems will contain individual points of use that require water flow rates at a fraction of that in the main loop. It would be extremely inappropriate to install point-of-use valves the same size as the main distribution loop. Point-of-use zero dead leg valves are available with "full-size" distribution loop fittings and reduced size "delivery" size fittings. For example, a $2 \times 2 \times \frac{1}{2}$ in. zero dead leg valve contains "flow" fittings of 2 in. with ½ in. diaphragm/weir discharge connection. The Basis of Design for a system should clearly outline the maximum anticipated flow rate for each point of use from the distribution loop. The delivery valve size should be selected based on information within the Basis of Design. Occasionally, systems will use flow orifices or "travel stops" on pneumatically actuated diaphragm valves to restrict the flow of water at a point of use. It is strongly suggested that this method of flow control, particularly the use of orifices, is inappropriate. For point-of-use flow rate control, proper valve size selection is more appropriate.

Throughout this chapter, dead legs from the distribution loop have been discussed. Ideally, all connections to distribution loops should not contain dead legs. An evaluation of distribution loop dead legs should not be limited to valves but include support equipment such as sensors and probes associated with instrumentation. Online conductivity and TOC monitoring for Purified Water and Water for Injection system distribution systems are common. Conductivity probes and "sample taps" for TOC analyzers should be installed in a manner that eliminates dead legs and the potential for back contamination of the loop. Online TOC analyzers should employ loop sample valves of zero dead leg diaphragm type. Other instrumentation, such as thermal wells associated with temperature elements and pressure gauges, should be combined with sanitary ferrules connected as close as possible to the recirculating water. In certain cases, considering the size of the distribution loop, it may be appropriate to directly weld a sanitary ferrule to a section of tubing to minimize a dead leg. The use of "short outlet" fittings should be considered. For plastic construction distribution loops, this issue again presents a challenge because of the limited availability of fittings. For example, for *socket-welded* PVDF or PVC distribution loops, installation of a sanitary ferrule conductivity probe (as opposed to a highly undesirable threaded probe) requires a transition fitting, which presents an undesirable dead leg.

Subloops may be provided from a main recirculating Purified Water or Water for Injection distribution system. For certain Purified Water applications, the recirculating loop from the storage tank may be maintained at an elevated temperature for microbial control, while one or more ambient (or "cold") temperature loops are configured from the main loop. This is also applicable for Water for Injection applications, in which subloops may be operated for a portion of the time at ambient (or cold) temperatures. A drawing depicting a typical subloop is shown in Figure 9.19. Within the subloop, water flows to a recirculation pump and heat exchanger to decrease the temperature of the water to a desired value. Water is recirculated from the effluent of the heat exchanger through the subloop, returning to the suction side of the subloop distribution pump. Points of use requiring ambient temperature (or cold) water are positioned from the subloop. Quite often, the operations requiring the ambient (or cold) temperature water (Purified Water or Water for Injection) are used for a fraction of each day. Subsequently, ambient (or cold) temperature water can be "displaced" from the subloop and replaced with elevated temperature water from the main recirculating loop, with the subloop heat exchanger chill water or cooling water turned off. This increases the temperature of the subloop performing sanitization or, in the case of Water for Injection, addresses operating criteria (FDA, 1986).

A second method of obtaining ambient temperature water or reduced temperature water from a recirculating hot distribution loop ($\geq 80°C$) employs point-of-use heat exchangers.

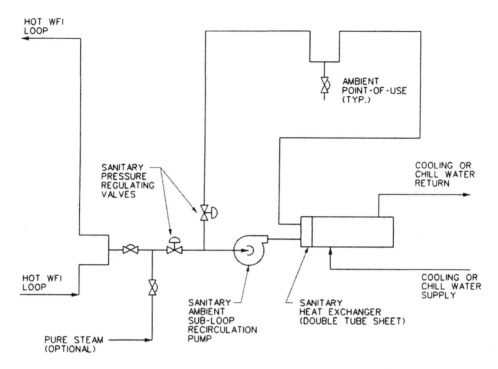

Figure 9.19 Ambient temperature subloop from hot recirculating Water for Injection loop. [All instrumentation, valves, and controls are not shown. A Pure Steam connection is optional for the sanitization of the ambient subloop. Hot Water for Injection, with cooling (or chill) water flow, may be used periodically (daily) for sanitization. All tubing, fittings, instrumentation connections, and so on are to be the sanitary type.]

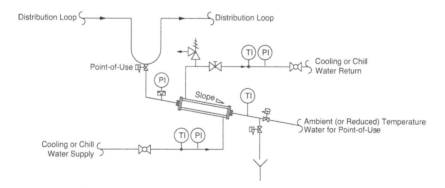

Figure 9.20 Sanitary point-of-use heat exchanger with accessories. (The isolation valve from the main hot recirculation loop is positioned as close as possible to the loop. The heat exchanger may be of sanitary design with double tube sheet. The heat exchanger shown is a straight tube. A U-tube heat exchanger may be used; however, it must be fully drainable, without water holdup. Discharge tubing from the heat exchanger may require eccentric reducer/fitting to ensure fully drainable status.)

Figure 9.20 depicts a point-of-use heat exchanger for a recirculating hot Purified Water or Water for Injection distribution loop. When ambient temperature water is required, the heat exchanger should initially be sanitized. This is accomplished by ensuring that the valve supplying water to the point of use is closed and that the valve diverting water-to-waste is open. It should be noted that certain state, local, or federal regulations may limit the temperature of water fed directly to drain to 140°F. To achieve this requirement, it may be necessary to blend the heated water associated with the sanitation operation with "tap water" to obtain the desired 140°F (or lower) temperature. The valve delivering water to the specific point of use from the main recirculating loop is open, delivering hot water through the heat exchanger, sanitizing both the heat exchanger and the distribution tubing. After a preestablished (and "validated") period of time, determined during PQ execution and from established SOPs, the cooling water flow to the heat exchanger is actuated. When the water temperature to the drain connection decreases to the desired delivery value, the valve diverting water to drain is closed. Water is then delivered to the point of use at the required temperature. The elapsed time period for "tempered" water use subsequent to completion of heat exchanger hot water sanitization must be determined for the specific application. It is suggested that a maximum time period of two to four hours be considered (and validated). Subsequent to the established time period or upon completion of heat exchanger use, the initial hot water sanitization step should be repeated and water drained from the heat exchanger. Pure Steam is often used for sanitization in lieu of hot water. On the basis of experience, the use of hot water sanitization not only produces superior microbial control than Pure Steam but also simplifies the operation.

While Figure 9.20 depicts a manual system, the valves may be automated to obtain the same results. Again, this can be an extremely effective method of delivering ambient temperature water when a fraction of points-of-use, in terms of number and/or flow rate, requires reduced temperature water, from a recirculating hot loop. As indicated, when the ambient temperature water delivery is complete, the cooling water flow should be stopped. The rinse-to-drain valve should open, flushing hot water to waste. The point-of-use valve from the main distribution system can be closed after a preestablished period of time. The waste-to-drain valve can remain open. This postdelivery operation, depending on the frequency of ambient temperature water demand, ensures that the tubing is not exposed to a constant wet environment since the hot water should at least partially evaporate from surfaces at the end of the operation. As indicated previously, it is inappropriate to allow hot stagnant water to remain in the heat exchanger and associated support tubing and valves when reduced temperature water is not required. The rinse-to-drain valve may be closed after a preset period of time after the postoperational sanitization operation. In this case, the heat exchanger and associated tubing will be filled with air as opposed to water. Obviously, it would be difficult to provide this system for a plastic piping material.

A final method of delivering reduced temperature water to points of use in hot systems is shown in Figure 9.21. As discussed in chapter 7 for ozonated Purified Water systems,

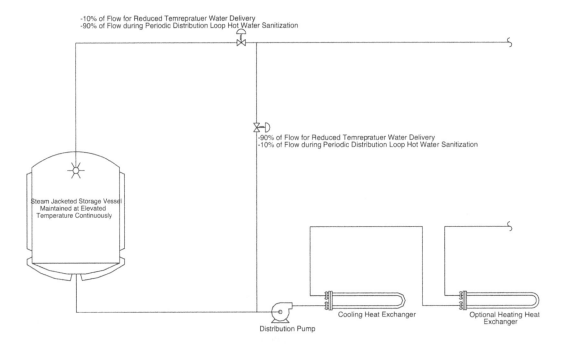

Figure 9.21 Continuous hot storage with ambient or hot recirculation.

point-of-use total viable bacteria and biofilm can be eliminated if water in the storage tank is entirely free of bacteria. This goal can be achieved for Purified Water and Water for Injection systems by operating with the stored water at a temperature greater than 80°C continuously. The hot Purified Water or Water for Injection flows through a distribution pump and to a cooling heat exchanger. The temperature of the water is reduced to a desired value, preferably not in the bacteria incubation range of 30 to 35°C. The cooled water passes through the loop, with draw-off at points of use as required. During periods when reduced temperature water is required, or for a time period not to exceed a suggested value of 20 consecutive hours each day, about 90% of the tempered water flow is delivered to tubing directing return flow to the suction side of the distribution pump. Approximately 10% of the water is directed through tubing to the storage tank. While the loop is maintained at a reduced temperature, feedwater is from the tank with water maintained at a temperature $\geq$80°C. Upon completion of operations requiring reduced temperature water or for at least four *consecutive* hours each day, the modulating valves in the distribution loop return tubing "switch" positions, directing 90% of the return water to the storage tank and 10% to the suction of the distribution pump. A heat exchanger in the return tubing provides heating to minimize temperature drop in the storage tank upon initiation of the loop "thermal sanitization" operation. The operation may be automated. It offers a very effective method for tempered water delivery while maintaining excellent microbial and biofilm control.

OPERATING AND MAINTENANCE CONSIDERATIONS

Periodic repassivation will be required for stainless steel distribution loops. The requirement may range from once every 12 to 18 months to once every 3 to 5 years or more, depending on the application. As discussed earlier, several factors will influence the degree of rouging and, subsequently, the frequency of repassivation for a distribution loop (Grant et al., 1997; Mangan 1991). For hot recirculating loops, thermally cycled loops, or loops that are periodically sanitized with ozone, yearly repassivation is suggested.

Generally, distribution loops operated at elevated temperatures will require repassivation more frequently than loops operated at ambient conditions or with periodic thermal sanitization. However, there are other important factors to consider, including the quality of the orbital welding; the mechanical finish of the stainless steel tubing; electropolishing of the stainless steel tubing; and "consistency" of the mechanical polish for the distribution system, storage system, distribution pump, and other components within the storage and distribution system (Bigelow and DiVasto, 1986). To determine the need for repassivation or derouging, the storage and distribution system should be periodically inspected at least once per year. The inspection frequency may be increased to once every six months if excessive rouging is noted. The inspection should include at least the following:

- The interior of the storage tank
- The storage tank spray ball assembly
- The section of distribution loop immediately downstream of the distribution pump
- Elbows or tubing tees downstream of the distribution pump
- Conductivity cells

Periodic inspection of the distribution system should be conducted. While it is acknowledged that portions of the distribution loop may not be readily accessible, every attempt should be made to inspect as much of the loop as possible. Stainless steel distribution tubing should be inspected once every one to two years. A suggested inspection frequency for plastic tubing/piping systems should be based on the "integrity" history of the system. Unfortunately, it is impossible to predict when a leak will occur in a plastic tubing/piping system. This particularly applies to socket-welded systems and systems with sanitary ferrules. If plastic tubing/piping systems are operating at pressures that are $\geq$70% of the recommended maximum operating pressure, or for hot or thermally cycled PVDF construction loops, the frequency of visual (exterior) inspection may be as great as once per week. Quite often, leaks from distribution loops, particularly plastic tubing/piping loops, will tend to evaporate. As a result, a "hands on" inspection may detect a leak that would not be visible by the presence of a "puddle" beneath the source of the leak. There may be indications of a leak, such as "water stains," particularly when tubing/piping is positioned behind sheetrock walls or suspended ceilings. Any water stains require investigation. The fact that the water stain does not appear to be "wet" does *not* indicate that a leak is not present.

Periodic inspection of the distribution loop may also reveal other important items critical to the successful operation of the distribution loop. This would include, for example, accelerated microbial growth on plastic tubing/piping systems in an area exposed to higher temperatures (visible), misalignment of tubing/piping due to periodic thermal cycling for sanitization, "shifting" of plastic tubing/piping loops, and movement of support members for plastic tubing/piping.

Leaks in distribution tubing/piping may be identified by visual inspection or located and identified as a result of an obvious increase in microbial levels as Purified Water or Water for Injection passes through the loop. During routine sampling of points of use, it is extremely critical to initiate a thorough sampling and bacterial monitoring program if a significant increase in total viable bacteria is noted at a particular point of use. This is of even greater concern when other points, generally upstream but also potentially downstream of the particular point of use, do not indicate similar elevated bacteria levels. (This is associated with data trending for points of use within the system.) Leaks may be associated with valve diaphragms and sanitary ferrule gaskets, as discussed earlier in this chapter. While leaks in stainless steel distribution systems, specifically at welds, are not common, experience indicates that they may occur particularly in systems that have not been properly maintained (repassivated or derouged). Furthermore, because of its apparent structural strength, larger diameter stainless steel tubing may incorrectly be used as a "stepping point" by personnel. It may also be used to provide support for smaller diameter piping. Obviously, both of these situations should be avoided.

Leak repair for plastic tubing/piping systems must be carefully evaluated. Repetitive leaks at sanitary ferrules may be associated with upstream or downstream stress on vertical or horizontal sections of tubing/piping. These stresses may be enhanced if the distribution loop is thermally cycled. The ultimate repair of repetitive leaks, occurring because of mechanical connections associated with plastic sanitary ferrules, could result in the replacement of sections of tubing/piping immediately upstream and downstream of the leak. Leaks at socket welds present a serious challenge. Often, the leaks are simply small "drips." This type of leak is frequently repaired by "backfill" of the tubing/piping using heat and "filler rod." In many cases, this is inappropriate. Leaks at heat-welded joints indicate that there is a crevice with a path from the interior of the tubing/piping to the atmosphere. The backfilling process may stop the flow of water, but not "fill" the crevice associated with the original leak (Fig. 9.22). It is suggested that leaks associated with heat-welded joints be repaired by qualified personnel. It is further suggested that it may be necessary to remove a section of tubing/piping, containing fittings, to achieve proper loop repair, free of crevices. Generally, the presence of crevices, remaining after

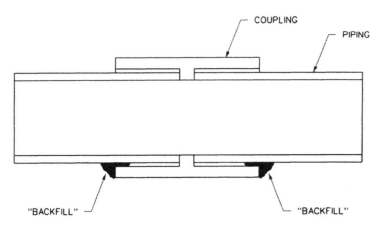

Figure 9.22 Example of "Backfill" for socket-welded plastic piping.

loop repair, will result in frequent sanitization, as determined by trending total viable bacteria levels as a function of time.

For systems using sanitary ferrules, it is important to use a periodic replacement program for sanitary ferrule gaskets. For systems that are thermally cycled periodically, it is suggested that either ethylene propylenediene monomer (EPDM) or Teflon envelope-type gaskets be used. EPDM gaskets exhibit excellent "flexibility" during expansion and contraction associated with thermal cycles. While Teflon gaskets are both hydrophilic and nonorganic, eluting Teflon in an envelope around EPDM provides a standard material for Water for Injection applications. To avoid potential leaks, it is suggested that Teflon envelope gaskets or solid Teflon gaskets only be "engaged" a single time. Finally, it is suggested that EPDM gasket material not be considered for hot Purified Water, Water for Injection, or ozonated Purified Water systems. While the literature indicates that EPDM is an acceptable material for hot or ozonated systems, experience indicates long-term oxidation, with resulting leaks, for both applications. On the other hand, EPDM, a relatively "soft" material, presents an excellent inner envelope material for Teflon envelope gaskets.

The program to replace sanitary ferrule gaskets should be conducted in parallel with a program to replace sanitary diaphragms in both manual and automatic diaphragm valves. In general, it is suggested that annual replacement of diaphragms be conducted if the system operates in a hot recirculating mode or if the diaphragms are either Teflon or Teflon backed with a "softer" material such as EPDM. The diaphragm, over a period of time, will become "scored" by the mating weir in the valve. Eventually, a leak may occur. During replacement of the diaphragm, the weir for each valve should also be inspected. Generally, there will not be any problems for stainless steel valves, but there may be a problem due to "depression" or "distortion" of a weir in a plastic valve. Plastic valve weirs should also be inspected for small cracks that could represent long-term serious problems. Finally, it is important to inspect the replaced diaphragm to ensure that foreign material is not present in the system, compressed between the weir and the diaphragm during normal operation. As discussed earlier, some stainless steel distribution systems may be passivated without the use of temporary strainers. Improper cleaning or incomplete passivation may introduce metallic impurities into the storage and distribution system. These impurities can be trapped between the diaphragm and weir of the valve and impinged into the diaphragm material. Stainless steel systems may also exhibit rouge impinged on the diaphragm. Often, inspection of a diaphragm may be used with other information to verify that repassivation or derouging of the stainless steel storage and distribution system is required.

Accessories installed in the distribution loop, such as pressure gauges, flow totalizers, mass flowmeters, temperature probes, conductivity cells, should be calibrated on a routine basis. A maximum time interval between calibrations is suggested as one year.

As discussed earlier, certain systems will be periodically sanitized with hot water. The sanitization frequency could be daily to once every one to three months, depending on the particular application and the desired microbial control within the storage and distribution system. It has been indicated that effective hot water sanitization can be achieved by ensuring that the return distribution loop temperature is less than 90 to 95°C. Effective sanitization can be achieved at this temperature when the sanitization process is conducted for two hours. To minimize heating and, more importantly, postsanitization cooldown, it is suggested that the initial step in the sanitization operation includes drain down of the storage system to a level consistent with maintaining the required net positive suction head (NPSH) of the distribution pump *and* allowing adequate water for flushing of each point-of-use valve with hot water during the sanitization cycle. As indicated in chapter 6, heat-up may be achieved using a heat jacket (single or multiple zone) positioned around the storage tank or an in-line heat exchanger installed in the distribution system. The sanitization cycle should ensure that the temperature of the water in the storage tank, and in the return section of tubing (or piping) from the distribution loop, is 90 to 95°C.

When sanitization at an elevated temperature is completed, it is highly desirable to return the water temperature within the storage and distribution system to its normal operating value (ambient temperature) as quickly as possible. It is desirable to minimize the time period that the water temperature is in the bacteria incubation range. Several different techniques may be employed to achieve the desired results. Cooling water or, preferably, chill water can be used with either a dedicated heat exchanger installed in the distribution loop or a single or multiple zone heating jacket surrounding the storage tank to reduce a small volume of water in the storage tank and distribution loop from the sanitizing temperature to ambient temperature. As an alternative, ambient temperature makeup water may be added to the storage tank, gradually decreasing the temperature of the water. Another cooldown technique employs filling the tank with ambient temperature water and simultaneously initiating cooldown using a heat exchanger with chill (or cooling) water. The length of time required for sanitization is "controlled" by the ability to cool down the hot water, rather than the time period required for heating the water or the time period at sanitizing temperature. Operating procedures, coupled with proper system design, should ensure that a complete sanitization operation, from and returning to normal operating conditions, should not exceed approximately six hours. The importance of minimizing the time required to complete the sanitization cycle cannot be overemphasized since it is a determining factor in establishing the sanitization frequency. Excellent microbial control can be maintained within systems that are sanitized on a routine basis. Extending sanitization frequency because of the time involved and its effect on "production" is often counterproductive, resulting in "emergency" sanitization as corrective action associated with exceeding established bacteria Alert and/or Action Limits.

As discussed in chapter 7, periodic sanitization of Purified Water storage and distribution system (and/or Drinking Water systems) can be conducted with ozone. These systems generally maintain a predetermined concentration of ozone in the storage system. The ozone is removed, downstream of the storage tank, by in-line ultraviolet units operating at a wavelength of 254 nm and an ultraviolet radiation intensity of about 80,000 to 100,000 $\mu W\ sec/cm^2$. Dissolved ozone monitors are positioned upstream and downstream of the in-line ultraviolet unit to confirm that ozone is present (feed) and removed (product) during normal operation. A dissolved ozone monitoring system is also provided in the return tubing from the distribution loop to control operation of the ozone generator during the sanitization operation, maintaining adequate dissolved ozone within the distribution loop. Obviously, during this operation, power to the in-line ultraviolet units is inhibited. In addition, point-of-use valves should not be opened (by operators) for processing applications. Additional information associated with the use of ozone, including periodic sanitization, is presented in chapter 7.

Periodic chemical sanitization requires greater operator attention when compared with hot water sanitization or sanitization with ozone. However, as indicated, the effectiveness of chemical sanitization with appropriate agent and procedure is not only destroys bacteria but also removes biofilm. The frequency of the chemical sanitization operation should be established for a particular system based on the characteristics of the system and the nature of the process and products. The sanitization operation with a disinfecting agent can take a moderate period of time to perform and completely remove the disinfecting agent. A procedure for effective chemical sanitization is presented earlier in the chapter.

Sanitary ferrule gaskets used in Pure Steam distribution systems should be of stainless steel impregnated Teflon (Rubber Fab, 2010). Gaskets constructed of various elastomers do not provide long-term physical stability.

REFERENCES

Adley C, Saieb F. Biofilm formation in high-purity water: Rastonia pickettii, a special case for analysis. Littleton, Colorado: Ultrapure Water 2005:14–17.

ASME BPE. Bioprocessing Equipment, Sections SD-4.11.4, SD-4.11.5, and SD-4.11.6, New York, NY: The American Society of Mechanical Engineers, 2009a:26–27.

ASME BPE. Bioprocessing Equipment, Section SD-3.11.1. New York, NY: The American Society of Mechanical Engineers, 2009b:17.

ASME BPE. Bioprocessing Equipment, Section DT-8. New York, NY: The American Society of Mechanical Engineers, 2009c:83.

ASME BPE. Bioprocessing Equipment, Part MJ, "Material Joining", New York, NY: The American Society of Mechanical Engineers, 2009d:108–119.

ASME BPE. Bioprocessing Equipment, Section MJ-7.2.3. New York, NY: The American Society of Mechanical Engineers, 2009e:116.

ASME BPE. Bioprocessing Equipment, Section M-7, "Inspection, Examination, and Testing". New York, NY: The American Society of Mechanical Engineers, 2009f:110–117.

ASTM. Reagent Grade Water. ASTM D 1193, Type I. Philadelphia, PA: American Society for Testing and Materials, 1993.

ASTM. Standard Practice for Cleaning, Descaling, and Passivation of Stainless Steel Parts, Equipment, and Systems, Designation A 380-06. West Conshohocken, PA: American Society for Testing and Materials, 2006a.

ASTM. Standard Specification for Reagent Grade Water, Designation D1193-06 Standard No. 7916. West Conshohocken, PA: American Society for Testing and Materials, 2006b.

Balmer KB, Larter M. Evaluation of chelant, acid and electropolishing for cleaning and passivating 316L stainless steel (SS) using auger spectroscopy. Pharm Eng 1993; 13(3):20–28.

Banes DP. Insights into rouge: definition, remediation, and monitoring, presented at Ultrapure Water Pharma, New Brunswick, NJ, May 22, 2010.

Bigelow PT, DiVasto RJ. Tubing welds and inspection techniques for WFI and other critical process fluids. Pharm Eng 1986; 6(1):21–24.

Chemtrol. Thermoplastic Piping Technical Manual, Chemtrol Thermoplastics Flow Solutions Manual No. C-CHTM-0401, a division of NIBCO, Elkhart, IN, 2001a:7.

Chemtrol. Thermoplastic Piping Technical Manual, Chemtrol Thermoplastics Flow Solutions Manual No. C-CHTM-0401, a division of NIBCO, Elkhart, IN, 2001b:10.

Coleman D, Evans RW. Fundamentals of passivation and passivity in the pharmaceutical industry. Pharm Eng 1990; 10(2):43–49.

Coleman DC, Evans R. Corrosion investigation of 316L stainless steel pharmaceutical WFI systems. Pharm Eng 1991; 11(4):9–13.

Collentro WV. Pharmaceutical Water System Fundamentals – Ion Removal – Reverse Osmosis, The Journal of Validation Technology. Duluth, MN: Institute of Validation Technology, Summer 2010; 16(3):66–75.

Costerton JW, Cheng KJ, Gessey GG, et al. Bacterial biofilms in nature and disease. Annu Rev Microbiol 1987; 41:435–464.

Costerton JW, Lewandowski Z, Caldwell DE, et al. Microbial biofilms. Annu Rev Microbiol 1995; 49:711–745.

Donlan RM. Biofims and Device-Associated Infections, Emerging Infectious Diseases. Atlanta, GO: Centers for Disease Control and Prevention, 2001; 7(2):Special Issue.

Enfield. Purity Sustained by Enfield. Enpure Pipe and Fittings, Catalog No. E983. Lake Bluff, IL: Enfield Industrial Corp, 1993.

Enpure™. High Purity Polypropylene System Manual, Industrial Technical Manual Series. Vol 3, 2nd ed., IPEX Enpure, Manual No. MNINEPIP070603, Mississauga, ON, Canada, 2007a:10.

Enpure™. High Purity Polypropylene System Manual, Industrial Technical Manual Series. Vol 3, 2nd ed., IPEX Enpure, Manual No. MNINEPIP070603, Mississauga, ON, Canada, 2007b:36.

Enpure™. High Purity Polypropylene System Manual, Industrial Technical Manual Series. Vol 3, 2nd ed., IPEX Enpure, Manual No. MNINEPIP070603, Mississauga, ON, Canada, 2007c:13.

EPA. Health risks from microbial growth and biofilms in drinking water distribution systems, United States Environmental Protection Agency, Office of Water (4601M), Office of Ground Water and Drinking Water, Distribution System White Paper, Washington, DC, June 17, 2002.

FDA. Water for Pharmaceutical Purposes. Inspection Technical Guide No. 46. Rockville, MD: Food and Drug Administration, Public Health Service, Department of Health and Human Services, 1986.

FDA. Guide to Inspections of High Purity Water Systems. Rockville, MD: Food and Drug Administration, Office of Regulatory Affairs, Office of Regulatory Operations, Division of Field Investigations, 1993.

Fleming HC, Wingender J. Relevance of microbial extracellular polymeric substances (EPS) – parts I and II – structure and ecological aspects / technical aspects. Water Sci Technol 2001; 43:1–8; 9–16.

George Fischer. Industrial plastic systems. In: Engineering Handbook. Vol. 5. Tuston, CA: George Fischer, Inc., 1995:8.1–8.47.

George Fischer. SYGEF PVDF Piping Systems, George Fischer Technical Bulletin, Section 4. Tustin, CA: George Fischer LLC, 2009a:4.8.

George Fischer. SYGEF PVDF Piping Systems, George Fischer Technical Bulletin, Section 4. Tustin, CA: George Fischer LLC, 2009b:4.19.

George Fischer. SYGEF PVDF Piping Systems, George Fischer Technical Bulletin, Section 4. Tustin, CA: George Fischer LLC, 2009c:4.21.

George Fischer. SYGEF PVDF Piping Systems, George Fischer Technical Bulletin, Section 4. Tustin, CA: George Fischer LLC, 2009d:4.23.

Grant A, Henon BK, Mansfeld F. Effects of purge gas purity and chelant passivation on the corrosion resistance of orbitally welded 316L stainless steel tubing. Pharm Eng 1997; 17(2):94–109.

Gray GC. Recirculation velocities in water for injection (WFI) distribution systems. Pharm Eng 1997; 17 (6):28–33.
Harfst WF. Selecting piping materials for high purity water systems. Ultrapure Water 1994; 11(4): 62–63.
Kilkeary JJ, Sowell T. New developments in passivation technology, presented at Interphex Pharmaceutical Conference, Jacob K. Javits Convention Center, New York, NY, April 21, 1999.
Lewis K. Riddle of biofilm, Antimicrobial Agents and Chemotherapy, American Society for Microbiology, 2001; 45(4):999–1007.
Mangan D. Metallurgical, manufacturing and surface finish requirements for high purity stainless steel system components. J Parenter Sci Technol 1991; 45(4):170–176.
Mazzola P, Martins A, Penna T. Identification of bacteria in drinking and purified water during the monitoring of a typical water purification system. BMC Public Health, 2:13 2002:1–11.
Mazzola P, Martins A, Penna T. Chemical resistance of the gram-negative bacteria to different sanitizers in a water purification system. BMC Infect Dis, 6:131 2006:1–11.
Meltzer TH, Jornitz MW. Pharmaceutical Filtration: The Management of Organism Removal, ISBN: 1-930114-77-X. Bethesda, MD: Parenteral Drug Association and River Grove, IL: Davis Healthcare International Publications, LLC, 2006.
Mittleman MW. Bacterial growth and biofouling control in Purified Water systems. In: Fleming HG, Gessey GG. Biofouling and Biocorrosion in Industrial Water Systems. Berlin, Germany: Springer and Verlang, 1991:133–154.
Mittleman MW. Microbial biofilms and bioburden generation in compendial water systems, presented at Ultrapure Water Pharma – 2010, New Brunswick, NJ, May 22, 2010.
MMWR Nosocomial Ralstonia pickettii Colonization Associated with Intrinsically Contaminated Saline Solution – Los Angeles, California,1998, Morbidity and Mortality Weekly Report, Atlanta, GA: United States Centers for Disease Control and Prevention, 1998; 47(14):285–286.
Potera C. Forging a link between biofilms and disease. Science 1999; 283:1837–1839.
Prakash B, Veeregowda BM, Krishnappa G. Biofilms: a study of stratergy of bacteria. Curr Sci 2003; 85:9; 10.
Pringle JH, Fletcher M, Ellwood DC. Selection of attachment mutants during continuous culture of Pseudomonas fluorescens and relationship attachment ability and surface composition. J Gen Microbiol 1983; 129:2557–2569.
Roberts AP, Pratten J, Wilson M, et al. Transfer of Conjuctive Transposon, Tn5397 in a Model Oral Biofilm. Delft, The Netherlands: Federation of European Microbiological Societies, 1999; 177(1):63–66.
Roll D. Current Methodologies & Chemistries Utilized in Effective Passivation Procedures", presented at the Interphex Pharmaceutical Conference, Philadelphia, PA: Pennsylvania Conference Center, April 15–17, 1997.
Roll D, Petrillo P. Rouge: monitoring, measuring, & maintenance in water and steam systems, presented at Ultrapure Water Pharma, New Brunswick, NJ, May 22, 2010.
Rubber Fab. Tuf-Steel, A Full Line of World Champion Gaskets, Technical Bulletin No. RF-160. Sparta, NJ: Rubber Fab Technologies Group, 2010.
Sani-Tech. Engineer and Design Catalogue Guide. 5th ed. Andover, NJ: Sani-Tech, Inc, 1996.
Theien FCM, O'Toole GA. Mechanism of biofilm resistance to antimicrobial agents. Trends Microbiol 2001; 9:34–39.
Vess RW, Anderson RL, Carr JH, et al. The colonization of solid PVC surfaces and the acquisition of resistance to germicides by water microorganisms. J Appl Bacteriol 1993; 19(2):215–221.

10 | Controls and instrumentation

There are numerous texts available that discuss controls and instrumentation. The purpose of this chapter is to provide a general overview of controls and instrumentation for pharmaceutical water systems. Chapters 11 and 13 will discuss specification preparation and validation of pharmaceutical water purification systems. The objective of this chapter is as follows:

- Present a general overview of control, monitoring, and instrumentation terminologies associated with specifications and drawings for water purification systems such as process and instrumentation diagrams (P&IDs).
- Present a list of instruments, sensors, indicators, switches, etc. employed in pharmaceutical water purification systems.
- Discuss both online conductivity and total organic carbon (TOC) measurements and the application of these measuring techniques to criteria set forth in the pharmacopeia.
- Discuss recommendations for water purification system central control panel processor "view screens" with suggested displays.

OVERVIEW

A thorough discussion of instrumentation, monitors, and controls for pharmaceutical water purification systems could generate significant material that is beyond the scope of this text. However, it would be inappropriate to omit an overview discussion of controls and instrumentation in a text dedicated to pharmaceutical water purification systems.

One of the primary purposes of this chapter is to provide a discussion of specific instrumentation, monitoring, and control issues frequently used/encountered in pharmaceutical water purification systems. In general, the discussion will be presented on a unit operation basis, similar to the discussion presented in previous chapters. By attempting to discuss the various control, monitoring, and instrumentation items for specific unit operations, the magnitude and detail required, with associated complexity, for a "system control philosophy" can be greatly simplified. Emphasis will be placed on the functionality of items to support the proper operation of the unit operation. In other words, controls, monitoring, and instrumentation will be treated as an accessory to a specific water purification unit operation.

Programmable logic controllers (PLCs) will briefly be discussed in this chapter. The discussion will be limited to the function of a PLC to receive inputs from various transmitters within the water purification system, provide control output to components, provide output to a facility central data collection system, and display/annunciate alarm conditions. Specific details describing the types of PLCs available, particularly systems with a central control function, are not included in this chapter. The documentation required for validation is presented in chapter 11.

A discussion of the International Society for Measurement and Control (ISA, formerly known as the Instrument Society of America) terminology will be presented. Every attempt will be made to limit the discussion to specific sensors, transmitters, controllers, and other appropriate items used for control and instrumentation in pharmaceutical water purification systems. In many cases, examples will demonstrate the use of this terminology for specific water purification unit operations.

This chapter will also discuss the importance of P&IDs for a water purification system, with reference to ISA terminology. As discussed in chapter 11, a detailed specification for a pharmaceutical water purification system, divided into subsections dealing with specific unit operations, should be accompanied by a P&ID (or multiple P&IDs). A P&ID integrates the individual sections of the specification and demonstrates interaction of the individual unit operations to provide a complete "system."

It would be inappropriate to discuss controls and instrumentation without addressing nontemperature-compensated conductivity monitoring systems and TOC analyzers (laboratory and online) for pharmaceutical applications.

Finally, the importance of a view screen with navigation to multiple displays will be discussed. A view screen can significantly increase operator knowledge and confidence by presenting "process flow diagram" displays with color-coded indication of valve, motor, status, etc.

SYMBOLS

The most fundamental general instrument or function symbol is that of a discrete field-mounted instrument, which is commonly used to represent a pressure or temperature gauge mounted directly in a section of tubing/piping. The field-mounted discrete instrument symbol is a simple circle connected to the tubing/piping, much as the pressure or temperature gauge would be directly mounted in the line (Fig. 10.1). The terminology employed is relatively simple. While some pharmaceutical manufacturing firms may have established symbol designation, the ISA designation is generally used. As an example, "PI" represents a "pressure indicator" and "TI" represents a "temperature indicator." To generate a proper P&ID, every instrument, sensor, transmitter, pump, automatic valve, etc. must be properly labeled. Subsequently, a pressure gauge may be labeled "PI-101," as shown in Figure 10.1. The three-digit code is arbitrarily selected. Because of the number of gauges that may be present in a system, a two-digit code, as a minimum, is generally required. There is another reason for selecting a three-digit code. While a matter of preference, the first letter in the three-digit number can assist operating personnel. The first digit of the code can indicate where the pressure gauge is located, such as the Purified Water or Water for Injection generating system or the Purified Water or Water for Injection storage and distribution system. The number "1" could indicate that the pressure gauge is located in the generating system, while gauges within the storage and distribution system would begin with the number "2."

Another symbol commonly used represents a pressure-sensing element, a temperature-sensing element (RTD, resistance temperature detector), or conductivity cell. The sensing elements would be mounted directly in-line, similar to that shown for a pressure sensor in Figure 10.2. A pressure-sensing element would be represented by "PE"; a temperature-sensing element is designated by "TE"; and a conductivity cell is represented by "AE," where "A" indicates an analytical sensing element. Again, the sensing element designation would be followed by a dash and a multiple-digit code. Since there are potentially many "analytical" elements, it is common practice to insert the specific parameter being detected at the lower right hand side of the symbol external to the symbol, as shown in Figure 10.2. This

PRESSURE GAUGE

TEMPERATURE GAUGE (WITH THERMOWELL)

PRESSURE WITH INSTRUMENT NUMBER

Figure 10.1 Sample gauges connected to piping line.

CONTROLS AND INSTRUMENTATION

PRESSSURE SENSING ELEMENT

CONDUCTIVITY SENSING ELEMENT

Figure 10.2 Typical locally mounted sensors.

figure shows conductivity sensor "AE-201." A sensing element cannot be a "stand-alone" item. It allows measurement capability that is provided to a transmitter, controller, and indicator.

In certain cases, an instrument may combine a sensing element with an integral transmitter and indicator. Transmitters may be used to accept a response from the sensing element, convert it to an analog "signal," generally 4 to 20 mA, and transmit it to an indicator, indicating controller, or input/output (I/O) module in the central control panel. Transmitters may also be used to accept a response from the sensing element, convert it to a discrete signal (e.g., open or closed), and transmit it to an indicator, indicating controller, or I/O module in the central control panel. Transmitters for pressure, temperature, and conductivity would be designated by "PT," "TT," and "AT," respectively. The symbol for a field-mounted sensor would be enclosed in a circle similar to that for pressure indicators and pressure elements. A line connects the circle for the sensing element to the circle for the transmitter. A sensing element and transmitter must be combined with some output source, such as an indicator, indicating controller or central control panel I/O module. If local indication, in addition to remote indication through the central control panel display screen is desired, an indicating-type transmitter (e.g.,"PIT" for pressure) may be used.

Indicators, transmitters, and indicating transmitters may be physically positioned remotely from the sensing element. The location of the indicator, transmitter, or indicating transmitter is generally referred to as a "primary location." The symbol for a pressure, temperature, and conductivity sensor indicating transmitter combination is shown in Figure 10.3. Generally, a sensing element and transmitter or indicating transmitter are coupled with a control function in the central control panel such as a programmed processor. For systems without a central control panel with programmed processor capability, indicating controllers may be employed. A pressure indicating controller, temperature-indicating controller, or conductivity indicating controller is designated by "PIC," "TIC," or "AIC," respectively. Since controllers or indicating controllers provide, as implied, a control function, there is generally a "local" output from the controller. The output could provide control to a back pressure regulating valve (PIC); a modulating steam valve heating a storage tank or heat exchanger (TIC); or a divert-to-drain valve for product water from a deionization system, RO unit, or distillation unit (AIC) for conductivity.

Quite often, indicating controllers are represented with adjacent symbols, designating high- and/or low-alarm conditions. As an example, "PAL" and "PAH" may be used to represent "pressure alarm low" and "pressure alarm high," respectively. Similarly, "TAL" or "TAH" can be used for "low- or high-temperature alarm conditions." Also, "AAH" could be used to represent high conductivity. This system may be modified for certain applications. As an example, as discussed in chapter 6, a four-point level sensing system may be employed for storage tanks. This system may use discrete signals from sensors (e/g., proximity switches) or

PRESSSURE SENSOR AND INDICATOR

PRESSSURE SENSOR, TRANSMITER AND INDICATING CONTROLLER

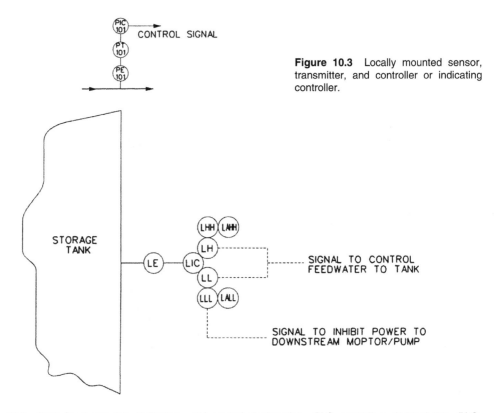

Figure 10.3 Locally mounted sensor, transmitter, and controller or indicating controller.

Figure 10.4 Tank level sensing, indicating, and control designation—PLC controller. *Abbreviation*: PLC, programmable logic controller.

analog signals from sensors and transmitters (e.g. differential pressure). In chapter 6, it was noted that "high-high" and "low-low" levels would indicate an unacceptable system excursion and energize an alarm, while "high" and "low" levels would be associated with the normal operating level "band" for the tank. For discrete sensors, alarm conditions may be provided by a "hard-wired" signal to an indicating light and audible alarm. For this particular arrangement, the high-high and low-low levels would be represented by symbols "LAHH" and "LALL." An example of alarm conditions associated with discrete sensors/controllers is shown in Figure 10.4. For analog sensor(s)/transmitters, the central control panel processor would display the alarm condition of a view screen and energize an alarm.

Prior to discussing specific additional symbols that are generally employed in P&IDs for pharmaceutical water purification systems, it is appropriate to define and demonstrate general instrument or function symbols as established by ISA. Figure 10.5 is a representation of "drawing balloons" used for discrete instruments, shared display or shared controls, computer functions, and programmable logic control. It should be noted that the only difference between field-mounted instruments and instruments mounted at a "primary" location is that the

CONTROLS AND INSTRUMENTATION

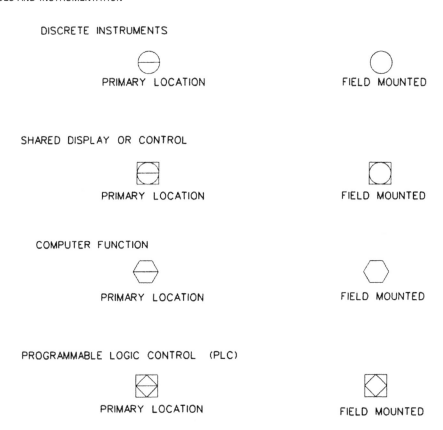

Figure 10.5 General instruments or function symbols. *Source*: ANSI/ISA 1986.

primary location instrumentation contains a solid horizontal line though the middle of the indicating symbol. The eight general instrument or function symbols provided in Figure 10.5 represent all applications that will be encountered in pharmaceutical water purification systems.

There are a few additional items that should be addressed. The solid horizontal line through the primary location symbol implies that the panel containing the component is normally accessible to operating personnel. Normally inaccessible or "behind the panel" devices (or functions) are depicted by a dotted (rather than solid) horizontal line. Some applications with an "auxiliary location" may contain instrumentation. This auxiliary location may be associated with a specific monitor that for some reason cannot be located within a central control panel. If an auxiliary location is used for instrumentation, a double horizontal line should appear through the appropriate symbol.

The following are example descriptions of discrete instruments, shared displays and shared controls, computer functions, and programmable logic control functions. A discrete instrument is a simple indicating device that merely provides an indication of a parameter, such as pressure. It has already been indicated that a discrete field-mounted instrument would be a pressure gauge. If pressure indication is desired on a pressure gauge mounted on the face of a panel, such as in some smaller flow rate RO applications, a diaphragm isolator coupled with a hydraulic transfer line could be used to transfer the pressure indication to the panel-mounted pressure gauge at a "primary location." An example of a shared display/shared control instrument is a conductivity meter that operates in conjunction with two field-mounted conductivity cells. A single conductivity meter would display values for each of the two locations. A computer function would be a "smart" transmitter. Generally, smart units, for monitoring an item such as pressure, are equipped with an integral sensor, a micro-based transmitter to provide a 4 to 20 mA signal that can be calibrated, and indicator. A PLC provides flexible input and output capability from field-mounted transmitters (or indicating transmitters). The analog or discrete input is used by a custom programmed processor to achieve a number of functions. For example, the custom programmed processor can be used to perform sanitization of a storage and distribution system automatically. Further, it can be programmed to limit the heat-up rate to an adjustable number of degrees per minute or hour. To avoid "exceeding/overshooting" the sanitization temperature set point, the programmed processor can automatically decrease the heat-up rate when the temperature approaches the desired sanitization value. Once the desired sanitization temperature is achieved, the processor can maintain the desired temperature for a preset sanitization time period. Finally, the custom programmed processor can control cooldown of the loop, using cooling water or chilled water, to a preset normal operating (postsanitization) temperature.

Table 10.1 presents several other symbols commonly encountered in pharmaceutical water purification systems with a brief description of the function/device.

Table 10.1 Typical Symbols for Pharmaceutical Water Purification, Storage, and Distribution Systems

Symbol	Description
Pilot light (indicating light) 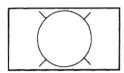	Used to indicate the status of an operation. Used on I/O or panels, motor starters, and local control panels. May also be used to indicate a "fault"/alarm condition
Diaphragm seal 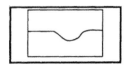	Used to eliminate dead legs in hot operating/sanitized systems and provide minimum dead leg for chemically sanitized systems
Interlock 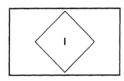	A method of tying one or more control functions together. As an example, an interlock may be used to inhibit the operation of an RO unit during backwash of upstream pretreatment components such as a multimedia filter or activated carbon unit
Relief valve 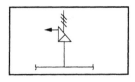	Installed at the top of a pressure vessel such as a multimedia filtration unit column. Also installed in the supply steam or cooling water piping to a heat exchanger
Rupture disc 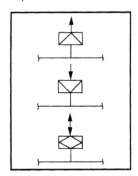	Symbols are used to indicate the function of a rupture disc, a positive method of mechanical relief for a sanitary vessel. The top symbol represents a disk that ruptures on positive pressure as indicated by the flow direction arrow. The middle symbol indicates a disk that ruptures only on vacuum. The lower symbol indicates a "compound-type" disk that ruptures on pressure and/or vacuum

CONTROLS AND INSTRUMENTATION

Table 10.1 (*Continued*)

Steam trap 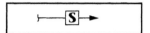	Used in conjunction with components requiring facility steam for heating such as the jacket on a sanitary storage tank or shell of a shell and tube type heat exchanger. Appears in condensate piping for the facility steam
Current to pneumatic converter 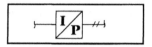	Symbol for a device that converts an analog output, generally 4–20 mA, to a proportional pneumatic output, generally 3–15 psig. The device is used for components such as distribution loop back pressure–modulating valves and facility steam modulating valves
Pneumatic signal 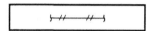	The "hashed" line symbol represents a proportional pneumatic signal to a field device such as a modulating valve. It would be used for the pneumatic tubing from an "I/P" converter
Valves 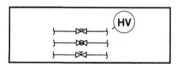	Some of the manual valve symbols used in pharmaceutical water systems. The top valve symbol represents a manual ball valve. The middle symbol represents a manual diaphragm-type valve. The bottom symbol represents a needle-type valve. The ball valve and diaphragm valve symbols may be combined with a pneumatic actuator to show a pneumatically operated valve

PROCESS AND INSTRUMENTATION DIAGRAM

The P&ID, which will be discussed in greater detail in chapter 11, is an important part of a detailed specification. It integrates the system controls, particularly when the specification is developed around individual unit operations. The P&ID demonstrates field-mounted sensing and monitoring devices that will provide an analog or discrete signal to an I/O module in central control panel. Further central control panel I/O modules, based on the processor program and additional I/O modules, will provide a control signal to pumps, solenoid valves, and component accessories for automatic system operation. The importance of establishing a numbering system for sensing elements, transmitters, etc. at the specification stage cannot be overemphasized. Furthermore, the documentation associated with each unit operation's components and accessories must be provided as part of any "turnover package." This should include Certificates of Calibration clearly stating that the calibration is traceable to a National Institute of Standards and Technology (NIST) source. It is strongly suggested that the Certificate of Calibration include the test report compiled for multiple-point calibration, as opposed to a document that states that the equipment has been calibrated to a manufacturer's standard. Finally, field instrumentation including sensors and transmitters must be properly "tagged." This could include etching on the actual device, a difficult to remove tag with permanent etching on a stainless steel or aluminum tag, or use of composite-type plastic tags.

Table 10.2 presents a summary of field devices used for pharmaceutical water systems. It should be noted that some systems may use components that are not listed in the table or use components for an application in a manner different than that stated.

CONTROL FUNCTIONS FOR PHARMACEUTICAL WATER PURIFICATION SYSTEMS

The majority of systems installed for "smaller" applications, such as research and development laboratories, smaller biotechnology facilities, pilot plant facilities, and small production facilities, may use discrete controls contained within a central control panel. Discrete

Table 10.2 Indicators and Other Control Elements Frequently Used in Pharmaceutical Water Systems with Potential Applications

Parameter: Pressure
Location: Various
Applications:
- Sanitary-type pressure gauges or bourdon tube–type pressure gauges with diaphragm isolators are positioned in the feedwater and product water piping/tubing to and from components.
- Certain components, such as reverse osmosis or ultrafiltration units, may require multiple pressure gauges for feed, product, waste, recycle, etc.
- Compound-type pressure gauges should be provided in the feedwater piping/tubing to pumps and at the top of storage tanks where vacuum conditions may exist.
- Pressure gauges should be provided in the product water piping/tubing from each pump.
- Pressure gauges should be provided in the feedwater and product water piping/tubing of any component where pressure drop is a concern such as a multimedia filtration unit, activated carbon, unit, heat exchanger, etc.
- While materials of construction for utility pressure gauges (steam, cooling water, etc.) may be of brass, steel or other materials, gauges employed throughout the pharmaceutical water system should be of 316 or 316L stainless steel material (sensor and case), appropriate diameter (2–3 in. face diameter minimum), and appropriate display range and increments.
- Pressure sensors may be employed in "pairs" for tank level detection.

Parameter: Temperature
Location: Various
Applications:
- Positioned downstream of a manually adjusted raw feedwater thermal blending valve system feeding a small capacity water purification system with reverse osmosis.
- Positioned in the face piping of a unit operation requiring regulated backwash flow, such as a multimedia filtration unit, to determine temperature, related viscosity of water, and required backwash flow rate.
- Positioned in the feedwater and product water piping/tubing to and from a heat exchanger.
- Positioned in the feedwater piping/tubing to a membrane process such as reverse osmosis or ultrafiltration.
- Positioned in the return piping/tubing from a distribution loop.
- Positioned in the feed and return piping for various utilities such as plant steam and condensate, cooling water, and/or chilled water.

Parameter: Storage tank level
Location: Positioned on storage tanks
Applications:
- Tank level measurement to control the flow of makeup water.
- Measures tank level to indicate extremely low conditions that may damage downstream distribution/recirculation pump as a result of cavitation. Automatically inhibits pump operation on extremely low tank level and energizes an audible and visual alarm.
- Measures tank level to provide an alarm on extremely high tank level that will result in overflow, flooding of hydrophobic vent filter, rupture disk relief, etc. Automatically energizes an audible and visual alarm.
- Level sensors may be proximity-type switches, "conductivity-type sensor, or differential pressure-type.
- Differential pressure-type systems should be considered for hot water (operating and periodic sanitization) and ozone applications.

Parameter: Flow rate
Location: Various
Applications:
- Positioned in feedwater piping/tubing to pretreatment unit operations to measure operating and backwash.
- Positioned in feedwater piping/tubing to water softening or deionization units to measure flow rate during operation, backwash, regeneration, and rinse.
- Directly measures at least two of the three flow streams for a membrane process such as reverse osmosis, preferably feedwater and wastewater, to establish, verify, and monitor operating and maintenance conditions.
- Measure the "recycle" and "recovery" flow rates for membrane-based systems such as reverse osmosis.
- Measures the waste flow rate for continuous electrodeionization units.
- Measures the flow rate in distribution loop piping/tubing.
- Measures the flow rate at points of use to control delivered "established" water volume (with automated valve) for a specific product/manufacturing operation.

Parameter: Temperature switch
Location: Downstream of raw water thermal blending systems, heat exchangers, and upstream of a membrane process such as reverse osmosis

Table 10.2 (*Continued*)

Applications:
- Positioned downstream of a raw water (hot and cold) thermal blending valve to avid damage to piping, valves, and membranes, particularly reverse osmosis membranes, associated with high temperature as a result of valve failure, loss of cold water flow, valve, or malfunction.
- Positioned in the feedwater piping/tubing to a reverse osmosis unit to inhibit unit operation on high feedwater temperature. High temperature (at operating pressure) will result in irreversible damage to membranes (includes "hot water sanitizable" membranes if experienced at normal operating pressures.
- Positioned downstream of heat exchangers (pretreatment, ion exchange loop, or distribution system) to provide a discrete output signal as a back-up to an analog signal to a temperature control system. The switch may be used in conjunction with a valve to terminate heating or cooling media flow.

Parameter: Pressure switch
Location: Suction piping/tubing to in-line pumps, and feedwater and product water piping/tubing of membrane-based processes, particularly reverse osmosis
Applications:
- Positioned in the feedwater piping/tubing to an in-line repressurization/recirculation pump, insuring proper feedwater pressure to avoid pump cavitation with resulting pump damage.
- Positioned in the feedwater piping/tubing to a reverse osmosis unit high-pressure pump, inhibiting unit operation to avoid pump damage.
- Positioned in the product water piping/tubing from a reverse osmosis unit to sense high pressure, an indication of catastrophic membrane failure/loss of integrity.
- Positioned in clean, oil-free facility air supply to control systems and pneumatically operated valves of air-to-open, spring-to-close type. Loss of instrument/control air (without loss of electrical power) should automatically terminate system operation and energize an audible and visual alarm.

Parameter: Proximity switch
Location: Level detection for unpigmented polyethylene or polypropylene tanks
Applications:
- Employed for nonintrusive level measurement in unpigmented polypropylene and unpigmented polyethylene storage tanks. Multiple probes may be used for design of a "four-point" (or greater) control system.
- Employed as a low-level sensors in systems using chemicals for reverse osmosis unit feed treatment such as reducing agents and/or antiscalants.
- Employed on brine storage tanks for water-softening units if tanks are translucent.
- Employed on raw water feed/break tanks to control the flow of raw makeup water flow to the system.
- Employed on translucent waste collection tanks where waste flow rate exceeds instantaneous drain capacity or nonpressurized drain is not available in the physical area of installation.
- Employed to collect higher purity or softened water waste streams such as reverse osmosis unit waste for collection and transfer for recovery to another facility applications.

Parameter: Flow totalizer
Location: Various
Applications:
- Positioned upstream of pretreatment systems to measure the total volume of raw water fed to the system.
- Positioned in the feed-water piping/tubing to a water-softening system to determine the amount of water processed. Volume must be "reset" or recorded subsequent to regeneration and may be used to determine regeneration frequency if feed-water hardness is fairly constant or measured frequently.
- Employed as part of a point-of-use flow rate delivery system to deliver a preset volume of water for a specific process/operation.

Parameter: Oxidation-reduction potential (ORP) meter
Location: Pretreatment system employing reduction of reducing agent for removal of raw water disinfecting agent. Located downstream of reducing agent injection
Applications:
- Positioned downstream of pretreatment system reducing agent injection (e.g., sodium bisulfite) to ensure that all raw water residual disinfecting agent and/or any upstream injected disinfection agent has completely removed.
- Positioned after introduction of disinfecting agent when raw water exhibits lack of adequate disinfecting agent and high bacteria levels. Introduction of disinfecting agent provides microbial control but must be monitored to indicate responsive volume based on fluctuation in feedwater quality.
- Installed in "Drinking Water" systems after injection of a bacteria control agent subsequent to treatment by unit operations such as multimedia filtration and/or water softening.

(*Continued*)

Table 10.2 Indicators and Other Control Elements Frequently Used in Pharmaceutical Water Systems with Potential Applications (*Continued*)

Parameter: Total hardness monitor
Location: Downstream of water softening system
Applications:
- Positioned to monitor water softening system product water total hardness. Detects total hardness when feedwater hardness concentration varies and also detects improper water softening unit regeneration and/or malfunction.
- Establish/initiates water softening unit regeneration and "switchover" for dual or multiple softener systems.
- Provides long term throughput volume as a result of loss of resin capacity with age and/or "iron fouling"

Parameter: Residual disinfectant monitor
Location: Pretreatment System
Applications:
- Positioned to verify the concentration of residual disinfecting agent in the raw water supply.
- Positioned to determine the concentration of residual disinfecting agent for systems with low, or lack of raw water disinfecting agent and high microbial levels and subsequent use of disinfecting agent injection systems.
- Installed in Drinking Water systems after injection of a bacteria control agent subsequent to treatment by unit operations such as multimedia filtration and/or water softening.

Parameter: Dissolved ozone monitor
Location: Within storage and distribution systems using dissolved ozone for microbial control
Applications:
- Positioned from piping/tubing to determine the dissolved ozone concentration immediately downstream of the storage tank.
- Positioned from piping/tubing to determine the dissolved ozone concentration downstream of the dissolved ozone destruct in-line ultraviolet unit to ensure that ozone is not present during operation but is present during periodic distribution loop sanitization.
- Positioned from the piping/tubing in the distribution loop "return" section to verify the presence and concentration of dissolved ozone during periodic sanitization.

Parameter: Conductivity
Location: After unit operations removing ions and in distribution loop supply and return piping/tubing
Applications:
- Employed with a greater "cell constant" to determine the conductivity of feedwater to a system, particularly when fluctuations in conductivity are common.
- Employed to measure product water ionic purity from a deionization system positioned after each ion removal unit operation. For example, in a cation-anion-cation system, conductivity should be measured after the anion unit (two-bed deionization system) and after the cation polisher.
- Employed to determine product purity from primary and polishing mixed bed deionization units.
- Positioned to determine product water purity from a single pass reverse osmosis unit or each pass of a double pass reverse osmosis unit.
- Employed to determine the product water purity from individual continuous electrodeionization units.
- Positioned in the product water tubing from a distillation unit to verify product water ionic purity.
- Employed to verify that non-ion removal unit operation, such as membrane filtration, ultrafiltration and/or in-line ultraviolet sanitization do not indicate an increase in conductivity.
- Positioned in the feedwater and return section of distribution loops to verify online conductivity and detect loop back contamination.
- Positioned in "sub-loops" from distribution loop to verify ionic quality.

Parameter: Conductivity ratio
Location: Distribution loop supply and return and reverse osmosis units
Applications:
- Major use for determination of percent rejection of ionic material through a reverse osmosis unit. Cells positioned in feedwater and product water piping/tubing to, and from the reverse osmosis unit. Meter displays each conductivity value (with corresponding temperature) and the percent rejection of ions. Ion rejection is generally determined directly by the meter not the central PLC.
- Positioned in distribution loop supply and return section of piping/tubing to verify that a notable increase in ions does not occur. An increase in conductivity may be associated with back contamination of the distribution loop at one or more points of use.

Table 10.2 (*Continued*)

Parameter: Differential pressure
Location: Filtration unit operations and tank level
Applications:
- Pressure sensors (elements) positioned in the feedwater and product water piping/tubing for prefilters and membrane filters to detect a pressure increase associated with accumulation of particulate matter. Note that extremely high levels of bacteria must be present to indicate a pressure drop in polishing/distribution loop applications.
- For back-flushable filtration units, such as a multimedia filtration unit, high differential pressure can indicate potential or pending breakthrough of particulate matter.
- Positioned in the feedwater and product water piping/tubing of an ultrafiltration unit to determine "transmembrane" pressure drop, critical to long-term successful operation of the unit.
- May be positioned in reverse osmosis feedwater, waste, and product water piping/tubing to determine pressure drop. However, this is generally performed by transmission of an analog pressure to the central PLC, and appropriate programming to establish maximum feed-to-waste and feed-to-product differential pressure values for specific unit operation.
- Differential pressure used for tank level determination as discussed earlier.

Parameter: Ultraviolet radiation intensity
Location: In-line ultraviolet units
Application:
- Employed on in-line ultraviolet units to indicate the ultraviolet radiation intensity. General indication but important for multiple lamp units and dissolved ozone destruct units.

Parameter: Load cell
Location: Base frame of storage tank
Applications:
- Employed as a nonintrusive method of determining storage tank level using the weight of water in the tank.
- Employed for point-of-use batching vessels to measure the "delivered" volume of water.

Parameter: Rupture disk integrity sensor
Location: Top of storage tanks
Applications:
- Employed for verification on the integrity of the rupture disk positioned on the top dome of a storage tank.
- Very infrequently used in distribution loop tubing to provide a positive means of mechanical relief of pressure for systems where the tubing is either operated in a hot or thermally cycled mode and has "in-line" valves. Closing of the valves could create a pressure or vacuum condition as a result of hydraulic pressure in a "solid" water section of tubing.
- May be used with discharge to a gaseous ozone vent tubing in an ozonated storage tank.
- The integrity strip is physically positioned in a full disk diameter physical arrangement with low electrical current providing a discrete output if integrity is lost.

Parameter: Pump motor starter
Location: Locally or pump motor control center
Applications:
- Positioned in the physical area of a pump to allow operating personnel to start/stop power to the pump locally.
- Generally of the "hands-off-auto" (HOA) type with "control leg," allowing automated execution of control functions such as inhibiting power on low upstream tank level or pressure.
- Occasionally positioned in a remote motor starter control center with electrical feed to a locally mounted "disconnect."
- Generally not mounted in PLC control panel without electric "field" shielding.
- Variable frequency drive power of electrical cycles (and subsequently pump motor speed/rotation) controlled through central PLC based on pressure or flow rate.

Parameter: Pump motor starter variable frequency drive
Location: Locally or pump motor control center
Applications:
- Variable frequency drive power of electrical cycles (and subsequently pump motor speed/rotation) controlled through central PLC based on pressure or flow rate.
- Employed in conjunction with a distribution pump system return piping/tubing pressure sensor or flow rate sensor for significant point-of-use flow rate requirements.

(*Continued*)

Table 10.2 Indicators and Other Control Elements Frequently Used in Pharmaceutical Water Systems with Potential Applications (*Continued*)

Parameter: TOC analyzer
Location: Distribution loop
Applications:
- Sample valve with flow to online unit generally positioned in distribution loop return piping/tubing monitors TOC by periodic sample collection and analysis.
- Dual monitoring locations can be established, one from the distribution loop feedwater tubing and one from the return tubing. Individual monitors or single monitor with valves may be used to provide results from both locations.
- Time delay for analysis should be considered. A change in online TOC value is not immediately displayed on the analyzer.
- Laboratory results for "grab" samples should be periodically compared with online meter reading.
- "System Suitability" must be performed periodically. Failure of System Suitability requires analyzer recalibration.
- Laboratory analysis of grab samples collected periodically throughout the system should be considered.

Abbreviations: PLC, programmable logic controller; TOC, total organic carbon.

controllers do not require the depth of validation required for a PLC-based control system. The adjustable set points available with discrete controllers require thorough operator knowledge of the control philosophy. This type of control system can present significant challenges when compared with a central PLC-based control system particularly with regard to flexibility. The function of a control system for a water purification system should be to ensure that the system is monitored and controlled as appropriate to provide water quality at a points of use. There is no mandate, from a regulatory standpoint, to use PLCs. The price of a PLC-based control system for smaller projects is low and provides desired "integration" of controls. While the "unit operation" philosophy emphasized in this text is important, integration of the individual unit operations, from a control standpoint, is critical. For smaller capacity applications, the use of discrete controls may allow operating and maintenance personnel to interface directly with the system. For other systems, particularly systems with multiple components and twenty-four hour per day operation, PLC-based systems are strongly suggested. With proper preventative maintenance, most PLC-based systems will operate without operator for extended periods of time without problems.

The suggested use on a PLC-based control system considers proper specification preparation, submittal review, and "turnover" package, discussed in chapter 11. An equipment supplier's "standard" control package may be appropriate for smaller capacity systems and systems that do not require integration of pretreatment, ion removal, storage, and distribution. An alternative to a single PLC-based central control panel, integrating the entire system from raw feedwater to points of use, is multiple manufacturer's standard PLC-based control panels with "communication" to a custom PLC-based central panel. A preferred alternative to this design is the use of locally positioned "input/output" panels controlled by a central PLC-based panel through control wiring. This alternative allows solenoid valves to be mounted in the physical area of components, minimizing the length of pneumatic tubing.

Figure 10.6 depicts a typical discrete monitoring and control section for the return loop tubing/piping of a conventional distribution loop. It demonstrates a locally mounted pressure gauge (PI) and temperature gauge (TI). A conductivity cell inserted into the tubing/piping (AE), providing an "analytical element" for a panel-mounted conductivity meter (AI). The conductivity meter energizes a panel-mounted audible alarm (and indicating light) on high conductivity (AAH). This alerts operating personnel to the fact that the conductivity in the return portion of the distribution loop has exceeded a preset value. In this particular case, the alarm condition does not provide any additional action, such as inhibiting the flow of water through the loop, or restricting draw-off of water at point of use. The established in-line high conductivity set point should be less than 1.3 µS/cm at 25°C or 1.1 µS/cm at 20°C for Purified Water and Water for Injection.

CONTROLS AND INSTRUMENTATION

Figure 10.6 Distribution system return line: representative monitoring, control, and alarms.

Figure 10.6 also demonstrates a combination temperature element (TE) and temperature transmitter (TT). The temperature transmitter provides a 4 to 20 mA signal to a panel-mounted TIC. This controller has preset high- and low-temperature alarm conditions, as shown, represented respectively as TAH and TAL. The alarm conditions activate an audible alarm on the panel and energize appropriately labeled lights on the panel. The temperature-indicating controller also provides a signal to a panel-mounted temperature recorder (TR) and provides a proportional signal (4–20 mA) to a current-to-pneumatic (I/P) converter, which provides a pneumatic signal (3–15 psig) to a modulating valve on the heating source to the system. The modulating valve is installed on a steam line to the heating jacket around the storage tank or the steam supply to a heat exchanger installed within the distribution loop. The locally mounted direct reading TI provides a method of verifying the temperature reading indicated by the temperature-indicating controller in the "remote" panel.

Finally, the distribution loop return tubing/piping is equipped with a back pressure regulating valve (modulating type) and an associated pressure monitoring and control system. This control function includes a pressure-sensing element (PE) and a pressure transmitter (PT). The transmitter provides a signal to a PIC positioned in the control panel. The high- and low-pressure set points are established respectively indicated as PAH and PAL. Either of these conditions would energize an audible alarm positioned on the control panel and appropriately labeled indicating lights. The pressure-indicating controller provides an electronic current signal (4–20 mA) to a current-to-pneumatic converter, providing a proportional pneumatic signal (3–15 psig) to a modulating-type back pressure regulating valve. The purpose of this detailed discussion of the simple control scheme for monitoring parameters in the return line from a distribution loop is to demonstrate how discrete controllers and instrumentation may be employed in water purification system.

Figure 10.7 presents a duplication of the control and monitoring function provided in Figure 10.6 for field devices and transmitters to a PLC-based central control panel. The conductivity, temperature, and pressure indication would typically be displayed on a panel-mounted display screen. "Navigation" to a "control set point" screen would display the high conductivity, low-pressure, high-pressure, low-temperature, and high-temperature values.

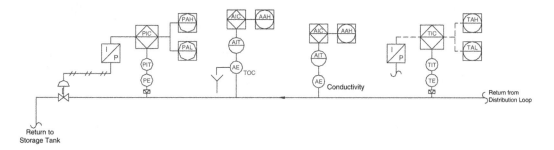

Figure 10.7 Distribution system return line: representative monitoring, control, and alarms with PLC.

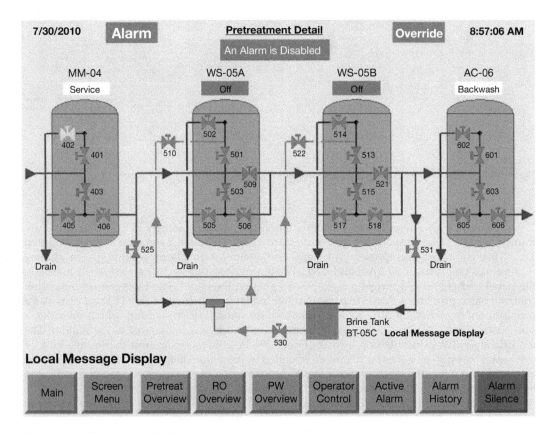

Figure 10.8 Pretreatment detail.

The values can be adjustable (at the highest screen access "code") or can be "fixed" by the custom processor program.

It is suggested that the benefits of employing a PLC central control panel with display screen can be demonstrated by "screen shots" of various sections of the system, screen navigation display, alarm set point display, and manual override display. A discussion of the indicated screen shots is presented below. Routinely, screens are presented in color. In general red "fill" for a water purification component or accessory such as a pump indicates that it is off or in the case of a valve, closed. Green-colored "fill" for a water purification component or accessory such as a pump indicates that it is on or in the case of a valve, open.

Figure 10.8 presents a screen shot for pretreatment components. The figures on the screen depict the individual pretreatment components; recirculation pump, cooling heat exchanger, multimedia filtration unit, activated carbon unit, in-line ultraviolet sanitization unit, two water softening units configured in series and feed to a reverse osmosis break tank. The system employs individual pneumatically operated diaphragm valves. Displayed items include the status of each pneumatically operated valve (green—open, red—closed), recirculating pump status (green—operating, red—off), and system make-up water flow to the RO Break Tank or return back to the suction side of the recirculation pump.

Figure 10.9 presents a screen shot for RO/CEDI Loop components. The figures on the screen depict the individual RO/CEDI components; RO Break Tank, repressurization pump, RO unit, post RO in-line ultraviolet sanitization unit, CEDI unit, post CEDI unit in-line ultraviolet sanitization unit, final filtration system and feed to a Purified Water storage tank. Displayed items include the status of each pump, RO Unit status, RO product water conductivity, post RO UV status, CEDI unit status, post CEDI unit UV status and system make-up water flow to the Purified Water Storage Tank or return back to the RO Break Tank.

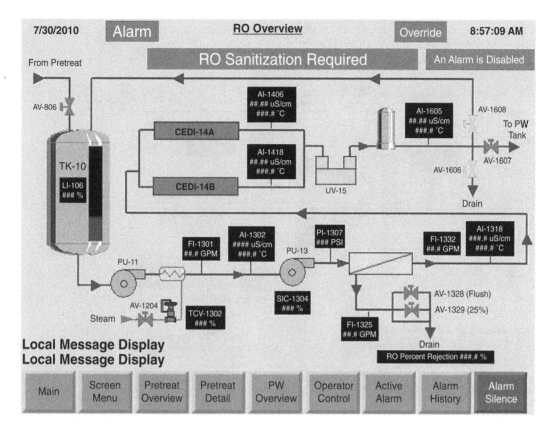

Figure 10.9 RO CEDI system.

Figure 10.10 presents a screen shot for the Purified Water storage and distribution system. The figures on the screen depict the individual storage and distribution components; storage tank with accessories, distribution pump, cooling heat exchanger, distribution loop, and return to the Purified Water storage tank. Displayed items include Purified Water storage tank level and temperature, distribution pump status, chill water flow status to the cooling heat exchanger, post heat exchange temperature, loop supply conductivity, loop return conductivity, loop return online TOC indication, and loop return pressure.

Figure 10.11 depicts the "alarm set point" screen. While certain alarm set point values may be "hard coded" in the PLC program, many are adjustable on this screen. As indicated earlier, it is suggested that a multiple pass code system be employed for access to particular screens such as the alarm set points. Access for operator level, supervisor level, and administration level should be considered.

Figure 10.12 depicts the "manual override" screen. This screen allows manual override operation of any pneumatic actuated automatic valve in an "auto" mode (normal), manually "open" mode, or manually "closed" mode. The screen also allows operation of an electrically powered component, such as a pump, in an auto mode (normal), manually "on" mode, or manually "off" mode. When *any* valve, pump, or other "controlled" item is in a manual override mode a "banner" indicating manual override will be displayed in the upper corner of every display screen.

Additional screen features that should be considered include "flashing red-colored banner" in the upper corner of the screen indicating an alarm that has been acknowledged but alarm condition still active, date, and time (military).

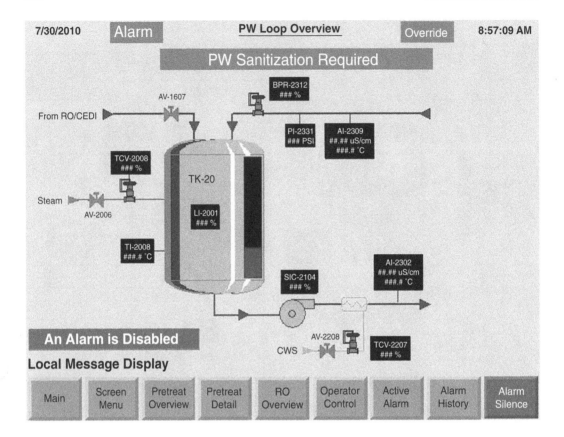

Figure 10.10 Purified Water storage and distribution system.

Data Acquisition

The central PLC-based control system should include provisions for output of all analog signals (4–20 mA) to a central facility data acquisition system. The PLC-based control system should also be capable of providing all discrete output conditions.

If the data acquisition system has the appropriate drivers, these data can be transmitted over a network connection.

Data Logger

Data logging by a management system is generally required for all manufacturing facilities. This technique is highly attractive at larger facilities that have larger water purification systems and a greater number of monitored items. While data logging provides an excellent means of obtaining a permanent record of monitored parameters within the system, there are some drawbacks. The presence of a data logger should not result in complacency on the part of operating personnel to visually monitor system performance. As a minimum, operating personnel should review and inspect system operation on a daily basis, noting critical parameters within the system. A daily logbook is suggested for monitoring critical parameters, even when a data collection system is used.

One of the important items that can be accomplished with a central PLC-based control system coupled to appropriate software and a data collection system is data trending. Specific software packages are available with the "water purification unit operation calculations" necessary to "custom trend" data for a particular application (Westwind Software 1988; La Firenza et al 1998). For example, a software package with proper input can automatically calculate RO unit performance (percent ion rejection, percent water recovery, etc.) and trend the data, alerting operating personnel when membrane-cleaning guidelines are approached.

CONTROLS AND INSTRUMENTATION

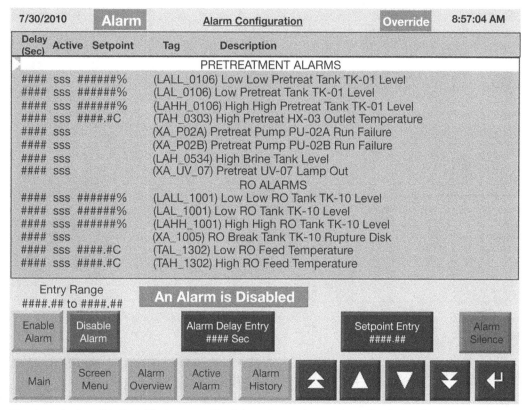

Figure 10.11 Alarm configuration.

Finally, it should be noted that the data logging system, particularly a system used to permanently record data, must be validated in accordance with acceptable criteria for record keeping.

PURIFIED WATER AND WATER FOR INJECTION CONDUCTIVITY MEASUREMENT

Physical Tests Section <645> of the current version of USP discusses the water conductivity measurement and specification for Purified Water and Water for Injection. Other pharmacopeia set forth a conductivity specification for compendial water in a generally *harmonized* manner.

The conductivity specification set forth in USP *Physical Tests* Section <645> was developed around the projected "end point" of historical "pass" or "fail" tests. The actual conductivity specification is associated with a chloride concentration of 0.47 mg/L (pH 5.0–6.5) and ammonium value of 0.3 mg/L (pH 6.6–7.0). The literature contains an excellent article discussing the conductance effect of each ion in developing this specification (Bevilacqua, 1996).

The conductivity measurement establishes a three-stage procedure. Stage 1 allows conductivity measurement online or off-line in a laboratory. Stages 2 and 3 measurements must be conducted in a laboratory.

Within the three-stage measurement technique, a sample is initially tested in accordance with stage 1. If the sample passes the stage 1 requirements, it is not necessary to proceed to stage 2 or stage 3. If the sample conductivity is greater than the stage 1 specification (at the sample temperature), stage 2 testing is required. If the conductivity is within the stage 2 limit, the sample passes; it is not then necessary to perform stage 3 testing. If the sample conductivity is greater than the value specified in stage 2, it is necessary to proceed to stage 3.

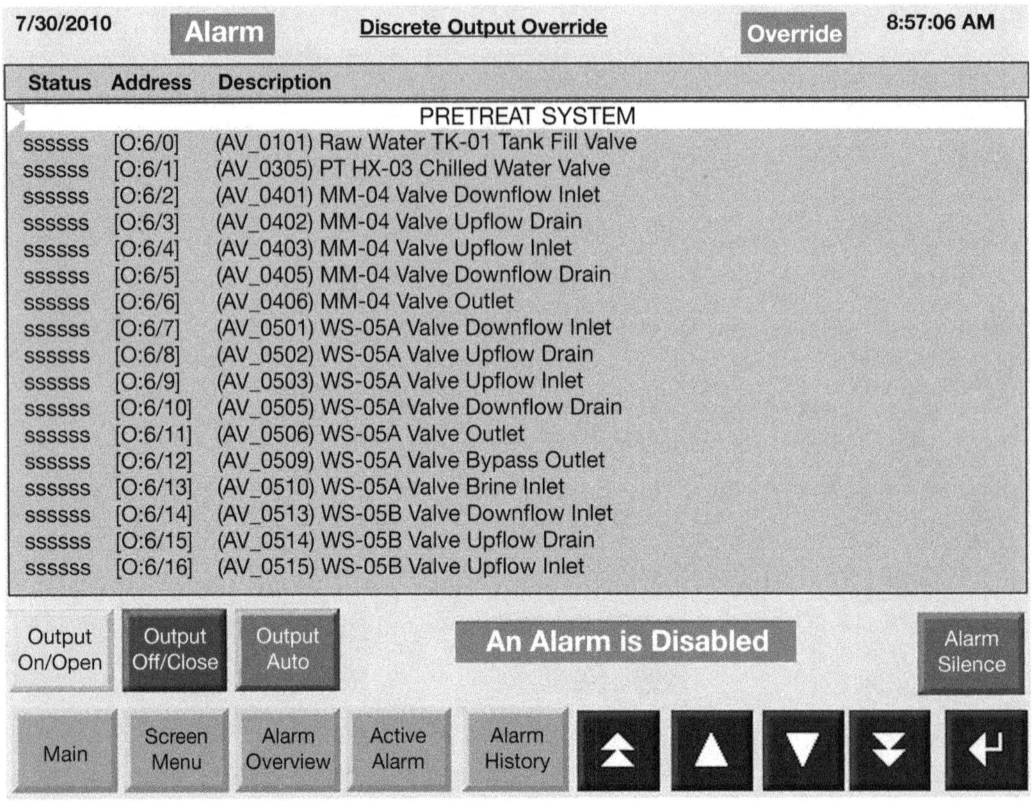

Figure 10.12 Output override.

The stage 1 maximum conductivity value at 25°C is 1.3 µS/cm. The maximum stage 2 conductivity at $25 \pm 1°C$ is 2.1 µS/cm. The maximum stage 3 conductivity value is a function of pH, varying between 2.1 and 4.7 µS/cm.

All conductivity values, both online and off-line, *must* be measured with a non-temperature-compensated conductivity meter.

It is strongly suggested that individuals responsible for the calibration and maintenance of conductivity monitoring systems for USP Purified Water or Water for Injection systems carefully review *Physical Tests* Section <645>. It is further suggested that online conductivity measurement is a much more reliable and accurate indication of water quality than laboratory analysis. As discussed on several occasions within this text, it is recommended that conductivity monitoring of Purified Water and Water for Injection distribution loops (online) should be performed with a single meter and two cells. One of the cells should be installed in the feed to the distribution loop and the second is installed in the return line from the distribution loop. It is also suggested that online conductivity measurement should be included as part of any data acquisition program. This allows operating personnel to determine if transient conditions within the storage and distribution system affect conductivity values within the distribution loop and, subsequently, at points-of-use.

Figure 4.20 provides several installation configurations for conductivity cells. The preferred mounting orientation for a conductivity cell is in the "run" portion of a tee, with feedwater "hitting" the cell directly, while maintaining recirculation through the branch side of the tee. Vertical mounting is not recommended, since it requires that the tubing/piping be free of any trapped air between the electrodes of the conductivity cell. Seven different arrangements of conductivity cell installations are demonstrated in Figure 4.20, of which only three are acceptable installation methods and four are unacceptable. To determine representative conductivity values, appropriate conductivity cell installation is required.

TOC MEASUREMENT

Physical Tests Section <643> of the current edition of USP presents the specification and criteria associated with measurement of TOC for USP Purified Water and Water for Injection applications.

Currently, the USP reference standard solutions for calibration are USP 1,4-Benzoquinone RS and USP Sucrose RS. The TOC monitoring system must have a manufacturer's specified limit of detection of 0.05 mg of carbon/L or lower. The response efficiency defined in *Physical Tests* Section <643> is $\geq 85\%$ and $\leq 115\%$ of the theoretical response to the standard solutions.

There are a number of TOC analyzers available for pharmaceutical water system applications. It is extremely important to select an analyzer capable of providing the desired monitoring capability. Questions associated with preference for online monitoring or laboratory analysis should be addressed. While many applications may use an online method, monitoring the TOC value in the return line from the distribution loop, it is suggested that this type of monitoring program, without periodic supplemental monitoring at points of use (actual "delivered water" to a manufacturing vessel, such as the end of a transfer hose), *could* be challenged from a regulatory standpoint. As discussed on several occasions within this text, the regulatory focus for water quality is associated with the delivered water to a process, application, etc. While it is fully acknowledged that the use of online monitoring provides ongoing indication of recirculating loop TOC values including any system transients, certain systems, particularly systems at larger manufacturing facilities with "platforms" using mass totalizing meters, should consider "supplemental" monitoring. As an example, periodic sampling from the end of transfer hoses should be considered since there is a possibility of contamination. While SOPs for platform activities should clearly define requirements to ensure proper handling and cleaning of hoses, field experience clearly indicates that this is extremely difficult to achieve in light of the multiple "batching" vessels, the length of the production cycle, the requirement for water at different points of use throughout the manufacturing cycle, the length of transfer hoses with their awkwardness for proper draining and storage, and the generally "wet" nature of the physical manufacturing area. The TOC-monitoring program for points of use should reflect the nature of operations at a particular facility including the potential contamination characteristics of the delivery systems through components such as flow/mass meters.

Another item that should be considered in selecting TOC analyzers is the ability to measure TOC values throughout the water purification system, from raw water to water at points of use. This is a critical factor for systems with a raw feedwater source from a surface supply. Some TOC analyzers, capable of measuring raw water TOC values, can be employed to determine the reduction in TOC levels through an activated carbon unit or a membrane process (reverse osmosis or ultrafiltration). Further, TOC analyzers may be used to determine the effects of other unit operations, such as ozonation or 185 nm ultraviolet units employed for destruction of residual disinfecting agent. In light of this discussion, the following presents a general summary of the basic principles, types of measuring techniques, and the limits of various TOC analyzers. It is not the intention to promote or endorse one technology over another for TOC monitoring. However, it is important that individuals selecting TOC monitors understand the method of operation and applicability for specific applications.

The majority of online TOC analyzers used for pharmaceutical water system applications employ an oxidation technique using ultraviolet radiation at a wave length of 185 nm and conductivity detection. Ideally, the 185 nm ultraviolet radiation oxidizes all organic material to carbon dioxide (in equilibrium with the bicarbonate and hydronium ion), which is detected by an increase in conductivity. Numerous enhancements to the basic oxidative and detection process have been and continue to be developed. Enhancements include, but are not limited to, use of differential conductivity, injection of acid to remove carbon dioxide associated with inorganic carbon, and addition of an oxidizing agent to increase oxidation of organic material. As indicated earlier, the selection of a particular TOC analyzer is a function of the inorganic and organic attributes of the water being measured. A particular analyzer may be applicable for "finished" Purified Water and/or Water for Injection TOC measurement but inappropriate for raw water and/or activated carbon unit product water TOC determination.

Another type of TOC analyzer, primarily used for laboratory applications, provides both total inorganic carbon and TOC values. Initially, total inorganic carbon is determined by acidification of the sample, converting all bicarbonate (and/or carbonate) to carbon dioxide. The carbon dioxide is collected on a molecular sieve, purged, and detected by a calibrated, nondispersive infrared detector. After determining the total inorganic carbon value, the sample is treated with a strong oxidizing agent, chemically oxidizing organic carbon in the sample at about 100°C. Organic carbon is converted to carbon dioxide, which is collected on a molecular sieve. The carbon dioxide is subsequently purged and detected by the calibrated, nondispersive infrared detector. This particular analyzer has significant versatility. The results are reliable for both high purity and raw water. The instrument is capable of determining TOC concentrations over a wide range by changing the size (volume) of the sample analyzed. There are no limitations with regard to the conductivity of the sample being analyzed, and there is no interference from THM compounds.

The final TOC analyzer to be discussed uses a combustion method and is essentially limited to laboratory applications. In this analytical scheme, the sample is catalytically combusted in the presence of an oxidizing atmosphere at user-selected temperatures between 680 and 950°C. Total carbon is determined by measuring the carbon dioxide resulting from the combustion of carbon in the sample with a nondispersive infrared detector. Total inorganic carbon is detected by simple sample acidification that converts carbonate and bicarbonate ions to carbon dioxide. The dissolved carbon dioxide from the inorganic radicals is then purged from the solution and measured by the nondispersive infrared detector. The actual TOC value is determined by subtracting the total inorganic carbon value from the total carbon value. Limitations associated with this analyzer relate to cost and the required high temperature for combustive destruction of organic material.

REFERENCES

ANSI/ISA. Instrumentation Symbols and Identification. ANSI/ISA S5.1-1984, approved November 5, 1986. Research Triangle Park, NC, USA: Instrument Society of America, ISBN 0-87664-844-8.

Bevilacqua AC. Calibration and performance of a conductivity system to meet USP 23. Ultrapure Water 1996; 13(8):25–34.

Godec R, O'Neill K, Hatte R. New technology for TOC analysis ion water. Ultrapure Water 1992; 9(9):17–22.

La Firenza S, Neto J, BuKay M. Improved monitoring at Hewlett-Packard by using a mobile barcode data collection system. Ultrapure Water 1998; 15(2):18–23.

McCurdy L. Implementing TOC testing for USP 23—a case study. Pharm Eng 1997; 17(6):96–104.

Westwind Software. Water System Detective, Version 1.7. San Jose, CA: Westwind Software, Inc., 1998.

11 | System design and specification guidelines

OVERVIEW
It is inappropriate to discuss the design, installation, validation, operation, and maintenance of pharmaceutical water purification systems without providing examples of required specifications and associated documentation. Specifications are an extremely important item because they provide the vehicle that delineates design and performance requirements to the water purification equipment manufacturer or system integrator. Some water purification equipment suppliers focus their sales and marketing activities on "standard products." Other manufacturers exhibit flexibility with regard to design and manufacturing criteria for individual components. The "ideal" specification should be prepared in a manner that clearly defines the unit operation specifics required for a project. "Standard equipment" manufacturers will either include "adders" for specified items that deviate from their standard product line or will take exception to specific sections of the specification in their proposal. The specification should, where possible, provide design details that are readily available from the majority of equipment manufacturers. Items that may be required for a specific component but would obviously increase the cost of the component disproportionately to the desired benefits should either be deleted or provided by "others" (mechanical contractor, installer, etc.). For example, an activated carbon removal port may be specified for an activated carbon unit. This is a critical item since it is required for the periodic replacement of activated carbon. It should be available from both standard equipment and "custom" manufacturers. On the other hand, it may be desirable to provide an exterior thermal insulation "package" for the unit, particularly if it is specified with hot water sanitization provisions. The insulation package is not an item readily available from most equipment manufacturers. Subsequently, it would be more cost-effective to install the insulation package (using a mechanical contractor) after the equipment is positioned "in-place." The insulation package is, for most installations, a necessary item, considering personnel protection. It should not be deleted. It should be included in the system specification by defining the "Scope of Supply." A project manager or construction manager can provide valuable assistance to ensure that items that are not standard from water purification equipment suppliers be included as part of the installed system.

This chapter discusses specific items that should be included in water purification system specifications and documentation. It also presents multiple attachments that provide general examples of required documents.

The chapter has been divided into the following "documentation and specification" subsections:

- General requirements
- Drawing requirements
- Instrumentation and control system requirements
- Operation and maintenance requirements
- Quality control/quality assurance requirements

GENERAL REQUIREMENTS
A *Basis of Design* should be prepared for the system. As implied, this document provides criteria that establish the requirements of the water system. As a minimum this should include the following:

- Classification of water required such as "Drinking Water," "Softened Water," "RO/DI Water," USP/EP/JP Purified Water, and USP/EP/JP Water for Injection.
- Description of the nature of demand at individual points of use such as manufacturing (ingredient), clean-in-place rinsing, and laboratory.

- Description of the physical points of use such as "hard piped," hoses, "transfer panel," heat exchanger, mass flow or totalizing meter, and manual diaphragm valve.
- List of each point of use, with identification number, maximum instantaneous demand (volume/unit time).
- List of each point of use, with identification number, maximum daily (or 8-hour shift) demand (volume).
- Point-of-use matrix showing the number of points that will be using water concurrently. This should include time periods when there will be no water demand from the distribution system.
- Point-of-use matrix with the delivered water temperature requirements for all operations at each point. This list should include the flow rate for different temperature applications if applicable.
- Total viable bacteria "Alert" and "Action" Limits for each classification of water including intended enumeration method.
- Feedwater source to the facility with analysis, historical data, and source (municipality or "private" supply).
- Available utilities (steam, raw feedwater flow rate, electrical, drain, ventilation, etc.).
- Local, state, county, province, or other regulations.
- Complete list of countries where product using water will be used.
- Any other facility requirements.

All pharmaceutical water purification systems require an *Equipment Specification*. This document demonstrates how the requirements, outlined in the Basis of Design, will be achieved. It is an important link in the documentation chain for ultimate system qualification or validation. The specification is prepared after conceptual and preliminary design options have been evaluated. It should clearly define "submittal" requirements for the successful water purification equipment supplier. The Basis of Design, detailed specification, and submittal material, coupled with a "turnover package" from the successful equipment supplier, will be used extensively during preparation for Installation Qualification (IQ) and Operation Qualification (OQ). The Basis of Design and detailed specification are referenced documents for both IQ and OQ. The "submittal package" forms the basis of a final turnover package, which should be consistent with the detailed specification and will be directly incorporated within the IQ, OQ, and standard operating procedures (SOPs), modified to the end user's format.

It is extremely beneficial to develop specification sections for individual water purification unit operations. This unit operation concept can be extended to preparation of IQ and "integrated" with all system components for preparation of the OQ, the topic of chapter 13. While there are certain functions of the system that are common to system operation, such as controls and instrumentation, the unit operation approach has several benefits. As discussed in earlier chapters, the design, operation, maintenance, and validation (or qualification) of a pharmaceutical water purification system weaves its way through numerous technical disciplines. It would be inappropriate to consider that *all* individuals involved in the process will have "in-depth" knowledge of water purification components, principles of operation, and other related technical matters. Subsequently, individuals can be literally overwhelmed with detailed design, operating, maintenance, and validation requirements of a water purification system. Conversely, the unit operation approach is a highly effective method of simplifying the technical aspects of components to a comprehensible level. As suggested in chapter 13, a training manual should be generated as part of the validation (or qualification) "package." Individual copies of this manual can be distributed to personnel during training sessions. The sessions should be structured for presentation to individual from a discipline cross section at a facility. The training sessions begin with the initial component in the system and proceeds through each component within the system. The training program is structured similar to earlier chapters in this book, outlining the basic theory, application, design considerations, operation, and maintenance considerations. Validation requirements, specifically verbatim compliance to SOPs, should also be discussed. When complete, a proper training session should explain why each component is used, why the components are configured in a specific arrangement with certain accessories, how each component should be operated and properly maintained, and the critical nature of the validation process.

A properly prepared specification should clearly define the design criteria for each component. Obvious criteria would include items such as column diameter and height. Less obvious criteria, but critical to a properly prepared specification, would include items such as the attributes of feedwater, product water sampling valves, the face diameter of pressure gauges, the range of pressure gauges, the materials of construction and type of pressure gauges, leveling provisions for legs on a column, column exterior paint requirements, motor horsepower, motor voltage, motor electrical power phases, and anticipated motor amperage requirements. In developing the detailed specification for a component, it is appropriate to consider a description of the component as if it were being assembled. A *Bill of Materials* associated with *equipment assembly drawings* should be listed as a submittal requirement in the specifications. If the level of detail, submittal requirements, and documentation are not clearly set forth in the specification, it will inevitably result in an "expansion of scope" from the water purification equipment manufacturer, with associated increase in the overall cost of the system. With few exceptions, there can never be too much detail in a properly prepared specification. The one issue that should be considered, however, is specification to the point where the individual unit operations are truly "custom components," containing details of options not readily available from the majority of equipment manufacturers as discussed earlier.

In addition to outlining the specific design requirements for *each* component, performance criteria for each component, as well as the system as a whole, should be clearly defined. This is a very important, specifically as it relates to the performance criteria for individual components. It is suggested that most pharmaceutical water purification system specifications, particularly "performance specifications," do not place enough emphasis on the performance of individual components. This generally results in a system that will initially meet established performance criteria but fails, in the long term, to provide the desired system performance. This is associated with the fact that system components, particularly pretreatment components, are not properly designed and adequate for the application. For example, an activated carbon unit with feedwater containing chloramines as a residual disinfectant (as opposed to residual chlorine) must be designed to operate at a maximum face velocity of about 3 gpm/ft^2 of cross-sectional bed area and a volumetric flow rate in the range of 0.5 to 0.75 gpm/ft^3 of activated carbon media, as discussed in chapter 3. If a system performance specification is provided, it is fully anticipated, based on the competitive nature of system procurement process, that both of the stated parameters for the activated carbon unit will be exceeded. The overall effects associated with chloramine breakthrough may not affect initial system operation. The system may operate for three to six months without any problems. Eventually, however, chloramine breakthrough will occur, potentially resulting in the "failure" of reverse osmosis (RO) membranes and ultimate system shutdown. If activated carbon unit product water residual chloramine levels are appropriately monitored, improper design will result in very frequent replacement of activated carbon media (e.g., once every 2–3 months). Activated carbon media replacement is not an easy maintenance operation to execute. On the other hand, a correctly sized unit, associated with a properly prepared specification for the activated carbon unit, would clearly define the performance criteria for a minimum time period of one year, a generally accepted standard. Failure to provide individual unit operation performance specifications is, perhaps, the greatest flaw in the development of a water purification system specification because it presents warranty issues that are often impossible to resolve since they are not obvious during initial system operation. A lack of performance criteria for each water purification component can generally be offset, as indicated above, by extensive system maintenance. If the water purification system is capable of meeting the "system performance criteria," regardless of the required maintenance, the water purification equipment manufacturer has technically met its obligation. This highly undesirable, and common situation, may be avoided by proper preparation of a detailed specification, including individual unit operation performance criteria.

The specification for each component, including accessories, should clearly state that appropriate documentation must be submitted for approval and/or provided with the final turnover package with delivery/receipt of equipment. This documentation would include, but not be limited to, the following:

- *Utility Matrix* containing *electrical requirements* voltage, phases, cycles, and amperage; *feedwater* flow rate and minimum and maximum temperature; *wastewater* flow rate and maximum temperature, *facility steam* pressure, thermodynamic quality and flow; *cooling/chill water* temperature, flow rate, and maximum temperature increase; maximum acceptable ambient temperature and relative humidity; *control and operating air* pressure, dryness, temperature, and volumetric demand.
- *American Society of Mechanical Engineers (ASME) documentation* including applicable code requirements, calculations, design pressure, test pressure, and test reports with ASME certified inspector signature.
- *Valve Matrix* presenting a list of every valve in the system, location of the valve in the system (unit operation), should be provided. The valve matrix should indicate the type, size, design, materials of construction, diaphragm/seat material, and actuator information (automatic-type) for each valve. Further, the valve matrix should include, as part of the description, the tag number correlating to the system P&ID.
- *Valve/Component Sequencing Chart and Functional Description* should be provided. The chart should summarize the sequence of valves and components for operation and associated operations such as component backwash and tank makeup.
- A *Bill of Materials* should be provided for each unit operation. The bill of materials should contain a list of all items used to assemble a component including, where possible, the manufacturer, part number, and serial number. The list should correspond to a reference on the appropriate *equipment assembly drawing*.

Appendix A to this chapter contains a generic specification for a hot water sanitizable activated carbon unit.

DRAWING REQUIREMENTS

A *process flow diagram (PFD)* should be provided. The PFD is a basic diagram depicting the major components in the system with accessories. Components are generally shown as "blocks." The flow path is indicated by lines with flow direction arrows. An example of a PFD for a simple RO/CEDI Purified Water system is depicted in Figure 11.1.

Piping (process) and instrumentation drawing(s) (P&ID) should be provided. The equipment manufacturer's P&ID(s) and the P&ID(s) provided as part of the specification should agree. Accessories, instrumentation, and format of the P&ID(s) should employ standards set for by the Instrument Society of America (ISA) (chap. 10). A "lead sheet" (for one or more system P&IDs) should be provided clearly indicating the "legend" for symbols used on all system P&IDs. If the pharmaceutical manufacturer uses an established legend for components and instrumentation that is not completely consistent with ISA terminology, this material should be provided to the water purification equipment manufacturer (or general contractor/system integrator) upon award of contract. Equipment labeling, valve number designations, instrument number designations, etc. should be clearly established on the final P&IDs, as discussed in chapter 12. Since the P&IDs provided with the detailed specification will probably contain a dotted line ("envelopes") that designates certain components, the P&IDs with complete instrument and valve numbering references can be supplied only by the equipment manufacturer or prepared by the pharmaceutical manufacturer or designated representative, using drawings provided by the water purification equipment manufacturer. It is strongly suggested that the company establishing and providing the valve and instrument tag numbers and physical tags (equipment manufacturer or pharmaceutical company) be defined within the specification.

General arrangement drawing(s) should be provided. The general arrangement drawings generally provide a two-dimensional arrangement of the water purification unit operations in the physical area available for installation. Obviously, the drawings are drawn to "scale." Generally, it is necessary to provide isometric drawings for items such as distribution loops with "drops" to individual points of use. General arrangement drawings provide valuable

SYSTEM DESIGN AND SPECIFICATION GUIDELINES

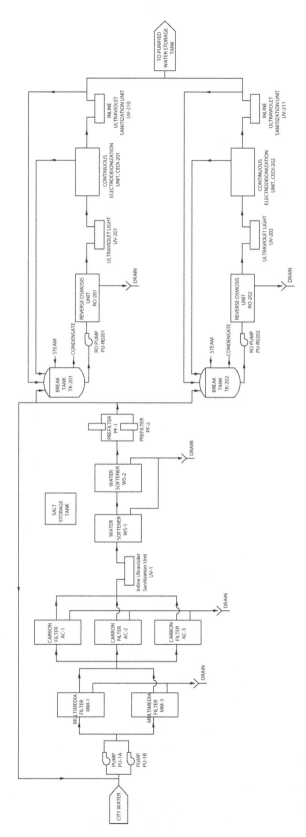

Figure 11.1 Process flow diagram: Purified Water generation system with RO/CEDI.

information for location of utilities, discussed in chapter 12. Further, the general arrangement drawings, submitted for review and approval, assist in resolution of space and access issues prior to "fixing" the location of support utilities. Potential equipment rigging issues can also be addressed. Finally, the general arrangement drawings should contain the operating weight of each component. Floor loading limitations can be reviewed prior to release of equipment for fabrication.

Equipment general assembly drawings should be provided. The drawings generally provide top, bottom, left side, right side, frontal, and rear view of individual unit operations with all accessories including valves and instrumentation. The drawings enhance information from the general arrangement drawings. The drawings are prepared to scale and depict the location of all "tie-ins" such as inlet, outlet, and drain.

CONTROL SYSTEM REQUIREMENTS

Control and instrumentation for pharmaceutical water purification system is presented in chapter 10. Specific documentation that should be provided for system controls and instrumentation include, but are not limited to, the following:

- Annotated source code
- Software development program
- Software test procedure
- I/O list(s)
- User's manual
- Screen shots
- Technical literature
- Hardware configuration drawing

OPERATION AND MAINTENANCE DOCUMENTATION REQUIREMENTS

- *Operating manuals* from all suppliers of equipment should be provided. It is strongly suggested that the operating manuals be developed into custom-prepared SOPs by the pharmaceutical manufacturer, system integrator, or contracted specialty organization.
- Depending on the scope of the project, primarily the volumetric capacity of the system, it may be appropriate to generate an *installation manual* for equipment. Criteria associated with installation are discussed in chapter 12.
- Recommended maintenance for system components and accessories should be provided as part of a manufacturer's *maintenance manual*. It is strongly suggested that the information in the manufacturer's maintenance manual(s) be developed into custom-prepared maintenance manual by the pharmaceutical manufacturer, system integrator, or contracted specialty organization.
- *Vendor "cut sheets"* for all components and accessories should be provided. The catalog cut sheets will be referenced in the IQ, discussed in chapter 13.
- *Data sheets* should be provided that supplement information in the cut sheets as appropriate.
- A *spare parts list* should be included. The list should be developed with reference to individual components. The list should provide the original manufacturer with model number and serial number. Spare parts should *not* contain a "coded" number to the equipment manufacturer if purchased from a supplier.

QUALITY CONTROL/ASSURANCE REQUIREMENTS

Many of the QC/QA requirements presented below are related to stainless steel and/or stainless steel welding. Appendix B contains a reference specification for a Purified Water stainless steel distribution loop including valves.

- A *hydrostatic or pneumatic test procedure* should be provided for verification of distribution and/or interconnecting tubing/pipe integrity. As indicated in chapter 9, pneumatic testing *should not* be performed to verify the integrity of any plastic

tubing/piping material. A hydraulic pressure "hold" test is preferred. The test procedure should establish the acceptance criteria for the pressure test. Since the pressure in tubing/piping completely filled with water will be significantly affected by temperature, the test procedure should clearly address temperature control.
- A *hydrostatic or pneumatic test report* should be provided. The report should include the original log sheet for testing with date, time (start and stop), pressure readings, temperature readings, and signature of the individual performing the test and qualified observer.
- *Material test reports* should be provided for all stainless steel surfaces in contact with water such as tubing, valve bodies, sanitary pressure gauges, and conductivity cells. The test reports should contain a complete chemical analysis of the stainless steel, the company generating the report, date of the report, traceable "heat number," and signature.
- *Certificates of Calibration* should be provided for all instruments. This includes sensors and monitors/indicators. Experience indicates that pharmaceutical facilities prefer to conduct field calibration of the installed instrumentation as part of the commissioning process (chap. 12). The "field calibration" does *not* negate the requirement to provide Certificates of Calibration from the instrument manufacturer. Finally, Certificates of Calibration must be traceable to a National Institute of Standards and Technology (NIST) source (chap. 10).
- A *Certificate of Analysis* should be provided for materials. This would include materials such as gaskets, valve diaphragms, RO membrane preservative, O-rings, and inert gas employed for orbital welding of stainless steel tubing.
- *Weld test coupons* should be provided. For orbital welding of stainless steel, the minimum number of coupons required is presented in the literature (ASME BPE, 2009), while Appendix B presents a slightly more intense test program. While infrequently performed, it is suggested that test coupons also be provided for socket welded and "bead and crevice-free" plastic welds [polypropylene, polyethylene, and/or polyvinylidene fluoride (PVDF)].
- *Stainless steel tubing mechanical polish and tests procedures* should be provided. The procedures should define the method of mechanically polishing stainless steel surfaces as well as the measurement techniques used to verify the finish.
- *Stainless steel tubing mechanical polish and tests reports* should be provided to very conformance with the procedures. The test report should be dated and signed.
- *Passivation procedures* should be provided. The passivation procedures should include the nature, concentration, and temperature of the prepassivation alkaline cleaning operation. The procedure should also include the chemical composition of the passivation solution, concentration of each chemical, temperature, and exposure time. Material Safety Data Sheets (MSDS) and Certificates of Analysis should be provided for all chemicals used.
- *Passivation test report* should be provided with results of Feroxyl Testing or other "owner"-approved alternative test method.
- *Riboflavin test results* should be provided demonstrating specified spray ball system "coverage" for storage tanks.

COMMERCIAL CONSIDERATIONS

The Scope of Supply for the water purification equipment manufacturer and others should be clearly defined. For example, a multimedia filter and an activated carbon unit are generally shipped and provided without media installation. The specification should clearly state who is responsible for installing the media. It is suggested that all internal components to various water purification unit operations be installed by the water purification equipment manufacturer. This suggestion is based on potential warranty issues. While the installation of support and filter media for a multimedia filtration unit would appear to be a relatively simple item, capable of being performed by either facility personnel or a subcontractor, this can present problems. If damage to the lower distributor in the multimedia filtration unit is caused directly by the installation operation (performed by others), the operation of downstream components may be affected, potentially "nullifying" any warranty provided by the water purification equipment supplier.

The equipment and performance warranty for each component should be clearly established in each section of the specification. A descriptive portion of the specification, provided in the individual component specifications, should clearly delineate the system warranty and, more importantly, the system performance warranty. Generally, warranties for system components and the system as a whole should not be less than 12 months. This 12-month period should not commence until the system is fully installed and operational. The warranty should *not* include "third-party" warranties. Specifically, this applies to accessories that are purchased by the water purification equipment manufacturer from others. This would include, for example, a recirculating pump or distribution pump and motor. The language within the specification should clearly state that failure of the pump and/or motor, repair or replacement, including labor, travel expenses, living expenses, etc., for a service person to correct the problem should be provided directly by the water purification equipment supplier, not the pump and/or motor manufacturer. Again, this is an extremely important item because, if not clearly defined, it would result in a system with multiple warranty contacts. Furthermore, it is suggested that this issue could result in potential voiding of the warranty by the equipment manufacturer if an incorrect item is installed by a third party.

Another important item to be addressed is the basis of a performance or component warranty. Some equipment manufacturers will agree to the warranty terms outlined in the detailed specification, while providing documentation that "qualifies" the parameters associated with the warranty. An excellent example of this situation relates to RO and electrodeionization units. While the manufacturer will indicate acceptance of warranty items outlined in a detailed specification, their post-contract award submittal package or "terms and conditions" will *qualify* the performance of the unit, stating that it is based on feedwater characteristics to the unit, which may not be achievable for the specified system. Quite often, a Silt Density Index (SDI) value, discussed in chapter 4, is specified prior to a RO system. The equipment manufacturer's warranty for the system (components, membranes, and performance) can imply that the SDI value of the feedwater must be below 3 to 5 under all conditions. If, as part of an analytical monitoring program, SDI values are measured less frequently than specified in the equipment manufacturers' terms and conditions, the warranty for the entire system may be "null and void." Proper wording within a specification can eliminate confusion or misunderstanding associated with this and similar items.

Criteria associated with system start-up (chap. 12) should clearly be defined within the detailed specification. Generally, start-up will occur after equipment arrival, unpacking, and connection of appropriate utilities (electrical, drain, air, waste, etc.). Frequently, equipment start-up is delayed. Many of these problems are related to services required for the components and "software" problems for systems employing a programmable logic controller (PLC). The extent of start-up provided by the water purification equipment manufacturer, in terms of the number of person-days allocated, is an issue since it is generally not defined in the specification. Even when the specification delineates that start-up will be required as part of the water purification equipment supplier's scope, there are often "scope expansion" issues associated with "delays" related to the connection of utilities. The two ways to address these issues are as follows:

- While often flawed by delays associated with facility services required by others, specifications may state that the water purification equipment supplier service personnel will remain at a facility until start-up is complete. This generally inflates the price of the entire system since start-up delays may be numerous and extremely time consuming. A start-up "call back" clause should be inserted in contract language to address this situation.
- A more appropriate method of defining the start-up expectations from the water purification equipment manufacturer is to define the number of days anticipated, and request, within the specification, a daily fee for providing qualified service personnel for additional "days." Further, the request for additional service time should include information associated with travel and living expenses. It is important to clearly establish the method of "invoicing" for the additional start-up time. For example, is travel time invoiced as service time, or is the service fee limited to the actual time that a service person is at a facility?

Start-up issues should include appropriate "acceptance criteria." Obviously, a definition is required to determine when start-up is complete, and system turnover is accepted by the pharmaceutical facility. The acceptance criteria should be the defining item for release of any monetary "retainer" for the project. In other words, terms of payment to the water purification equipment supplier should include a suggested 10% retainer until the system is operating in accordance with provisions stated in the acceptance criteria. The retainer provides incentive for everyone to complete the project as quickly as possible while complying with clearly defined criteria set forth in the specification.

Shipping requirements should be clearly defined within the specification. It is strongly suggested that the water purification equipment manufacturer or accessory provider "maintain" ownership of the equipment until it arrives at the pharmaceutical manufacturer's facility. This is accomplished by stating the "freight-on-board" (FOB) location as the pharmaceutical facility. Note that stating that the freight charges will be the responsibility of the supplier is *not* adequate. Unless the FOB point is specified as indicated, the equipment becomes the property of the pharmaceutical manufacturer once it leaves the equipment manufacturer's or supplier's facility. It is suggested that the most important item associated with shipping relates to the receipt of material at the pharmaceutical facility, unloading and unpacking the equipment, inspecting the equipment for visual damage during shipment, and reporting the damage *immediately* to the manufacture and/or supplier. Where practical, equipment should be shipped in dedicated trailers or containers on an "exclusive" use basis. The specification should clearly state that the water purification equipment manufacturer or component supplier should provide an authorized individual to verify that shipping damage has not occurred upon receipt of the equipment at the pharmaceutical manufacturer's facility. Essentially all transportation companies will not accept liability for equipment damage identified greater than 24 to 48 hours after receipt of the material at a facility. While the water purification equipment may arrive days, weeks, or even months before it will be installed at a facility, immediate inspection is very important. If there is damage to a component that requires replacement of the component, system start-up could be significantly delayed if the damage is not identified until the system is being unpacked and installed, potentially at a much later date than received.

As discussed previously, a turnover package, including documentation required for preparing the validation protocols, should be included with the system. The "acceptance criteria" outlined earlier should include receipt of an acceptable turnover package. This package should not be "accepted" without a thorough review, by the appropriate individuals at the pharmaceutical facility, specifically individuals preparing the validation (or qualification) documents. It is much easier to obtain "missing" documentation from a water purification equipment supplier prior to the final release of any monetary retainer than after release. To justify the water purification equipment manufacturer's position, however, the required information is more readily available at the time of, or slightly after, shipment from their facility.

The detailed specification should also define "manufacturing milestones" for the components of the system. The specification should state that the pharmaceutical manufacturer, and/or a designated representative, shall reserve the right to inspect the equipment at designated "milestones." Periodic inspection throughout the manufacturing process can eliminate needless delays associated with minor items during system start-up. Obviously, it is much easier to install an alternative sample valve or pressure gauge, for example, at the equipment manufacturer's facility, as opposed to correcting the system during start-up at the pharmaceutical manufacturer's facility. Further, periodic inspections may reveal items that were inadvertently overlooked during preparation of the detailed specification. This could include, for example, equipment dimensions that would restrict installation, or unplanned movement of one or more components from the unloading area at a facility to the area of installation.

In addition to periodic inspections of the equipment components at the manufacturer's facility, a final inspection should also be clearly defined within the specification. This final inspection should be performed by the individuals who will be involved in unpacking and inspecting the equipment when it is received at the pharmaceutical manufacturer's facility. During the final inspection, the individuals who will be performing the start-up of the system from the water purification equipment manufacturer's organization should be physically present. These items can be performed with a Factory Acceptance Test (FAT).

REFERENCE
ASME BPE. Bioprocessing Equipment, Section MJ-7.2.3, New York, NY: The American Society of Mechanical Engineers, October 20, 2009:116.

APPENDIX A: ACTIVATED CARBON UNIT

USP Purified Water System
ABC Pharmaceutical Company
Any Town, U.S.A.

XYZ Engineering & Design Company
123 Maple Street
Another Town, U.S.A.

Revision History			
Rev	Author	Description	Date
A	A. Person	Submitted Client Review	
B	A. Person	Issued for Bid	

Job Number: XXXXX-01   **Job Description:** USP Purified Water System Project

Written and Released for: Bid Solicitation

Item #	Spec Category	Specified Attribute	Additional Comments

General
1. Tag number(s): AC-102 • N/A
2. Model number(s): To be determined • N/A
3. Quantity: One (1)
4. Sanitary vessel: X Yes ☐ No • N/A
5. Type: Vertical • N/A
6. Orientation: Cylindrical • N/A
7. Top type: Domed • Flanged & Dished
8. Bottom type: Domed • Flanged & Dished
9. Applicable vessel codes:
    a. ASTM—American Society for Testing and Materials, A-l 82
        i. A967-05 Standard Specification for Chemical Passivation Treatment for Stainless Steel Parts
        ii. A-380-99 Standard Practice for Cleaning, Descaling, and Passivation of Stainless Steel Parts, Equipment, and Systems
    b. CFR Title 29, Part 1910—OSHA Safety and Health Standards
    c. American Society of Mechanical Engineers
        i. Materials of Construction—Section II
        ii. Welding and Brazing Qualification—Section VIII
        iii. Vessel Design and Construction—Section VIII
    d. ANSI—American National Standards Institute
    e. AWS—American Welding Standards
10. Stamping: The vessel must be stamped and/or labeled with the following items
    a. Serial number
    b. Design/test pressures
    c. Design/test temperatures
    d. ASME
    e. National Board Number

SYSTEM DESIGN AND SPECIFICATION GUIDELINES 397

Job Number: XXXXX-01	Job Description: USP Purified Water System Project		
Written and Released for: Bid Solicitation			
Item #	Spec Category	Specified Attribute	Additional Comments
11.	Applicable codes and requirements	a. United States Department of Health and Human Services—Food and Drug Administration, CFR Title 21, Part 210 & 211. b. IAMES—International Association of Milk, Food and Environmental Sanitarians—3A Sanitary Standards c. ASME/ANSI F2.1—Food, Drug and Beverage Equipment. d. HIMA—Health Industry Manufacturers' Association, Document 3, Volume 4, 1982. e. USP—United States Pharmacopoeia, Class VI Test for Extractables f. Seismic Zone—Designed for Seismic Zone 2A—Calculations not required	
Submittals			
12.	Vessel certificates: The following must be submitted explicitly on the Certificate of Compliance or through an explicit certificate with each vessel:		
		X U-1A Form X Model number X Serial number X Wetted material of construction X Jacket material X Tank insulation material ☐ Tubing insulation material	• Face piping to be designed to be field insulated with 2-in. thick fiberglass insulation with PVC jacket by others
		X Surface finish X Polish X Passivation ☐ Other:	
13.	Welders certifications	X Yes ☐ No	• Required for all individuals performing welding for vessels or face tubing
14.	Drawings	X Yes ☐ No	• Provide detailed drawing with elevation view, plan view, typical ferrule weldment detail, other critical weld details, materials of fabrication, vessel insulation details, weld finish information, code design requirements, bolt down location detail, include shipping and operating weights on drawing • Detailed list of all components used
Vessel parameters			
15.	Vessel contents		
		30 ft^3 of activated carbon	• Acid-washed activated carbon, 12 × 40 mesh, Calgon Corporation—Centaur • Freeboard = 100%
		Support media	• Anthracite • Must cover minimum of 4 in. above bottom distributor

(Continued)

Job Number: XXXXX-01		Job Description: USP Purified Water System Project	
Written and Released for: Bid Solicitation			
Item #	Spec Category	Specified Attribute	Additional Comments
16.	Vessel design parameters		
	Design temperature	250°F	• Per ASME Code
	Design pressure	150 psig/full vacuum	• Per ASME Code
	Pressure drop	>5 psid and <12 psid including face tubing/piping and internals	• Measured using service inlet and service outlet pressure gauges during normal operation
17.	Vessel operating parameters		
	a. Temperature range	220°F maximum 65°F nominal	
	b. Pressure range	60–80 psig nominal	
	c. Flow range	25 gpm normal, 50 gpm maximum	
18.	Vessel material	Series 316L stainless steel	• Material certification required
19.	Surface finish	Mill finish—welds ground smooth	• Internal and external
20.	Insulation/jacket	Vessel to be insulated with sheathing on shell only. Vessel heads to be insulated with approved material without sheathing	• Verification documentation required • 2-in. thick chloride-free insulation with series 304L stainless steel jacket. Jacket external finish must be <15Ra. Mirror sheathing riveted in-place
21.	Other		• The flow of water through the unit will be from left to right • The activated carbon unit will be provided as a stand-alone unit for field mounting directly to the floor of the facility, by others. The column should contain four (4) support legs, 90° apart. Each leg should terminate in an elevation adjustable support pad. Each support pad should contain four individual holes, also 90° apart, for securing each leg to the floor.
Dimensions			
22.	Inside diameter	To be determined per ASME Code Calculation	• Per ASME Code
23.	Outside diameter	42 in.	• Not including insulation and sheathing
24.	Straight side height	96 in.	• N/A
25.	Overall skid height	<12 ft	• Including relief valve
26.	Discharge height (center to ground)	To be determined per ASME Code Calculation	• Submittal approval required
Vessel fittings and details			
27.	Top head fittings		
	a. Sanitary Tri-Clamps	1½-in. sanitary ferrule	• Pressure relief valve fitting—vendor to confirm size
		2-in. sanitary ferrule	• Top inlet/backwash outlet
		1-in. sanitary ferrule	• Vent
	b. Other	• Manway	• Upper column straight side or domed top position clearly outlined in proposal • 14 in. × 18 in. elliptical • Non sanitary strong back style • Per ASME Code
		• Lifting lugs	• Vessel to be equipped with two (2) lifting lugs on the top head configured in a plane parallel to the face tubing/piping

Job Number: XXXXX-01	Job Description: USP Purified Water System Project	
Written and Released for: Bid Solicitation		
Item #	Spec Category	Specified Attribute	Additional Comments
---	---	---	---
28.	Bottom tank fittings a. Sanitary Tri-Clamps	2-in. sanitary ferrule	• Bottom outlet/backwash inlet
29.	Miscellaneous tank fittings a. Miscellaneous	• 8-in. ANSI flange with mating blind flange	• Media removal and lower distributor access port positioned approximately 12 in. above bottom straight side at 90° location from face tubing/piping
		• 8-in. ANSI flange with mating blind flange	• Media removal and lower distributor access port positioned approximately 12 in. above bottom straight side at 90° location from face piping, 180° from other flange
		• 1½-in. sanitary ferrule	• Side mounted location for thermal well, about 24 in. below top. Vendor to confirm size
30.	Accessories a. Bottom distributor	• Series 316L stainless steel	• Header-lateral or hub-radial design, 2 in. drilled pipe with series 316 stainless steel screen (Johnson Screen or approved equal)
	b. Upper distributor	• Series 316L stainless steel	• Vendor standard design to eliminate media loss during backwash
	c. Gaskets	• EPDM	• Heat compatible • FDA-approved material
31.	Face piping (tubing) a. Flow rates	• 25 gpm nominal • 50 gpm maximum • 35 gpm hot water sanitization	• Maximum flow rate occurs during rinse of softener—Infrequent condition
	b. Material of construction	• Series 316 L stainless steel	• Orbitally welded
	c. Size	• 2 in. sanitary tubing	• N/A
	d. Internal finish	• Ra ≤ 30 microinches	• Internal only, exterior may be mill finish
	e. Insulation	• Provided by others	• Piping to be designed for field insulation with 1 in. insulation
32.	Valves	• Automatic valves, ACAV-109	• Series 316L stainless steel 1½ in. sanitary diaphragms with pneumatic actuators
		• Automatic valves, ACAV-104, ACAV-105, ACAV-106, ACAV-111, ACAV-114	• Series 316L stainless steel 2 in. sanitary diaphragms with pneumatic actuators
		• Manual feedwater and product water isolation valves, ACHV-102, ACAV-117	• Series 316L stainless steel 2 in. sanitary diaphragm valves
		• Manual ambient backwash flow rate control valve, ACHV-119, and rinse flow regulating valve, ACHV-112	• Series 316L stainless steel 2 in. sanitary diaphragm valves
		• Manual bottom drain valve, ACHV-118	• Series 316L stainless steel 2 in. sanitary diaphragm valves

(Continued)

Job Number: XXXXX-01		Job Description: USP Purified Water System Project	
Written and Released for: Bid Solicitation			
Item #	Spec Category	Specified Attribute	Additional Comments
		• Relief valve ACPSV-118	• Nonsanitary design is acceptable (connection to tank must be sanitary) • Series 316L stainless steel construction • ASME Code stamp (with calculations required to justify sizing) • Relief valve tubing to be series 316L stainless steel directed to within 12 in. of the base of the column by vendor
		• Sample valves—Feedwater and product water, ACSV-110, ACSV-116	• Series 316L stainless steel ½-in. sanitary diaphragm with EPDM diaphragms and sanitary ferrule connections • Valves to be connected to a short bull, minimal dead leg tee
		• Vent valve, ACHV-107	• Series 316L stainless steel 1-in. sanitary diaphragm valve directed to within 12 in. of the base of the column by vendor
33.	Instrumentation		
		• Feedwater and product water pressure elements and indicating transmitters, AC-PE/PIT-103, AC-PE/PIT-115 • Column temperature element and indicating transmitter AC-TE/TIT-113	• 2-in. sanitary ferrule connection • Feedwater and product water location per P&ID • Single units with sensor, digital indicator, and transmitter • Sanitary connection to thermal well in *upper* straight side of the column
		• Sight glasses in waste line from unit, ACSG-120	• Units to be John C. Ernst Co. #255 • 2 in. size • Series 316 stainless steel • Borosilicate glass with minimum length $4\,3/4$ in., 360° visibility, 100 psig maximum operating pressure, 300° F maximum operating temperature, with Teflon gasket material.
34.	Controls		

a. A single I/O panel will be provided for the activated carbon. The I/O panel will contain the solenoid valves for operation of each pneumatic valve. Further, the I/O panel, through a communications cable with the central PLC for the entire pretreatment system will not only provide a signal for valve control but will also receive (and transmit) signals for communication with field devices for the activated carbon.

| | b. I/O panel | • NEMA 4X stainless steel | • Hoffmann enclosure
• Mounted on vessel
• Dimensions: TBD |
| | c. Pneumatic tubing | • ¼-in. white polyethylene | • Tubing to be neatly routed and bundled in conduit where possible. |

SYSTEM DESIGN AND SPECIFICATION GUIDELINES 401

Job Number: XXXXX-01	Job Description: USP Purified Water System Project	
Written and Released for: Bid Solicitation		
Item #	Spec Category	Specified Attribute	Additional Comments

35. Additional comments:
   a. Unit to be prepiped. The face piping (tubing) and columns should be pressure tested and passivated prior to shipment. The face piping (tubing) should be disconnected from the column, clearly labeled to identify the column associated with the tubing, and shipped in a wooden crate separate from the column.
   b. All open sections of tubing or column connections should be sealed with tape or other material to eliminate introduction of impurities during shipment.
   c. Wooden support frames should be provided for the column such that the column can be shipped in a horizontal position without damage. Any damage to the column and/or face piping during transportation will be the responsibility of the manufacturer.
   d. The outer sheathing on the column should be protected from minor scratching during shipment by use of "shrink wrapped" plastic or other suitable material.
   e. Documentation should include, but not be limited to, general arrangement drawing, piping & instrumentation drawing, ASME U form and code calculations, recommended spare parts list, passivation procedure, passivation log, hydrostatic test procedure, hydrostatic test logs, welding procedures, welding procedure specification, welding procedure qualification record, welder performance qualification, tubing weld log, tubing weld maps, orbital welding print out for each weld, material certificates, video tape inspection (on DVD) of the inspection of the interior of each weld (360°) (100% of welds), weld coupons for each tubing size for everyday orbital welding was performed, column weld maps, column weld logs, column certificates of material, post passivation Feroxyl Test Report, MSDS sheets for carbon, and QA/QC manual. Additional documentation required per the vendor data requirements (VDR) form.
   f. Units to operate in downflow direction during normal service and upflow during ambient backwash and sanitization. Ambient backwash to be conducted at a flow rate 1.5 times the service flow rate. Sanitization flow to be conducted at 1.5 times the service flow rate.
   g. Interconnecting tubing and wiring between feedwater pump skid, this unit, and the post-activated carbon UV units to be provided by others. All field connections to terminate in sanitary ferrules consistent with the piping schedule.
   h. The top of each column should be equipped with two lifting lugs for rigging and positioning of the column.
   i. Media to be shipped separately in bags or 5 cubic foot drums. Proper crating should be provided to ensure that the media containers remain intact during shipment.

APPENDIX B: USP PURIFIED WATER STAINLESS STEEL DISTRIBUTION TUBING AND FITTINGS

USP Purified Water Distribution System

USP Purified Water System
ABC Pharmaceutical Company
Any Town, U.S.A.

XYZ Engineering & Design Company
123 Maple Street
Another Town, U.S.A.

Revision History			
Rev	Author	Description	Date
A	A. Person	Submitted Client Review	
B	A. Person	Issued for Bid	

A. Scope of Supply

1. This section of the specification provides a general procedure for fabrication, welding, and installation of the series 316L stainless steel tubing for the USP Purified Water system at the facility. The intent of this section of the specification is to provide general guidelines that will be followed by the tubing installer. Material contained in this section of the specification provides guidelines that will be supplemented by detailed procedures provided by the tubing installer.

B. Codes, Standards, and References

1. Definition
Owner—Designated ABC Pharmaceutical personnel or their appointed representative.

2. Codes
   a. The installed stainless steel tubing will meet the appropriate sections of the current critical code items, outlined below.
      i. ASME—American Society of Mechanical Engineers
         - Materials of Construction—Section II
         - Welding and Brazing Qualification—Section VIII
         - ASME BPE-2009 Guide for Bioprocessing Equipment and all applicable Addenda
      ii. ASTM—American Society for Testing and Materials
      iii. ANSI—American National Standards Institute
      iv. AWS—American Welding Society
      v. CFR Title 29, Part 1910—OSHA Safety and Health Standards

3. *Regulatory Requirements*
   a. All welding, material handling, inspection, and passivation should be consistent with current Good Manufacturing Practices (cGMP).
   b. The welding and inspection program with documentation should be consistent with the requirements of the U.S. Food and Drug Administration, for review of the final validated system, as a USP Purified Water system.

4. *Submittals*
   a. The tubing installer will provide all information associated with sanitary stainless steel orbital welding, including, but not limited to:
      i. A quality assurance/quality control manual.
      ii. A list of certified welders employed directly by the tubing installer that could potentially work at the site, including a copy of welder's certificates and training.
      iii. A list, including model and serial no., of all orbital welding machines directly owned by the tubing installers.
   b. At the conclusion of the installation, the tubing installer will provide "weld map" drawings, corresponding orbital welder printout, and boroscopic inspection results including a videotape copy.
   c. Prior to the commencement of daily work, the tubing installer will provide "weld coupons" to demonstrate weld quality and integrity for review by the owner. These coupons should be dated, categorized, and submitted to the owner at the end of each workweek. Should relocation of an orbital welding machine occur during the course of the day, an additional coupon be generated.
   d. All supporting documentation for the above, including copies of each welders' ASME certification.
   e. Boroscopic inspection of 100% of welds. Tubing installer should provide a video DVD record of the inspection.

5. *Certification Standards*
   a. Certifications should be in accordance with the following standards:
      i. ASME SA-249/ASME A-270
      ii. ASME SA-249/ASME A-92
      iii. ASTM A-270-90
      iv. ASME SECTION IX, Welder Qualification

C. Conflicting Requirements
If section requirements conflict with other specifications or with industrial standards, confer with the owner before proceeding with affected operations.

D. Warranty
The tubing installer will warranty the soundness of welding performed under this section of the specification. If there is a failure in a weld, the tubing installer will be responsible for repair of the weld, including but not limited to all associated documentation and subsequent actions required for repairing the weld.

E. Quality Assurance
As outlined above, the tubing installer will provide a quality manual defining the complete quality assurance and quality control procedures used for tubing installation.

F. Exceptions
Any exceptions to these specifications should be submitted to the owner for approval prior to use.

G. Performance and Manufacturing Criteria

1. *General*
   a. Tubing will be installed and welded per the tubing installer's quality assurance and quality control manual, which will include detailed welding procedures as well as inspection, material handling, cleaning, preparation, testing (hydraulic), and passivation.
   b. General guidelines provided in this specification will be followed.

2. *Procedure Criteria*
   a. Purge gas utilized for the project will be supplied with a complete "Certificate of Analysis" (COA) including oxygen concentration and moisture concentration. Purge gas may be utilized from a liquefied source or high-purity gas cylinder to ensure proper quality.
   b. The inert gas provided for the project will be used for shielding gas, acting as an envelope surrounding the weld pool exterior and as purge gas, on the interior of the tube being welded to eliminate air from contacting the weld pool. Both shielding and purge gas will consist of argon with a quality of at least 99.999% (five nines). The moisture in the purge gas will be less than 2 ppm. The oxygen concentration in the purge gas will also be less than 2 ppm.
   c. A copy of the welder-training program for each of the individuals performing welding should be provided.
   d. The method of purging welds should be defined by the tubing installer.
   e. The procedures for cutting tubing will not introduce any foreign substances (such as dirt, oil, or grease) or create a situation where the tubing is "crushed" or damaged in any other way to cause it to be "out-of-round." All tools used for cutting and preparation of the stainless steel surface will be nonferrous, dedicated exclusively to 316L stainless steel material, and segregated from tools utilized for cutting other material.
   f. Abrasive wheels may *not* be used for cutting tubing under any circumstances.
   g. The cutting wheels, blades, or bands utilized for cutting tubing will be replaced on a regular basis to avoid creating out-of-round conditions.
   h. All cut tubing will be "faced" utilizing tools with a hardened steel bit. These tools will be exclusively purchased for facing 316L stainless steel tubing of the diameters required for this project. The use of an abrasive wheel or belt to face tubing is *not* acceptable. The facing operation will remove a minimum of 1/32 in. from the cut end. Cut ends will be square with a tolerance of plus or minus 10% of the wall thickness of the tubing.
   i. If deburring of the tubing is required, it will be performed with hardened tool steel blade equipment specifically manufactured for deburring of 316L stainless steel. These tools should consist of either a pencil reamer or deburring tool. In general, steel files will not be utilized for this application.
   j. Tubing ends will be cleaned of residual contaminants, including grease, oil, dust or shavings utilizing disposable, approved wiping pads, and isopropyl alcohol. After the material is clean, the welder should not touch the clean surface with his hands. Further, after this operation, no visible foreign matter will be evident on the surface to be welded.
   k. Individual log sheets will be provided for each weld.
   l. Finish welds will have 100% wall penetration with smooth internal surface. With the exception of "closure welds," there will be no blind welds.

3. *Tubing, Piping, Materials Criteria*
   a. Material
      i. Tubing
         - 316L (low carbon) stainless steel tubing. ASTM A269 or A270, but with 0.005% to 0.017% maximum sulfur content and ASTM A269 dimensional tolerance, bright annealed with square ends for automatic butt welding.

ii. Fittings
- Fittings should be manufactured from the same material, wall thickness, dimensional tolerance and finish as specified for tubing. Fittings should be long tangent type for automatic weld fittings (AWF) type for automatic machine welding.

iii. Joints
- Only automatic machine butt-welded and Tri-Clamp-type joints should be provided. Tri-Clamp-type joints should be used when connecting to valves, equipment, and accessories with Tri-Clamp-type ends. Otherwise, joints will be automatic machine butt-welded. Exceptions should be shown on the drawings.
- Threaded joints are not permitted.

iv. Ferrules
- Ferrules should be manufactured from the same material; 0.005% to 0.017% maximum sulfur content, wall thickness, dimensional tolerance and finish as specified for tubing. Ferrules should be welding type with one end for Tri-Clamp-type connection as manufactured by Tri-Clover, Inc.

v. Clamps
- Tri-Clamp type for use in connecting distribution loop tubing to valves, equipment, and accessories with Tri-Clamp-type ends.
- ½ in. and ¾ in. 304 stainless steel, two-segment, heavy construction with metal wing nut, 1500 psig at 70°F, 1200 psig at 250°F, Tri-Clover, Inc.—13 MHHS-3/44.
- 1 in. through 6 in. 304 stainless steel, two-segment, heavy construction with metal wing nut, 150 psig at 70°F, 75 psig at 250°F, Tri-Clover, Inc.—13 MHHM. Provide 13 MHHS three-segment clamps with metal wing nut at connections that will be fastened and unfastened frequently such as product tank connections.

b. Tubing and piping mechanical criteria
  i. Material for the sanitary tubing system will be 316L stainless steel. The tubing is to be orbitally welded and fitted with sanitary Tri-Clamps as required. There will be no threaded connections in this system. Clamp connections should be minimized.
  ii. Tubing should be welded tube, bright solution annealed (ID and OD).
  iii. The sanitary tubing outside diameter will have a tolerance of ±0.010 in.
  iv. Tubing wall thickness allowance:
  - ¼ to 3/8 in, Tubing is 0.035 in. ±10%.
  - ½ to 4 in. Tubing is 0.065 in. ±10%.
  v. Tubing to have a minimum slope of 1/8 in./linear ft to allow for draining through the use-point valves and low point valves. Slope to be verified by level and direction noted on the isometric tubing drawings.
  vi. Tubing surface finish
  - Internal: <25 Ra (maximum)
  - External: <35 Ra (maximum)

c. Tubing should conform to the following standards:
  i. ASME SA-249
  ii. ASTM A269/A270

d. Hygienic connections
  i. Connections to hygienic tubing should be Tri-Clover style. Gasket material should be designed for the nominal tubing size.

e. System valves
  i. Valves should match material type; inner and outer surface finish of the connecting tubing.
  ii. To prevent damage to any gaskets, seals, valve plugs and other nonmetallic components, the valves may be disassembled before welding. If diaphragm valves are supplied with the weld extensions, they may not require disassembly before welding.

iii. Sanitary diaphragm valves
- Valve bodies: 0.5 to 2 in. valve bodies should be forged 316L stainless steel. Valve bodies should have controlled sulfur set forth by the ASME 2007 Table DT-3. Butt weld ends should be 16 gauge 0.065 wall thickness. Butt weld ends should be integral to the valve body and extended to accommodate orbital welding. Tangent butt weld ends should be per ASME BPE-2009 fittings table DT-4. Valve body interior should be polished to match interior surface finish of tubing.
- Manual bonnets: 0.5 to 2 in. manual valve bonnets should be 316 stainless Steel with glass-filled PAS handwheel. Adjustable travel stop should be provided to prolong diaphragm life.
- Actuators: 0.5 to 2 in. automated valves should be air-to-open spring-to-close (unless specifically noted as air-to-close). Bonnet and actuator should be constructed of PAS. Actuators should be suitable of closing against a line pressure of 80 psig at 0% delta P conditions. Optional features should include adjustable limit opening stop.
- Certification: 0.5 to 2 in. valves should be provided with certified material test reports for the valve body. 0.5 to 2 in. Valve diaphragms should be provided with a Certificate of Compliance to USP Class VI Biological Reactivity.

iv. Diaphragms
- EPDM: 0.5 to 2 in. valve diaphragms should be compliant to FDA 21CFR Part 177.2600 and USP Class VI.
- PTFE: 0.5 to 2 in. valve diaphragms should be PTFE with EPDM backing compliant to FDA 21CFR Part 177.1550 (a) USP Class VI (for use in ozonated systems).

v. Point-of-use valves
- Valve type: Manual diaphragm—U-bend. Valves must be zero-static or zero dead leg design prefabricated. Must comply with Section 2.3.1.3.1.

vi. Back pressure modulating valves
- Valve type: Radial diaphragm back pressure regulating
- Valve control: External via pressure transmitter, fail open
- Valve body material of construction: 316L stainless steel
- Valve diaphragm material of construction: EPDM or TFE backed EPDM for exposure to low concentrations of continuous dissolved ozone if applicable
- Manufacturer: ABC Valve Company or equal

f. Gasket and O-ring material
i. All gaskets in the sanitary tubing system must be of EPDM or Viton material of construction.
ii. Materials should comply with USP Class VI Certification and Cytotoxicity, and CFR, Title 21, Sections 177.1550 and 177.2600
iii. Certificates are required verifying compliance with regulatory requirements, traceability lot and batch and certification lot and batch.
iv. All gaskets should be postcured to minimize extractables. All gaskets and compounds must meet the CFR and USP requirements and have a certificate to verify compliance.

g. Tubing clamps and hangers
i. Clamps should be designed to be secure, maintenance-free
ii. Clamps in contact with hygienic process tubing must incorporate a "smooth bore" contact surface between clamp and tube diameter
iii. Clamp material must be polypropylene or owner-approved material suitable for the application.
iv. Hangers should be stainless steel and adjustable via bolt connection.

4. Insulation
    a. General
        i. Continuous insulation
           Insulation should be continuous through openings and sleeves except:
           - Insulation may be omitted on hot tubing/piping (between 20 and 100°C) that passes through fire-rated and nonrated walls, floors, and other partitions.
           - Where otherwise shown on drawings.
        ii. Quality assurance
           - Installation of insulation systems to comply with manufacturer's published instructions.
           - Comply with manufacturer's published instructions for ventilation and safety procedures.
           - Unless specified otherwise, the insulation system (including all components) should have a composite noncombustible fire and smoke hazard system rating and label as tested by ASTM E84, NFPA 255, and UL 723 not exceeding Flame Spread 25, Fuel Contributed 50, and Smoke Developed 50.
    b. Piping insulation materials
        i. Insulation for hot piping
           - This paragraph applies to insulation for all hot systems, including tubing/piping systems that are heat sanitized.
           - For stainless steel tubing/piping, insulation will be XYZ Mfg. Co. tube insulation or approved equal. Insulating cement to be ABC Chemical Co. inorganic-based, asbestos-free. Use molded tube fitting of the same make whenever available. Otherwise use mitered sections of tube insulation.
           - Products in contact with stainless steel tubing systems must be suitable for use with austenitic stainless steel.
           - For hot water piping, insulation thickness should be 1.5-in. thick with PVC jacket.
           - For steam piping, insulation thickness should be 2.0-in. thick with PVC jacket.
        ii. Installation
           - Fitting insulation fabricated from matching tube/pipe insulation should be tightly fitted against adjoining tube/pipe insulation and sealed with adhesive.
           - Sliding insulation around elbows or turns, as a substitute for fitting insulation, should not be permitted.

5. *System Controls and Instrumentation Connections*
    a. All field instruments will be connected to the system using sanitary Tri-Clamp connections.
    b. Pressure gauges and transducers will be connected to short outlet tees.
    c. Temperature transmitters will be connected to short outlet tees.

6. *Sanitary Service Requirements*
    a. The stainless steel tubing, valves, and support items will be installed in accordance with sanitary requirements as outlined in cGMP (21CFR 210, 21CFR 211).

7. *Passivation*
    a. All surfaces of tubing/piping equipment and valves in a sanitary stainless steel tubing/piping system should be cleaned and passivated.
    b. Passivation will be performed after installation per approved passivation procedures and will be the responsibility of the tubing installer to coordinate and perform. The tubing installer should submit the passivation procedures and safety manual for review and approval by the owner before work commences.
    c. MSDS should be provided to the owner for all chemicals used.

    d. The tubing installer should submit a detailed passivation schedule for the owner to approve prior to the start of work.
    e. The passivation procedure should conform to the owner's and chemical manufacturer's requirements for safety and handling of chemicals.
    f. A Feroxyl Test, or owner-approved equivalent test, is required to verify proper passivation.

8. Hydrostatic Pressure Testing
    a. Pressure testing will be the responsibility of the tubing installer to be conducted per owner-approved hydrostatic pressure test procedures. The hydrostatic test procedure will be submitted to the owner for review and approval.
    b. The hydrostatic test procedure should conform to the owner's requirements for safety.

H. Execution

1. Storage, Handling, and Delivery

Storage, handling, and delivery of the stainless steel tubing, fittings, and related items are defined in detail in Section 3.3.2 of this specification.

2. Installation

The tubing installer will ensure that the tubing is installed to meet all of the criteria outlined in this tubing specification.

3. Material Acceptance and Quality Control
    a. Upon completion of installation of stainless steel tubing, an owner-approved hydraulic test demonstrating that the system, as installed, is intact will be performed (by the tubing installer). Hydrostatic testing will be conducted at a pressure approximately 50% in excess of the anticipated maximum operating pressure.
    b. The tubing installer will establish a restricted area (limited access area) at the job site. All stainless steel material will be received into this area. The tubing installer will maintain control of all material entering or leaving the area. All stainless steel material in the area will be in sealed plastic bags or, in the case of tubing, tightly capped at any ends.
    c. When stainless steel tubing, fittings, or other components are ordered by the tubing installer, the purchase order must clearly state that all material (individual sections of tubing or individual fittings) will be "stamped" with a mill "heat number" accompanied with associated analysis of compliance, "Mill Certificate." Prior to arrival of any material at the Facility, the Mill Certification Analysis must be available. The owner will reserve the right (with the material supplier) to "reject" tubing, fittings, or other stainless steel components if the Mill Certification is not available upon receipt (arrival) of the material at the job site.
    d. The tubing installer will review shipments of stainless steel components with a "packing list" provided by the supplier with the shipment. The tubing installer will verify that all material contain heat numbers matching the Mill Certificates. A log will be maintained by the tubing installer of all incoming components with respective heat numbers. Heat number should be etched to components.
    e. Material released from the controlled area will only be authorized by the tubing installer's designated individual(s).
    f. Material released from the controlled area should be installed in the system within 48 hours. Material not used after 48 hours should be returned to the controlled area (in a clean, bagged, or capped condition), within one shift after the 48-hour time period.
    g. If the tubing installer releases a section of tubing that will be cut into two or more sections, the individual will provide approximate dimensions of each section of tubing. Prior to release, the tubing installer's authorized material release person will etch duplicate (triplicate, etc.) heat numbers into the midpoints of the estimated tubing lengths.

h. The project may utilize seamed or welded stainless steel tubing. Seamed tubing will be received in a manner similar to other stainless steel material, to the controlled area. However, the material will be positioned in a clearly labeled "hold area." To avoid inadvertent access to this material, this operation will be completed within 48 hours of receipt of the tubing, depending on the "volume" of the shipment. The tubing installer will reserve the right (with the supplier) to reject tubing considered unacceptable by the qualified inspector. Inspected material, meeting criteria acceptable to the inspector, will be released from the hold area, recapped, and placed in the controlled area inventory for release.
i. All tubing provided by the tubing supplier to the tubing installer must be accompanied by certificate of surface finish and provided along with Mill Certificates.

4. Weld Inspection Procedure
   a. All welds will, as a minimum, meet the examination and testing requirements of ASME B31.3.
   b. The visual inspection will include an examination of joint details prior to welding, inspection for defects observed during welding (as accessible), and undercut, overlay, overheating, and reinforcement dimensions after welding.
   c. The tubing installer's inspector will verify that the interior of the weld is a smooth, continuous surface that will not provide a location for bacteria to attach and replicate, once the system is operational. The boroscopic inspection is a multipurpose test, verifying soundness of the weld (visually), but also verifying continuity of the interior weld surface, absence of holes or pits, and smoothness required to ensure that bacteria proliferation will not occur at the weld.
   d. A weld will be rejected for any of the conditions presented below.
      i. Observation of *any* pinholes.
      ii. Irregular or erratic weld bead appearance.
      iii. Observation of thermal cracking.
      iv. Observation of dark brown, black, or rainbow color around weld seam.
      v. The maximum weld bead width should not be less than 50% of the maximum weld bead and the maximum weld bead meander should be 25% of the weld bead width, measured as a deviation from the weld centerline in accordance with ASME BPE-2010 Guide for Bioprocessing Equipment
      vi. Observation of stainless steel oxidation products.
      vii. Misalignment of welded section of tubing.
      viii. Weld "overrun" due to improper operation of orbital welder.
      ix. Restart of orbital welder or welding "skips" around perimeter.
      x. Noticeable convex or concave area of weld above base stainless steel material.
   e. Welds identified as unacceptable to the tubing installer's inspector should be refitted with new tubing and/or fittings, as required. "Second Pass" welding to correct an unacceptable weld will only be performed with approval of the inspector, and the owner.
   f. Welds that are unable to be inspected via boroscope due to their location should be categorized, and the corresponding data from the orbital weld should be noted on the inspection log.

5. Documentation and Control
   a. The tubing installer will have a documented quality assurance/quality control manual.
      i. While the tubing installer is working on the job, a copy of the quality assurance/quality control manual will be available for reference.
   b. The tubing installer will also provide documentation stating that all procedures, welding machines, and individual welders performing the work meet the appropriate specification, and have been qualified in accordance with the ASME Code, Section IX.

c. A valid qualification certificate for each individual performing welding will be provided. The documentation will include the welder's name, social security number, "stencil number," and current (valid) certificate.
d. At the beginning of a workday (or shift, as applicable), each welder must provide the tubing installer's inspector with appropriate test weld(s). The test weld(s) will be performed on the diameter tubing planned for welding that day (or shift). If multiple diameter tubing sizes are welded by the welder, representative test welds will be provided for each diameter scheduled. The welds will be labeled by number, date, and welder stencil number. Welds may be identified by number only, with the date and welder number noted in the corresponding daily weld log. Each welder must obtain approval from the tubing installer's shift supervisor, based on inspection of the weld(s) by the inspector, before proceeding to perform *any* welding that day (shift). Test welds samples will be stored in a secured area by the tubing installer and retained by the owner upon completion of the installation.
e. If the inspector rejects the quality of a welder's test weld, the inspector will witness the performance of a second test weld. If accepted by the inspector, the welder may proceed with work for the project. If the quality of the second weld is not acceptable, the inspector will immediately notify the tubing installer's project manager/supervisor. Under *no* circumstances will the welder be allowed to continue work on the project. Further, the welding machine used for providing the test welds will also be removed from service until the nature of the problem is resolved. Requalification/recertification of the welder (and/or welding machine) will be required before the welder's name (and/or welding machine) is placed back on the tubing installer's roster of welder's (and/or welding machines), for the project. The situation described above also applies when an "installed" unacceptable weld is detected by the inspector.
f. Prior to inspecting a weld, the inspector will receive the individual weld data for the weld from the welder. The boroscopic inspection results will be entered on a separate section of the data sheet for that specific weld. At the end of each day (shift), the inspector will ensure that these data sheets are provided to the project manager/supervisor. The original data sheets will be retained by the owner. If for some reason an owner's representative is unavailable, the tubing installer should retain the records and submit to the owner at the next possible shift. Copies of the data sheets will be retained by the tubing installer.
g. At the completion of each day (shift), the tubing installer's project manager/supervisor will provide a copy of an isometric drawing showing the elevation of *all* welds performed during the day (shift). Heat numbers for tubing, fittings, and valves will be included in the drawing. The drawing copies will be stored in a secured area. Periodically, the drawings will be used (by the tubing installer) to develop larger drawings, as appropriate, for sections of installed tubing. However, the original elevation drawings will be retained for record purposes, and will be retained by the owner for system validation at the conclusion of the tubing installation.
h. Upon completion of the project, the tubing installer will provide electronic AutoCAD isometric drawings for the entire installed tubing.

6. Project Completion
a. Upon completion of the project, the tubing installer should have a "project closeout" meeting with the owner.
b. At this time, three (3) copies of a complete turnover package including all of the information and documentation outlined in this specification should be provided.

12 | System installation, start-up, and commissioning

OVERVIEW
System installation, start-up, and commissioning are important to the successful long term operation of a pharmaceutical water purification system. Throughout this text, emphasis has been placed on proper component design, component and system specification, as well as component documentation. Incorrect component installation can result in failure of one or more components to operate in an improper manner, ultimately effecting proper operation of the entire system. Further, many small but important system operating issues may be noted during start-up. As an example, proper system start-up may identify a minor but important change to the system control logic that could have a significant effect on proper operation. Finally, proper system commissioning not only verifies correct installation and operation of components during all sequences but also provides an opportunity to verify items in the IQ and OQ prior to "official" execution. This can significantly reduce or eliminate IQ and OQ execution "deviations."

Proper installation should address the following:

- Facility considerations
- Utility considerations
- Technical considerations

Each installation item of concern is discussed below.

SYSTEM INSTALLATION: FACILITY CONSIDERATIONS
The physical characteristic of the physical area of the water system installation area, including dimensions must be evaluated. The area should not be far from the points of use since distribution tubing must transverse the area from the storage tank to the manufacturing area requiring water. However, the location should also consider access to utilities such as raw water, facility steam, chill water, cooling water, drains, electrical service, and clean, dry oil-free air. The installation area should be enclosed with limited access where possible. Finally, the area should not be in an environment with airborne contamination since non particulate contamination, such as volatile organic compounds will pass through the hydrophobic vent filtration systems during storage tank drawdown.

Floor loading for the water purification components should be carefully reviewed. Many water purification components contain support "pads." The area and number of the support pads determines the floor loading. During equipment specification it is possible to outline support pad size if the floor loading appears marginal. If floor loading is inadequate, the installation area may be reconsidered. As an alternative, particularly for older facilities where floor loading is either unavailable or questionable, the area for equipment installation can be excavated and replaced with appropriate reinforced concrete to meet the floor loading requirements. Occasionally, tanks and/or equipment are positioned on mezzanines. A tank may also have an access mezzanine. To address tank mezzanine loading requirements a "support band" may be positioned around the center of a tank to distribute the weight of the tank around the perimeter. This technique may also be used for large volume tanks that cannot be installed on a single floor because of height restrictions. The support band requirements must be established and outlined in the specification for the tank.

Personnel, maintenance equipment, and replacement equipment access to the installation area should also be considered. Personnel access should consider the entry/egress route from an operator's office to the equipment. If possible, the route should not transverse buildings that are not physically connected (outdoor passage). Further, the route should not transverse clean (contamination controlled/classified) areas that require personnel to done gowns and remove gowns as part of normal access. Routine maintenance will include items such as addition and removal of RO membranes, activated carbon media, water softening resin, etc. Access should not only allow transfer of these items but consider the method of transfer and nature of the items such as containers, potential for leaks of material or water, weight, required transport equipment (pallet jack, dolly, fork lift, etc.) and physical dimensions. Throughout the life cycle of the equipment, it is possible that a component may fail. The ability to remove and replace a component should be considered. This may include "removal wall sections" to an equipment staging area.

The physical area for installation of a pharmaceutical water system is frequently marginal. Emphasis is placed on manufacturing area with minimization of facility support equipment areas. Subsequently, water systems may be installed in physical areas with other facility equipment. This equipment can include boilers, air compressors, condensate receivers, HVAC equipment, vacuum pumps, waste collection tanks, waste neutralization/treatment equipment, etc. Newer facilities may have clearly defined dedicated areas for equipment installation. Installation of pharmaceutical water purification equipment into a "shared area" presents challenges. As an example, air compressors or vacuum pumps may release oil into the environment. The oil will not be removed by a hydrophobic vent filter on a storage tank or condensing unit of a distillation unit. Any oil introduced to stored purified water and/or water for injection or introduced to condensed water for injection *would*, from a strict regulatory/compliance standpoint, be considered as an "added substance." Since an added substance is prohibited by the *official monographs* for both purified water and water for injection, measurement for the presence/absence of oil or a method of eliminating oil from the air to the hydrophobic vent filter must be considered. It should be noted that the chemical attributes for purified water and water for injection (conductivity, TOC, and nitrates for EP) may be within specification. However, the potential presence of oil requires a monitoring program as indicated. If oil is detected, the "intake" location for the hydrophobic vent filter(s) can be routed to withdraw air from an adjacent "oil-free" area.

Prior to finalizing system specifications, a comprehensive program for rigging components into their installed position should be established. This "rigging plan" should consider any existing equipment, access, temporary removal of obstruction (where possible) and the size/type of the rigging equipment required. It is suggested that it may be appropriate to construct a "model" of equipment for critical (tight-fitting) rigging applications. Field experience indicates that the use of cardboard of inexpensive schedule 40 PVC piping models can be of great assistance for rigging involving movement in multiple directions through doorways and corridors.

The location of safety showers and eye wash stations should be established. These safety components should be positioned in areas of higher "risk" such as activated carbon units and water softening units where media will be periodically replaced. Raw water should be "hard piped" to both eye wash stations and safety showers. For certain applications, portable eye wash stations may be used. The portable units are recommended for higher risk areas where access to raw water is a challenge.

It is suggested that tubing/piping be insulated to the greatest extent possible for colder raw water or hot water lines. Cold raw water, chill water, and cooling water lines *will* "sweat" during summer months and/or during periods of high relative humidity. The floor area beneath a pharmaceutical water purification system should be free of water. This allows detection of leaks from the system. Obviously, tubing/piping containing elevated temperature water or steam should be insulated for personnel protection. Experience indicates that 1-in-thick fiberglass insulation with a PVC jacket is adequate for raw water, chill water, cooling water, and hot water tubing/piping. 2-in-thick fiberglass insulation with a PVC jacket is recommended for all steam tubing/piping.

Color coded tubing/piping identification labels with flow direction indication should be provided. The labels may be custom purchased. It is suggested that labels not only assist personnel understanding system operation but identify tubing/piping section for "interface" personnel such as individuals collecting water samples. At least one label should appear in vertical or horizontal sections of tubing/piping longer than about five feet. The font size on the label should legible to operating personnel from floor level when feasible.

Floor drains may exist in the installation area or may be installed for a new system. Floor drains require a "pitch" in floor level for proper draining. The floor drain system should be capable of minimizing "puddles" of water in improperly sloped areas. Field experience indicates that many older facility drain systems will require "upgrade" for installation of a new water purification system. Upgrade may be required because the floor area around a drain actually is sloped away from the drain. This appears to be associated with long-term settling of the floor around the drain while the steel drain piping connected to the drain retains the physical elevation of the drain. Further, it is strongly suggested that the drain piping capacity be determined by a field test not drawing pipe diameter size. Drain piping may corrode, collapse, and or become clogged with material over time. This may require excavation and replacement of drain piping. As an alternative, high flow rate operations such as multimedia filtration unit backwash, a collection tank may be specified, procured and installed. The collected backwash and rinse water can be released to drain through an adjustable valve (e.g., butterfly type), at a much lower flow rate than the backwash flow rate. However, the addition of the tank may consume physical space required for water purification components. Finally, discharge water is regulated. The maximum allowable temperature of the "combined" waste from a facility is 140°F. This may not be an issue at larger facilities with high total water use. However, smaller facilities may require a waste heat exchanger (chill water cooling) to reduce the temperature of waste water from operations such as activated carbon unit or RO/CEDI loop hot water sanitization. Dilution of water with cold raw water is not recommended and violates many local and state restrictions.

As indicated previously, the water purification system installation should be executed such that there is restricted access to the equipment. This may require installation of removable walls and doors. Unauthorized access to the water purification system compromises the "control" of an "ingredient," purified water and/or water for injection. To facilitate access for approved personnel "card scan" systems may be employed. These systems can also be used, as part of a facility-wide access system, to record the individual entering the area as well as the date and time of entry/exit.

As discussed earlier in this chapter, chemical fumes or vapors in the area of installation are a concern. The environmental conditions may not only affect water quality (through hydrophobic vent filtration systems) but also the long term operation of critical components. As an example, most pump motors are totally enclosed fan cooled (TEFC). Vapors or dust in the environment will pass through the fan on the motor ultimately damaging bearings and seals. Further, environmental contaminants may also clog the cooling air through a "vortex-tube" cooler frequently used for power panels with heat generating items such as CEDI transformers.

Lighting in the installation area should allow observation of all equipment, particularly manual valves. Periodic maintenance items such as RO membrane "rotation" and activated carbon media replacement result in water flow to the floor. Adequate lighting minimizes the use of "extension cords" to perform maintenance with associated increase in personnel safety.

Painting and preparation of the installation area should be performed prior to arrival of equipment. A chemically resistant epoxy-type paint should be considered for floors.

Upon completion of equipment installation, removal metal barriers and guiderails may be appropriate to protect equipment. While "classical" yellow-colored guiderails may be used, custom fabricated stainless steel tubing can be welded, with fittings, to produce a visually attractive removable barrier.

The installation area will contain an overhead sprinkler system. If a "false" ceiling is installed, sprinkler head relocation may be required. Further, regulations associated with the distance from a sprinkler head to any "obstruction" must also be reviewed. Relocation and/or addition of sprinkler heads may be required.

Once the exact component dimensions are know, installation of "curbing" and/or "equipment pads" may be considered. As indicated earlier in this chapter, the area beneath the equipment should be dry. While preventative maintenance will introduce water to the floor beneath certain components, the water can be contained or directed to a drain and removed with proper post maintenance "housekeeping." Curbing tends to "crack" over a period of time, principally at edges and corners. Equipment pads increase the height of equipment above manufacturer's "standard design" and also are subject to long-term cracking requiring maintenance. Subsequently, it is suggested that components can generally be mounted directly to a well prepared floor using adjustable "leveling" legs (specified).

SYSTEM INSTALLATION: UTILITY CONSIDERATIONS

The location and condition of the steam supply to the installation area should be field verified (for systems requiring steam). Field verification should include pressure, flow, and thermodynamic steam quality. The condition of the steam piping should be determined by exterior examination of the piping and collection of steam condensate free of particulate matter (iron oxides).

Connection of components requiring steam should employ appropriate accessories. The supply steam piping to a component, depicted in Figure 12.1, should contain a manual isolation valve, pneumatically actuated ball valve, pressure gauge, pressure regulator, steam modulating control valve (fail close), pressure gauge, temperature indicator, and relief valve. A "drip leg" containing an isolation valve, strainer with drain valve, and steam trap should be connected between the regulated supply steam pressure and the steam condensate line from the component. The drip leg, as implied, assists in the removal of condensate from the supply steam during periods of steam demand. The steam relief valve should be sized for the application with calculations retained as part of the project file. It is strongly suggested that regulations require individual pressure relief valves for each component requiring steam. Preferably, the discharge from each relief valve should be directed upward, through the roof. The penetration through the roof should be appropriately sealed. The relief line should terminate in a double elbow configuration with "bird screen" positioned around the end of the piping to avoid accumulation of debris. To avoid water "backflow" from rain/snow/ice, the discharge of the relief valve, near the water purification component should contain a "drip pan." The drip pan collects water that may enter the relief valve piping and delivers it to a collection "cup." The cup can be fitted with a small diameter section of piping to deliver water to drain. The steam condensate piping, also shown in Figure 12.1, should contain a steam trap, temperature gauge, pressure gauge, and isolation valve. The discharge from the condensate

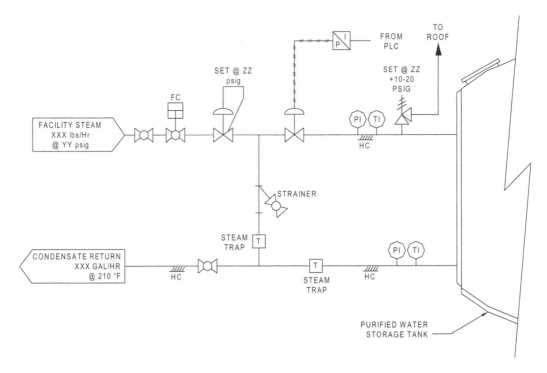

Figure 12.1 Steam and condensate piping for storage tank heating jacket.

line should be directed to a local condensate receiver for return to the facility boiler condensate return/feed water system. For applications with small or infrequent steam demand such as a steam-heated jacket on a hydrophobic vent filter housing or monthly hot water sanitization of a RO/CEDI loop, condensate return may not be technically justified. The condensate may be directed to drain. However, raw water should be simultaneously introduced to the drain to reduce the temperature of the condensate (previously indicated temperature discharge limit) and protect personnel. Steam and condensate piping should be insulated using 2-in-thick fiberglass material with PVC outer cover.

The compressed air requirements for installation should consider both control air for opening/closing pneumatically actuated valves through solenoid valves and higher volume use such as vortex coolers for electronic panels. It is suggested that the compressed air supply be dry, clean of debris, and oil free. Filters should be employed in the compressed air line to remove the indicated contaminants. Air regulation should be provided at each major use point in the system. Each regulator should contain a filter and pressure gauge. An isolation valve should be positioned prior to each regulator to allow for maintenance. An important item to consider is water purification system status in the event of an electrical power failure. Ideally, all water purification system components including storage, distribution, and controls should have a back-up emergency generator electrical supply. However, this emergency electrical supply must include a central or dedicate air compressor. The system includes an *electrically* powered control system with *pneumatically* actuated valves. While valves are specified to fail in a "safe" position, there are some valves, such as the automatic feed water valve to a RO unit, that will fail in a safe closed position upon loss of air but must be opened when electrical power is available and the RO unit operational. In other words, for proper system operation both compressed air and electrical power are required. As a safety feature to avoid catastrophic component failure if electrical power is available without compressed air, a pressure switch should be positioned in the compressed air supply line. If air pressure is not detected, a discrete control signal to the central control panel will inhibit control function (comparable to depressing the "Emergency Stop" button on the face of the panel) until the compressed air pressure is restored.

Installation area ventilation, including humidity and temperature control is highly desirable. Bacteria proliferation increases significantly with increasing temperature. Further, central control panel components [processor, input/output (I/O) cards, etc.] have specified high temperature operating conditions. Components in other panels, such as the CEDI power cabinet transformers, generate heat. It is highly desirable to maintain ambient temperature in the installation area less than 100°F. If this condition cannot be achieved, cabinet coolers (e.g., vortex type) should be installed to cool panels. Further, the use of inline heat exchangers to provide microbial control of stored and recirculated water should be considered. HVAC requirements are frequently specified in regulations based on the "volume" of the installation area. HVAC system should evenly distribute air flow to the water purification components, considering heat generation from individual components and periodic maintenance items such as hot water sanitization.

Raw feed water piping to the installed system should employ a back flow preventor. Again, this is a regulatory requirement. For recirculating pretreatment systems, as discussed in chapter 3, use of a raw water "break tank" may be appropriate.

SYSTEM INSTALLATION: TECHNICAL CONSIDERATIONS

Electrical components should be provided with an Underwriters Laboratory (UL) stamp. Most electrical components manufactured or assembled in Europe will contain a "Conformité Européenne" (CE) stamp indicating that the product complies with the European Union health, environmental, and safety regulations. Unfortunately, in the United States, some state, local, or insurance agency requirements limit acceptance of CE stamped components, requiring the UL stamp. This requirement should be identified in the component specifications as well as the general requirements for a project. It is possible to have a component that is CE stamped inspected by UL and UL stamped. However, the process is tedious, expensive, and sometimes destructive. On the basis of experience, appropriate UL stamping of *all* components that may require an UL stamp is strongly suggested.

Electrical power requirements should have been established as part of the specification and submittal process. The following items must be addressed prior to and during equipment installation:

- Electrical power requirements for each panel including motor starters. This includes voltage, single or three phase, cycles, and amperage.
- The location of the nearest motor control center (MCC) for electrical power feed to the installation area.
- Acceptable conduit for power and control wiring. This may range from flexible conduit to "water-tight" threaded galvanized steel conduit.
- Conduit diameter.
- UL requirements as discussed earlier in this chapter.
- Emergency generator provisions as discussed earlier in this chapter.
- Local, state/province, county, and federal electrical requirements.
- For power panels employing vortex coolers to remove heat, it is important that conduit be "sealed" at the panel. If proper seal is not provided, cooling air will flow through conduit deceasing the cooling air available to the cabinet.

Control and power panels should be supplied and installed such that they meet a specified National Electrical Manufactures Association (NEMA, 2008) classification. For indoor installations, the environment of the installation area as well as activities in the area of the panels/cabinets must be considered. NEMA designations consider protection of cabinet interiors from contaminants such as dust, water, vapors, corrosive materials, non corrosive liquids, steam, and explosive substances. The specified NEMA classification (Type) should consider worst possible conditions.

It is suggested that all control wiring to and from cabinets/field devices be positioned in conduit. Field wiring should be labeled at the field device and in the cabinet. Label identification should correspond with project drawings. Field experience indicates that certain analog signals from components (e.g., ultraviolet unit intensity meters and certain electrolytic ozone generator power indicators) require "isolators" to avoid "feedback." Often, these items are not identified by equipment suppliers. Since these items can take several days to identify during system commissioning, they should be evaluate for each component prior to installation.

The discharge location for tank rupture discs should be positioned to eliminate operator exposure to pressurized water, particularly if the water is at an elevated temperature, upon pressure relief. "Solid" hydraulic system relief is rapid but dramatic. Adequate space should be provided in an inaccessible or limited access area at the base of the tank. The diameter of the rupture disc discharge tubing should be one to two sizes larger than the diameter of the rupture disc. For example, a 3" diameter rupture disc should "exhaust/relieve" into a 3 ½" or 4" diameter section of tubing.

For systems with a PLC-based central control panel and remote I/O panels, solenoid valves may be mounted on the local I/O panels. The I/O panels should be positioned as close as possible to the component(s) associated with the panel. Solenoid valves should be provided with individual local manual "override" capability. While the "output override" capability of the central control panel should be used for "manual" override, certain maintenance operations may require use of the local override provisions. It is suggested that the physical method of solenoid valve override should not employ a push button–type device but a device requiring a simple tool such as a screwdriver. Experience indicates that push buttons may be depressed (manual override) inadvertently. This will result in improper component operation that will *not* be detected or displayed by the central control panel. Finally, the type of pneumatic tubing employed from individual solenoid valves to pneumatic valve actuators and components receiving a controlled [3–15 pounds per square inch gauge (psig) controlled air signal] should be discussed. The use of "hard-piped" tubing of copper or stainless steel material is not recommended. Copper tubing requires continuous support to avoid bending with time and components operation, oxidizes to produce a blue/green color, and technical may require galvanic isolation. Further, physically positioned c it is not aesthetically attractive.

Stainless steel tubing requires significant installation effort to provide visually appealing horizontal and vertical "runs" and lacks required flexibility for connection to solenoid valves and actuators. It is suggested that polyethylene tubing be considered. To avoid a "spaghetti-like" appearance, the tubing can be positioned in electrical conduit routed through a central area close to the exterior surface of the "controlled" component. While small sections of installed tubing may be visible, they do not interfere with maintenance operations when compared with "hard-piped" stainless steel or copper. Further, connection to solenoid valves and valve actuators is enhanced by the flexible nature of the tubing. Flexible polyethylene tubing should not be considered for applications where direct contact with hot surface may be possible. Stainless steel tubing should be used for these applications.

Recovery of softened or RO wastewater (during RO/CEDI loop recirculation) should be considered. Collection tank(s) with air breaks and overflow directed to drain may be part of the reclaim system. The reclaimed water may be used for applications such as cooling tower makeup, facility boiler feed water, and closed loop cooling application makeup. For facilities with daily volumetric discharge limitations, the use of RO waste recovery and treatment (makeup mode) may be appropriate. RO waste may be processed through an "industrial-grade" RO unit with product recovered for applications indicated above. Conservatively, at least 50% of RO waste during makeup (non-recirculating) application may be recovered.

SYSTEM START-UP AND COMMISSIONING CONSIDERATIONS

As built drawings for water purification system interconnecting tubing should be generated. The drawings should contain dimensions or have a reference point and drawn to scale. Stainless Steel tubing downstream of the RO unit, as a minimum, should contain reference to "heat numbers" and orbital weld numbers, similar to purified water or water for injection 316L stainless steel distribution loops. It is suggested that as built drawings are not required for utility piping such as steam, compressed air, drains, etc. However, for reference, it is suggested that the routing of utility piping be shown on dimensioned "single-line" drawings where practical.

Pressure testing of interconnecting tubing/piping should be performed. The requirement and suggested test methods have been discussed previously in chapter 10.

A functional test of PLC-based operating sequence for each component should be conducted. This test should include each normal operation for the component. As an example, testing of a multimedia filtration unit should include normal operation, backwash, settle, post backwash rinse-to-drain cycles and automatic return to normal operation. The position of each automatic valve should be verified during cycle testing. Further, function of support components such as the speed of an upstream pump, flow meter indication, pressure gauge indication, etc., should be conducted. Using the "Operator Override" display screen, the manual opening and manual closing of each automatic valve on the multimedia filtration unit should be verified. Manual operation of multimedia filtration unit accessories, such as an upstream repressurization/recirculation pump, should be verified in the manual on and manual off mode. When complete, I/O testing should verify that the PLC controls every automatic valve and accessory in the intended manner. The indicated testing will be used to correct any PLC program discrepancies and will be duplicated during execution of the Operation Qualification for a validated system. It may be used to satisfy the requirements of a "Commissioning Operational Qualification" for a "qualified" system. While it is suggested that Installation Qualification items can be verified during commissioning for a validated system, it is suggested that OQ related commissioning testing should be verified during execution of the OQ. Frequently, Factory Acceptance Testing (FAT) and Site Acceptance Testing (SAT) are used to satisfy validated system OQ test requirements. Prior to execution of the OQ for a validated system, the system Standard Operating Procedures (SOPs) must be prepared, reviewed, and approved. Generally approved SOPs are not available when the FAT and/or SAT are performed. The suggested "repetition" of testing is based on the importance of SOPs to OQ execution for validated systems.

Instrument calibration, discussed in chapter 10, should be performed. In addition, the "zero" value and "span" of electronic components and pneumatic signals to accessories such as modulating valves, should be verified. For analog signals, the zero valve should correspond to a 4-mA signal. The full scale value/position should correspond to a 20-mA signal.

As discussed earlier, passivation of interconnecting tubing should be performed. Passivation should include an initial hot alkaline cleaning step and final Feroxyl Test.

As indicated earlier, Commissioning IQ, Commissioning OQ, and Commissioning PQ documents should be considered. The intent of the Commissioning IQ may be to provide necessary documentation for "qualified" systems or to eliminate reexecution/verification of items during IQ execution for a validated system. The intent of the Commissioning OQ may be to provide necessary documentation for qualified systems or to decrease the possibility of deviations during OQ execution for a validated system. The Commissioning PQ may be used for system operation with sampling to verify anticipated point-of-use water quality prior to initiation of formal IQ, OQ, and PQ execution for validated systems. For qualified systems, the PQ may be used as the sole document for verification of point-of-use water quality. The recommended sequence presented above represents a conservative approach to system validation. Variation from this approach may be performed at the discretion of the system "owner."

A SAT of the system, including control function should be performed. This test verifies that all system unit operations are functioning properly. Ant programming, component, or accessory issues should be addressed and corrected.

The Commissioning Installation Qualification (CIQ) should be executed. As indicated previously documentation and execution of the CIQ may be used as an IQ for a qualified system. Further, when executed in accordance with criteria set for in a validated system IQ (complete documentation, execution member requirements), the data may be used as a reference for sign-off in the actual IQ.

Formal system start-up should be performed. Upon completion of system start-up, all components should function as if in a normal operating mode.

Open completion of system start-up the Installation Operation Qualification (IOQ) can be performed. Execution of the IOQ verifies proper system operation for a qualified system. Further, control verification during execution of the IOQ should allow validated system execution of the OQ to be performed without "deviations."

Upon execution of the IOQ and IOQ, initial sanitization of the system may be performed if required. This could include pretreatment components, RO/CEDI loop (or RO/rechargeable canister system), and storage and distribution system.

Prior to proceeding further with formal validated system execution as discussed in chapter 13, it is suggested that about a week of intra component and point-of-use testing be performed in accordance with a schedule in an Installation Performance Qualification (IPQ). Testing at this stage could exhibit an issue, affecting anticipated water quality that can easily be addressed prior to execution of a validated system IQ, OQ, and PQ. Further, this document, perhaps with extended execution time period (e.g., two to four weeks) can be used to meet acceptance criteria for a qualified system.

REFERENCE
NEMA. Enclosures for Electrical Equipment (1,000 Volts Maximum), NEMA Standard Publication 250-2008. Rosslyn, VA: National Electrical Manufacturer's Association, 2008.

13 | System validation

INTRODUCTION

Validation is a requirement for USP/EP/JP Purified Water and Water for Injection systems. Many pharmaceutical water purification systems have also established internal "qualification" programs, with the same general structure as a validation program. Qualification as a non compendial system has been discussed in chapters 11 and 12. Throughout this text, numerous references have been made to the fact that validation (or qualification) is "woven" throughout any pharmaceutical water purification system project. Beginning with the Basis of Design/Validation Master Plan and proceeding through PQ with Summary Report, validation clearly demonstrates that the system performs and meets all of the stated objectives outlined in the Basis of Design/User Requirement Specification (URS) (Collentro and Angelucci, 1992). It has been suggested that proper validation of a water purification system involves integrating specific requirements throughout the project. The validation process will vary with the type and nature of the system, installation, and facility as well as "format/procedures" established by the "owner." A summary of milestones that *may* be considered for this process *could* include but not be limited to the following:

- Basis of Design/URS
- Risk Assessment
- Validation Master Plan
- Conceptual design
- Conceptual design review
- Preparation of detailed specification
- Review of detailed specification
- Final revision of detailed specification and associated drawings, such as a P&ID
- Request for water purification equipment supplier's proposals
- Review of water purification equipment supplier's proposals
- Formal documentation of proposal review
- Water purification equipment supplier "kickoff" meeting
- Review of water purification equipment supplier's "submittal package"
- Periodic inspection, with meetings, of water purification equipment supplier's progress at indicated "milestones"
- Final inspection of equipment at water purification equipment supplier's facility including Factory Acceptance Test (FAT)
- Receipt of equipment and "turnover package" from water purification equipment supplier
- Review of turnover package, including drawings and Bill of Materials for components
- Equipment installation
- Performance of Site Acceptance Test
- Preparation of IQ
- Execution of IQ
- Preparation and approval of OQ, training manual, and maintenance manual
- OQ execution
- System start-up
- Preparation of PQ
- Initiation of PQ
- Ongoing review of data generated during PQ execution
- Compilation of data generated during PQ execution
- Preparation of a validation Summary Report
- Preparation of a suggested "perpetual" analytical monitoring program for the system

It is extremely important to consider the fact that validation (or qualification) of a pharmaceutical water system cannot commence once the water purification equipment is received at a facility, installed, and started up. The process begins during preparation of the Basis of Design, the initial step in defining the requirements for the system, and proceeds throughout the project, with specified definition of required documentation. Further, it is important to remember that multiple disciplines from a pharmaceutical manufacturing firm, such as facility engineering, quality control/assurance, validation, maintenance, analytical laboratory, controls and instrumentation, administration, manufacturing, and other appropriate disciplines, must be involved *throughout* the project. The input of each discipline is extremely important during execution of a successful project.

Finally, it is suggested that successful validation of a pharmaceutical water purification system must include participation from individuals with an understanding of the various unit operations. It is impossible to generate an IQ or OQ document for a water purification operation unless the individual(s) preparing the document(s) has a basic understanding of the function and operation of the specific water purification components. When one reviews the myriad of regulatory citations for water purification systems, it is obvious that the majority of these citations relate to inadequate documentation, record keeping, and a lack of system understanding, which is reflected in preparation of the individual validation documents, specifically SOPs and operating/maintenance logs.

GENERAL DISCUSSION

The literature contains numerous references clearly defining the purpose of the validation process (Neal, 1997; King, 1997). In addition, specific articles associated with the validation of pharmaceutical water purification systems have also been published (Elms and Green, 1995a, 1995b; Collentro, 1995; Roczniak, 1995; Fessenden, 1996). In general, a broad definition of the validation process would state that it provides necessary documentation to establish that a system performs, on a continuous basis, in the manner defined in the Basis of Design. As such, proper system validation, with required SOPs, operating logs and maintenance logs, will insure that water quality, at individual points of use, is within "specification" 100% of the time. This is accomplished by ensuring that the description of the system, within the IQ, is detailed to the point where it is impossible to change *anything* that will affect the ultimate performance of the system without appropriate "Change Control." components and accessories, including models and serial numbers are precisely documented insuring that "replacement" can only be performed with an identical (like-for-like) component. If a specific component or accessory is no longer manufactured, the Change Control process, to be discussed later, provides a method of insuring that an "equivalent" component is used. Throughout this chapter, the word *validation* can be interchanged with the word *qualification* and used for non compendial systems where "control" is desired.

As indicated, the format used to develop a validation protocol for a water purification system will vary from company to company. Some organizations, as an example, prefer to exclude model numbers and serial numbers of accessories particularly within the IQ. This validation method may allow individuals executing the documents to "fill in the blanks" with regard to items such as valve numbers, instrument numbers, model numbers, and serial numbers. An alternative, discussed in chapter 12, is to reference and attach CIQ documentation to the IQ.

It is suggested that the validation process define, in detail, the specifics for all components, accessories controls, operation, record keeping, and maintenance. Rigorous validation provides documentation that minimizes the opportunity to substitute accessories and support components for a unit operation, deviate from procedures, or improperly maintain the system. As a result, the more rigorous validation documents inherently ensure that product water quality, at points of use, meets the appropriate specifications 100% of the time. It is suggested that this rigorous method of preparing validation documents, with appropriate training, will eliminate regulatory challenge to the system, procedures, documentation, and philosophy.

It appears appropriate to establish the steps in the validation process. In previous chapters, the validation requirements throughout the entire water purification project were

SYSTEM VALIDATION

presented. Within the individual steps for this process specific reference was made to critical validation documents, which have either been previously discussed or are defined below.

Validation Master Plan

A Validation Master Plan is used to summarize the expectation for system validation. A typical Validation Master Plan may include the following:

- Approval page
- Revision history
- References to pharmacopeia, cGMPs, and owner internal documentation
- Purpose
- Scope
- Definition of terms
- Objectives
- General requirements
- Responsibilities
- Project description
- Documentation requirements

User Requirement Specification

A URS is used to summarize the final product water requirements. Generally this document is brief, with references to individual pharmacopeia. However, the document *could* include the following:

- Approval page
- System overview
- Pharmacopeia and cGMP requirements
- Design requirements
- Process requirements
- Instrumentation and control requirements
- Calibration requirements
- Monitoring and sampling requirements
- Document requirements

Risk Assessment Document

A Risk Assessment Document should be generated. This document presents an evaluation of the probability, detectability, and severity of potential failure of components, including controls, on point-of-use water quality. A suggested format for a Risk Assessment Document would include the URS item, potential risk, "product" risk, ultimate consequence of the risk, explanation of the risk, immediate action to mitigate the risk, and routine monitoring procedure that assist in identification of the risk. An example of a URS item for a Purified Water system that could be evaluated is failure of a pre RO water softening unit to regenerate. A Risk Assessment Document "line item" for this situation would include the following:

- URS item: Purified Water conductivity.
- Potential risk: Failure of a RO unit pretreatment system water softening unit to regenerate.
- Product risk: Potential.
- Ultimate consequences of the risk: Very low.
- Explanation of the risk: The water softening units are positioned upstream of the RO unit, part of the pretreatment system. If a water softening unit failed to regenerate, multivalent cations such as calcium and magnesium would be present in the feed water to the RO unit. The presence of multivalent cations in RO feed water would result in scaling of the RO membranes. Product water flow rate through the RO unit

would decrease. The feed water pressure to the RO unit would increase. Both RO low product water flow rate and RO high pressure will generate a system alarm.
- Immediate action to mitigate the risk: Low RO product water flow rate (alarm), high RO feed water pressure (alarm) and high water softening unit product water total hardness analyzer alarm.
- Routine monitoring: Backup total hardness test kit.

Factory Acceptance Test
A Factory Acceptance Test Document can be prepared as discussed in chapter 12. The document is submitted for approval and comment. The approved document is executed at the equipment manufacturer's facility prior to shipment. IQ-related items in the FAT may be used for reference during execution of a validated system IQ as discussed in chapter 12. Individual items that *may* be included during execution of the FAT document include, but are not limited to, the following:

- Approval page with revision log
- Documentation verification items
  - Process documents
  - Mechanical assembly documents
  - Subassembly drawings
  - Component documents
  - Electrical cabinet assembly drawings
  - Electrical cabinet schematic drawings
  - Control drawings including annotated ladder logic drawings
  - Stainless steel mill certificates
  - Orbital weld logs
  - P&IDs
  - Orbital weld boroscopic inspection
  - Orbital weld test coupons
  - Passivation test results
  - Certificates of Analysis
  - Instrument supplier calibration sheets
  - Certificates of Conformance
  - Catalog cut sheets
- Field tests
  - Review and verification of all P&IDs
  - Mechanical assembly verification
  - Component verification
  - Remote I/O panel hardware verification
  - Central control panel hardware verification
  - Field device verification
  - Point-to-point electrical wiring verification
  - Remote I/O panel electrical wiring verification
  - Central control panel electrical wiring verification
  - Discrete input testing
  - Discrete output testing
  - Analog input testing
  - Analog output testing
  - Software version verification
  - Display screen graphics review
  - Display screen format review
  - Operational sequence testing
  - Alarm response verification
  - "Event" response verification
  - Password level and code testing

- Loss of electrical power testing
- Loss of control air testing
- PLC logic verification

Installation Qualification

While each section of the overall validation protocol is important, the IQ contains the greatest volume of material and provides the most definitive description of system components and accessories. It is the defining validation document, providing a solid basis for the preparation of subsequent validation documents, such as the SOPs, OQ, and the maintenance manual.

Later in this chapter, an IQ protocol for a USP Water for Injection Storage Tank will be presented and discussed in detail. However, it is important to initially discuss a suggested method for preparing an IQ document for a pharmaceutical water purification system.

As discussed earlier, for IQ preparation it is strongly suggested that a pharmaceutical water purification system be divided into multiple individual unit operations. In chapter 11, a method for developing individual specifications for individual water purification unit operations was presented. This technique combines the individual unit operation specifications into a project specification. Reference was made in chapter 11 to the fact that this same "unit operation" principle would be extremely valuable during IQ preparation. The IQ document should consist of multiple subsections of individual IQs for specific unit operations. The "boundaries" in terms of feed water and product water, as well as any utility requirements, must be clearly established for each unit operation. Once the boundaries are defined, an "envelope" can be established around the unit operation, and a highly detailed description of the unit operation accessories can be developed. The logical IQ for a unit operation would begin at the feed water boundary to the unit operation, proceed through the unit operation, and terminate with the product water boundary from the unit operation. Reference to upstream and downstream components precisely defines the position or location of accessories within the specific unit operation. An IQ is a verbal description, in multiple line items, of the unit operation's components and accessories. It attempts to describe in words a detailed "videotape" of all components and accessories, including inaccessible or non visible items, such as the interior of tubing/piping and columns. When completed, the IQ for a particular unit operation should *not* allow operating personnel to remove, replace, change, modify, or in any other way alter components or accessories without creating a conflict between the multiple line item description within the IQ and the "altered" component. Any alterations must be addressed by Change Control.

Again, it should be emphasized that the rather lengthy and highly descriptive nature of the IQ is critical to long-term successful operation of a system. If a specific unit operation's component accessories are properly operated in accordance with SOPs, maintained in accordance with the maintenance manual, and not changed, the quality of water at points of use should not change. As-built drawings, provided by the water purification equipment manufacturer; catalog cut sheets and information for components and accessories; component and accessory maintenance and operating manuals; MSDSs for each component, accessory, or material; and other pertinent information, as discussed in chapter 11, should be included within the IQ. It is the section of the validation protocol that provides the "encyclopedia" of information for all components and accessories within the system. It would be extremely inappropriate to position a lengthy troubleshooting manual for a particular component or accessory within SOPs or maintenance procedures. If the SOPs are too long, it is suggested that individuals will not follow them verbatim or ignore them entirely. The SOPs should not contain material other than that implied by their name, *standard* operating procedures. Material in the maintenance manual should be limited to periodic routine preventative maintenance, not unscheduled repair or replacement. Again, the logical place to position specific, detailed component and accessory information within the validation documents is the IQ. Support documentation may be inserted at the end of the IQ, as an appendix, containing lengthy material for the numerous accessories used in fabrication of water purification unit operations.

As discussed in chapter 12 and earlier in this chapter, verification of certain items during execution of a documented FAT may be used (within the indicated appendix) as a reference to verify applicable IQ line items.

Standard Operating Procedures

System SOPs should be prepared, reviewed, and approved by appropriate owner personnel prior to execution of the OQ. The SOPs establish the precise method of system operation. Since the OQ verifies system operation, it is critical that parameters be defined prior to execution.

SOPs for the system should concisely define the steps necessary to operate the system properly. SOPs should not include maintenance procedures, system troubleshooting, and/or any references to operations that are not *"standard."* SOPs require verbatim compliance. Subsequently, the documents should be brief, concise, and highly specific. Lengthy SOPs, containing "nonstandard" information, such as maintenance requirements, troubleshooting techniques, and calibration procedures, make it extremely difficult for operating personnel to focus on the portion of the document required for normal system operation.

In general, SOPs should provide a sequential list of operations that must be performed, such as opening and closing valves, to achieve system start-up, normal operating functions, and system shutdown. It is suggested that maintenance items, such as periodic sanitization of an activated carbon unit, periodic chemical sanitization of a RO/CEDI loop, and periodic chemical sanitization of a storage and distribution system, be included in the maintenance manual, not the SOPs. Conversely, critical items necessary to the successful operation of the system, such as housekeeping practices at points of use to avoid back contamination of the storage and distribution system, *should* be included as part of the SOPs. Water purification equipment manufacturer information, included in the "turnover package" as a "system operating manual," should be reviewed and *incorporated* as appropriate, during the development of SOPs. It is strongly suggested that "raw" generic manufacturer's operating procedures should not be used as the system SOPs. The operating manual from the water purification equipment supplier may also define parameters that should be included as part of a routine analytical monitoring program and other items that should be included in the maintenance manual. The individuals involved in document preparation should ensure that all information included within the turnover package is incorporated within the validation documents for the system. The majority of this information will be placed in the encyclopedia for the system—the IQ. However, other portions of the operating manual will be included in other documents, such as the SOPs.

Operational Qualification

The general purpose of OQ is to verify normal operating conditions for the system including items such as periodic backwash of a multimedia filter and regeneration of a water softening unit. Further, the OQ should create, by adequate description, *all* potential "transient" or "accident" conditions, verifying that the system, as designed and installed, will respond, resulting in shutdown and/or alarm. In the IQ discussion presented earlier in this chapter, it was emphasized that individual IQs for each unit operation should be prepared and integrated into the actual *system* IQ. Unlike IQ, the OQ is must address the interaction of various unit operations within the water purification system. A representative OQ protocol for a Water for Injection storage system is presented and discussed later in this chapter. A properly prepared OQ will ensure that routine operations requiring intracomponent coordination function properly and are consistent with the SOPs for the system.

The following example demonstrates the intracomponent relationship in the OQ. Consider a system with a multimedia filtration unit, an activated carbon unit, and dual water softening pretreatment section feeding a RO/CEDI loop. Two examples of "unusual" conditions for this system would be low feed water pressure or high feed water temperature to the RO unit. Either condition should activate an alarm on the system control panel and inhibit operation of the RO/CEDI loop. Subsequently, each transient/accident conditions would be clearly defined, as separate execution items within the OQ. However, as indicated, a normal operating condition that should also be verified during OQ execution for this system would be backwash of the multimedia filtration unit or activated carbon unit. When backwash of either unit occurs, which may have been initiated automatically, feed water *may* not be available to the downstream RO unit. A properly designed system should inhibit operation of the RO unit during backwash of either the multimedia filtration unit or activated carbon unit. Individual line items within the OQ document should verify that the RO unit will not function when either of these unit operations (multimedia filtration unit or activated carbon unit) is in a backwash mode, assuming a "single-component" pretreatment train.

There are numerous normal operating conditions, similar to those indicated above, that should be included within a properly prepared OQ for a pharmaceutical water purification system. While it is extremely important to verify that abnormal conditions will shutdown components and energize audible alarms with appropriate indicating/status lights on the central control panel display screen, routine operations requiring control interface must also be verified. Quite often, potential normal operating "conflicts" are not recognized and documented. In the system referenced earlier, assume that the dual water softeners are arranged for series operation. The lead water softening unit is regenerated on the basis of throughput (water processed since the last regeneration cycle), while the trailing water softener is regenerated at a preset time, such as once every Saturday morning. Dual water softening units are provided because it takes about 2.5 to 3 hours to regenerate and rinse a water softening unit. During this time period, feed water may be required to the RO unit. The intent of providing dual units would be to insure that there is makeup water flow to the RO unit. The OQ should verify that regeneration of both water softening units does not occur simultaneously. Many OQs for the system described would not include a line item to verify this condition, and it is possible that the central control panel is not programmed to avoid simultaneous regeneration. This is complicated by the fact that a system may operate for several months before the simultaneous regeneration scenario occurs. Careful evaluation of intracomponent control and functionality must be verified as part of the OQ for the water purification system.

Finally, it is important to indicate that "created" transient or accident conditions, simulated during OQ execution, should not affect the quality of system product water. Intentional chemical, bacteria, and bacterial endotoxin contamination of the system should *never* be performed during OQ execution.

Performance Qualification

The PQ for a pharmaceutical water purification system consists of an intense sampling and analysis program conducted after execution of the IQ and OQ for the system. Generally, it is suggested that this program be conducted for a minimum period of four weeks. During this time period, samples should be obtained from each point of use within the system at least once per working day. Further, samples obtained from the feed water to the system, as well as the product of all components within the USP Purified Water or Water for Injection system, should be obtained at *least* once per week. Later in this chapter, a typical PQ protocol for a USP Water for Injection system is presented and specific details discussed. However, there are some important issues that should be addressed regarding proper PQ preparation and execution.

As previously indicated, all points of use in the USP Purified Water or Water for Injection system should be sampled at least once per working day. For larger facilities with a significant number of points of use, "sample rotation" *may* be considered but is discouraged. Is sample rotation is used, the PQ should not present a sampling schedule that is identical (from week to week) for a suggested six- to eight-week period. When preparing a PQ with sample rotation it is desirable to establish the sample collection using a schedule that collects samples on Monday for the first week of sampling, Tuesday for the second week of sampling, Wednesday for the third week of sampling, and so on. This rotating method of preparing the PQ schedule insures that the same point of use is not sampled on the same day of the week, each week. The time of day for sampling should also be varied. For facilities that operate one shift per day, samples should be collected, on subsequent weeks, in the morning, midday, and afternoon. For facilities operating three shifts per day, seven days per week, samples should be collected throughout the seven-day work week on all three shifts. The logic for this sampling schedule should be obvious. The purpose of PQ is to verify, through appropriate analytical techniques, that the chemical, bacteria, and bacterial endotoxin (for Water for Injection systems) levels meet the Purified Water (or Water for Injection) *Official Monograph* specifications *and* that the established total viable bacteria Alert and Action Limits are not exceeded. As discussed earlier, the total viable bacteria Alert and Action Limits for Purified Water system are "product and process dependent." Subsequently, the Alert and Action Limits *may* vary for each system. Total viable bacteria Alert and Action Limits are presented in the *General Information* section of USP. It is extremely important to understand that the *General Information* section, as implied by its title, provides "support" information. Purified Water and Water for Injection total viable bacteria Action Limits presented in the USP *General Information* section are, as indicated, for general informational purposes only.

Once the sampling schedule has been established, it is appropriate to consider the nature of the required analysis. The first item that will be addressed relates to the chemical parameters stated in the individual monographs for USP Purified Water and Water for Injection. For USP, the chemical specifications include conductivity and TOC. For EP, the chemical specifications include conductivity, TOC, and nitrates. It should be pointed out that several pharmaceutical companies, manufacturing products that will be distributed in other parts of the world which may require analysis that do not appear in USP. It is important that pharmaceutical manufacturers clearly recognize the requirement to perform analysis in accordance with the criteria for the part of the world where the product will be distributed.

It is suggested that point-of-use monitoring for conductivity and TOC employ online monitors/analyzers. Systems should be provided with online conductivity measurement of feed water to a distribution loop and return water from the distribution loop as well as an online TOC analyzer to monitor the return water from the distribution loop. During execution of the PQ "grab" samples for conductivity and TOC should be collected and analyzed from individual points of use on a daily basis. While it is suggested that online conductivity and TOC monitoring should be the primary indicator after successful PQ execution, the intense PQ monitoring program provides an excellent method of comparing grab and online data. Regulatory investigators focus on "delivered" water quality at points of use. There are numerous situations, discussed in detail in chapter 9, where transfer hoses and other required support accessories (e.g., volume or mass totalizers) are used to provide water to multiple manufacturing applications from points of use. One cannot guarantee that the water quality from accessories such as mass totalizing meters with transfer hoses will be the same as the water quality in the return line from the distribution loop. Subsequently, it is suggested that a less intense point-of-use monitoring program, for chemical parameters, be established subsequent to successful execution of the PQ. The extent of this post PQ sampling and analysis program will, in general, vary depending on the number and nature of the points of use. However, it is suggested that the program provide sampling and chemical analysis of each distribution loop point of use at least once per month.

During PQ execution, samples should be collected for total viable bacteria. As indicated, each point of use should be sampled at daily and evaluated using an appropriate microbial monitoring technique. The USP *General Information* section provides *suggested* methodology for determining total viable bacteria in Drinking Water, Purified Water, and Water for Injection. It is strongly suggested that most USP Purified Water systems, particularly systems where bacterial control is critical, should use a sample volume and enumeration method consistent with established Alert and Action Limits for the system/application. One of the primary reasons for suggesting testing of larger volume samples using Membrane Filtration relates to the fact that a heterotrophic plate count of a 1-mL sample will generally (for systems where bacteria control is important) provide a result of <1 cfu/mL. As discussed earlier, it is impossible to "trend" data unless an absolute number is obtained by the analytical technique employed.

When samples are obtained for bacteria monitoring, the conditions should simulate the conditions encountered during routine use of the recirculating water. Specifically, a sample point should not be rinsed to drain for a preestablished period (e.g., 10–15 min) unless water will be rinsed to drain for the same time period every time water is used from the point of use. Rinsing prior to sampling for analysis must replicate the conditions used during manufacturing. Undoubtedly, "sample rinsing" of duration greater than normal operation rinsing will result in a regulatory challenge of all microbial data. Furthermore, PQ documents often reference the use of sample valve cleaning with an isopropyl alcohol solution. Again, unless points of use are pre washed with isopropyl alcohol (before each manufacturing operation), this cleaning operation could also be challenged by regulatory investigators. In addition total viable bacteria samples must be free of any residual isopropyl alcohol to provide a representative result. TOC samples collected immediately prior to collection of bacteria samples can be analyzed for TOC which will indicate if residual isopropyl alcohol is present.

The suggested sample volume for USP Water for Injection systems and USP Purified Water systems where bacteria control are important, has been addressed above. It is beyond the scope of this text to address sample size, culture media, incubation time period, and incubation temperature set forth in individual pharmacopeias. However, on the basis of field experience, suggested total viable bacteria enumeration conditions are presented.

For Purified Water systems where bacteria Alert and Action Limits are >10 cfu/mL *and* exclusion of specific pathogens is not a concern, the suggested methodology *may* be Membrane Filtration of a 1.0-mL sample diluted with 99 mL of Sterile Water, PCA culture media, 48- to 72-hour incubation time period, and 30°C to 35°C incubation temperature. If the absence of specific pathogens is a concern (*Pseudomonas aeruginosa, Escherichia coli,* etc.) specific tests set for in the literature (Eaton et al., 2005) should be employed in addition to total viable bacteria measurement. The suggested sample volume for the indicated pathogens is 100 mL.

For Purified Water systems where bacteria Alert and Action Limits are ≤1 cfu/mL *and* exclusion of specific pathogens is a concern, the suggested methodology *may* be Membrane Filtration of a 100-mL (or larger) sample, PCA or R2A culture media, 72- to 120-hour incubation time period, and 30°C to 35°C incubation temperature. Note that in "nutrient-starved" environments more representative total viable bacteria results may be obtained at an incubation temperature of 22°C with R2A culture media. If the absence of pathogens is a concern any observed colonies should be identified. If available, a Riboprinter should be used to provide definitive organism identification.

For Water for Injection systems the suggested methodology *should* be Membrane Filtration of a 100-mL (or larger) sample, R2A culture media, 72- to 120-hour incubation time period, and 30°C to 35°C incubation temperature. Note that in nutrient-starved environments more representative total viable bacteria results may be obtained at an incubation temperature of 22°C with R2A culture media. Any bacteria colonies should be identified using a Riboprinter if available.

During PQ execution, it is important to evaluate the effect of bacteria results with a change in the incubation temperature, incubation time period and culture media. The suggested enumeration method for USP Water for Injection total viable bacteria testing has been presented. However, during PQ execution, one duplicate USP Water for Injection sample each day should be incubated at 30°C to 35°C, while the other sample is incubated and counted at 3, 5, 8, and 10 days. Experience indicates that additional bacteria colonies are not observed after five days. Another set of duplicate samples should be collected for enumeration using PCA culture media and R2A culture media. Experience appears to indicate that R2A culture media, with an incubation temperature of 22°C and an incubation time of 72 hours, provide representative results that should not be exceeded, to any great extent, by any other culture media and incubation time period and temperature. It is further suggested that a similar test program be employed for USP Purified Water systems where bacteria control is a concern. This evaluation allows personnel to evaluate the most appropriate culture media, incubation time period and incubation temperature. This evaluation should *not* be made solely on the method providing maximum "recovery" of bacteria but should consider important operation consideration such as the amount of product generated during the incubation time period. The selected enumeration may be "coupled" with rapid microbiological methods (Riley, 2004) to produce a responsive and technically justifiable program.

Many pharmaceutical companies, particularly with smaller water systems and limited laboratory capability, will use the services of a contract laboratory for bacteria determination. If a contract laboratory is used, it is critical that the proper chain of custody be established. The chain of custody is achieved by assigning a unique number to each sample provided to the contract laboratory. The numbers should be sequential. An adequate number of digits should be used to avoid repetition. An alphanumeric code may be also used to avoid the duplication of numbers. Samples provided to a contract laboratory should be stored and shipped at ~2°C to 4°C. The bacteria enumeration technique *must* be initiated within 24 hours of sample collection.

A final item that must be addressed regarding bacteria determination during the PQ relates to the species of bacteria present. As indicated earlier, it is suggested that any colonies of bacteria observed during PQ execution for an USP Water for Injection system samples should be sub cultured and speciated. Gram-negative bacteria should not be present in USP Water for Injection samples. Organism identification by Riboprinter is attractive if available. If unavailable, the literature provides information for selection of an identification technique (Sutton, 2007). During PQ execution for an USP Purified Water system where bacteria control is critical, it is suggested that colonies exhibiting the same morphology be sub cultured and speciated. If *"undesirable"* organisms are detected for critical products (topical solutions, inhalants, antacids, ophthalmics, etc.), system evaluation should be initiated to determine their source. Corrective action, such as an increase in the sanitization frequency of the storage and distribution system, may be required. Further, reinitiation of the PQ may be appropriate once the source of contamination has been identified and/or operating procedures, such as sanitization frequency, changed.

When the intense monitoring portion of the PQ has been successfully completed, a routine bacteria monitoring program should be established. This program, collecting samples from points of use and water system intracomponent locations, should be structured around the characteristics of the distribution loop, the critical nature of microbial control for the application, and the nature of the manufacturing process. It is suggested, as a minimum, that each point of use be sampled for bacteria monitoring once per week. For USP Water for Injection systems or critical (from a bacteria control standpoint) USP Purified Water systems, point-of-use sampling and bacteria determination may be more frequently. Regulatory requirements (FDA, 1993) indicate that the PQ last for a one year time period to capture seasonal and climatic changes in feed water. However, for a compendial water system, point-of-use monitoring should be performed throughout the life of the system.

Bacterial endotoxin monitoring should be conducted for all samples collected from points of use (with the same frequency as bacteria monitoring) for USP Water for Injection systems. Several USP Purified Water systems will employ an "internal" bacterial endotoxin specification. Generally, this specification is the same as the bacterial endotoxin limit for Water for Injection, 0.25 endotoxin units/mL.

As indicated earlier, during PQ execution, samples should be obtained from raw feed water (to the water purification system) and product water from each unit operation in the water purification system. Consistent with the requirements stated in the individual monographs for USP Purified Water and Water for Injection, feed water must meet the NPDWR defined by the EPA (or applicable equivalent), as discussed earlier in this text. There are a number of regulated organic compounds defined by the NPDWR. Municipal water treatment facilities will generally have an established monitoring program for organic compounds, verifying that they are within established limits. For private water supplies to a pharmaceutical facility, it may be necessary to use the services of a contract laboratory that is familiar with the extent of organic analysis required for Drinking Water. This is particularly true if the feed water source is from a groundwater supply where contamination from numerous sources, such as fertilizer, pesticides, or industrial pollutants, may be possible.

Analysis for inorganic impurities in raw water should include items specifically contained in the NPDWR, as well as other items that could affect the performance of downstream water purification unit operations. A suggested analytical monitoring program for inorganic compounds is discussed in the PQ protocol presented later in this chapter. Microbial monitoring should include an analysis for Total Coliform and total viable bacteria by heterotrophic plate count (Eaton et al., 1995). In addition to raw water samples, intra-component samples should be obtained and appropriately analyzed. Bacteria levels in intracomponent samples prior to a membrane process such as RO, should be determined by heterotrophic plate count. Bacteria levels in samples obtained after bacteria "specific" components, such as RO or ultrafiltration, may, more appropriately, be determined by Membrane Filtration of a 100-mL sample.

The nature of chemical analysis conducted in the feed water and product water to and from each unit operation within the water purification system is established by the functional of the component. For example, feed water to an activated carbon unit (assuming that the raw feed water is from a surface water source or groundwater source influenced by a surface source) should be analyzed for residual disinfectant concentration, TOC, and TSS. Product water from the activated carbon unit should be sampled and analyzed for the same components. Feed water and product water chemical values and total viable bacteria levels determined by heterotrophic plate count for the activated carbon unit should be plotted as a function of time. This monitoring program will establish the normal operating conditions and maintenance requirements for the unit, such as required hot water sanitization frequency (based on product water bacteria levels) and activated carbon media replacement (based on an increase in product water TOC levels beyond a preestablished value or "breakthrough" of monochloramine). During PQ execution it is suggested that intracomponent monitoring be performed weekly. Subsequent to PQ execution it is suggested that intracomponent monitoring be performed monthly. For certain components, monitoring may be required more frequently on the basis of the potential effects of excursions in unit operations on point-of-use water quality. For example, if high TOC values, the presence of residual disinfectant, and/or high microbial levels in the product water from an activated carbon unit will affect the performance of a downstream single- or double-pass RO unit, periodic monitoring, as frequently as once per day, should be performed to insure that the RO membranes are not adversely affected. Obviously, a loss of integrity of the RO system membranes for a Purified Water application will eventually affect bacteria levels at individual points of use within the distribution system.

In conclusion, the PQ not only verifies that a system operates in a manner consistent with the design parameters but also provides two other highly valuable objectives. It verifies that component selection and design will, throughout the life of the system, function in an integrated manner producing water, at points of use, meeting specified requirements (chemical, bacteria, and, where applicable, bacterial endotoxin). The PQ also serves as a valuable tool for establishing the requirements for routine monitoring, specifically between components in the water purification system, as well as guidelines for the concentration of specific impurities (including bacteria) from individual components, which are necessary to insure proper system operation.

Maintenance Manual

The maintenance manual should contain all required short- and long-term maintenance items. For a Purified Water system, this may include replacement procedures for support and filter media in the multimedia filtration unit, activated carbon unit media replacement, activated carbon unit periodic hot water sanitization, dual water softening system ion exchange resin replacement, single- or double-pass RO membrane cleaning provisions, pump maintenance criteria, storage system and distribution loop sanitization procedure, etc. A detailed preventative maintenance schedule should be included for the execution of each item. To minimize confusion associated with required routine maintenance operations, it is suggested that the schedule clearly define all maintenance items that should be performed in a chronological sequence, beginning with the date of system start-up. Instrument calibration, generally performed annually, should be included as part of the maintenance procedures. Initially, it may be appropriate to "stagger" calibration requirements, performing certain items after six months, thus avoiding significant calibration tasks at one specific time period. On the other hand, it may be helpful to establish the required calibration items during an "annual shutdown," when there may be greater access to system components and accessories requiring calibration as well as a longer time period to execute.

The maintenance manual should address applicable items presented within the turnover package provided by the water purification equipment manufacturer. While specific maintenance items associated with components "directly" manufactured by the supplier may be clearly defined, additional maintenance procedures, such as those for "purchased" accessories for various unit operations, should also be reviewed and incorporated. The maintenance manual should include support items that were provided directly by the pharmaceutical manufacturer (or subcontractor) but are an integral part of the system. This would include, for example, mass totalizing meters for certain batching operations and delivering hoses at points of use. The maintenance manual for the system should, if properly followed, insure that the system operates in accordance with the criteria set forth in the Basis of Design.

"Failure conditions" associated with a component (not due to improper system maintenance) should be addressed. However, the procedures for replacing components and accessories, specifically details such as model and serial numbers, should be consistent with material in the appropriate section of the IQ. Where applicable, Change Control documentation may be required for replacement of a component and/or accessory.

Training Manual

The training manual is a document prepared for qualification of individuals who will operate and maintain the system as well as engineering, supervisory, management, laboratory, quality assurance, and validation personnel. It should provide an overview of the Basis of Design, including the function of each water purification unit operation in the system. It is strongly suggested that "controlled" copies of the training manual be provided to operating and maintenance personnel (as well as others who may interface with the system), during one or more videotaped training sessions. It is further suggested that the training sessions, including questions and answers from personnel, be included as part of the training manual. The videotape and training manual may be used to provide periodic operator and maintenance personnel "retraining" as appropriate. The training manual may also contain appropriate articles from the literature, presenting, in greater depth than discussed in the training session, the purpose and function of a particular water purification unit operation. It is also suggested that enclosures to the training manual contain a copy of additional support information, specifically regulatory guidelines, such as the *Guide to Inspection of High Purity Water Systems* (FDA, 1993). The training session should include a "hands-on" review of the actual system.

Summary Report

After the IQ and OQ have been successfully executed, the intense sampling and monitoring phase of the PQ can be initiated. Once all data are available from the intense sampling phase of PQ execution, a Performance Qualification Summary Report can be generated. This report tabulates data from the PQ and discusses the results. A summary section, positioned at the beginning of the report, explains the format of the report and discusses the chemical, bacteria, and, if appropriate, bacterial endotoxin results for all samples collected during PQ execution. On the basis of these results, the report should define a routine analytical monitoring program to insure that system point-of-use data verify that Purified Water or Water for Injection is continuously available to end users. The methodology used for microbial monitoring should be clearly defined for both water purification system and point-of-use samples. The ability of the system to maintain point-of-use bacteria levels below established Alert and Action Limits should be demonstrated. The frequency of required maintenance to insure that point-of-use chemical, bacteria, and, where applicable, bacterial endotoxin specifications are maintained should be clearly defined.

The body of the PQ Summary Report should contain information associated with documented "chain of custody" if the services of a contract laboratory were used. The actual PQ protocol should also be included as part of the report. Original laboratory data sheets for grab conductivity and TOC measurements, as well as observed online conductivity and TOC indications during sample collection should be included. Data from the system logbooks, clearly tabulating all samples collected during PQ execution should be chronologically presented. Point-of-use results for conductivity, TOC, bacteria, and, if applicable, bacterial endotoxins, should be chronologically tabulated. The PQ Summary Report should clearly demonstrate that the system, as operated and tested during PQ execution, is capable of meeting the criteria defined in the Basis of Design for the system. As discussed previously, the Basis of Design is developed around specific requirements for the product and process used to manufacture a single product or multiple products at a pharmaceutical facility.

A "final" PQ Summary Report should be prepared after one year of operation. The data in the report should support the intense sampling period PQ Summary Report, verifying system ability to meet point-of-use chemical and microbial parameters with seasonal and climatic changes in the feed water during a one year time period.

Execution of Validation

As indicated in the discussion above, validation documents, specifically the IQ, OQ, and PQ, require execution. Occasionally, a Purified Water or Water for Injection system may be installed on a "turnkey" basis. As part of the turnkey nature, IQ and OQ execution may be requested by the system "owner." On certain occasions, PQ execution may also be requested. It is suggested that it is inappropriate for a company (such as a water purification equipment manufacturer) supplying the turnkey system and validation documents to execute the validation documents. The primary basis for this suggestion is that the intent of the validation process is to insure that the owner of the Purified Water or Water for Injection system understands all of the information set forth in the IQ and OQ documents and, as a minimum, initials/signs protocol line items during execution. The owner must also verify, during PQ execution, that the system performs as outlined in the Basis of Design and that sample collection is performed in accordance with criteria outlined in the PQ. There may be a conflict for a firm providing the system to execute validation documents, particularly the PQ. Execution of validation documents by a "contracted firm" *may* be challenged by regulatory authorities. If the "execution" process is not challenged, regulatory personnel will most likely attempt to verify that all personnel fully understand all issues associated with the validation execution. It is suggested that this requirement can be achieved by the indicated owner sign-off for protocol line items verifying proper execution.

IQ execution is an extremely important vehicle to insure that operating and maintenance personnel develop a comprehensive understanding of the water purification system. It appears inappropriate to delegate this responsibility to a contract firm. On the contrary, it is further suggested that IQ and OQ execution, by appropriate individuals employed directly by the pharmaceutical manufacturer, be videotaped. Pharmaceutical personnel execution of validation protocols by the owner demonstrates that there is a thorough understanding on the part of the pharmaceutical manufacturing firm of system operation and installation criteria. The videotape may be used as an important training tool, supplementing material in the training manual.

Change Control

Earlier in this chapter, reference was made to Change Control, a process for changing, revising, or in any other way modifying validation documents in response to a change in system components, control, operation, maintenance, testing, etc. Change Control must be initiated and submitted to the disciplines originally reviewing and approving the validation documents for the system. As the requested change is reviewed by the various disciplines, the overall effect of the requested modification on the performance of the system (supplying water to all points of use) must be evaluated. If Change Control for a specific modification has no effect on system performance, it is still necessary to document the change. An example of a modification that would fall under this category is the replacement of a faulty pressure gauge with another pressure gauge from a different manufacturer, but of identical construction, range, accuracy, and so on as the original gauge. Change Control *is* required because the IQ specifically states that the pressure gauge in question was manufactured by a particular company with a unique model number and serial number. The individuals responsible for reviewing and approving the initial validation documents must review the request to use an alternate pressure gauge supplied by a different manufacturer. If it is determined that the attributes of the "new" gauge are similar to those of the "faulty" gauge, in terms of overall effect on the system, Change Control can be processed with minimum testing to verify acceptance. As a minimum, this subtle change would require modification to the original IQ, verifying the change in manufacturer, model number, serial number, and any other pertinent information.

Other items requiring Change Control could have a significant effect on system operation and, ultimately, on water quality at points of use. For example, a validated Purified Water system contains a sanitary stainless steel storage tank, a distribution pump, and a distribution loop. Malfunction of a fork lift truck at the facility could result in significant damage to the sanitary storage tank, rupturing the tank or the outer heating jacket. A review of the damage indicates that the tank cannot be repaired and must be replaced. However, the delivery time for a new tank is excessive (>10–16 weeks). The pharmaceutical manufacturing firm operates multiple facilities and determines that a similar but not identical tank is available from another facility. The tank has not been used for a period of time. Fortunately, the tank had been used as a Purified Water storage tank prior to decommissioning of that system. When the previously used tank is installed into the system, accessory and support components at the top of the tank must be modified to accommodate the different orientation of fittings. The size of the tank is somewhat larger than the irreversibly damaged tank. Modifications to the control system are required. Derouging of the tank and subsequent repassivation of the storage tank and distribution loop are required.

The situation described above would obviously require Change Control. The extent of the modifications to the system could have a significant impact on water quality at points of use. As a minimum, new sections of the IQ must be prepared for the reused tank. A revised OQ must be prepared. The revised IQ and OQ must be reexecuted for the storage tank. It is strongly suggested that the nature of the change, because of its "proximity" to points of use, would require complete reexecution of a new PQ. Appropriate changes would be required to the SOPs and the maintenance manual for the system. The training manual may also be affected. A new PQ Summary Report must be prepared after data are accumulated during execution of the new PQ. Information associated with the revised component and accessories, including control changes, and PQ data must be "dovetailed" into the existing validation documents for the balance of the system.

Change Control is *very* important. The training program should clearly inform all personnel directly or indirectly interfacing with the water purification system that *any* change to equipment, accessories, components, procedures, etc., must be approved through the Change Control process. Personnel must clearly understand that any modifications initiated without Change Control can ultimately *invalidate* the system.

Types of Validation
It would be inappropriate to conclude a general discussion associated with the validation of pharmaceutical water purification systems without discussing the timing of the validation process relative to system installation and use of product water for manufacturing operations. In general, three terms may be used to describe the timing of the validation process.

Retrospective Validation
A system that has been operating for a period of time and is not designated as a validated system, for example, a "qualified" high purity water system, may require "redesignation" as a Purified Water system. This is not an uncommon occurrence. Occasionally, a pharmaceutical manufacturing firm may determine that an application requires Purified Water as opposed to "Drinking Water" or "Deionized Water." However, the specific manufacturer has been using this water purification system for a number of years to provide water for the application. The system is properly maintained delivering product water to points of use and meets the chemical specifications for Purified Water including internally established Alert and Action Limits for total viable bacteria. Appropriate validation documents can be generated. It can take considerable effort to provide the required documents for the IQ and OQ. A Basis of Design, specification, and appropriate drawings must be generated. The validation process, outlined earlier in this chapter, must be completed through the PQ Summary Report. The validation process for this particular application is referred to as a "retrospective" validation, since it is being performed after the system has been installed and operating. Once validated, assuming that the Summary Report clearly indicates functionality as a Purified Water system, product water at points of use can be appropriately designated as Purified Water. This process is *not* intended to indicate that water produced prior to the validation process was Purified Water. In other words, designation of a system as a validated Purified Water system is *not retrospective*.

Prospective Validation
The most appropriate way to conduct validation of a pharmaceutical water system is to perform all of the steps indicated earlier in this chapter, from preparation of a Basis of Design through compilation and preparation of a PQ Summary Report. Validation would be complete

before water is used as Purified Water or Water for Injection. This manner of validating a system is referred to as "prospective" because all of the validation steps, including verifying that water quality at points of use, meet Purified Water or Water for Injection requirements, were established before the water is used as Purified Water or Water for Injection for any applications at the facility. During PQ execution for prospective validation, it is important that "simulated" water demand conditions are created. In other words, water should be withdrawn from the system, preferably at points of use, and directed to drain at the intended flow rate, projected volumetric demand, and frequency.

Concurrent Validation
The final method of water system validation is occasionally used. A "concurrent" validation uses water from the system as Purified Water or Water for Injection before entirely completing execution of the PQ and preparing the PQ Summary Report. Water used prior to completion of the validation process is used "at risk." It is strongly suggested that product generated during this process be quarantined. In this validation method, all documents, from the Basis of Design through the OQ, including SOPs, the maintenance manual, and the training manual, have been generated. The IQ and OQ protocols have been executed. The PQ protocol has been prepared, approved, and partially executed. Generally, when concurrent validation is performed, the pharmaceutical manufacturing facility will not use water from individual points of use until at least two full weeks of PQ data are available. Once it is demonstrated that product water quality at points of use meet the chemical, bacteria, and, where applicable, bacterial endotoxin specifications, the water *may* be used, at risk, as Purified Water or Water for Injection. As indicated, there is risk associated with this method of validation. This risk can be *minimized* by quarantining the product until PQ execution is fully complete. The concurrent validation process is completed by continuing the PQ for a recommended 30-day period and preparing the PQ Summary Report. The decision to perform concurrent validation is generally limited to Purified Water systems. Another situation that could be considered for concurrent validation is a USP Purified Water system where the stored and recirculated water is constantly maintained at a temperature >80°C.

Revalidation
A final item that should be briefly discussed concerns system revalidation. The frequency of system revalidation is related to the number and extent of Change Control documents generated for a given system. At some point in time, drawing revisions and markups to IQs, OQs, SOPs, and maintenance manuals become difficult to manage and comprehend. Certainly, the nature of documentation is somewhat expanded by the Change Control process. While there is no "set" time period established for system revalidation, a major change to a portion of the system may prompt revalidation. For example, many older Purified Water systems have been "upgraded" over the past 5 to 10 years by incorporating continuous electrodeionization. Over this same time period, many control systems have been updated to incorporate the use of PLCs, thus providing flexibility with regard to system monitoring and "operator-friendly" control functions. Extensive changes are required to install continuous electrodeionization systems and/or controls. These changes will, by the nature of the Change Control process, prompt system revalidation. Subsequently, revalidation will increase as water purification, control, and storage and distribution system technology evolve.

INSTALLATION QUALIFICATION PROTOCOL: USP WATER FOR INJECTION STORAGE TANK WITH ACCESSORIES
Experience indicates that the most effective method of demonstrating the contents of an IQ for a water purification system is by example. Workshops and seminar sessions are presented each year discussing the validation of water purification systems. This technique provides personnel with an excellent background of the validation process but does not, in general, specifically present an actual IQ. While it should be emphasized that "template" validation should not be employed, it is appropriate to review a step-by-step procedure for a specific component in a pharmaceutical water system. The examples in chapter 11 for specification

preparation discussed Purified Water system components. This chapter will focus on a USP Water for Injection component—a storage tank with accessories. Appendix A presents a representative section of an IQ for a USP Water for Injection storage tank. The major items within the example are discussed below.

Certificate of Installation
The first page of each section of the IQ document should contain a Certificate of Installation signed by designated individuals at the pharmaceutical facility, including such disciplines as technical services, facilities engineering, and quality control. This list can be expanded to include other disciplines discussed earlier in this chapter. As a general note, it should be pointed out that a "header" for each section of the IQ be used. Further, a "footer" is also appropriate, particularly with regard to establishing the latest revision of the document. Finally, it should be noted that the Certificate of Installation contains a note with regard to Change Control and potential requalification/ revalidation of the system.

Purpose
A general one paragraph statement outlining the purpose of each section of the IQ should be provided. This summarizes the format of the IQ sections.

Equipment Rationale
The Basis of Design for the system has outlined the function of each component. The equipment rationale provides an overview of the component presented in the IQ. In this particular example, the equipment rationale states that the Water for Injection storage tank and accessories provide a depressurized location for the collection of water from the upstream multiple-effect distillation unit. Further, the tank provides a reservoir of water to support manufacturing operations and a point for delivering water to a recirculation pump and distribution loop and returning to the tank. Accessories provided on the tank, such as a hydrophobic vent filter, compound-type rupture disk assembly, spray ball, back pressure regulating valve, level control system, and temperature control system, are also mentioned. The equipment rationale provides a brief but informative section integrating the design criteria (established by detailed specifications) with the validation requirements.

Protocol Format Approval
The approval section of the IQ document provides a "sign-off" requirement indicating that individuals overseeing the validation of the Water for Injection system have reviewed the IQ protocol *prior* to execution. It is important that both the cover sheet for the Certificate of Installation and protocol format approval are signed by the appropriately designated individuals prior to execution. Signing the first page of the IQ for the Water for Injection storage tank merely indicates that the lead disciplines have reviewed execution of the document. Signing the protocol format approval indicates that the lead disciplines have thoroughly read and reviewed the document and concur that the document is adequate as an IQ for the Water for Injection storage tank.

Conclusion
The conclusion section states that the tank selected is consistent with the system design requirements, equipment specifications, and the design criteria set forth in the Basis of Design for the system. This conclusion "weaves" the validation process through the project from the Basis of Design to the IQ.

Components and Accessories—Inspection Items
The "Components and Accessories" section tabulates the major items that will be delineated, with description, in the IQ. For example, item 14 in the example IQ, "tank fittings size and location," could require several line items within the detailed component inspection for the tank. The inspection list summarizes the myriad of items that must be addressed as part of the IQ for the Water for Injection storage tank.

Component Inspection

The "Component Inspection" section delineates, in detail, the specific items associated with the Water for Injection storage tank, the component being addressed in this section of the IQ for the project. The first column of the table is designated "Parameter," which states the specific item to be inspected. A second column of the table indicates the "Test Method," which, for the majority of items, will be "visual." The final column of the table is labeled "Actual Results." Any deviation from the specific information contained in the Parameter column should be clearly documented in the Actual Results column.

This section demonstrates the broad variation encountered in preparation of the document. For example, the second line item in the sample IQ states that the individual executing the document should "verify, using tank nameplate information, that the tank is serial no. 12345." It is suggested that a proper IQ for any component can only be prepared by conducting a thorough inspection of the components during preparation of the document. In fact, the individual(s) preparing the document *should* simulate execution of the document during this preparation process. This individual at the pharmaceutical facility, who should be cognizant of the water purification system, should verify the specific information contained within the IQ.

As discussed earlier, there are numerous ways of preparing validation documents. A less desirable manner of preparing the IQ is to simply leave the actual serial number blank, requiring the individual who executes the document to fill in the blanks. This could literally be interpreted as preparation of the IQ during its execution. Further, many organizations will prepare an IQ from documentation in the turnover package provided by equipment manufacturers and suppliers, without ever visiting the facility, inspecting the water purification system and support components, and verifying that the material received from the manufacturers and suppliers is correct. Validation document preparation, specifically preparation of the IQ and OQ documents, should be a hands-on activity. Critical items defining the nature of the components and accessories can only be established by onsite visual inspection of the various components.

Individual line items within the "Component Inspection" section require verification of the physical dimensions and volume of the tank, the maximum working pressure, available heat transfer area on the sidewall and bottom (two zones) of the tank, external shroud and tank materials of construction, operating pressure for the heating jacket around the tank, internal mechanical polish for the tank, the overall dimensions of the tank and outer shroud, use of chloride-free insulation between the tank and outer shroud, the "National Board number" for the unfired pressure vessel, inspection of the top-mounted sanitary manway, the presence of lifting lugs on the top of the tank, clarification of support legs and the diameter of the legs, description, size, and location of tank fittings, the hydrostatic test pressure for the tank, and the minimum metal design pressure for the tank.

Accessory Inspection

The "Accessory Inspection" section of the example IQ provides a list of items of tank accessories. Some of the accessories are the hydrophobic vent filter (housing and membrane filter), the compound-type rupture disk, vent tubing from the compound rupture disk, the divert valve on the supply line on the upstream multiple-effect distillation unit, feed water tubing from the upstream multiple-effect distillation unit (with indicated slope), the back pressure regulating system, including a pressure sensor installed on the return tubing to the storage tank from the distribution loop, tank level sensing devices with support equipment, tank temperature sensors, monitors, and support equipment, and the dual zone heating jacket supply steam modulating valve with accessories.

Control System Inspection

The "Control System Inspection" section discusses, in detail, the individual items to interface with the sensors for temperature, level, pressure, etc., discussed in the Accessory Inspection section. Items included in this section for the Water for Injection storage tank are the level-indicating transmitter, the temperature-indicating transmitter, the return tubing pressure transmitter, the current-to-pneumatic converter for the back pressure regulating valve, the facility steam modulating supply valve to the tank heating jacket, the divert-to-waste valve in the feed water line to the tank from the upstream multiple-effect distillation unit, and the solenoid valves to support the divert system.

Instrumentation Inspection
Instrumentation inspection section identifies all sensors and transmitters for the Water for Injection storage system. This section should also include identification of the calibration reference standard. It should also distinguish between "critical" instrumentation required for routine system monitoring and "noncritical" instrumentation. An example of critical instrumentation would include the temperature or level sensing elements, transmitters, and indicating controllers. An example of a noncritical instrumentation would include pressure and temperature indicators installed in the supply steam line to the heating jacket of the tank. Items discussed in this section include, but are not limited to, the tank level sensor/transmitter (differential with two sensors), the tank differential level-indicating transmitter, the tank temperature sensor/transmitter, and the distribution loop return tubing pressure sensor/transmitter.

Utilities Inspection
The utilities inspection section presents utilities for the tank, such as supply steam for the heating jacket. The specific items discussed in this section include, but are not limited to, the facility steam modulating supply valve to the heating jacket, the condensate return from the heating jacket, the regulated [2–3 pounds per square inch gauge (psig)] supply steam to the heating jacket for the hydrophobic vent filter, the condensate return from the hydrophobic vent filter heating jacket, the waste line for the divert-to-waste valve from the upstream multiple-effect distillation unit, facility air for the divert-to-waste valve associated with the upstream multiple-effect distillation unit, and a regulated low voltage electrical supply to the Water for Injection system control panel.

Documentation Verification
The "Documentation Verification" section outlines all of the support documents provided as part of the "turnover" package from the component supplier, accessory manufacturer's information and accessory "cut sheets." Additional information should be obtained, as appropriate, to insure that the IQ for the Water for Injection storage tank, is, as discussed previously, the encyclopedia for the tank. The documents enclosed as part of this section should contain a title page corresponding to the individual line item designations tabulated under "document or drawing reference." The enclosures *will* be several pages longer than the actual IQ for the tank, but they provide a single source, in a logical location, for supporting long-term operating, maintenance, troubleshooting and replacement of items for the Water for Injection storage tank and its accessories.

The items presented in this section of the IQ include instructions for manual diaphragm control valves, the actuator installation and operating maintenance manual for control valves, the maintenance manuals for the hand wheel operated diaphragm valve weir and straightway and diaphragm valves, the instruction manual for the electronic sanitary temperature transmitter, the differential level transmitter, the sanitary electronic pressure transmitter, the service manual for air-actuated diaphragm valves and the loop back pressure regulating valve, manufacturer's information for the sanitary hydrophobic vent filter housing, "sterilization" procedure for the hydrophobic vent filter assembly, the manufacturer's selection guide for the hydrophobic vent filter, sanitary fittings, the final assembly drawing for the 1500-gal storage tank, the Certificates of Calibration for tank level sensors/transmitters, the tank level-indicating transmitter, the tank temperature sensor/transmitter, and the distribution loop tubing pressure sensor/transmitter.

OPERATIONAL QUALIFICATION PROTOCOL: USP WATER FOR INJECTION STORAGE TANK WITH ACCESSORIES
As discussed earlier in this chapter, the OQ for a particular unit operation demonstrates that the system operates in a normal mode including routine maintenance steps such as multimedia filtration unit backwash and is capable of responding to unacceptable transients.

Unlike an IQ that can be distinctly prepared for a specific unit operation, an OQ will generally interface with upstream and downstream components and, more importantly, the central control panel or multiple control panels for the system. Similar to the IQ for the Water for Injection storage tank, the most effective method of demonstrating the suggested requirements of an OQ is to present an actual OQ. Appendix B contains an OQ for the USP Water for Injection storage tank discussed above. The format of the OQ may be modified, as required, for consistency with other validation protocols for a particular pharmaceutical manufacturing company. However, as discussed earlier, preparation of the document should be generated at the manufacturer's facility. It is highly possible that set points and limits established for control and instrumentation documentation provided by the equipment supplier/installer will not be consistent with the actual values encountered during OQ execution. The OQ, more than any other validation document, provides a verification of system monitoring and control capability exceeding that performed during system start-up. A summary of OQ sections for the USP Water for Injection storage tank is presented below.

Certificate of Operation

The Certificate of Operation is similar to the Certificate of Installation. This sign-off format insures that designated individuals (by signing and dating the Certificate of Operation) verify that the material contained within the document has been properly executed. Furthermore, the Certificate of Operation refers to Change Control and the potential requirement for requalification/ revalidation of the system if significant modifications to the operational characteristics of the system are implemented.

Purpose

The purpose of the OQ for the Water for Injection storage tank is defined. It clearly states that the OQ will verify that the system operates as intended, activate specific alarms when the system malfunctions, and indicate/react to an unacceptable excursion during system operation.

Equipment Rationale

The "Equipment Rationale" section defines the function of the particular unit operation, in this case the Water for Injection storage tank, within the system. This equipment rationale is identical to the equipment rationale for IQ.

Protocol Format Approval

The protocol format approval in the OQ document serves the same purpose as that in the IQ document. It ensures that the designated individuals responsible for verifying that the OQ has been properly prepared have reviewed the OQ prior to execution, are familiar with the its contents, and concur with the content of the document.

Conclusion

The conclusion reiterates that the purpose of the OQ document is to ensure that components and support accessories operate in an integrated manner from a controls standpoint during normal operation and unacceptable transients or abnormal conditions that generate an alarm alerting operating personnel that an unacceptable condition exists. This OQ, when integrated with other OQs for the entire system, insures that the quality of water at points of use will not be compromised by a transient condition in one or more unit operations.

Equipment Description and List of Components

Reference to pertinent documents that will be used during OQ execution is stated in the "Equipment Description and List of Components" section. This includes reference to the P&ID for the storage and distribution system. In addition, the electrical drawing for the control panel would also be used during OQ execution. Finally, references are made to manufacturer support information, which should be contained in the IQ, SOPs, and maintenance manual for the system.

Critical Elements
Critical accessories required for proper execution of the OQ for the Water for Injection storage tank are indicated. For example, critical elements would include the individual indicating lamps on the central control panel (or display panel icons), temperature control system for the storage and distribution system, level control system for the storage tank, alarm system on the central control panel, and appropriate system electrical interlocks contained within the central control panel for the system. As emphasized earlier, reference to these critical elements demonstrates the integration of individual OQs when compared with the structured individual unit operation nature of the "example" IQ.

Critical Parameters and Data Sheets
The various steps necessary to create a transient condition are presented in the "Critical Parameters" section to insure that the system responds in an appropriate fashion. The body of this section is presented as a three-column table. The first column, Parameter, represents a critical step in creating the unacceptable transient condition or a description of how the transient condition will ultimately be noted by operating personnel. The second column, Test Method, describes the method for creating the transient condition. The final column, Actual Results, would be initialed by the individual(s) from the pharmaceutical manufacturing firm executing the OQ for the storage tank. Initials (preferably a signature) and date are required in this column. Each of the steps required to create the transient condition must be verified within the Actual Results column.

The first objective of OQ is to verify the function of the level control system for the tank. As discussed in chapter 6, the Water for Injection storage tank is equipped with a four-point level control system. The four level set points, from the top to the bottom of the tank are indicated as "high-high," "high," "low," and "low-low." The multiple-effect distillation unit contains a dedicated control panel, provided as a "package" with the unit. The face of the multiple-effect distillation unit control panel has a "manual/auto" switch (display screen indication) that is very helpful for OQ execution, since the "auto" control function can be "defeated." With the switch for the multiple-effect distillation unit in the "manual" position, the tank level is increased to the high level. An indicating light on the central control panel/display screen icon verifies that the high level situation has been encountered. The level is further increased to the high-high level. The OQ requires verification that an indicating light/display screen icon on the central control panel, an audible alarm, and a red-colored strobe light are all activated when the high-high tank level is reached. Subsequent to this operation, the stored Water for Injection in the tank is intentionally drained (not completely) through a valve at the base of the tank as discussed in chapter 9. During the draining process, individual OQ line items are provided to verify the fact that the high-high and high level indication and, as appropriate, alarm conditions, "clear." Tank draining proceeds until the low level is reached. An OQ line item verifies that an indicating light/display screen icon is energized when this level is reached. Tank draining continues until the low-low level is reached—an excursion condition. This condition energizes an indicating light/display screen icon on the central control panel, a red strobe light, and an audible alarm. A line item of the OQ verifies that the downstream Water for Injection distribution pump is automatically deenergized at a low-low tank level. At this point, the tank drain valve is closed, and the multiple-effect distillation unit is allowed to refill the tank. As the level increases above the low-low set point, the alarm conditions should clear. Further, as the water level continues to increase, the low-low level-indicating light/display screen icon should be deenergized, and the Water for Injection distribution pump reenergized.

The next portion of the OQ verifies the proper performance of the temperature-indicating control system by adjusting the set points to produce alarm conditions. While it may appear inappropriate to adjust set points to achieve alarm conditions, in this particular case, the high temperature set point is 95°C, very close to boiling conditions. It would be extremely inappropriate to allow the tank to be heated to boiling conditions. A small change in the set point, to a value slightly less than 95°C, can be used to verify the functionality of the temperature control system. The temperature excursions are noted on a temperature recorder or data logger. The high tank temperature condition is further noted by activation of a control light/display screen icon on the central control panel, a red-colored strobe light, and an audible alarm. A similar procedure is used to verify low distribution loop return temperature, a value monitored from a temperature probe installed in the return distribution tubing upstream of the back pressure regulating valve.

Finally, the OQ verifies the proper operation of the integrity strip (electrical continuity strip) within the compound rupture disk assembly, by manually disconnecting the electrical line from the central control panel to the "mating connection" from the compound rupture disk. Disconnecting the electrical junction simulates disk failure and generates an alarm. The condition is noted by activation of a light/display screen icon on the central control panel, a red strobe light, and an audible alarm.

The transient conditions described within the OQ for the Water for Injection storage tank demonstrate that a visual indicating light/display screen icon, an audible alarm, and a red-colored strobe light are all activated when an unacceptable system transient conditions occurs. This is the purpose of the OQ for this component.

PERFORMANCE QUALIFICATION PROTOCOL: USP WATER FOR INJECTION STORAGE TANK WITH ACCESSORIES

Specific details associated with preparation of the PQ were presented earlier in this chapter. Appendix C is representative of a PQ for an USP Water for Injection storage and distribution system with limited points of use. The material in the appendix attempts to demonstrate the format, sampling criteria, analytical criteria, data logging, and schedule preparation for the PQ. It should be pointed out that Appendix C contains only the first week of a four-week sampling schedule. Because of the limited number of sample points, the balance of the sampling schedule has not been included. Critical items that should be considered in preparation of the PQ are presented below.

Protocol Format Approval

Protocol format approval for PQ is identical to that for the IQ and OQ for the USP Water for Injection storage tank. Designated individuals from the pharmaceutical manufacturing firm are responsible for reviewing the PQ, verifying that the proposed format is acceptable to the disciplines of technical services, facility engineering, and quality control.

Summary

The "Summary" section of the PQ for the Water for Injection storage tank provides a summary of the sampling program. It states that the program will consist of a four-week intense monitoring program. Since the distribution loop contains only three points of use, samples will be obtained from each point of use daily. Samples are only obtained during working days, Mondays through Fridays. In addition, samples will be collected in the supply portion of the distribution loop as well as the return portion of the loop. Once per week, samples will be collected of the feed water to the multiple-effect distillation unit and product water from the multiple-effect distillation unit. The summary states that a10 cfu/100 mL total viable bacteria Action Limit applies for all Water for Injection samples. It is also stated that *any* bacteria detected in Water for Injection samples will be sub cultured and speciated. The samples collected from points of use within the Water for Injection distribution loop will be designated as "WFI" samples. Samples obtained from the feed water to the multiple-effect distillation unit, including product water from the distillation unit, will be designated as "WGS" samples.

Chemical Testing—WFI Samples

Chemical testing is outlined in accordance with the criteria discussed earlier in the text for Water for Injection systems. The chemical parameters measured are TOC, conductivity and nitrates. The chemical tests are conducted in the laboratory for individual points-of-use samples. Since there are a limited number of points of use, online measurement of TOC was not employed. The return line tubing from the recirculating loop contains a conductivity sensor connected to a non temperature compensated conductivity meter. At the conclusion of the PQ, individual conductivity values obtained from points of use were compared with continuously recorded non temperature compensated conductivity values from the Water for Injection return tubing conductivity sensing element.

Bacteria Testing—WFI Samples

The bacteria testing sampling procedure, shipment criteria, analysis format, and required documentation are outlined in "Bacteria Testing—WFI Samples." The sampling procedure describes proper collection of a sample. As written, this PQ does not consider a flush of the external surfaces of point-of-use valves or other sampling points, such as the end of hoses. It should be noted that there is a minimal time period for flowing water to drain prior to collecting a sample. The specified time period is only 10 seconds. This relatively short time period is literally specified to allow sampling personnel to collect the sample. It does not imply that flushing is required to obtain a "representative" sample. Extensive flushing, prior to sample collection, could be challenged by regulatory investigators unless water is flushed from each point of use for the same time period indicated in the PQ sampling procedure. Flushing, prior to sampling, is further discouraged because it indicates that there is something inherently wrong with the system.

It is stated that samples that will not be analyzed within a reasonable time period, suggested as 30 minutes, be refrigerated (dedicated area) at $\sim 2°C$ to $4°C$. Further, all bacteria enumeration methods should be initiated within 16 hours of sampling. If the services of a contract laboratory are used, bacteria enumeration should *not* be initiated >24 hours after sampling. Criteria for shipping the samples are outlined. The samples should be transferred in a portable cooler with "ice packs" to maintain temperature at $2°C$ to $4°C$. The sample temperature should not exceed $10°C$ during shipment. Further, the sample temperature should not decrease below $1°C$. The bacteria monitoring technique specified for the PQ of the USP Water for Injection system is Membrane Filtration of a 100-mL sample with incubation in R2A culture media for 72 hours at $22°C \pm 2°C$. The procedure indicates that at least one sample per week should be analyzed to provide results observed after 5-, 8-, and 10-day incubation time periods. The procedure also states that any bacteria colonies detected should be sub cultured and speciated.

If the samples for bacteria determination are sent to a contract laboratory, a chain of custody must be completed to properly identifying the samples with a sequential logbook number, as discussed earlier. Data Sheet B provides a typical sample bottle label. This label contains appropriate information, such as the sequential logbook number assigned to the sample, the date of sampling, the time of sampling, the sample location, and the initials or signature of the individual collecting the sample. Results from the contract laboratory should be provided in a formal report, with appropriate support documentation and signature of a designated individual at the contract laboratory supervising the bacteria determination technique. It may take as long as two to three weeks to obtain a formal report from the contract laboratory. Subsequently, it is strongly suggested that the contract laboratory, as part of their "Scope of Supply," provide results as they become available. A copy of the results should be sent via a facsimile machine or electronically to the pharmaceutical manufacturing firm. Further, the Alert and Action Limits for total viable bacteria should be provided to the contract laboratory. If a sample indicates the presence of bacteria above the Alert Limit, the designated individual at the pharmaceutical manufacturing facility should be immediately contacted by the contract laboratory. On rare occasions, bacteria colonies may be observed after an incubation period of <72 hours, even as short as 24 hours. The contract laboratory should observe, without compromising the 72-hour results, bacteria colonies on a daily basis. Identification of excessive bacteria growth prior to the 72-hour incubation time period provides the pharmaceutical manufacturing firm with extremely valuable information, which should prompt a full investigation of the situation. This simple but highly responsive program should obviously be employed if bacteria monitoring is conducted directly at the pharmaceutical manufacturing facility. This program should be within the SOPs and employed for samples collected after PQ execution.

If the services of a contract laboratory are used, the pharmaceutical manufacturing firm, or a qualified designated representative, should periodically inspect the facility. The inspection should, as a minimum, evaluate the procedures for sample receipt, the method of insuring proper chain of custody, bacteria monitoring techniques, the segregation of samples with high anticipated bacteria levels (from other clients), the qualification program for the technician performing the analysis, laboratory supervisor, administrative personnel qualification, SOPs; and the established quality assurance/quality control program. Subsequent inspections of the contract laboratory should be conducted at annually.

Bacterial Endotoxin Testing

The sampling procedure, shipment requirements (where applicable), the analytical method, and required documentation are presented in the section entitled "Bacterial Endotoxin Testing." The sampling procedures are similar to the bacteria sample collection procedures, except for the size of the sample container. The shipment criteria, when the services of a contract laboratory are used, are similar to those for bacteria samples. The method of analysis may vary depending on the specific preferences of the pharmaceutical manufacturing facility or, in the case of a contract laboratory, available technique. The bacterial endotoxin detection technique should provide a quantitative result, with a minimum sensitivity at least one order of magnitude less than the 0.25 endotoxin unit/mL specification outlined within the *Official Monograph* for Water for Injection. Using a technique that provides a "pass" or "fail" result at the 0.25 endotoxin unit/mL specification is highly discouraged, since it does not provide a quantitative number required for data trending and determination of long-term system performance. The documentation and qualification for the contract laboratory are similar as for bacteria testing.

Chemical Testing—WGS Samples

The requirements for the chemical testing of WGS samples, the feed water and product water to and from the multiple-effect distillation unit, are presented in the "Chemical Testing" section of the PQ for the Water for Injection storage tank. The first part indicates the frequency of sample collection, which is established as weekly during PQ execution. The feed water to the pretreatment system feeding the multiple-effect distillation unit should be analyzed for several parameters. Some of the parameters are not necessarily included in the NPDWR as defined by the EPA, but are critical in establishing the quality of the feed water, by identifying the requirements to improve water quality. Further, the feed water to the multiple-effect distillation unit has been treated by a system similar to that of a typical Purified Water system. As part of the lengthy list of analysis, elements that could create potential scale-forming precipitates have been included. Specifically, this would include aluminum, barium, calcium, iron, magnesium, manganese, and lead. The analysis for reactive silica is important because it is volatile and *may* be carried over with the steam generated within the individual "effects" of the multiple-effect distillation unit (chap. 5). In addition to the analysis required for feed water to the multiple-effect distillation, the analytical requirements for distillation unit product water for chemical constituents are outlined (TOC and conductivity).

Bacteria Testing—WGS Samples

The sampling procedures for WGS samples is very similar to the sampling procedures for point-of-use WFI samples. One of the subtle differences relates to the "flush-to-drain" time prior to collecting a sample. The specified time for WGS samples as long as 5 to 10 minutes. The additional flush time has been selected on the basis of the fact that it reflects the operation of the multiple-effect distillation unit. For example, during routine operation of the distillation unit, a signal is received from the USP Water for Injection storage tank through the control panel. The product water from the multiple-effect distillation unit will be diverted to waste for a preestablished time period, generally 5 to 10 minutes, before flowing to the tank. Thus, specifying a 5- to 10-minute "rinse-to-drain" period prior to sample collection is appropriate.

The shipment and transport requirements of WGS samples are similar to those for WFI samples. The analysis section is also similar. However, the procedure limits sub culturing and speciation of any observed bacteria to samples obtained from the product water of the multiple-effect distillation unit, not the feed water to the unit. It is important to note that the pretreated water to the multiple-effect distillation unit is from a pretreatment system where microbial control is maintained. Generally, bacteria monitoring of feed water samples to a multiple-effect distillation unit will employ a heterotrophic plate count, not Membrane

Filtration of a 100-mL sample with R2A culture media at 22°C for 72 hours. The heterotrophic plate count can be performed with PCA media, incubation temperature of 30°C to 35°C and incubation time period of 48 hours. For this particular PQ, the bacteria monitoring technique of the feed water to the multiple-effect distillation unit is based on the anticipated bacteria levels and the ability to "trend" the levels.

Documentation of this testing is similar to the documentation required for other sections of the PQ—the chemical, bacteria, and bacterial endotoxin testing for point-of-use WFI samples.

Typical Logbook Pages, Log Sheets, and Sample Bottle Labels

The enclosures to the PQ protocol are representative sections of support information. A supplemental sampling and monitoring program is demonstrated. A typical log sheet for WFI samples with sequential numbers is provided. Note that the actual logbook would consist of a hardbound book with string binding and pre printed sequential page numbering to provide a permanent record of all samples collected with their results. The nature of the book clearly demonstrates that no pages have been removed or altered in any way. Corrections are made by simply drawing one line through an entry and initialing and dating the line. A similar logbook page for the WGS samples is also provided. Data Sheet A is the conductivity log for product water conductivity monitoring from the multiple-effect distillation unit. In this particular system, the conductivity monitoring system for the Water for Injection storage and distribution loop was equipped with a continuous recorder. However, the conductivity monitor for the product water from the multiple-effect distillation unit was not equipped with a recorder. Subsequently, daily recording of conductivity is suggested. Data Sheet B depicts a typical sample bottle label.

Sample Locations

The PQ should provide a summary of sample locations for both the Water for Injection distribution loop and distillation unit feed water and product water samples. A list of four sampling points for the Water for Injection distribution loop is provided. The "type" of sample refers to either "direct" or "sample valve." In this particular case, a point-of-use valve and the distribution loop supply valve are available for direct collection of water at sample points WFI-304 and WFI-310. However, two production vessel points of use are "hard piped" directly to the vessel. Subsequently, a sample valve is installed from the point-of-use valves for each production point of use. These valves are referred to as a "sterile access port type" (chap. 9). The samples obtained from sample valves WFI-305 and WFI-307 are representative of the water being provided to production vessels 1 and 2, respectively. Finally, information is provided for WGS samples, referred to as "WGS-201" and "WGS-202," feed water and product water to and from the multiple-effect distillation unit, respectively.

Sampling Schedule

The final page of Appendix C is the first week's sampling schedule. It should be noted, as discussed earlier, that samples are collected from individual points of use on a daily basis. Since operation is conducted over a five-day work week (Monday through Friday), samples are collected on each of the five working days. The length of the sampling schedule is four weeks, providing approximately 20 sample days. The WGS samples are collected weekly.

REFERENCES

Collentro WV. Proper validation of a water purification system: am inherently flawed process? J Validation Technol 1995; 1(3):17–23.

Collentro WV, Angelucci L. Coordinating validation requirements for pharmaceutical water purification systems. Pharm Technol 1992; 16(9):68–78.

Eaton A, Clesceri LS, Rice E, Greenberg A, eds. Standard Methods for the Examination of Water and Waste Water. 21st ed. American Public Health Association, American Water Works Association, and Water Environmental Federation, 2005.

Eaton AD, Clesceri LS, Greenberg AE. Standard Methods for the Examination of Water and Wastewater. 19th ed. American Public Health Association, American Water Works Association, and Water Environment Federation. Washington, D.C.: American Public Health Association, 1995.

Elms B, Green C. Water systems: the basics, part 1: design as a prelude to validation. J Validation Technol 1995a; 1(2):36–42.

Elms B, Green C. Water systems: the basics, part 2: validation and maintenance. J Validation Technol 1995b; 1(3):716.

FDA. Guide to Inspection of High Purity Water Systems. Rockville: Food and Drug Administration, Office of Regulatory Affairs, Office of Regional Operations, Division of Field Investigations, 1993.

Fessenden B. A guide to water for the pharmaceutical industry: part 3—purification and validation techniques. J Validation Technol 1996; 2(2):132–148.

King J. A practical approach to equipment validation. J Validation Technol 1997; 4(1):84–89.

Neal C Jr. Validation 101: back to basics. J Validation Technol 1997; 3(13):281–283.

Riley BS. Rapid Microbiology Methods in the Pharmaceutical Industry, American Pharmaceutical Review. Vol. 7, Issue 2. Indianapolis, IN: Russell Publishing, 2004:28–31.

Roczniak BJ. A guide to validating a purified water or water for injection system. J Validation Technol 1995; 1(3):24–27.

Sutton S. How do you decide which microbial identification is best. Pharm Microbiol Forum 2007; 13(1):4–12.

APPENDIX A: INSTALLATION QUALIFICATION FOR AN USP WATER FOR INJECTION STORAGE TANK

ABC Pharmaceutical Company
Anytown, U.S.A.
Installation Qualification
USP Water for Injection Storage Tank

CERTIFICATE OF INSTALLATION

The USP Water for Injection storage tank, tag number TK-1, is installed in a manner consistent with the requirements set forth in this document.

Group	Name	Signature	Date
Technical Services			
Facilities Engineering			
Quality Control			

Note: Change alert/requalification

Requalification may be required when changes to a system or piece of equipment have been made. All future changes/modifications should be recorded in an appendix to this protocol/report. This will form the engineering record for the component.

Equipment Installation Qualification Index

Subject

I. Installation Qualification Protocol and Report
   A. Purpose
   B. Equipment Rationale
   C. Protocol Format Approval
   D. Conclusion

II. Installation Qualification—Execution
   A. Component and Accessories—Inspection Items
   B. Component Inspection
   C. Accessories Inspection
   D. Control System Inspection
   E. Instrumentation Inspection
   F. Utilities Inspection
   G. Document Verification

I. INSTALLATION QUALIFICATION PROTOCOL AND REPORT

A. Purpose

The purpose of section II of this protocol is to verify that the selection, installation, and accessories of the USP Water for Injection storage tank (tag number TK-1) are consistent with the specified tank for the project. Physical characteristics, specified for the USP Water for Injection storage tank will be verified and appropriately recorded. The presence of specified accessories, including isolation valves, sample valves, and instrumentation, will also be verified. The control function shall be in accordance with custom logic, outlined in the specification. When completed, the IQ for the USP Water for Injection storage tank shall verify that the system, as installed, is consistent with all criteria outlined in the Basis of Design, project specification, and tank supplier's proposal.

B. Equipment Rationale

The USP Water for Injection storage tank (tag number TK-1) has been selected and positioned in the system downstream of the multiple-effect distillation unit but prior to the USP Water for Injection distribution pump(s) (tag nos. P-1 and P-2) to provide a location for storage of Water for Injection and a number of additional critical functions. The tank is of vertical cylindrical design with a domed top and an inverted dome base. The tank is provided with an external heating jacket, insulation, and an outer stainless steel "sheath."

The tank provides a reservoir for storage of USP Water for Injection from the upstream multiple-effect distillation unit. During normal operation, the flow rate requirement for USP Water for Injection for production applications may exceed the makeup capability of the multiple-effect distillation unit. The 1500-gal (nominal) capacity of the USP Water for Injection storage tank allows flexibility with regard to all production application requirements. Further, a depressurized point for the product water from the multiple-effect distillation unit must be provided.

Another function of the USP Water for Injection storage tank is to provide a method of recirculating and maintaining USP Water for Injection temperature >80°C (176°F) to eliminate concerns associated with bacteria proliferation in the storage and distribution system. All common species of bacteria are destroyed when the USP Water for Injection temperature is maintained at $\geq 80°C$. Obviously, "dead legs" must be avoided, since they can produce potential "thermal sinks" where stagnant areas of USP Water for Injection can exist with a "local" temperature <80°C.

The USP Water for Injection storage tank is equipped with appropriate accessories, such as a makeup valve (positioned on the multiple-effect distillation unit), hydrophobic vent filter with heating jacket, compound-type rupture disk, a spray ball, back pressure regulating valve on the return tubing line, level control system, and temperature control system. In general, control details, including components, for the USP Water for Injection storage tank are contained in the central control panel for the USP Water for Injection system.

C. Protocol Format Approval

It is the responsibility of individuals from technical services, facilities engineering, and quality control to review and approve sections I and II of this protocol prior to execution. The review of data gathered after implementation of the protocol and final approval is their responsibility.

Group	Name	Signature	Date
Technical services			
Facilities engineering			
Quality control			

D. Conclusion

1. Section I: The purpose, rationale, and format of the IQ for the USP Water for Injection storage tank (tag number TK-1) properly justifies component selection.

2. Section II: The inspection of items, defined in detail in the execution section of this document, establishes that the USP Water for Injection storage tank (tag number TK-1) complies with the requirements outlined in the Basis of Design for the system and in the tank supplier's proposal.

II. INSTALLATION QUALIFICATION—EXECUTION
A. Component and Accessories—Inspection Items
1. Manufacturer
2. Serial number
3. Tank capacity
4. Tank diameter and straight side height
5. Tank wall thickness
6. National Board certification number
7. Maximum allowable internal working pressure—vessel
8. Maximum allowable external working pressure—vessel
9. Hydrostatic test pressure—vessel
10. Minimum design temperature—vessel
11. Maximum allowable working pressure—steam heating jacket
12. Hydrostatic test pressure—steam heating jacket
13. Tank materials of construction
14. Tank fittings—size and location
15. Tank internal finish
16. Tank external finish
17. Level control sensors—type, model number, and manufacturer
18. Level control sensors—mounting and location
19. Hydrophobic vent filter housing—manufacturer, model number, and type
20. Hydrophobic vent filter housing—materials of construction
21. Hydrophobic vent filter membrane—manufacturer, type, sealing mechanism
22. Control wiring—level sensor
23. Spray ball assembly
24. Control interface
25. Feed water sample valve
26. Tank discharge sample valve
27. Operating and support manuals

B. Component Inspection

Parameter	Test method	Actual results
Verify that the USP Water for Injection storage tank (tag number TK-1) of company X, model number X has a stated capacity of 1500 gal.	Visual	
Verify, utilizing tank nameplate information, that the tank is serial number 1234.	Visual	
Verify, utilizing tank manufacturer's information, that the 1500-gal tank is of vertical cylindrical type.	Visual	
Verify, utilizing tank manufacturer's information, that the tank has X in. inside diameter × Y in. straight side (tangent to tangent) × Z in. overall height to the top of the tank including support legs.	Visual	
Verify, utilizing tank manufacturer's information, that the tank is designed for a maximum internal working pressure of 30 psig and full vacuum at 300°F.	Visual	
Verify, utilizing tank manufacturer's information, that the sidewall of the tank is equipped with XXX ft^2 of channel wall heat transfer, and that the bottom of the tank contains XX ft^2 of dimple surface heat transfer, connected in a single heating zone, with 1.25-in. NPT inlet and outlet connections.	Visual	
Verify, utilizing tank manufacturer's information, that the heat transfer surfaces, excluding the tank exterior wall, are of 304 stainless steel construction.	Visual	

(Continued)

(*Continued*)

Parameter	Test method	Actual results
Verify, utilizing tank manufacturer's data, that the heat transfer surface is designed for operation at facility steam pressure ≤125 psig at 360°F.	Visual	
Verify, utilizing tank manufacturer's information, that the tank is constructed of 316L stainless steel.	Visual	
Verify, utilizing tank manufacturer's information, that the interior of the tank has been mechanically polished to a Ra = 15–25 finish.	Visual	
Verify, utilizing tank manufacturer's information, that the exterior of the tank, specifically the sidewall and dished bottom, have been mechanically polished to a Ra = 30–35 finish.	Visual	
Verify, utilizing tank manufacturer's information, that all tank interior welds have been mechanically polished to a Ra = 15–25 finish and that all exterior welds have been mechanically polished to a Ra = 30–35 finish.	Visual	
Verify, utilizing tank manufacturer's information, that exterior sections of all heat transfer surfaces are covered with XXX coating.	Visual	
Verify, utilizing tank manufacturer's information, that the sidewall and bottom (inverted) dish of the tank are encased by a 304 stainless steel "shroud" approximately XX in. outside diameter (straight side).	Visual	
Verify, utilizing tank manufacturer's information, that the space between the outer shroud and the tank contains 2-in.-thick chloride-free insulation.	Visual	
Verify that the tank contains a nameplate indicating that the tank; was designed, manufactured, and tested in accordance with the ASME Code for "Unfired Pressure Vessels," with National Board number XXXX.	Visual	
Verify, by inspection and tank manufacturer's information, that the top of the tank is equipped with an 18-in. diameter circular manway with hinged cover, Viton O-ring gasket, and swivel bolt hold-owns with hand knobs.	Visual	
Verify, by inspection and tank manufacturer's literature, that the top of the tank is equipped with two lifting lugs of 316 stainless steel material, positioned 180° apart.	Visual	
Verify that the tank is supported by 4 legs, with base plates, 3-in. IPS, schedule 40, 304 stainless steel pipe, each XX in. long.	Visual	
Verify, by inspection and tank manufacturer's information, that the top of the tank is equipped with a center-mounted 4-in. sanitary ferrule equipped with a 2.5-in. 316L stainless steel removable spray ball, capable of providing 360° water "coverage."	Visual	
Verify, by inspection and tank manufacturer's information, that the top of the tank is equipped with 3 individual 1.5-in. sanitary ferrules with orientation as shown on tank manufacturer's drawing, and center lines on an XX-in. radius from the center of the domed tank top.	Visual	
Verify, by inspection and tank manufacturer's information, that the top of the tank is equipped with 2 individual 2-in. sanitary ferrules with orientation as shown on tank manufacturer's drawing, with center lines on a XX-in. radius from the center of the domed tank top.	Visual	
Verify, by inspection and tank manufacturer's information, that the inverted domed bottom of the tank is equipped with a 2-in. center-positioned, sanitary ferrule for tank discharge to the downstream USP Water for Injection distribution pump(s).	Visual	
Verify, by inspection and tank manufacturer's information, that the top of the tank is equipped with a 4-in. sanitary ferrule with orientation as shown on tank manufacturer's drawing and center lines on an XX-in. radius from the center of the domed tank top.	Visual	
Verify, by inspection and tank manufacturer's information, that the lower straight side of the tank is equipped with an "alcove" containing 2 individual 1.5-in. sanitary ferrules.	Visual	
Verify by tank data plate inspection that the vessel was hydrostatically tested at a pressure of 45 psig.	Visual	
Verify by tank data plate inspection that the minimum metal design temperature for the vessel is 32°F.	Visual	

C. Accessories Inspection

Parameter	Test method	Actual results
Verify that a hydrophobic vent filter, tag number VF-1, is connected to a 1.5-in. sanitary ferrule on the top of the USP Water for Injection storage tank.	Visual	
Verify that the hydrophobic vent filter housing is XXX part number XX, with an external heating jacket.	Visual	
Verify that the hydrophobic vent filter housing is connected to a 1.5-in. sanitary ferrule on the domed top of the tank, utilizing a gasket and clamp.	Visual	
Verify that the hydrophobic vent filter housing is approximately XX in. in diameter × CC in. high equipped with a 1.5-in. sanitary ferrule inlet and outlet connections.	Visual	
Verify that the hydrophobic vent filter housing is equipped with a heating jacket with 0.5-in. FPT top side-mounted inlet fitting and 0.5-in. FPT bottom side-mounted outlet fitting.	Visual	
Verify utilizing manufacturer's information, that the hydrophobic vent filter housing is designed to contain a 20-in.-long, single-open-end, Code 7 (226 double O rings with locking tabs) membrane cartridge with fined end.	Visual	
Verify, utilizing hydrophobic filter housing manufacturer's information, That all metallic surfaces in contact with filtered air are 316L stainless steel.	Visual	
Verify that the filter housing is installed such that the inlet flow direction is downward from the atmosphere, into the USP Water for Injection storage tank.	Visual	
Verify, utilizing installation records and hydrophobic membrane filter manufacturer's information, that the filter is XXX part number XXX.	Visual	
Verify, utilizing hydrophobic membrane filter manufacturer's information, that the membrane material is PTFE with polypropylene support, drainage, and hardware.	Visual	
Verify that the 4-in. (noncenter) sanitary ferrule on the domed top of the tank is connected to a compound-type rupture disk, tag number PSE-1.	Visual	
Verify that the rupture disk holder is XXX model number XX, with a 3-in. disk diameter, with 4-in. sanitary ferrule inlet and outlet connections.	Visual	
Verify that a section of stainless steel tubing (and fittings) is connected to the discharge fitting of the rupture disk, directing any liquid or gas associated with "rupture" to the area at the base of the tank.	Visual	
Verify, utilizing rupture disk manufacturer's information, that the disk is constructed of Teflon® with a 316 stainless steel "girdle."	Visual	
Verify that the feed water tubing to the USP Water for Injection storage tank, from the upstream multiple-effect distillation unit, is equipped with a pneumatically actuated, three-way diaphragm type valve (tag number AV-1) positioned downstream of sample valve XX, but upstream of the USP Water for Injection storage tank.	Visual	
Verify that the feed water tubing to the USP Water for Injection storage tank from the multiple-effect distillation unit is slopped to ensure complete draining of water after terminating the distillation unit operation.	Visual	
Verify that the USP Water for Injection distribution loop return tubing to the USP Water for Injection storage tank contains a pressure sensor/transmitter (tag number PT-1) positioned downstream of temperature sensor TE-4 and upstream of back pressure regulating valve PCV-1.	Visual	
Verify that the pressure transmitter is XX model number XX, with a 1.5-in. sanitary ferrule connection.	Visual	
Verify that the pressure transmitter provides a proportional control signal to the USP Water for Injection distribution loop back pressure regulating valve (PCV-1), through the USP Water for Injection system central control panel.	Visual	
Verify that the USP Water for Injection distribution loop return tubing contains a back pressure regulating valve, tag number PCV-1, positioned downstream of pressure transmitter PT-1 and prior to the USP Water for Injection storage tank return tubing, connected to the top of the tank by a 4-in. sanitary ferrule (connected to the spray ball assembly).	Visual	

(Continued)

(*Continued*)

Parameter	Test method	Actual results
Verify that the USP Water for Injection distribution loop return tubing back pressure regulating valve is XX model number XX.	Visual	
Verify that back pressure regulating valve, PCV-1, is equipped with 1.5-in. sanitary ferrule inlet and discharge connection.	Visual	
Verify that the USP Water for Injection storage tank is equipped with a level sensor, tag number LE-1A, positioned in a 1.5-in. sanitary ferrule at the lower straight side of the tank, and a second level sensor, tag number LE-1B, positioned in a 2-in. sanitary ferrule, reduced to 1.5-in. positioned on the domed top of the tank.	Visual	
Verify that the USP Water for Injection storage tank level transmitter is XX model number XX.	Visual	
Verify that the USP Water for Injection storage tank is equipped with a temperature sensor, tag number TE-1, positioned in an XX thermowell, tag number TW-1, mounted on a 1.5-in. sanitary ferrule located on the lower straight side of the tank.	Visual	
Verify that the USP Water for Injection storage tank temperature thermowell, TW-1, is XX model number XX.	Visual	
Verify that temperature-indicating transmitter, TIT-1, is model number XX.	Visual	
Verify that a proportional control signal is transferred from the USP Water for Injection storage tank temperature sensor/transmitter (TTK-1) through the USP Water for Injection system central control panel to a modulating type steam supply valve (TCV-1) to the heating jacket surrounding the tank.	Visual	
Verify that the heating jacket surrounding the tank (and inverted domed base) is supplied steam through a steam modulating valve, tag number TCV-1, connected to the 0.5-in. supply line located at the upper straight side of the tank.	Visual	
Verify that steam modulating valve, TCV-1, is XX model number XX.	Visual	
Verify that the heating jacket surrounding the USP Water for Injection storage tank is equipped with a supply steam condensate trap, positioned in the effluent piping from the steam jacket (1.25-in. MPT fitting), at the inverted domed base of the tank.	Visual	
Verify that the 2-in. sanitary ferrule discharge from the USP Water for Injection storage tank valve DV-1 is connected to a section of 2-in. tubing feeding the downstream USP Water for Injection distribution pump, tag number P-1.	Visual	

D. Control System Inspection

Parameter	Test method	Actual results
Verify, by direct inspection and manufacturer's literature, that an analog signal is available from USP Water for Injection storage tank level-indicating transmitter (LITK-1) to the USP Water for Injection system central control panel.	Visual	
Verify, by direct inspection and manufacturer's literature, that an analog signal is available from USP Water for Injection storage tank temperature-indicating transmitter (TTK-1) to the USP Water for Injection system central control panel.	Visual	
Verify, by direct inspection and manufacturer's literature, that an analog signal is available from USP Water for Injection storage loop return tubing pressure transmitter (PTK-1) to the USP Water for Injection system central control panel.	Visual	

(*Continued*)

SYSTEM VALIDATION

(*Continued*)

Parameter	Test method	Actual results
Verify, by direct inspection and manufacturer's literature, that a proportional pneumatic supply is available from the USP Water for Injection system central control panel to USP Water for Injection distribution loop back pressure regulating valve (PCV-1).	Visual	
Verify, by direct inspection and manufacturer's literature, that a proportional pneumatic supply is available from the USP Water for Injection system central control panel to USP Water for Injection storage tank heating jacket steam supply valve (TCV-1).	Visual	
Verify, by visual inspection, that a pneumatic line from the USP Water for Injection system central control panel (direct or through a remote solenoid valve) is connected to the USP Water for Injection storage tank feed water divert-to-waste valve (AV-1).	Visual	

E. Instrumentation Inspection

Note: All instrumentation is critical.

Parameter	Test method	Actual results
USP Water for Injection storage tank level sensor/transmitter (LTK-1A) with current calibration sticker.	Visual	
Record tank level sensor/transmitter (LTK-1A) calibration identification number.	Record	
USP Water for Injection storage tank level sensor/transmitter (LTK-1B) with current calibration sticker.	Visual	
Record tank level sensor/transmitter (LTK-1B) calibration identification number.	Record	
USP Water for Injection storage tank differential level-indicating transmitter (LITK-1) with current calibration sticker.	Visual	
Record tank differential level- indicating transmitter (LITK-1) calibration identification number.	Record	
USP Water for Injection storage tank temperature sensor/transmitter (TTK-1) with current calibration sticker.	Visual	
Record tank temperature sensor/transmitter (TTK-1) calibration identification number.	Record	
USP Water for Injection distribution loop return tubing pressure sensor/transmitter (PTK-1) with current calibration sticker.	Visual	
Record distribution loop return tubing pressure sensor/transmitter (PTK-1) calibration identification number.	Record	

F. Utilities Inspection

Parameter	Test method	Actual results
Verify that facility steam is connected to the USP Water for Injection storage tank heating jacket modulating steam supply valve (TCV-1).	Visual	
Verify facility condensate return provisions are available for receiving the steam condensate from the USP Water for Injection storage tank heating jacket.	Visual	

(*Continued*)

(*Continued*)

Parameter	Test method	Actual results
Verify that facility steam, regulated to a 2–3 psig pressure, is available to the steam fitting at the top of the hydrophobic vent filter housing (heating jacket).	Visual	
Verify that a method of removing facility steam condensate from the bottom of the hydrophobic vent filter housing (heating jacket) is available.	Visual	
Verify that a depressurized waste line is available to receive water from the multiple-effect distillation unit divert-to-waste valve (AV-1).	Visual	
Verify that facility air is available, through solenoid valves or the USP Water for Injection system central control panel, to automatic and modulating pneumatically controlled valves.	Visual	
Verify that a regulated low voltage supply is available, through the USP Water for Injection system central control panel to all control transmitters.	Visual	

G. Documentation Verification

Document	Rev.	Date	Source
Instruction manual for diaphragm control valves			
Diaphragm actuator installation, operation, and maintenance manual			
Diaphragm valve hand wheel operated diaphragm valves weir and straightway maintenance manual			
Diaphragm valve maintenance manual			
Custom equipment: instruction and parts list			
Instruction manual: electronic sanitary temperature transmitter			
Instruction manual: model XXX differential level transmitter			
Service manual number XXX: air-actuated valves series XXX			
Instruction manual: model XX sanitary electronic pressure transmitter			
XXX sanitary T-type filter housing service instructions			
Sterilization procedures for stainless steel sterilizing membrane filter assemblies that utilize replaceable filter elements			
XXX selection guide: model XXX filter housings			
Sanitary valve service instructions			
XX fittings: catalog			
Air-actuated valve service manual			
Model number XXX			
Final assembly 1500 gal × X in.			
ID vertical, insulated pressure vessel			
Certificate of Calibration: tank level sensor/transmitter (LTK-1A)			
Certificate of Calibration: tank level sensor/transmitter (LTK-1B)			
Certificate of Calibration: tank level-indicating transmitter (LITK-1)			
Certificate of Calibration: tank temperature sensor/transmitter (TTK-1)			
Certificate of Calibration: distribution loop return tubing pressure sensor/transmitter (PTK-1)			

APPENDIX B: OPERATIONAL QUALIFICATION FOR AN USP WATER FOR INJECTION STORAGE TANK

ABC Pharmaceutical Company
Anytown, U.S.A.
Installation Qualification
USP Water for Injection Storage Tank

CERTIFICATE OF OPERATION

The Water for Injection storage tank, tag number TK-1, operates in a manner consistent with the requirements set forth in this document. The storage tank and accessories, including monitors and controls, responds to transient conditions, energizing appropriate alarms, and alerting operating personnel of the condition.

Group	Name	Signature	Date
Technical services			
Facilities engineering			
Quality control			

Note: Change alert/requalification

Requalification may be required when changes to a system or piece of equipment have been made. All future changes/modifications should be recorded in an appendix to this protocol/report. This will form the engineering record for the component.

Equipment Operation Qualification Index

Subject

I. Installation Qualification Protocol and Report
   A. Purpose
   B. Equipment Rationale
   C. Protocol Format Approval
   D. Conclusion
   E. Equipment Description and List of Components

II. Operational Qualification—Execution
   A. Critical Elements
   B. Critical Parameters and Data Sheets

I. OPERATIONAL QUALIFICATION PROTOCOL AND REPORT

A. Purpose

The purpose of section II of this protocol is to verify that the operation, function, control, and alarm criteria of the USP Water for Injection storage tank (tag number TK-1) are consistent with the design, control, instrumentation, and monitoring requirements for the project. Critical factors associated with the operation of the unit will be reviewed. Out-of-specification situations requiring system control (and alarm) response will be created.

    The purpose of the steps in this OQ is to insure that each component, including controls, interface to produce USP Water for Injection. In the event of malfunction, the controls activate an alarm, alerting operating personnel to the condition, and/or stop the recirculating flow of Water for Injection to the points of use.

B. Equipment Rationale

The USP Water for Injection storage tank (tag number TK-1) has been selected and positioned in the system downstream of the multiple-effect distillation unit but prior to the USP Water for Injection distribution pump(s) (tag nos. P-1 and P-2) to provide a location for storage of Water for Injection and a number of additional critical functions. The tank is of vertical cylindrical design with a domed top and an inverted dome base. The tank is provided with an external heating jacket, insulation, and an outer stainless steel "sheath."

The tank provides a reservoir for storage of USP Water for Injection from the upstream multiple-effect distillation unit. During normal operation, the flow rate requirement for USP Water for Injection for production applications may exceed the makeup capability of the multiple-effect distillation unit. The 1500-gal (nominal) capacity of the USP Water for Injection storage tank allows flexibility with regard to all production application requirements. Further, a depressurized point for the product water from the multiple-effect distillation unit must be provided.

Another function of the USP Water for Injection storage tank is to provide a method of recirculating and maintaining USP Water for Injection temperature >80°C (176°F) to eliminate concerns associated with bacteria proliferation in the storage and distribution system. All common species of bacteria are destroyed when the USP Water for Injection temperature is maintained at ≥80°C. Obviously, "dead legs" must be avoided, since they can produce potential "thermal sinks" where stagnant areas of USP Water for Injection can exist with a "local" temperature <80°C.

The USP Water for Injection storage tank is equipped with appropriate accessories, such as a makeup valve (positioned on the multiple-effect distillation unit), hydrophobic vent filter with heating jacket, compound-type rupture disk, a spray ball, back pressure regulating valve on the return tubing line, level control system, and temperature control system. In general, control details, including components, for the USP Water for Injection storage tank are contained in the central control panel for the USP Water for Injection system.

C. Protocol Format Approval

It is the responsibility of individuals from technical services, facilities engineering, and quality control to review and approve sections I and II of this protocol prior to execution. The review of data gathered after implementation of the protocol and final approval is their responsibility.

Group	Name	Signature	Date
Technical services			
Facilities engineering			
Quality control			

D. Conclusion

1. Section I: The purpose, rationale, and format of the OQ for the USP Water for Injection storage tank (tag number TK-1) insures that the component operates in an integrated manner, capable of reacting to alarm/transient conditions.
2. Section II: The inspection of parameters identified in detail in the execution section of this document establishes that the Water for Injection storage tank (tag number TK-1) and accessories operate in accordance with the SOPs for the system, the Basis of Design, and the system specification.

E. Equipment Description and List of Components

A description of the USP Water for Injection storage tank (tag number TK-1) is contained in the specification for the project. A P&ID (diagram number P&ID-123-04) provides information associated with system interface and the position of the USP Water for Injection storage tank. The IQ for the USP Water for Injection storage tank contains the manufacturer's support information. SOPs provide an outline of routine operation. The maintenance manual provides procedures required for long-term successful system operation.

II. OPERATIONAL QUALIFICATION—EXECUTION
A. Critical Elements
The elements to be checked are those deemed to impact the storage and accessory function significantly.
1. Indicator lamp operation
2. Temperature control system
3. Level control system
4. Alarm system
5. System electrical interlock

B. Critical Parameters and Data Sheets
Note: Certain transient conditions may cause multiple alarms while executing the OQs. Each alarm will be verified one at a time.

Parameter	Test method	Actual results
Ensure that the multiple-effect distillation unit control panel is operating in the auto mode by pressing the "auto" switch.	Manual	
Verify that the auto (amber) light illuminates.	Visual	
Insure that the Water for Injection storage and distribution system control panel is "on" by pressing the "power" (red) switch.	Manual	
Verify that the power (red) light illuminates.	Visual	
Verify, utilizing the Water for Injection storage tank level indicator that the Water for Injection storage tank level is at or above the "high-level" set point.	Visual	
Verify that the facility steam, dedicated for Water for Injection storage tank heating, is supplied by opening manual Isolation valve (tag. number VSS-1) and noting the pressure on pressure gauge PI-7.	Manual	
Record the "high-high TK-1 level" set point on the tank level indicator (tag number LI-1).	Manual	
If the Water for Injection storage tank water level is not at the high-level alarm, operate the multiple-effect distillation unit in the "manual" mode to increase the level to the high-level set point.	Manual	
When the Water for Injection storage tank water level reaches the "high" level, return the multiple-effect distillation unit to the auto mode.	Manual	
Verify that the control light "distillate tank full" (amber) on the multiple-effect distillation unit control panel is on.	Visual	
Verify that the multiple-effect distillation Unit is in the "standby" mode.	Visual	
With the multiple-effect distillation unit operating in manual mode, by pressing and holding down the manual switch, let the Water for Injection storage tank fill to the high-high TK-1 level.	Manual	
Verify that the high-high TK-1 level indicating light (white) is on.	Visual	
Verify that the control panel strobe light (red) is activated.	Visual	
Verify that the audible alarm horn is activated.	Audible	
Silence the alarm by pressing the acknowledge button ("ACK") on the Water for Injection storage and distribution system central control panel.	Manual	
Verify that the panel strobe light (red) and the high-high TK-1 level indicating light remain energized after pressing the acknowledge button (ACK).	Visual	
Turn off the multiple-effect distillation unit by pressing the switch "off" (black) on the multiple-effect distillation unit control panel.	Manual	
Drain the Water for Injection storage tank by opening drain valve WFI-2.	Manual	
Record the Water for Injection storage tank level indicator reading when the high-high TK-1 level light is deenergized.	Record	
Continue draining the Water for Injection storage tank. Record the Water for Injection storage tank level indicator reading when the "high TK-1 level" light is deenergized.	Record	
Continue draining the Water for Injection storage tank. Record the Water for Injection storage tank level indicator reading when the "low TK-1 level" light is energized.	Record	

(Continued)

(*Continued*)

Parameter	Test method	Actual results
Continue draining the Water for Injection storage tank. Record the Water for Injection storage tank level indicator reading when the "low-low TK-1 level" light is energized.	Record	
Close the Water for Injection storage tank drain valve (WFI-2).		
Verify that the low-low TK-1 level indicating light (white) is on.	Visual	
Verify that the panel strobe light (red) is activated.	Visual	
Verify that the audible alarm horn is activated.	Audible	
Silence the alarm by pressing the acknowledge switch (ACK) on the Water for Injection storage and distribution system central control panel.	Manual	
Verify that Water for Injection distribution pump P-1 is deactivated.	Visual	
The multiple-effect distillation unit resumes distillate production when the black switch auto is pressed on the multiple-effect distillation unit central control panel.	Manual	
Verify that the control light on the multiple-effect distillation unit "distillate tank full" (amber) is off.	Visual	
Continue operating the multiple-effect distillation unit until the level in the Water for Injection storage tank increases to the high TK-1 level.	Manual	
Record the "low TK-1 temperature" set point limit on the AA temperature recorder (tag number TRC-1) located on the Water for Injection storage and distribution system central control panel.	Manual	
Record "high TK-1 temperature" set point limit on the AA temperature recorder (tag number TRC-1) located on the Water for Injection storage and distribution system central control panel.	Record	
Adjust the low TK-1 temperature set point on the AA temperature recorder (tag number TRC-1) above the actual displayed tank temperature but not greater than 95°C.	Manual	
Verify that the panel strobe light (red) is activated.	Visual	
Verify that the audible alarm horn is activated.	Audible	
Silence the alarm by pressing the acknowledge switch (ACK) on the Water for Injection storage and distribution system central control panel.	Manual	
Verify that the steam control valve (tag number TCV-1) located on the facility steam inlet line to the Water for Injection storage tank (tag number TK-1) is "open" by noting the location of the valve stem position indicator.	Manual	
Return the low TK-1 temperature set point on the AA temperature recorder (tag number TRC-1) to the previously recorded value.	Manual	
Adjust the high TK-1 temperature set point on the AA temperature recorder (tag number TRC-1) below the actual displayed tank temperature.	Manual	
Verify that the high TK-1 temperature control light (white) is on.	Visual	
Verify that the panel strobe light (red) is activated.	Visual	
Verify that the audible alarm horn is activated.	Audible	
Silence the alarm by pressing the acknowledge switch (ACK) on the Water for Injection storage and distribution system central control panel.	Manual	
Verify that the steam control valve (tag number TCV-1) located on the facility steam inlet line to the Water for Injection storage tank (tag number TK 1) is "closed" by noting the location of the valve stem position indicator.	Visual	
Return the high TK-1 temperature set point on the AA temperature recorder (tag number TRC-1) to the previously recorded value.	Manual	
Record the "low return temperature" set point on the AA temperature recorder (tag number TRC-1).	Record	
Adjust the low return temperature set point on the AA temperature recorder (tag number TRC-1) above the actual temperature.	Manual	
Verify that the low return temperature control light (white) is on.	Visual	
Verify that the panel strobe light (red) is activated.	Visual	
Verify that the audible alarm horn is activated.	Audible	

(*Continued*)

SYSTEM VALIDATION

(*Continued*)

Parameter	Test method	Actual results
Silence the alarm by pressing the acknowledge switch (ACK) on the Water for Injection storage and distribution system central control panel.	Manual	
Verify that the steam control valve (tag number TCV-1) located on the facility steam inlet line to the Water for Injection storage tank (tag number TK-1) is open by noting the location of the valve stem position indicator.	Visual	
Return the low return temperature set point on the AA temperature recorder (tag number TRC-1) to the previously recorded value.	Manual	
Manually break the electrical connection between the electrical connection on the storage tank rupture disk and the connection to the storage and distribution system central control panel.	Manual	
Verify that the "rupture disk" control light (white) is on.	Visual	
Verify that the panel strobe light (red) is activated.	Visual	
Verify that the audible alarm horn is activated.	Audible	
Silence the alarm by pressing the acknowledge switch (ACK) on the Water for Injection storage and distribution system central control panel. Reconnect the electrical connection.	Manual	

APPENDIX C: PERFORMANCE QUALIFICATION FOR AN USP WATER FOR INJECTION SYSTEM

ABC Pharmaceutical Company
Anytown, U.S.A.
Installation Qualification
USP Water for Injection Storage Tank

PERFORMANCE QUALIFICATION INDEX

Subject

I. Protocol Format Approval
II. Summary
III. Testing—WFI Samples
   A. Chemical Testing—WFI Samples
   B. Bacteria Testing—WFI Samples
   C. Bacterial Endotoxin Testing
IV. Testing—WGS Samples
   A. Chemical Testing—WGS Samples
   B. Bacteria Testing—WGS Samples
Enclosures
     1. Typical Page from a WFI Sample Logbook
     2. Typical Page from a WGS Sample Logbook
     3. Data Sheet A: Multiple-Effect Distillation Unit Product Water Conductivity Log
     4. Data Sheet B: Typical Sample Bottle Label
V. Sample Locations
   A. Water for Injection System Distribution Loop
   B. Distillation Unit
VI. Sampling Schedule

I. PROTOCOL FORMAT APPROVAL
It is the responsibility of individuals from technical services, facilities engineering, and quality control to review and approve this protocol prior to execution.

Group	Name	Signature	Date
Technical services			
Facilities engineering			
Quality control			

II. SUMMARY
The PQ will consist of a four-week intense monitoring program. During this period, samples will be obtained from each point of use in the WFI distribution loop once a day, every working day of the week. In addition, a sample of the feed water to the loop will be obtained each working day during the four-week period. Samples from the loop supply and points of use are designated as Water for Injection (WFI) samples. Once a week, samples are collected of the feed water and product water to and from the multiple-effect distillation unit. These samples are referred to as Water for Injection Generating System (WGS) samples.

All WFI samples will be analyzed for conductivity, TOC, nitrates, bacterial endotoxins, and total viable bacteria. Microbial analysis for WFI samples shall be by Membrane Filtration of a 100-mL sample. Any bacteria detected will be sub cultured and speciated. After successfully completing the four-week intense monitoring program, a revised, less intense monitoring program will be initiated.

All WFI samples shall be recorded in the "WFI sample logbook" and assigned a sequential number. An example of a typical logbook page is presented as Enclosure 1 (page 14). All WGS samples shall be recorded in the "WGS sample logbook" and assigned a sequential number. An example of a typical logbook page is presented as Enclosure 2 (page 15). Assigned sample numbers shall appear on any contract laboratory results, thus insuring proper "chain of custody."

III. TESTING—WFI SAMPLES
A. Chemical Testing—WFI Samples
1. ABC Pharmaceutical Company personnel shall record the online conductivity and temperature readings from the distribution loop return tubing cell/meter prior to collecting samples. The value should be documented, along with other required information on Enclosure 3, Data Sheet A (page 16).
2. Samples must be properly labeled. Enclosure 4, Data Sheet B (page 17) is an example of a typical label with the required information clearly demonstrated.
3. The following chemical tests will be performed by the ABC Pharmaceutical Company or a contracted testing laboratory:
   a. TOC per USP, EP, JP, or applicable pharmacopeia
   b. Conductivity per USP, EP, JP, or applicable pharmacopeia
   c. Nitrates per EP or applicable pharmacopeia as appropriate

B. Bacteria Testing—WFI Samples
1. Sampling Procedure
   a. Samples for microbiological evaluation should be collected in appropriately prepared bottles.
   b. ABC Pharmaceutical Company personnel should remove the cap from the bottle immediately prior to obtaining a sample.
   c. The sample side of the cap should *not* come in contact with any surface, including the fingers or hands of the individual obtaining the sample.
   d. The cap may be set on the top of a clean surface, top side against the surface.
   e. Open the sampling/use point valve about one turn, or until there is a steady stream of water.

f. Allow water to flow to drain for approximately 10 seconds.
g. Obtain the sample by filling the bottle.
h. Do not allow the bottle or the water in the bottle to come in contact with the valve.
i. Fill the bottle, without overflowing the bottle.
j. Remove the bottle from the sample stream and place the cap on the bottle as quickly as possible.
k. Tighten the cap securely.
l. Close the valve.
m. Fill in the information on the face of the bottle, as required.
n. Collect all sample bottles per the schedule and store in a dedicated refrigerated area (2–4°C) if bacteria analysis will not be initiated within 30 minutes after sample collection.
o. Insure that bacteria analysis is initiated within 16 hours after sampling. If the services of a contracted laboratory are utilized, the bacteria analysis *must* be initiated within 24 hours after sampling.

2. Shipment
    a. If a contract laboratory is utilized, samples shall be sent by express service to arrive at the contract laboratory as soon as possible after sampling. Samples should be treated and incubation initiated within 24 hours of sampling.
    b. Sample container transfer boxes should include provisions for maintaining the sample temperature at approximately 2°C to 4°C during shipment.
    c. The sample temperature shall not exceed 10°C at any time after initial refrigeration.
    d. Samples shall not be frozen or cooled to a temperature $\leq 1°C$.

3. Analysis
    a. The sample volume for all WFI samples shall be 100 mL.
    b. Analysis should be by Membrane Filtration (0.45-μm filter disk) utilizing R2A culture media.
    c. The incubation time period shall be 72 hours. At least one sample per week shall be analyzed to provide results observed after 5-, 8-, and 10-day incubation periods.
    d. The incubation temperature shall be $22°C \pm 2°C$.
    e. Results shall be reported in cfu/100 mL.
    f. Subculturing and speciation of any bacteria colonies shall be performed for all WFI samples.

4. Documentation
    a. If a contract analytical laboratory is utilized for bacteria determination, the laboratory shall prepare a formal report of the results.
    b. The contract laboratory shall have a documented quality control and assurance program.
    c. The contract laboratory shall have established SOPs that are acceptable to the U.S. FDA.
    d. Analytical procedures and test methods shall be validated to FDA standards.
    e. All documentation must demonstrate proper chain of custody.

C. Bacterial Endotoxin Testing
1. Sampling Procedure
    a. Samples for bacterial endotoxin evaluation should be collected in appropriately prepared bottles. Sample bottles shall be 15 mL, sterile, apyrogenic, and of material approved for collection and storage of bacterial endotoxin samples.
    b. ABC Pharmaceutical Company personnel should remove the cap from the bottle immediately prior to obtaining a sample.
    c. The sample side of the cap should not come in contact with any surface, including the fingers or hands of the individual obtaining the sample.
    d. The cap may be set on the top of a clean surface, top side against the surface.

e. Open the sampling/use point valve about one turn, or until there is a steady stream of water.
f. Allow water to flow to drain for approximately 10 seconds.
g. Obtain the sample by filling the bottle.
h. Do not allow the bottle or the water in the bottle to come in contact with the valve.
i. Fill the bottle, without overflowing the bottle.
j. Remove the bottle from the sample stream and place the cap on the bottle as quickly as possible.
k. Tighten the cap securely.
l. Close the valve.
m. Fill in the information on the face of the bottle, as required.
n. Collect all sample bottles per the schedule and store in a dedicated refrigerated area (2–4°C) if bacterial analysis will not be initiated within 30 minutes after sample collection.
o. Ensure that bacterial endotoxin analysis is initiated within 16 hours after sampling. If the services of a contracted laboratory are utilized, the bacterial analysis *must* be initiated within 24 hours after sampling.

2. Shipment (If Applicable)
a. If a contract laboratory is utilized, samples shall be sent by express service to arrive at the contract laboratory as soon as possible after sampling. Samples should be treated and incubation initiated within 24 hours of sampling.
b. Sample container transfer boxes should include provisions for maintaining the sample temperature at approximately 2°C to 4°C during shipment.
c. The sample temperature shall not exceed 10°C at any time after initial refrigeration.
d. Samples shall not be frozen or cooled to a temperature $\leq 1°C$.

3. Analysis
a. The sample volume for all WFI bacterial endotoxin samples shall be about 2.5 mL.
b. Analysis should be by the LAL gel clot method, with reference to the 1987 U.S. FDA *Guideline for Validation of the Limulus Amebocyte Lysate Test as an End Product Endotoxin Test for Human and Animal Parenteral Drugs, Products, and Medical Devices.*
c. Results shall be reported in endotoxin units/mL (EU/mL or IU/mL).

4. Documentation
a. If a contract analytical laboratory is utilized for bacterial endotoxin determination, the laboratory shall prepare a formal report of the results.
b. The contract laboratory shall have a documented quality control and assurance program.
c. The contract laboratory shall have established SOPs that are acceptable to the U.S. FDA.
d. Analytical procedures and test methods shall be validated to FDA standards.
e. All documentation must demonstrate proper chain of custody.

IV. TESTING—WGS SAMPLES
A. Chemical Testing—WGS Samples
1. Description
a. All sample dates containing the symbol WGS indicate that samples from the Water for Injection Generating System will be obtained. The samples include feed water to and product water from the multiple-effect distillation unit.
b. The ABC Pharmaceutical Company or a contracted analytical laboratory shall supply the appropriate number and size of sample bottles for the specified analysis scheme.

2. Analytical Scheme
a. Analysis of the multiple-effect distillation unit feed water is as follows:
   ○ pH
   ○ Total alkalinity

- Conductivity
- Nitrate as nitrogen
- Chloride
- Free and total chlorine
- Total hardness
- Total silica
- Total iron

b. Analysis of the multiple-effect distillation unit product water is as follows:
- TOC
- Conductivity

B. Bacteria Testing—WGS Samples

1. Sampling Procedure
   a. Samples for microbiological evaluation should be collected in appropriately prepared bottles.
   b. ABC Pharmaceutical Company personnel should remove the cap from the bottle immediately prior to obtaining a sample.
   c. The sample side of the cap should *not* come in contact with any surface, including the fingers or hands of the individual obtaining the sample.
   d. The cap may be set on the top of a clean surface, top side against the surface.
   e. Open the sampling/use point valve about one turn, or until there is a steady stream of water.
   f. Allow water to flow to drain for approximately 5 to 10 minutes.
   g. Obtain the sample by filling the bottle.
   h. Do not allow the bottle or the water in the bottle to come in contact with the valve.
   i. Fill the bottle, without overflowing the bottle.
   j. Remove the bottle from the sample stream and place the cap on the bottle as quickly as possible.
   k. Tighten the cap securely.
   l. Close the valve.
   m. Fill in the information on the face of the bottle, as required.
   n. Collect all sample bottles per the schedule and store in a dedicated refrigerated area (2–4°C) if bacteria analysis will not be initiated within 30 minutes after sample collection.
   o. Ensure that bacteria analysis is initiated within 16 hours after sampling. If the services of a contracted laboratory are utilized, the bacteria analysis *must* be initiated within 24 hours after sampling.

2. Shipment (If Applicable)
   a. If a contract laboratory is utilized, samples shall be sent by express service to arrive at the contract laboratory as soon as possible after sampling. Samples should be treated and incubation initiated within 24 hours of sampling.
   b. Sample container transfer boxes should include provisions for maintaining the sample temperature at approximately 2°C to 4°C during shipment.
   c. The sample temperature shall not exceed 10°C at any time after initial refrigeration.
   d. Samples shall not be frozen or cooled to a temperature $\leq 1°C$.

3. Analysis
   a. The sample volume for all WGS samples shall be 100 mL.
   b. Analysis should be by heterotrophic plate count (1-mL, PCA culture media, 30–35°C incubation temperature, and 48-hour incubation time period) for the multiple-effect distillation unit feed water and Membrane Filtration (100-mL, R2A culture media, 22°C $\pm$ 2°C incubation temperature, and 72-hour incubation time period).
   c. Results shall be reported in cfu/mL (multiple-effect distillation unit feed water) or cfu/100 mL (multiple-effect distillation unit product water).
   d. Subculturing for species of bacteria will be required for any multiple-effect distillation unit product water samples indicating the presence of bacteria.

4. Documentation
   a. The contracted laboratory utilized for bacteria determination shall prepare a formal report of the results.
   b. The contract laboratory shall have a documented quality control and assurance program.
   c. The contract laboratory shall have established SOPs that are acceptable to the U.S. FDA.
   d. Analytical procedures and test methods shall be validated to FDA standards.
   e. All documentation must demonstrate proper chain of custody.

ENCLOSURE 1: TYPICAL PAGE FROM A WFI SAMPLE LOGBOOK

Sample number	Date	Time	Identification	Initials	Results
WFI0001					
WFI0002					
WFI0003					
WFI0004					
WFI0005					
WFI0006					
WFI0007					
WFI0008					
WFI0009					
WFI0010					
WFI0011					
WFI0012					
WFI0013					
WFI0014					
WFI0015					
WFI0016					
WFI0017					
WFI0018					
WFI0019					
WFI0020					
WFI0021					

ENCLOSURE 2: TYPICAL PAGE FROM A WGS SAMPLE LOGBOOK

Sample number	Date	Time	Identification	Initials	Results
WGS0001					
WGS0002					
WGS0003					
WGS0004					
WGS0005					
WGS0006					
WGS0007					
WGS0008					
WGS0009					
WGS0010					
WGS0011					
WGS0012					
WGS0013					
WGS0014					
WGS0015					
WGS0016					
WGS0017					
WGS0018					
WGS0019					
WGS0020					
WGS0021					

ENCLOSURE 3: DATA SHEET A: WATER FOR INJECTION DISTRIBUTION LOOP RETURN ONLINE CONDUCTIVITY AND TEMPERATURE

Date	Time	Location	Temperature	Conductivity

ENCLOSURE 4: DATA SHEET B: TYPICAL SAMPLE BOTTLE LABEL

The ABC Pharmaceutical Company number

Date ──────────── Time ────────────

Location ────────────────────────

Sample collected by ────────────

V. SAMPLE LOCATIONS
A. Water for Injection Distribution Loop

Identification number	Description	Type
WFI-304	Distribution pump effluent sample valve	Direct
WFI-305	Production vessel number 1	Sample valve
WFI-307	Production vessel number 2	Sample valve
WFI-310	Production utensil wash area	Direct

B. Distillation Unit

Identification number	Description	Type
WGS-201	Distillation unit feed water	Sample valve
WGS-202	Distillation unit product water	Divert valve

VI. SAMPLING SCHEDULE—WEEK NUMBER 1

Monday 01/01/2011	Tuesday 01/02/2011	Wednesday 01/03/2011	Thursday 01/04/2011	Friday 01/05/2011
WFI-304	WFI-304	WFI-304	WFI-304	WFI-304
WFI-305	WFI-305	WFI-305	WFI-305	WFI-305
WFI-307	WFI-307	WFI-307	WFI-307	WFI-307
WFI-310	WFI-310	WFI-310	WFI-310	WFI-310
		WGS-201		
		WGS-202		

Index

Page numbers followed by f indicate figure and those followed by t indicate table.

Accessories, for reverse osmosis, 126–127
Acrylic acid, 133
Acrylic resins, 135–136, 136t, 144
ACS. *See* American Chemical Society (ACS)
Action Levels, 8
Action Limits. *See* Alert and Action Limits
Activated carbon units
  ASME code vessel in, 66
  backflushable of, 64
  backwash flow rate, 66
  cartridge-type, 64
  catalytic carbon, 65
  channeling through, 69
  design criteria for, 64–68
  hot water sanitation of, 64f
  operating and maintenance considerations, 68–70
  in pharmaceutical water purification systems, 68
  post hot water sanitization total viable bacteria levels—steam and hot water, 59t
  recirculation and, 115–116
  reverse osmosis and, 104, 108–109, 127
  specifications, 396–401
  theory and application, 58–64
  unit–effluent microbial levels, 63f
Actuators, automatic, 264
Added substances, 10, 34, 247
Additives, to reverse osmosis units, 109–110
Alarm conditions
  multiple-effect distillation units, 202, 204
  vapor compression distillation units, 211
Alarm set point, 381, 383f
Alert and Action Limits, 223, 225, 364, 425–426, 427
  for USP Purified Water system, 221
  for Water for Injection, 221
Alert Levels, 7–8
Alkalinity, 133
Aluminum
  acid washing and, 109
  alum treatment and, 110
  water softening and, 112
American Chemical Society (ACS), 293
American Society for Testing and Materials (ASTM), 293
  "type E-1" semiconductor water specifications, 300t

American Society of Mechanical Engineers (ASME), 48, 197, 390
  Code for Unfired Pressure Vessels storage tanks, 227–230
American Type Culture Collection (ATCC)-19146, 298
Amines, 195
  filming, 196
Ammonia, 20, 183
Ammonium, selectivity coefficient for, 137t
Anode, 260, 278
Antacids, 317
Antiscalants, 103, 109
ASME. *See* American Society of Mechanical Engineers (ASME)
ASME BPE, 337, 402, 406, 409
ASTM. *See* American Society for Testing and Materials (ASTM)
ASTM Standard F 838-05 (ASTM, 2005), 298
ASTM Type 1 water, 350, 363t
ATCC-19146. *See* American Type Culture Collection (ATCC)-19146
Attributes, Pure Steam, 183t
Auxiliary location
  instrumentation and, 371

Back contamination, 318, 326, 350, 351, 355
Backfill, 362, 362f
Backflushing, 71, 72f
Bacteria, 244
  action and alert limits, 7–9
  biofilm formation by
    in Purified Water system, 221
    sanitization and, 224
  endotoxin specification, 9
  L-forms of, 299
  ozone and, 256, 262
  removal, ASTM standards, 298
  sanitization and, 224
Bacterial endotoxins, 23, 181
  analysis of, 224
  CEDI unit, 170
  challenge test, 202
  ultrafiltration to remove, 307

Ball-type valves, 57
Ball valves, 115, 123
Bank-to-bank chemical cleaning/sanitization, 311–312, 311f
Basis of Design, 226–227
Berkholderia cepacia, 288
Biofilm, 326–334, 327f, 329f, 329t, 330f, 331t–334t, 332f
  characteristics of, 327–328
Blowdown tubing, vapor compression distillation units, 207
Boiling inside tubes
  multiple-effect distillation units
    external boilers, 202–205, 203f
    without external boilers, 199–202, 199f
  vapor compression distillation unit with pre-evaporation degasification, 205–209, 206f
Boiling on the shell side and no external evaporation, 197–199, 198f
Booster pump, 122, 128
BP. See British Pharmacopeia (BP)
Brine stream, 160
British Pharmacopeia (BP), 1
British Thermal Units/hour (BTU/hr), 43
Bromate, 28
Bromodichloromethane, 28
Bromoform, 28
BTU/hr. See British Thermal Units/hour (BTU/hr)
Buffer solutions, 40
Burkholderia cepacia, 80, 328

Calcium
  alkalinity and, 133
  selectivity coefficient for, 137t
  in water softening, 136
Calcium carbonate ($CaCO_3$), 14, 103, 111–112, 113–114, 129
  balance, 113, 114, 137–138
  equivalents, 137
Calcium ions, 70
Calibration
  conductivity cell, 148
  of two-bed deionization instrumentation, 148
Canisters, rechargeable mixed bed, 292
  commercially available standard rechargeable, 293t
  design considerations, 293–296, 294f
  operating and maintenance considerations, 296–297
  techniques, 296–297
  theory and application, 293
  used as, 295
Carbon dioxide (CO_2), 17, 19, 183
Cartridge filtration
  design considerations, 89–90
  operating and maintenance considerations, 90–91
  theory and application, 88–89
Casing drain, pumps, 285

Catalytic carbon, 65
Cathode, 260, 278
Cation polishers
  design considerations, 152
  operating, and maintenance considerations, 152
  for purified Water, 152
  theory and application, 152
Cation resin
  bed expansion, 73f
  oxidation of, 75
CEDI. See Continuous electrodeionization (CEDI)
Cellulose acetate, 100–101
Cellulose triacetate, 100
CEN. See European Committee for Standardization (CEN)
Centrifugal pyrogen separation system, double-jacket patented, 200–201, 200f
Certificate of Analysis, 248, 393
Certificate of Calibration, 278, 373, 393
Certificate of Completion, 230
CE stamp. See Conformité Européenne (CE) stamp
CFI. See Cross-flow index (CFI)
CGMPs, 227
Check valve, 126, 144
Chemical injection systems, 34–35
Chemical sanitization, 324–326, 326f
Chloramines, removal of, 117–118
Chloride pitting corrosion, 185
Chloride stress corrosion, 185, 186f
Chlorination, of municipal water, 26
Chlorine, removal of, 117–118
Chlorite, 28
Chloroform, 28, 183, 207
Chromium, plastic storage tanks, 226
CIP. See Clean-in place (CIP)
CIQ. See Commissioning Installation Qualification (CIQ)
Cleaning, of membranes, 103, 105, 110–112, 120, 125
Clean-in place (CIP), 2
Clumping, 154
Colloidal removal, electrodialysis reversal system, 165
Colloidal silica, 183
Colloids, 25–26
  membrane fouling by, 103, 104–105
Commissioning, 417–418
Commissioning Installation Qualification (CIQ), 418
Component installation. See Installation
Compressed air, for installation, 415
Concentration polarization, 97, 102, 102f, 104
Concurrent validation, 433
Condensate feedback feedwater system, 195–196, 196f
  purification, 196–197, 197f
    operating and maintenance considerations, 216
  use of, 195–196
Condensers, in low-velocity, single-effect distillation units
  design, 190–191, 191f
  materials of construction, 190

INDEX

Condensing units, 214–215
  operating and maintenance considerations, 219–220
Conductivity
  bicarbonate ion concentration and, 114
  in cation polishers, 152
  mixed bed deionization and, 153, 155
  monitoring of
    in feedwater piping, 128
    in two-bed deionization, 143, 144, 145, 147
  nitrogen and, 248
  for Purified Water, 248, 383, 384
  vs. resistivity, 16
Conductivity cells, 15, 277–278, 378
  calibration of, 148
  proper and improper positioning of, 143, 143f
  in RO system, 123, 126
Conductivity indicating controller, 369
Conductivity sensor, vapor compression distillation units, 209
Conformité Européenne (CE) stamp, 415
Continuous electrodeionization (CEDI), 159, 207, 221, 227, 243, 276, 276f
  bacterial endotoxin levels, 170
  chamber with resin, 167f
  design considerations, 170–177
  historical perspectives, 167
  operating and maintenance considerations, 177–179
  operating dynamics of, 168, 169f
  reverse osmosis and, 380, 381f
  RO/CEDI systems, 170–173, 171f
  theory and application, 167–170
  unit
    with instrumentation, 176–177, 176f
    ion exchange capacity, 178–179
    manufacturer's feedwater recommendations, 177, 178t
    manufacturer's performance data, 177, 178t
    microbial control preservative, 179
    product water conductivity, 178
    product water flow rate, 179
    replacement, 179
Control and instrumentation, 367–386
  conductivity measurement, 383, 384
  documentation for, 392
  functions of, 373, 378–382
  overview, 367–368
  parameters/applications, 373–378
  P&ID for, 373
  symbols for, 368–373, 371f, 372t
  TOC analyzers, 385–386
Cooldown, of storage system, 233, 234t
Corona discharge, 258
Countercurrent-regenerated deionization units
  design considerations, 151
  operating and maintenance considerations, 151–152
  for Purified Water, 148
  resin bed during, 149, 150f, 151
  theory and application, 148–151

Cross-flow index (CFI), 108
Current density, defined, 163
Cytotoxicity testing
  membrane filtration, 301, 302f

Data acquisition, 382
Data logger, 382, 383
Data trending, in RO system, 128
Dead-ended distribution systems, 318, 319f
Dead leg
  in cleaning process, 129–130
  defined, 129
  distribution loop, 324, 325f
  in microbial control, 115–116
Degasification
  in vapor compression distillation units
    post-evaporation, 209–212, 210f
    pre-evaporation, 205–209, 206f
Deionization, 168, 222. See also Continuous electrodeionization (CEDI)
  affinity for, 136, 137t
  of feedwater, 185–186
  representative equations for, 136, 137f
Deionized water stream, 160
Demineralization, 136
Diaphragm isolators
  with pressure gauges, 115
  for RO system, 118, 123
  in two-bed deionization system, 141
Diaphragms, 276–277
Diaphragm valves, in two-bed deionization systems, 142
Dibromomonochloromethane, 28
Differential pressure, 307
Differential-type pressure sensing system, 238
Dip tube, 274–275
  configuration design, 274f
  dissolved ozone, discharge location, 275, 275f
Dip-type cell, 15
  classical conductivity of, 16
Discharge tubing, distribution/recirculation pumps, 285
Disinfection by-products, 28–29
Dissolved gases
  nonreactive, 18
  reactive, 19–20
Dissolved ozone generation, 272
  multiple-cell, 261f
  theory and application, 259–262
Distillate piping, vapor compression distillation units, 206
Distillation, frequency of, 222
Distillation and pure steam generation
  applications, 184–187
  deionization of feedwater, 185–186
  distillation units
    bacterial endotoxin challenge test, 202
    condensing, 214–215
    Factory Acceptance Test, 202
    multiple-effect, 197–205

[Distillation and pure steam generation distillation units]
  operating and maintenance considerations, 216–220
  operating pressures, 202
    single-effect, 188–197
    vapor compression, 205–214
  enthalpy, 182
  percent conversion, 182
  principles of, 181
    associated with phase change, 181–182
    purity of steam during, 182–183
    theory, 181
    volatile organic impurities, 183
    water impurities, 183
Distribution loop, 221, 223
  microbial control within, 263
  sanitization of, 224, 270
Distribution/recirculation pumps, 242
  casing drain, 285
  curves at different speeds, 281, 282f
  design considerations, 283–286
  discharge tubing, 285
  double mechanical seal with flush, 284f
  dual distribution pumps with check valves, 286, 286f
  efficiency, 282
  installed spare pump, process flow diagram of, 285, 286f
  lubrication fitting, 287
  NPSH, 282
    calculation, 283t
  operating and maintenance considerations, 287
  sanitary centrifugal pump curve, 280, 281f
  seal material, 284
  selection, 280
  speed, 282
  theory and application, 280–283
Distribution systems, 317–364
  dead-ended, 318, 319f
  design considerations, 334–361
    for distribution tubing/piping materials, 322t, 345–361, 348t, 349f–350f, 351t–351t, 354f, 356f, 359f
    heat-welded plastic distribution, 342–344, 342f–343f
    plastic distribution systems with sanitary ferrule connections, 344–345, 344t, 345f–347f
    seamless plastic systems, 322t, 340–342, 341f
    socket-welded plastic distribution, 342–344, 342f–343f
    stainless steel loops, 335–340, 337t–338t
  ladder-type, 318, 319f
  operating and maintenance considerations, 361–364
    inspection, 361–362
  serpentine-type, 318, 319
  theory and application, 318–326, 319f, 320t–323t, 324f–326f
    biofilm, 326–334, 327f, 329f, 329t, 330f, 331t–334t, 332f
    materials of construction, 318

Distribution tubing, sample valve for, 352, 354f
Distributors, for two-bed deionization, 140–141, 141f
Divert-to-waste system, 126
Double block and bleed system, 355
Double-elbow system, 71
Double-jacket patented centrifugal pyrogen separation system, 200–201, 200f
Double-pass reverse osmosis, 96
Drains, 413
Drawing specifications, 390–392
  equipment general assembly drawings, 392
  general arrangement, 390, 392
  PFD, 390
  P&ID, 390
Drinking Water
  chemical specifications of, 2, 3–6
  contaminants for, 3–6
  distribution loops for, 318–319, 319f
  ozonation of, 223, 364
  storage tank for, 223
  total viable bacteria level for, 221
  volume requirements for, 222
  welding requirements for, 343
Drip pans, 46
Dual distribution pumps, with check valves, 286, 286f
Dual treatment units, progressive piping for, 73f

Electrical components, technical considerations for, 415
Electrical power, 416
Electrochemical oxidation potential (EOP), 256, 257t
Electrochemical separation process, 159
Electrodialysis
  anion and cation exchange membranes (ion flow), 160, 161–162, 161f
  cell arrangement with flow paths, 160, 161f
  colloidal/organic/microbial fouling, 163
  historical perspectives, 159
  limitations, 163
  membrane arrangement, 160, 160f
  operating and unit assembly criteria for, 161–162
  polarization, 162–163
  theory and application, 159–163
Electrodialysis reversal
  design considerations, 165–166
  flow path with four-way valves, 164, 165f
  foulants removal from anion exchange membranes, 164, 165f
  ion removal and polarity reversal, 164, 164f
  limitations, 166–167
  pretreatment requirements, 165–166
Electrolytic ozone generation, 259–262, 260f, 261f
  design considerations, 272–275
  manufacturer's data, 273
  selection of, technical considerations for, 262
Electropolishing, 235, 235f, 339
  defined, 339
  iron and, 339

INDEX

Energy, required to increase water temperature, 181–182
Enthalpy, 182
  liquid water, 182
  of steam, 182
EOP. *See* Electrochemical oxidation potential (EOP)
EP. *See* European Pharmacopeia (EP)
EPA. *See* U.S. Environmental Protection Agency's (EPA)
EPDM. *See* Ethylene propylenediene monomer (EPDM)
EPS. *See* Extracellular polymeric substance (EPS)
Equipment general assembly drawings, 392
Equipment installation. *See* Installation
Equipment pads, 413–414
Equivalent length, of distribution loop, 320
Escherichia coli, 22
Ethylene propylenediene monomer (EPDM), 254, 263, 276–277, 363
European Committee for Standardization (CEN), 182
European Pharmacopeia (EP), 1
Evaporator, vapor compression distillation units, 206
External boilers
  boiling inside tubes and, 202–205, 203f
  boiling inside tubes and no, 199–202, 199f
  high-velocity single-effect distillation units with, 193–195, 194f
Extracellular polymeric substance (EPS), 327

Face velocity, 51
Facility considerations, for installation, 411–414
  equipment pads, 413–414
  floor drains, 413
  floor loading, 411
  lighting, 413
  maintenance, 412
  personnel access, 411–412
  restricted access, 413
Factory Acceptance Test (FAT), 202, 417, 422–423
Falling film evaporative process, 192
Fast flushing, in ultrafiltration, 307, 309–311, 310f
FAT. *See* Factory Acceptance Test (FAT)
Feed channel spacers, 129
Feedwater
  analysis of, 100t, 113–114
  in cleaning process, 110–112
  deionization of, 185–186
  flows through interstage heat exchangers in multiple-effect distillation unit, 204
  impurities in, 183
  microbial control in, 115–116
  parameters, 9–10
  silica in, 183
  spacers, 98, 100, 100f, 103
  vapor compression distillation units, 207
Feedwater pump, RO high-pressure, 121–123, 121t
Fiberboard, 74
Fiberglass-reinforced polyester columns, 140
Filler rod, 362

Film composite polyamide membrane, 98, 101–102, 117–118
Filming amines, 196
Finn-Aqua America Inc., 201
Fixed beds deionization. *See* Countercurrent-regenerated deionization units
Flat gasket sealing mechanism, 119
Flat sheet membrane, 100, 100f
Floor drains, 413
Floor loading, 411
Flowmeter, 354, 355, 363
Flux, defined, 96
Foreign substance and impurity, in Purified Water, 222
Foreign substances, 11
Foulants, 103
  removal from anion exchange membranes by polarity reversal, 164, 165f
Fouling, colloidal/organic/microbial, 163
Four-effect distillation unit with boiling on the shell side and no external evaporation, 197–199, 198f
Freeboard space, 151, 154
Front-end pluggage, 104
Fulvic acid, 24

Gas chromatography, 10, 324
Gaseous ozone generation, 262
  design considerations, 266–272, 267f
  manufacturer's data for, 267t
  theory and application, 258–259, 258f
  thermal ozone destruct units, 271t
Gaseous ozone response, 265
Gases, reactive, 183
Gaskets, 358, 363, 364
Gaussian distribution, 306
Gelular resins, 134–135, 134f
Gelular-type resins, 77
General arrangement drawings, 390, 392
Giardia, 22
Gooseneck faucets, 351
Grains, 138
Gram-negative bacteria, 23, 68, 221
Gram's stain, 23
GWR. *See* U.S. EPA Ground Water Rule (GWR)

HAA5. *See* Haloacetic acids (HAA5)
Haloacetic acids (HAA5), 26
Head loss, 347
Health Industry Manufacturers Association (HIMA) bacteria challenge test, 298
Heat content, 182
Heat exchangers, 184, 369
  cyclic temperature control, 46
  design criteria for, 44–48
  in distribution system, 330, 358–361, 359f, 360f
  location of, 45
  multiple-effect distillation unit, 204
  operating and maintenance considerations, 48–49

[Heat exchangers]
   point-of-use, 330, 331
   sizing of, 46
   system with controls, 45f
   theory and application of, 43–44
   trim cooling application, 47
   vapor compression distillation units, 206–207, 208
Heating coil, 343
Heat transfer surface area
   dimpled arrangement, 231, 232f, 233t
   half-pipe arrangement, 231, 232f, 233t
   size of, 230
Heat-welded plastic distribution. *See* Socket-welded plastic distribution
High-velocity, single-effect distillation units, 192, 193f
   with an "external" boiler, 193–195, 194f
      advantages, 195
      components, 193–194, 194f
HIMA. *See* Health Industry Manufacturers Association (HIMA) bacteria challenge test
Hollow fiber ultrafiltration membrane, 307. *See also* Ultrafiltration
   heat-up limitations, 312t
   selection, 312
   techniques for cleaning, 309–310
Hot water sanitization, 224–225, 237, 317, 320, 328, 357, 360, 363, 387
Hot water solenoid, 43
Humic acid, 24
Hydrochloric acid, 146t
Hydrogen, 260
Hydrogen peroxide, 105, 111, 129, 276, 317, 325, 328, 330, 332t
Hydrogen sulfide, electrodialysis reversal system, 165
Hydrolysis process, in cellulostic membranes, 100
Hydronium ion, 65
Hydronium ions, 19, 162–163
Hydrophilic membrane filtration, 297
Hydrophobic vent filtration system, 246–248, 270, 278
   integrity testing of, 253
   sizing, calculations for, 248, 249f
Hydrostatic/pneumatic test
   procedure, 392–393
   report, 393
Hydroxyl ions, 162–163

Inhalants, 317
In-line repressurization pump, 318, 346
In-line ultraviolet radiation
   ozonation, 223, 364
Inline ultraviolet units
   design considerations, 81–82, 289–291
   double block and bleed tubing and valve arrangement, 290, 291f
   effectiveness, 290
   guidelines, 288–289
   microbial control with, 116
   observation of, 290

[Inline ultraviolet units]
   operating and maintenance considerations, 82–83, 292
   recirculation with, 116
   storage system for, 116
   theory and application, 287–289
   theory and application of, 79–81
   ultraviolet sanitation unit "light trap," 291, 291f
Inorganic material, 96, 109
Inspection
   of distribution systems, 361–362
   of storage tanks, 254, 275
Installation, 411–417
   facility considerations, 411–414
   technical considerations, 415–417
   utility considerations, 414–415
Installation Operation Qualification (IOQ), 418
Installation Performance Qualification (IPQ), 418
Installation qualification (IQ), 423
   Change Control and, 423, 431, 432
   commissioning, 418
   execution, 431
   plastic extractables and, 324
   preparation/initiation, 423
   retrospective validation and, 432
   revalidation and, 432
   for USP Water for Injection storage tank, 433–436
Integrity testing, of membrane filters, 304–305, 305f
International Society for Measurement and Control (ISA), 367, 368, 370, 390
Inverse solubility, 45
Ion exchange
   applications of, 138–139
   calculations, 137–138
   capacity units, 138
   defined, 136
   ion removal technique by, 133–158
   resins
      acrylic, 135–136, 136t
      bed depth for, 140
      description, 133–136
      gelular, 134–135, 134f
      macroreticular, 134–135
      physical factors affecting, 138
      strong acid cation exchange, 133
      strong base anion exchange, 133–134
      styrenic, 135–136, 136t
      weak acid cation exchange, 133
      weak base anion exchange, 134
   theory of, 136–139
   water softening, 139
Ion exchange membranes, in electrodialysis, 160, 161–162, 161f
Ion exchange vessel, 56, 56f
Ionic material, 14–18
   classification of, 14
   level of, 17
Ionizations, 15
   degree of, 17

Ion removal
    continuous electrodeionization.
        See Continuous electrodeionization
        (CEDI)
    electrodialysis, 159–163
    electrodialysis reversal, 164–167
    polishing
        design considerations, 293–296
        operating and maintenance considerations, 296–297
        theory and application, 293
Ions, nature of, 17
IOQ. See Installation Operation Qualification (IOQ)
IPA. See Isopropyl alcohol (IPA)
IPQ. See Installation Performance Qualification (IPQ)
Iron
    electropolishing and, 339
    plastic storage tanks, 226
Iron removal, in electrodialysis reversal system, 165
ISA. See International Society for Measurement and Control (ISA)
Isolation valve, for heat exchanger, 359
Isometric drawings, 340
Isopropyl alcohol (IPA), 90

Japanese Pharmacopeia (JP), 1
Joule's heat, 43, 224, 231, 240
JP. See Japanese Pharmacopeia (JP)

Labels, for tubing/piping identification, 412
Ladder-type distribution systems, 318, 319f
Langelier saturation index (LSI), 105, 106t, 128
Latent heat, 182, 184
Leak, distribution systems, 362
    repair, 362, 362f
Level sensing device, 238–239
L-forms, of bacteria, 299
Lifting lugs, 242
Lined steel column exchange vessels, 140
Lipopolysaccharide, 23
Lithium, selectivity coefficient for, 137t
Load cell, 38, 354
Loop polishing, 276, 276f
Loop sanitization, 264, 266
Low-velocity, single-effect distillation
        units, 188–192, 188f
    condenser materials of construction and
        design, 190–191, 191f
    conductivity sensor, 191
    distillate collection system, 190
    divert-to-waste system, 191
    evaporator section, 189
    system design, 188, 188f, 191f
    tubing, 191–192
    vertical cylindrical section, 189
LSI. See Langelier saturation index (LSI)
316L stainless steel, used in distillation units, 185, 189, 190

Macroporous resins, 79, 134–135, 135f
Macroreticular resins, 134–135, 135f
Magnesium
    alkalinity and, 133
    selectivity coefficient for, 137t
    in water softening, 136
Magnesium carbonate ($MgCO_3$), 15
Magnesium ions, 70
Maintenance manual, 429–430
Manganese, removal, in electrodialysis reversal system, 165
Manual override, 381, 384f
Manway, diameter of, 236
Mass spectrometry (MS), 10, 324
Mass totalizing meters, 350
Material test reports, 393
Mechanical polish, 328, 338, 339, 361
Membrane contactors, 96
Membrane filtration
    aspects, 299–300
    bacteria and, 222
    bacterial penetration and, 298–299
    Code 7 filter housing, 301, 303, 303f
    cytotoxicity testing, 301, 302f
    design considerations, 301–304
    evaluating the use of, 300–301
    filter configuration, 301
    flow rating, 303
    general experience, 304
    hydrophilic, 297
    operating and maintenance considerations, 304–306
    periodic nonintrusive integrity testing, 304–305, 305f
    point-of-use, 297–298
    scope of, 298
    theory and application, 298–301
    "T-style" housing, 301
    "Validation Guide," 301
Membrane flux
    defined, 96
    equations for, 97
Membrane fouling, 103, 104–105
    antiscalants and, 103, 109
    colloidal, 103, 104–105
    feedwater spacers for, 100, 103
    flux and, 97
    measurement of, 105–109, 109t
    microbial, 97, 105
    symptoms of, 103
Membranes
    cellulose acetate, 95, 100–101
    composite, 93, 101–102, 117–118
    replacement of, 103, 109, 115, 127–128
    for reverse osmosis. See Reverse osmosis (RO)
    semipermeable, 93–95
Membrane scaling, 103–104
Microbial membrane fouling, 97, 105
Microfiltration, 95
Microorganisms, 20–23
    CT factors, 36

Microswitches, 143
Mill heat numbers, 337, 340
Mixed bed deionization units
　anion resins for, 153, 154
　backwashing of, 153–154
　cation resins for, 153–154
　conductivity and, 153, 155
　distributors for, 154
　pH and, 158
　rechargeable canisters in, 155–158
　rechargeable units
　　design considerations, 155–158
　　operating and maintenance considerations, 155–158
　regenerative units
　　design considerations, 154
　　operating and maintenance considerations, 154–155
　theory and application, 153–154
Molybdenum, plastic storage tanks, 226
Monochloramine, 65, 183
MS. *See* Mass spectrometry (MS)
Multicomponent system, 60f
Multiple-effect distillation units, 185, 197–205
　alarm conditions, 202, 204
　boiling inside tubes
　　and external boilers, 202–205, 203f
　　and no external boiler, 199–202, 199f
　boiling on shell side and no external evaporator, 197–199, 198f
　capacity
　　as a function of facility steam supply pressure, 205t
　　versus steam consumption for units with three to six effects, 205t
　divert-to-waste system, 204, 205
　heating requirements, 184
　instrument (local and remote) and control requirements, 201–202
　operating and maintenance considerations, 217
　operating pressures, 202
　presence of silica in water, 183
　unit condenser-closed loop cooling system, 201, 201f
Multistage centrifugal pumps, 121–122, 121t, 128
Multivalent cations, 127

NaCl. *See* Sodium chloride (NaCl)
Nanofiltration, 95–96
National Committee of Clinical Laboratory Standards (NCCLS), 293
National Electrical Manufacturers Association (NEMA), 416
National Institute of Standards and Technology (NIST), 83, 290, 373
National Primary Drinking Water Regulations (NPDWR), 2, 13, 318
Naturally occurring organic material (NOM), 2, 13, 34, 183
　adsorption of, 60
　thermal decomposition of, 185

NCCLS. *See* National Committee of Clinical Laboratory Standards (NCCLS)
NEMA. *See* National Electrical Manufacturers Association (NEMA)
Net positive suction head (NPSH), 225, 230, 282, 363
　calculation, 283t
Nickel, plastic storage tanks, 226
NIST. *See* National Institute of Standards and Technology (NIST)
Nitrogen
　blanketing, 247, 248, 249, 250f
　and conductivity, 248
Nitrogen oxides
　ozone generator and, 259
NOM. *See* Naturally occurring organic material (NOM)
Nonreactive dissolved gases, 18
Nonreactive gases, 96
NPDWR. *See* National Primary Drinking Water Regulations (NPDWR)
NPSH. *See* Net positive suction head (NPSH)

Occupational Safety and Health Administration (OSHA), 256
Operational qualification (OQ), 424–425
　Change Control and, 432
　commissioning, 417, 418
　concurrent validation and, 433
　execution, 431
　revalidation and, 433
　for USP Water for Injection storage tank, 436–439
Orbital welding, 336, 337
　limitations of, 338
Organic material, 23–24
　from activated carbon by softened water
　　elution of, 62f
Organic removal, electrodialysis reversal system, 165
Organic scavenging resins
　design considerations, 78
　operating and maintenance considerations, 78–79
　theory and application, 76–78
O-ring sealing mechanism, 119, 127–128, 130
OSHA. *See* Occupational Safety and Health Administration (OSHA)
Osmosis, defined, 93, 94f
Osmotic pressure, 93–95
Out-gassing, of ozone, 261
Oxidation-reduction potential (ORP) analyzer, 128
Oxygen, 258, 260
　feed system, 267, 267f
　generator, 259
　　capacity of, 267
　　impurities in, 259
Ozonation, for Drinking Water, 364
Ozone
　bacteria and, 256, 262
　design considerations, 262–266

INDEX

[Ozone
  design considerations]
    electrolytic generation, 272–275
    gaseous generation, 266–272
  Drinking Water and, 223
  effects of personnel exposure to, 257t
  of in-line ultraviolet radiation, 223
  monitors. *See* Ozone monitors
  operating and maintenance
    considerations, 275–278
  out-gassing of, 261
  as oxidizing agent, 10
  sanitization and, 317, 324, 325, 331, 361, 364
  short circuiting of, 268
  storage tank and, 223
  theory and application, 256–258
    dissolved generation, 259–262
    gasous generation, 258–259
Ozone monitors, sidestream-dissolved, 265–266
  performance of, 275

Particulate matter, 13–14
  definition of, 13
  electrodialysis reversal system, 165
  Tyndall effect of, 14
Particulate removal filters
  design criteria for, 51–57
  operating and maintenance considerations, 57–58
  theory and application, 49–51, 50f
Passivation, 235, 329f–330f, 330, 331f, 339
  bacteria excursion after, 329f, 330f
  procedures, 393
  test report, 393
PCA. *See* Plate count agar (PCA)
Peracetic acid, 129, 317, 325, 328, 330, 332t, 333
Percent rejection of ions, 94–95, 97–98
Performance qualification (PQ), 223, 425–429
  commissioning, 418
  concurrent validation, 433
  of distribution system, 331, 333, 334t–336t
  execution, 431
  prospective validation, 430
  purpose of, 425
  summary report of, 430
  for USP Water for Injection storage tank, 439–442
PFD. *See* Process flow diagram (PFD)
PH
  activated carbon and, 127
  in cation polishers, 152
  cellulostic membrane sensitivity to, 100
  in cleaning operation, 112, 128–129
  in double-pass reverse osmosis, 96
  flux and, 96
  hydrolysis and, 100
  membrane fouling and, 103, 104
  membrane scaling, 103
  mixed bed deionization units and, 158
  silica and, 103
  thin-film composite membranes and, 101–102
  in two-bed deionization system, 145

Pharmaceutical waters, definition of, 1–2
Phase change, of water, 181–182
P&ID. *See* Process and instrumentation diagram (P&ID)
Piping/tubing
  distribution, 224
  plastic
    diameter of, 320, 321t, 324, 324f–325f
    materials, design considerations for, 322t, 345–361, 348t, 349f–350f, 351t–351t, 354f, 356f, 359f
    pressure rating for, 321, 322t
    temperature limitations for, 321–322, 322t
    thermal coefficient of expansion, 323, 323t
    in reverse osmosis, 123, 124–125
Plastic distribution systems with sanitary ferrule connections, 344–345, 344t, 345f–347f
Plate count agar (PCA), 21, 328
PLC. *See* Programmable logic controller (PLC)
Pneumatically actuated valves, in two-bed deionization systems, 142
Point-of-use membrane filtration, 297–298
Point-of-use valves, 324, 350, 352, 355, 356, 360, 364
Polarization, 162–163
Polishing components
  distribution pumps, 280–287
  inline ultraviolet units, 287–292
  membrane filtration, 297–306
  polishing ion removal, 292–297
  ultrafiltration, 306–315
Polyester, 98
Polyethylene, 74
Polymers, 34, 98, 109–112
Polypropylene, 74, 247
  for distribution loop, 318
  for distribution piping, 322–323, 322t
  fitting slippage, 343–344
  pressure rating for, 321, 322t
  for spiral wound RO membranes, 98
  for storage tanks, 250–253, 251f, 252f
  welding parameters for, 343, 343f
Polysulfone membrane material, 98, 101
Polyvinyl chloride (PVC)
  for distribution loops, 318, 323–324
  for piping, 343
  socket welding of, 343, 343f
  thermal coefficient of expansion for, 323
Polyvinylidene fluoride (PVDF), 170, 226, 250–251
  for distribution loop, 318
  for distribution piping, 322, 322t, 323
  hot water sanitization of, 318
  pressure rating for, 321, 322t
  temperature and, 318, 322–323, 322t, 323t
Post-evaporation degasification, in vapor compression distillation units, 209–212, 210f
Potassium, selectivity coefficient for, 137t
PQ. *See* Performance qualification (PQ)
Pre-evaporation degasification, in vapor compression distillation units, 205–209, 206f
Prefiltration, 118–119, 119f, 127–128

Pressure gauges, 115, 368, 371, 378
  diaphragm isolators with, 115
  for distribution system, 358, 363
  for RO system, 118, 123
  sanitary, 244
  for two-bed deionization systems, 141
Pressure hold test, 341
Pressure indicating controller, 369
Pressure indicators/sensors, 368
  vapor compression distillation units, 211
Pressure regulating valve, 369
Pressure-sensing element, 368, 379
Pressure sensors, 370f
  in RO system, 126–127
Pressure swing adsorption (PSA) system, 259, 259f, 267
  oxygen generation cycle, 268f
Pressure transmitter (PT), 379
Pressure vessels, 98–99, 99f, 120, 124, 227, 228f–229f
Pretreatment systems
  chemical injection systems, 34–35
  overview, 34
Primary location, instrumentation and, 369
Process and instrumentation diagram (P&ID), 367–368, 373, 378, 390
Process flow diagram (PFD), 390, 391f
Profilometer, 235
Programmable logic controller (PLC), 143, 367, 371–372, 382
Prospective validation, 432–433
PSA system. *See* Pressure swing adsorption (PSA) system
Pseudomonas aeruginosa, 256, 257, 299, 328
Pseudomonas cepacia, 299
Pseudomonas species, 100
Psig Pure Steam, 206
PT. *See* Pressure transmitter (PT)
Pulsation dampers, 128
Pumps
  distribution/recirculation
    casing drain, 285
    curves at different speeds, 281, 282f
    design considerations, 283–286
    discharge tubing, 285
    double mechanical seal with flush, 284f
    efficiency, 282
    NPSH, 282, 283t
    operating and maintenance considerations, 287
    sanitary centrifugal pump curve, 280, 281f
    seal material, 284
    selection, 280
    speed, 282
    theory and application, 280–283
Pure Steam, 224, 246
  attributes, 183t
  capacity, 215f
  generators for
    silica and, 147–148
  pressure, 213
  quality of, 182

[Pure Steam]
  for sanitization, 339
  superheat, 213
  thermodynamic quality ("dryness"), 213
Pure Steam generators, 212–214. *See also* Distillation and pure steam generation
  condensing unit and, 214–215
  feedwater tank and pump, 186, 187f
  method for relieving overpressurization of, 213
  noncondensable gases, 212–213
  operating and maintenance considerations, 219
  parameters, 214
  presence of silica in water, 183
  superheat, 213
  thermodynamic quality ("dryness"), 213
Purified Water, 96, 100, 133
  acrylic resins for, 135–136
  cation polishers for, 152
  chemical sanitization, 224
  chemical specifications of, 2, 3–6
  conductivity for, 248, 383, 384
  countercurrent-regenerated deionization units for, 148
  distribution systems for, 317, 318, 319, 324, 338, 348, 351, 352, 355, 363
  during loop sanitization, 264
  flow diagram of, 268–270, 269f, 270f
  foreign substance and impurity in, 222
  hydrophobic vent filtration systems used in, 246–247
  macroreticular strong acid cation exchange for, 134, 135
  macroreticular strong base anion exchange for, 134, 135
  microswitches for, 143
  nitrogen blanketing for, 248, 249, 250f
  rechargeable canisters for, 157
  spargers and, 268–270. 269f, 270f
  storage and distribution system, 265, 381, 382f
  storage tank
    interior of, inspection of, 275
    vent system for ozonated, 270–272, 271f
  TOC for, 358
  total viable bacteria control for, 256
  two-bed deionization for, 140, 145
PVC. *See* Polyvinyl chloride (PVC)
PVC jacket, 412
PVDF. *See* Polyvinylidene fluoride (PVDF)
Pyrex®, 314

Quality control/assurance (QC/QA) specifications, 392–393

R2A, culture media, 288
Ralstonia pickettii, 257, 299, 328
Raw feed water, design parameters for, 36–37
Raw water, impurities in, 183
  ionic material, 14–18
  overview, 13
  particulate matter, 13–14
Raw water total viable bacteria levels, 21

INDEX 473

Reactive dissolved gases, 19–20
Reactive gases, 96, 183
Reactive silica, 183
   electrodialysis reversal system, 165
Reagent Grade Water, 353t
Rechargeable canister, 153, 155–157, 155f, 156f
Recirculation
   activated carbon and, 115–116
   in-line ultraviolet with, 116
   and repressurization pumps
      design considerations, 84–87
      operating and maintenance
         considerations, 87–88
      theory and application, 83–84
   in two-bed deionization systems, 143–144,
      144f, 145
   in Water for Injection systems, 221, 223
Regenerant salt storage system, 74
Regeneration
   distributors, 151
   fixed beds deionization, 148–149, 150f
   mixed bed deionization, 153–154
   with sodium hydroxide, 144, 146, 147t
   throughput for, 138, 144
   in two-bed deionization systems,
      142, 144–145, 146
Relief valve, 54
   for distribution loop, 357
Removal port, 140
Residual chlorine, electrodialysis reversal
   system, 165
Residual disinfectant, 26–28
   in feed water
      design parameters for, 36–37
      maintenance considerations of, 37–38
      operating considerations of, 37–38
Residual disinfecting agent
   analyzers, 128, 146
   design criteria for, 38–39
   in distribution system, 326, 345
   feedback systems for, 39
   measurement of, 37
   operating and maintenance considerations, 39
   theory and application, 38
Resin bed, 149, 150f, 151
Resin fouling, 135, 145, 146
Resin fragmentation, 138
Resin removal port, 140
Resins
   acrylic, 135–136, 136t
   bed depth for, 140
   gelular, 134–135, 134f
   macroreticular, 134–135
   physical factors affecting, 138
   strong acid cation exchange, 133
   strong base anion exchange, 133–134
   styrenic, 135–136, 136t
   weak acid cation exchange, 133
   weak base anion exchange, 134
Resin traps, 142, 143
   with backwash provisions, 72f

Resistivity
   in cation polishers, 152
   in countercurrent-regenerated deionization
      units, 149
   monitoring of
      in two-bed deionization system, 144, 147
   vs. conductivity, 16
Resistivity cells, 15
Retrospective validation, 432
Return factor, 319–320, 357
Reverse fast flushing, in ultrafiltration, 313
Reverse osmosis (RO), 2, 39–40, 93–131, 167, 207,
      380. See also Sodium hydroxide injection,
      during reverse osmosis
   activated carbon and, 104, 108–109
   additives to, 109–110
   basic theory, 93–95, 94f
   chemically sanitized, 116, 117f
   continuous electrodeionization and, 380, 381f
   design considerations for, 112–115
      accessories for, 126–127
      booster pump for. See Booster pump
      cleaning system for, 125
      high-pressure feedwater pump for,
         121–123, 121t
      membranes for, 123
      microbial control in feedwater system,
         115–116
      monitoring parameters in, 125–126
      piping/tubing for, 124–125
      polishing system design, 116, 117f, 118t
      prefiltration for, 118–119, 119f
      pressure vessels for, 98–99, 99f, 120, 124
      residual chlorine/chloramines, 117–118
      sampling provisions for, 119–121, 120f
   double-pass, 96
   effect of water temperature on, 41t
   flux equations for, 96–97
   hot water sanitized, 116, 117f
   integrity of, 221
   ion removal by, 93–131
   material size and, 95, 96
   membrane array, 113, 113f, 129
   membranes for, 36, 95–96, 98–102, 123
      autopsy of, 103
      brackish water, 96, 99
      cleaning of, 110–112, 120, 125
      composition, 98–102
      film composite polyamide, 98, 101–102, 117–118
      flat sheet, 100, 100f
      flow characteristics, 96–97
      fouling of, 98, 103–104. See also Membrane
         fouling
      hollow fiber, 98
      ion rejection capability of, 94–95, 97–98
      plate and frame, 98
      polypropylene, 98
      size of, 95
      softening of, 95
      spiral wound, 98–102, 99f
      tubular, 98

[Reverse osmosis (RO)]
  operating and maintenance considerations for, 127–128
    cleaning, 128–130
    trending/monitoring, 128
    troubleshooting, 130–131
  operating temperature for, 127
  for Purified Water, 221, 225, 226, 227, 243
  recirculating design of, 41t
Reynolds number, 319, 320, 320t
Riboflavin test results, 393
RiboPrinter, 331, 333t, 334t
Rinse-to-drain, 76
Rinse-to-drain valve, 118, 126, 127, 360
Risk Assessment Document, 421–422
RO. See Reverse osmosis (RO)
RO/CEDI systems, 170–173, 171f
  break tank discharge tubing, 171–172, 171f
  hot water sanitized, 173, 173f
  RO/CEDI loop, 324, 328, 330
    hot water sanitization items, 174–175
    including RO break tank, 172, 172f
    tank accessories, 170–174
    typical chemically sanitized, 171f
Root mean square, 338
Rouging, 237, 254, 324, 337, 339, 361
Rupture discs, 245, 254
Rupture disk, for sanitary applications, 357

Salt passage, percentage of, 97
Salt permeation rate. See Percent rejection of ions
Sample valves, 352, 354f
  ball valve as, 115
  check valve with, 126
  for two-bed deionization systems, 143
Sampling, of RO system, 119–121, 120f
Sanitary ferrule connections
  on diaphragm valves, 363
  in distribution system, 344–345, 344t, 345f–347f
  improper support of, 344
  leaks in, 362
  medical grade–reinforced silicone, 355
  orbital welding and, 335
  plastics and, 344–345, 345f–346f
Sanitization, 278
  chemical, 324–326, 326f
  of distribution loop, 224, 270
  hot water, 224–225, 237, 317, 320, 328, 357, 360, 363, 387
  loop, Purified Water during, 264
  with ozone, 317, 324, 325, 331, 361, 364
  storage tank and, 224
SAT. See Site Acceptance Testing (SAT)
Scale inhibitors, 104, 109–112
Scanning electron microscope (SEM), 235
Scope of Supply, 393–395
SDI. See Silt Density Index (SDI); Silt density index (SDI)
Seal material, distribution/recirculation pumps, 284

Seamless plastic distribution systems, 322t, 340–342, 341f
SEM. See Scanning electron microscope (SEM)
Semipermeable membrane, 93–95, 260, 278
Sensible heat, 182
Sensors, 369f, 370, 370f
Serpentine-type distribution systems, 318, 319
Short circuit, of ozone, 268
Short cycling, 148
Sight glasses, 140, 154
Silica
  deionization of, 147–148
  in feedwater, 183
  measurement of, 147–148
  membrane fouling/scaling by, 103, 104–105
  Pure Steam generation and. See Pure Steam
  removal of, 96
  vapor-liquid disengaging sections and, 101
Silicates, in membrane cleaning, 103
Silt density index (SDI), 34, 394
  calculation of, 107, 107t
  color and, 107
  electrodialysis reversal system, 165
  equipment for, 106f
  first stage differential pressure, 108f
  microbial levels and, 107, 108f
  procedure for, 105, 107
  for reverse osmosis, 113, 119, 120, 128
Single-effect distillation units, 188–197
  components, 184
  condensate feedback, 195–196, 196f
  condensate feedback purification, 196–197, 197f
    operating and maintenance considerations, 216
  heating requirements for, 184
  heating steam pressure for, 192
  high-velocity units, 192, 193f
    with an "external" boiler, 193–195, 194f
  low-velocity units, 188–192, 188f, 191f
  operating and maintenance considerations, 216
  supply steam flow for, as a function of distillate water flow, 184t
  vent filtration systems, 190
Site Acceptance Testing (SAT), 417
Six-pipe diameter rule, 116
Socket-welded plastic distribution, 342–344, 342f–343f
Sodium, selectivity coefficient for, 137t
Sodium bisulfite, 39
Sodium chloride (NaCl), 14
Sodium hydroxide
  regenerant chemical characteristics, 147t
  regeneration with, 144, 146
Sodium hydroxide injection, during reverse osmosis
  design criteria for, 40
  operating and maintenance considerations, 40
  theory and application, 39–40
Sodium hypochlorite, in raw feed water, 36
Softening membrane process, 95
Solenoid valves, 57
  in two-bed deionization systems, 142

Solubility product, 14
SOP. *See* Standard Operating Procedures (SOP)
Spargers, Purified Water system and, 268–270, 269f, 270f
Specifications/documentation
  control system, 392
  drawing, 390–392
  general, 387–390
  operations and maintenance, 392
  QC/QA, 392–393
Spiral-type retaining ring, 124
Spiral wound RO membranes, 98–102, 99f
  cellulose acetate, 100–101
  thin-film composite, 101–102, 117–118
Split-type ring, 124
Spray ball system, 233, 236–238, 237f–238f
Stainless steel
  for check valve, 126
  distribution loops, 335–340, 337t–338t
  distribution tubing and fittings, specifications, 402–410
  for filter housings, 118, 119
  for impellers, 121, 122
  316L, 230
  for piping, 124–125
  for pressure vessels, 124
  resin traps with, 142
Stainless steel tubing mechanical polish and tests
  procedures, 393
  reports, 393
Standard Operating Procedures (SOP), 2, 39, 331, 355, 417
  for housekeeping practices, 225
  for platform activities, 385
  for revalidation, 424
  for system validation, 424
Steam, quality of, 182
Steam pressure, 46
Storage systems, 221–254. *See also* Storage tank
  cooldown of, 233, 234t
  design considerations, 226
    ASME Code for Unfired Pressure Vessels, 227–230
    conductivity, 248
    finish, 233, 234–236, 234t
    fittings, 242–243, 263
    heat transfer, 230–233
    hydrophobic vent filtration, 246–248
    material of construction, 230, 249–253
    sensing/control devices for, 238–241
    tank access, 236
    tank dimensions, 227
    tank orientation, 230
    tank pressure, 244–246
    tank size, 226–227
    tank support, 242
    utility parameters, 241
  segregation of water purification system from, 221, 222f
  sizing of, 222
  theory and application, 221–226

Storage tank
  access, 236
  conductivity, 248
  dimensions, 227
  finish, 233–236, 234t
  fittings for, 242–243, 263
  four-point level sensing system for, 369–370, 370f
  heat exchange, 224
  heat transfer, 230–233
  hydrophobic vent filtration, 246–248
  inspection of, 254
  material of construction, 230, 249–253
  operating and maintenance factors, 253–254
  orientation of, 230
  ozonation and, 223
  polypropylene for, 250–253, 251f, 252f
  pressure, 244–246
  Pure Steam and, 224
  purpose of, 221
  and sanitization, 224
  sensing/control devices for, 238–241
  size of, 222, 226–227
  specifications prepared for, 230
  support, 242
  thermal cycling of, 227, 230
Strong acid cation exchange resin, 133
Strong base anion exchange resins, 133–134
Styrenic resins, 135–136, 136t
Superheat, Pure Steam, 182, 213
Supply pressure, electrodialysis reversal system, 165
Surface Water Treatment Rule, 13
System Basis of Design. *See* Basis of Design
System installation. *See* Installation

Tank heating jacket, 223, 224, 231, 232, 240, 241
TDS. *See* Total dissolved solid (TDS)
Technical considerations, for installation, 415–417
Teflon, 190, 198, 236, 254, 263, 269, 277, 328, 363, 364
Temperature, electrodialysis reversal system, 165
Temperature element, 379, 379f
Temperature gauge, 368, 368f
Temperature-indicating control system, 240, 369
Temperature indicators/sensors, 211
Temperature indicator (TI), 368, 378
Temperature-sensing element, 368
Temperature switch, 122, 123
Temperature transmitter (TT), 379, 379f
Thermal blending systems
  design criteria for, 41–42, 42f
  operating and maintenance considerations, 42–43
  theory and application for, 40–41
Thermal blending valve, 41
  construction of, 42
  flow restriction of, 42
  operation of, 43
  pretreatment components, 42
Thermal decomposition, of NOM, 185
Thermowell, 239
THM. *See* Trihalomethanes (THM)

Throttling valve, 123
TI. *See* Temperature indicator (TI)
TOC. *See* Total organic carbon (TOC)
Total coliform bacteria, 23
Total Coliform Rule, 26
Total dissolved solid (TDS)
 as calcium carbonate, 138
 for two-bed deionization systems, 140, 145
Total hardness, electrodialysis reversal system, 165
Total Hardness Monitor, 75
Total heat, 182
Total organic carbon (TOC), 2, 35, 185, 253, 278
 analyzers for, 368, 385–386
Total suspended solids (TSS), 57, 195
Training manual, 430
Transmembrane pressure, defined, 307
Transmitters, 369–370
Trihalomethanes (THM), 26
 as volatile organic impurities, 183, 207
Trim cooling application, during heat exchangers, 47, 48f
TSS. *See* Total suspended solids (TSS)
TT. *See* Temperature transmitter (TT)
Tubing/piping
 discharge, in distribution/recirculation pumps, 285
 low-velocity, single-effect distillation units, 191–192
 plastic
  diameter of, 320, 321t, 324, 324f–325f
  materials, design considerations for, 322t, 345–361, 348t, 349f–350f, 351t–351t, 354f, 356f, 359f
  pressure rating for, 321, 322t
  temperature limitations for, 321–322, 322t
  thermal coefficient of expansion, 323, 323t
 in reverse osmosis, 124–125
Turbidity, electrodialysis reversal system, 165
Two-bed deionization systems
 anion resin for, 140
 backwashing of, 141–142, 146–147, 147t
 calibration of, 148
 conductivity for, 141, 143, 144, 145
 design considerations for, 140–145
 distributors for, 140–141, 141f
 heat exchanger in, 144
 operating and maintenance considerations, 145–148
 pressure gauges for, 141
 for Purified Water, 140, 145
 recirculation in, 143–144, 144f, 145
 regeneration in, 142, 144–145, 146
 relief valve for, 145
 sample valves for, 143
 theory and application, 139–140
 total dissolved solid for, 140, 145

U-1 (Form), 227, 228f–229f
UL. *See* Underwriters Laboratory (UL) stamp
Ultrafiltration, 95, 221, 226, 248
 advantages, 307
 bacteria and, 222
 "bank-to-bank" chemical cleaning/sanitization, 311–312, 311f
 complexed organic inorganic ions removal by, 308–309, 308t
 decrease in conductivity associated with post mixed bed deionization unit, 308t
 design considerations, 312–314
 "fast flushing," 307, 309–311, 310f
 historical perspectives, 307
 membrane "pore size" for, 306
 microbial control, 306
 operating and maintenance considerations, 314–315
 reverse fast flushing, 313, 315
 theory and application, 307–312
 transmembrane pressure drop, 307, 313
 TOC reduction as function of, 309, 309t
Ultrafiltration product water flow rate
 effect of temperature, 41t
Underwriters Laboratory (UL) stamp, 415
Unfired Pressure Vessels, 51
 ASME code for, 55
Unit condenser-closed loop cooling system, 201, 201f
United States Pharmacopeial Convention, 1
URS. *See* User Requirement Specification (URS)
U.S. Environmental Protection Agency's (EPA), 2
U.S. EPA Ground Water Rule (GWR), 26
U.S. EPA National Secondary Water Regulation concentration, 74
U.S. Pharmacopeia (USP), 1
 bacteriostatic water for injection, 7
 sterile purified water, 6
 sterile water for inhalation, 7
 sterile water for injection, 7
 sterile water for irrigation, 7
 water for injection, 6–7
User Requirement Specification (URS), 421
USP. *See* U.S. Pharmacopeia (USP)
USP Glycerin, 130
USP Purified Water. *See* Purified Water
Utility considerations, for installation, 414–415

Validation
 change control, 431–432
 execution, 431
 FAT, 422–423
 installation qualification. *See* Installation qualification (IQ)
 operational qualification. *See* Operational qualification (OQ)
 performance qualification. *See* Performance qualification (PQ)
 SOP, 424
 summary report, 430
 types of, 432–433
 URS, 421

INDEX

Validation Guide, 301
Validation Master Plan, 421
van der Waal's forces, 60
Vapor compression distillation units, 205–214
  alarm conditions, 211
  boiling inside tubes: pre-evaporation degasification, 205–209, 206f
  boiling outside tubes: degasification after evaporation, 209–212, 210f
  features for, 206–207
    distillate piping, 206
    evaporator, 206
    feedwater and blowdown tubing, 207
    heat exchangers, 206–207
    standard design, 210
  instrumentation, 211
  operating and maintenance considerations, 218–219
  operation of, information associated with, 207–209
  overview, 185
  pressure indicators/sensors, 211
  Pure Steam generators, 212–214
  temperature indicators/sensors, 211
Vapor-liquid disengaging device
  in high-velocity, single-effect distillation units with an "external" boiler, 194
  in low-velocity, single-effect distillation units, 189–190
Vapor-liquid disengaging process, 182–183
Vapor pressure, 246
Variable frequency drive (VFD), 225, 254, 264, 270, 281, 348
Vent filtration systems
  single-effect distillation units, 190
Ventilation, of installation area, 415
Vent system, for ozonated Purified Water storage tank, 270–272, 271f
Vertical cylindrical vessels, 51
Vessels, ASME data plate for, 52
VFD. *See* Variable frequency drive (VFD)
Vinyl ester columns, 140
Viton, 254, 263, 276
Volatile organic impurities, 183, 207

Warranties, 393–394
Waste-to-drain valve, 360
Water, latent heat of vaporization for, 44t

Water for Injection (WFI), 221
  Alert and Action Limits for, 221
  ambient temperature, 223
  bacterial endotoxin testing for, 441
  bacterial testing for, 440–441
  chemical test method, 96
  heat transfer, 230–233
  hydrophobic vent filtration systems used in, 246–247
  membrane filtration for, 427
  method of production, 11
  nitrogen blanketing for, 248, 249, 250f
  storage and distribution systems, 223–224
  storage tank qualification
    installation qualification (IQ), 433–436, 443–450
    operational qualification (OQ), 436–439, 451–455
    performance qualification (PQ), 439–442, 455–462
Water (liquid)-free steam, 189
Water purification unit, 15
Water softening, 104, 114, 127
  aluminum and, 112
  calcium and, 139
  magnesium and, 139
  for Purified Water, 139
  reverse osmosis and, 139
  for vapor compression distillation, 139
Water softening units
  design for, 70–75, 139
  operating and maintenance considerations, 75–76, 139
  recirculation and, 116
  regeneration of, 126
  theory and application, 139
  theory and application of, 70
Water vapor, accumulation of, 247
Weak acid cation exchange resins, 133
Weak base anion exchange resins, 134
Welding
  of 316L stainless steel, 230
  orbital, 336, 337
    limitations of, 338
Weld numbers, 340
Weld test coupons, 393
WFI. *See* Water for Injection (WFI)

Zero dead leg, 324, 325f, 356, 356f
  valve, 243, 243f

CPSIA information can be obtained
at www.ICGtesting.com
Printed in the USA
LVHW101553241218
600958LV00009B/120/P